DISCARD

THE ROYAL INSTITUTION
LIBRARY OF SCIENCE

(being the Friday Evening Discourses in Physical Sciences held
at the Royal Institution: 1851–1939)

PHYSICAL SCIENCES

Volume 3

THE ROYAL INSTITUTION
LIBRARY OF SCIENCE

(being the Friday Evening Discourses in Physical Sciences
held at the Royal Institution: 1851–1939)

PHYSICAL SCIENCES
Volume 3

EDITED BY

SIR WILLIAM LAWRENCE BRAGG,
C.H., O.B.E., M.C., F.R.S., M.A. (Cantab)
Professor Emeritus and formerly Director of the
Royal Institution of Great Britain and Fullerian Professor of Chemistry

AND

PROFESSOR GEORGE PORTER,
F.R.S., M.A., Sc.D. (Cantab)
Director of the Royal Institution of Great Britain and
Fullerian Professor of Chemistry

APPLIED SCIENCE PUBLISHERS LTD
LONDON

APPLIED SCIENCE PUBLISHERS LTD.
RIPPLE ROAD, BARKING, ESSEX, ENGLAND.

IBSN 0 85334 615 1

© COPYRIGHT IN THIS FORM 1970
APPLIED SCIENCE PUBLISHERS LTD.

All rights reserved. No part of this publication may be reproduced, stored in a retrieval system, or transmitted in any form, or by any means, electronic, mechanical, photocopying, recording, or otherwise, without the prior written permission of the original publisher, The Royal Institution, London, England.

Printed photo litho in Great Britain by Page Bros. (Norwich) Ltd., Norwich and London

CONTENTS

1878

The Explanation of certain Acoustical Phenomena—Lord Rayleigh	1
Experiments in Electro-Photometry—James Dewar	8
A Nocturne in Black and Yellow—W. Spottiswoode	11
The Effects of Stress on the Magnetism of Iron, Cobalt, and Nickel—Sir William Thomson	18
The Liquefaction of Gases—James Dewar	21

1879

A New Chemical Industry, established by M. Camille Vincent—H. E. Roscoe	28
The Sorting Demon of Maxwell—Sir William Thomson	36
Molecular Physics in High Vacua—William Crookes	38
The Optical Study of the Elasticity of Solid Bodies—A. Cornu	60
Spectroscopic Investigation—James Dewar	63

1880

Investigations at High Temperatures—James Dewar	80

1881

The Phenomena of the Electric Discharge with 14,400 Chloride of Silver Cells—Warren De La Rue	91
The Teachings of Modern Spectroscopy—Arthur Schuster	121
Elasticity viewed as possibly a Mode of Motion—Sir William Thomson	136
Selenium and its applications to the Photophone and Telephotography—Shelford Bidwell	138

The Conversion of Radiant Heat into Sound—J. Tyndall	150
Indigo, and its Artificial Production—H. E. Roscoe	159
Origin and Identity of Spectra—James Dewar	174

1882

Spectrum Analysis in the infra red of the Spectrum—Captain W. de W. Abney	207
Matter and Magneto-Electric Action—W. Spottiswoode	216

1883

The Size of Atoms—Sir William Thomson	227
The Ultra-Violet Spectra of the Elements—George D. Liveing	257
Thoughts on Radiation, Theoretical and Practical—J. Tyndall	268

1884

Theory of Magnetism—D. E. Hughes	281
The Two Manners of Motion of Water—Osborne Reynolds	291
Volta-Electric and Magneto-Electric Induction—Willoughby Smith	300
Researches on Liquefied Gases—James Dewar	312

1885

Liquid Films—A. W. Rücker	318

1886

Capillary Attraction—Sir William Thomson	325
Experiments showing Dilatancy, a property of Granular Material, possibly connected with Gravitation—Osborne Reynolds	350
On Recent Progress in the Coal-tar Industry—Sir Henry E. Roscoe	360

On Dissociation Temperatures with special reference to
   Pyrotechnical questions—Frederick Siemens . . . 377
Suspended Crystallisation—John Millar Thomson . 394

1887

Genesis of the Elements—William Crookes . . . 403
Sunlight Colours—Captain W. de W. Abney . . 427
On the Colours of Thin Plates—Lord Rayleigh . . 438
Light as an Analytic Agent—James Dewar . . . 440
The Rolling Contact of Bodies—H. S. Hele Shaw . 451

1888

Diffraction of Sound—Lord Rayleigh 454
Electrical Influence Machines—James Wimshurst . . 466
Phosphorescence and Ozone—James Dewar . . . 472

1889

Electrostatic Measurement—Sir William Thomson . 476
Electrical Stress—A. W. Rücker 478
The Discharge of a Leyden Jar—Oliver Lodge . . 481
Iridescent Crystals—Lord Rayleigh 493
Aluminium—Sir Henry Roscoe 496
Optical Properties of Oxygen and Ozone—James Dewar. 510
Optical Torque—Silvanus P. Thompson . . . 516
An Attempt to apply to Chemistry one of the Principles of
   Newton's Natural Philosophy—D. Mendeléeff . . 540
Quartz Fibres—C. V. Boys 560

Friday, March 15, 1878.

The Duke of Northumberland, D.C.L. President, in the Chair.

The Lord Rayleigh, M.A. F.R.S. *M.R.I.*

The Explanation of certain Acoustical Phenomena.

Musical sounds have their origin in the vibrations of material systems. In many cases, e. g. the pianoforte, the vibrations are free, and are then necessarily of short duration. In other cases, e. g. organ pipes and instruments of the violin class, the vibrations are maintained, which can only happen when the vibrating body is in connection with a source of energy, capable of compensating the loss caused by friction and generation of aerial waves. The theory of free vibrations is tolerably complete, but the explanations hitherto given of maintained vibrations are generally inadequate, and in most cases altogether illusory.

In consequence of its connection with a source of energy, a vibrating body is subject to certain forces, whose nature and effects are to be estimated. These forces are divisible into two groups. The first group operate upon the periodic time of the vibration, i. e. upon the pitch of the resulting note, and their effect may be in either direction. The second group of forces do not alter the pitch, but either encourage or discourage the vibration. In the first case only can the vibration be maintained; so that for the explanation of any maintained vibration, it is necessary to examine the character of the second group of forces sufficiently to discover whether their effect is favourable or unfavourable. In illustration of these remarks, the simple case of a common pendulum was considered. The effect of a small periodic horizontal impulse is in general both to alter the periodic time and the amplitude of vibration. If the impulse (supposed to be always in the same direction) acts when the pendulum passes through its lowest position, the force belongs to the second group. It leaves the periodic time unaltered, and encourages or discourages the vibration according as the direction of the pendulum's motion is the same or the opposite of that of the impulse. If, on the other hand, the impulse acts when the pendulum is at one or other of the limits of its swing, the effect is solely on the periodic time, and the vibration is neither encouraged nor discouraged. In order to encourage, i. e. practically in order to maintain a vibration, it is necessary that the forces should not depend solely upon the position of the vibrating body. Thus, in the case of the pendulum, if a small impulse in a given direction acts upon it every time that it passes through its lowest position, the vibration is not maintained, the advantage gained as the pendulum makes a passage in the same direction as that in which the impulse acts being exactly neutralized on the return passage, when the motion is in the opposite direction.

As an example of the application of these principles the maintenance of an electric tuning fork was discussed. If the magnetic forces depended only upon the position of the fork, the vibration could not be maintained. It appears therefore that the explanations usually given do not touch the real point at all. The fact that the vibrations are maintained is a proof that the forces do not depend solely upon the position of the fork. The causes of deviation are two—the self-induction of the electric currents, and the adhesion of the mercury to the wire whose motion makes and breaks the contact. On both accounts the magnetic forces are more powerful in the latter than in the earlier part of the contact, although the position of the fork is the same; and it is on this *difference* that the possibility of maintenance depends. Of course the arrangement must be such that the retardation of force *encourages* the vibration, and the arrangement which in fact encourages the vibration would have had the opposite effect, if the nature of electric currents had been such that they were more powerful during the earlier than during the later stages of a contact.

In order to bring the subject within the limits of a lecture, one class of maintained vibrations was selected for discussion, that, namely, of which *heat* is the motive power. The best understood example of this kind of maintenance is that afforded by Trevelyan's bars, or rockers. A heated brass or copper bar, so shaped as to rock readily from one point of support to another, is laid upon a cold block of lead. The communication of heat through the point of support expands the lead lying immediately below in such a manner that the rocker receives a small impulse. During the interruption of the contact the communicated heat has time to disperse itself in some degree into the mass of lead, and it is not difficult to see that the impulse is of a kind to encourage the motion. But the most interesting vibrations of this class are those in which the vibrating body consists of a mass of air more or less completely confined.

If heat be periodically communicated to, and abstracted from, a mass of air vibrating (for example) in a cylinder bounded by a piston, the effect produced will depend upon the phase of the vibration at which the transfer of heat takes place. If heat be given to the air at the moment of greatest condensation, or taken from it at the moment of greatest rarefaction, the vibration is encouraged. On the other hand, if heat be given at the moment of greatest rarefaction, or abstracted at the moment of greatest condensation, the vibration is discouraged. The latter effect takes place of itself, when the rapidity of alternation is neither very great nor very small, in consequence of radiation; for when air is condensed it becomes hotter, and communicates heat to surrounding bodies. The two extreme cases are exceptional, though for different reasons. In the first, which corresponds to the suppositions of Laplace's theory of the propagation of sound, there is not sufficient time for a sensible transfer to be effected. In the second, the temperature remains nearly constant, and the loss of heat occurs during the *process* of condensation, and not when the condensation is effected. This case corresponds to Newton's theory of the velocity of sound. When the transfer of heat takes place at the moments of greatest condensation or of greatest rarefaction, the pitch is not affected.

If the air be at its normal density at the moment when the transfer of heat takes place, the vibration is neither encouraged nor discouraged, but the pitch is altered. Thus the pitch is *raised*, if heat be communicated to the air a quarter period *before* the phase of greatest condensation, and the pitch is *lowered* if the heat be communicated a quarter period *after* the phase of greatest condensation.

In general both kinds of effects are produced by a periodic transfer of heat. The pitch is altered, and the vibrations are either encouraged or discouraged. But there is no effect of the second kind if the air concerned be at a loop, i. e. a place where the density does not vary, nor if the communication of heat be the same at any stage of rarefaction, as in the corresponding stage of condensation.

The first example of aerial vibrations maintained by heat was found in a phenomenon which has often been observed by glass-blowers, and was made the subject of a systematic investigation by Dr. Sondhauss. When a bulb about three-quarters of an inch in diameter is blown at the end of a somewhat narrow tube, 5 or 6 inches in length, a sound is sometimes heard proceeding from the heated glass. It was proved by Sondhauss that a vibration of the glass itself is no essential part of the phenomenon, and the same observer was very successful in discovering the connection between the *pitch* of the note and the dimensions of the apparatus. But no explanation (worthy of the name) of the production of sound has been given.

For the sake of simplicity, a simple tube, hot at the closed end and getting gradually cooler towards the open end, was first considered. At a quarter of a period *before* the phase of greatest condensation (which occurs almost simultaneously at all parts of the column) the air is moving inwards, i. e. towards the closed end, and therefore is passing from colder to hotter parts of the tube; but the heat received at this moment (of normal density) has no effect either in encouraging or discouraging the vibration. The same would be true of the entire operation of the heat, if the adjustment of temperature were instantaneous, so that there was never any sensible difference between the temperatures of the air and of the neighbouring parts of the tube. But in fact the adjustment of temperature takes *time*, and thus the temperature of the air deviates from that of the neighbouring parts of the tube, inclining towards the temperature of that part of the tube *from* which the air has just come. From this it follows that at the phase of greatest condensation heat is received by the air, and at the phase of greatest rarefaction is given up from it, and thus there is a tendency to maintain the vibrations. It must not be forgotten, however, that apart from transfer of heat altogether, the condensed air is hotter than the rarefied air, and that in order that the whole effect of heat may be on the side of encouragement, it is necessary that previous to condensation the air should pass not merely towards a hotter part of the tube, but towards a part of the tube which is hotter than the air will be when it arrives there. On this account a great range of temperature is necessary for the maintenance of vibration, and even with a great range the influence of the transfer of heat is necessarily unfavourable at the closed end, where the motion is very small. This is probably the reason of the advantage of a bulb. It is obvious that if the *open* end of the tube were heated, the effect of the transfer of heat would be even more unfavourable than in the case of a temperature uniform throughout.

The sounds emitted by a jet of hydrogen, burning in an open tube, were noticed soon after the discovery of the gas, and have been the subject of several elaborate inquiries. The fact that the notes are substantially the same as those which may be elicited from the tube in other ways, e. g. by blowing, was announced by Chladni. Faraday proved that other gases were competent to take the place of hydrogen, though not without disadvantage. But it is to Sondhauss that we owe the most detailed examination of the circumstances under which the sound is produced. His experiments prove the importance of the part taken by the column of gas in the tube which supplies the jet. For example, sound cannot be obtained with a supply tube which is plugged with cotton in the neighbourhood of the jet, although no difference can be detected by the eye between the flame thus obtained and others which are competent to excite sound. When the supply tube is unobstructed, the sounds obtainable are limited as to pitch, often dividing themselves into detached groups. In the intervals between the groups no coaxing will induce a maintained sound, and it may be added that, for a part of the interval at any rate, the influence of the flame is inimical, so that a vibration started by a blow is damped more rapidly than if the jet were not ignited.

Partly in consequence of the peculiar behaviour of flames and partly for other reasons, the thorough explanation of these phenomena is a matter of some difficulty; but there can be no doubt that they fall under the head of vibrations maintained by heat, the heat being communicated periodically to the mass of air confined in the sounding tube at a place where, in the course of a vibration, the pressure varies. Although some authors have shown an inclination to lay stress upon the effects of the current of air passing through the tube, the sounds can readily be produced, not only when there is no through draught, but even when the flame is so situated that there is no sensible periodic motion of the air in its neighbourhood. In the course of the lecture a globe intended for burning phosphorus in oxygen gas was used as a resonator, and, when excited by a hydrogen flame well removed from the neck, gave a pure tone of about 95 vibrations per second.

In consequence of the variable pressure within the resonator, the issue of gas, and therefore the development of heat, varies during the vibration. The question is under what circumstances the variation is of the kind necessary for the maintenance of the vibration. If we were to suppose, as we might at first be inclined to do, that the issue of gas is greatest when the pressure in the resonator is least, and that the phase of greatest development of heat coincides with that of the greatest issue of gas, we should have the condition of things the most unfavourable of all to the persistence of the vibration. It is not difficult, however, to see that both suppositions are incorrect. In the supply tube (supposed to be unplugged, and of not too small bore) stationary, or approximately stationary, vibrations are excited, whose phase is either the same or the opposite of that of the vibration in the resonator. If the length of the supply tube from the burner to the open end in the gas-generating flask be less than a quarter of the wave length in hydrogen of the actual vibration, the greatest issue of gas *precedes* by a quarter period the phase of greatest condensation; so that, if the development of heat is *retarded* somewhat in comparison with the issue of gas, a state of things exists *favourable* to the maintenance of the sound. Some such retardation is inevitable, because a jet of inflammable gas can burn only at the outside; but in many cases a still more potent cause may be found, in the fact that during the retreat of the gas in the supply tube small quantities of air may enter from the interior of the resonator, whose expulsion must be effected before the inflammable gas can again begin to escape.

If the length of the supply tube amounts to exactly one quarter of the wave length, the stationary vibration within it will be of such a character that a node is formed at the burner, the variable part of the pressure just inside the burner being the same as in the interior of the resonator. Under these circumstances there is nothing to make the flow of gas, or the development of heat, variable, and therefore the vibration cannot be maintained. This particular case is free from some of the difficulties which attach themselves to the general problem, and the conclusion is in accordance with Sondhauss' observations.

When the supply tube is somewhat longer than a quarter of the wave, the motion of the gas is materially different from that first described. Instead of preceding, the greatest outward flow of gas *follows* at a quarter period interval the phase of greatest condensation, and therefore if the development of heat be somewhat retarded, the whole effect is unfavourable. This state of things continues to prevail, as the supply tube is lengthened, until the length of half a wave is reached, after which the motion again changes sign, so as to restore the possibility of maintenance. Although the size of the flame and its position in the tube (or neck of resonator) are not without influence, this sketch of the theory is sufficient to explain the fact, formulated by Dr. Sondhauss, that the principal element in the question is the length of the supply tube.

The next example of the production of sound by heat, shown in the lecture, was a very interesting phenomenon discovered by Rijke. When a piece of fine metallic gauze, stretching across the lower part of a tube, open at both ends and held vertically, is heated by a gas flame placed under it, a sound of considerable power, and lasting for several seconds, is observed almost immediately *after* the removal of the flame. Differing in this respect from the case of sonorous flames, the generation of sound was found by Rijke to be closely connected with the formation of a through draught, which impinges upon the heated gauze. In this form of the experiment the heat is soon abstracted, and then the sound ceases; but by keeping the gauze hot by the current from a powerful galvanic battery, Rijke was able to obtain the prolongation of the sound for an indefinite period. In any case from the point of view of the lecture the sound is to be regarded as a *maintained* sound.

In accordance with the general views already explained, we have to examine the character of the variable communication of heat from the gauze to the air. So far as the communication is affected directly by variations of pressure or density, the influence is unfavourable, inasmuch as the air will receive less heat from the gauze when its own temperature is raised by condensation. The maintenance depends upon the variable transfer of heat due to the varying *motions* of the air through the gauze, this motion being compounded of a uniform motion upwards with a motion, alternately upwards and downwards, due to the vibration. In the lower half of the tube these motions conspire a quarter period *before* the phase of greatest condensation, and oppose one another a quarter period after that phase. The rate of transfer of heat will depend mainly upon the temperature of the air in contact with the gauze, being greatest when that temperature is lowest. Perhaps the easiest way to trace the mode of action is to begin with the case of a simple vibration without a steady current. Under these circumstances the whole of the air which comes in contact with the metal, in the course of a complete period, becomes heated; and after this state of things is established, there is comparatively little further transfer of heat. The effect of superposing a small steady upwards current is now easily recognized. At the limit of the inwards motion, i. e. at the phase of greatest condensation, a small quantity of air comes into contact with the metal, which has not done so before, and is accordingly cool; and the heat communicated to this quantity of air acts in the most favourable manner for the maintenance of the vibration.

A quite different result ensues if the gauze be placed in the *upper* half of the tube. In this case the fresh air will come into the field at the moment of greatest rarefaction, when the communication of heat has an unfavourable instead of a favourable effect. The principal note of the tube therefore cannot be sounded.

A complementary phenomenon discovered by Bosscha and Riess may be explained upon the same principles. If a current of *hot* air impinge upon *cold* gauze, sound is produced; but in order to obtain the principal note of the tube the gauze must be in the upper, and not as before in the lower, half of the tube. An experiment due to Riess was shown in which the sound is maintained indefinitely. The upper part of a brass tube is kept cool by water contained in a tin vessel, through the bottom of which the tube passes. In this way the gauze remains comparatively cool, although exposed to the heat of a gas flame situated an inch or two below it. The experiment sometimes succeeds better when the draught is checked by a plate of wood placed somewhat closely over the top of the tube.

Both in Rijke's and Riess' experiments the variable transfer of heat depends upon the motion of vibration, while the effect of the transfer depends upon the variation of pressure. The gauze must therefore be placed where both effects are sensible, i. e. neither near a node nor near a loop. About a quarter of the length of the tube, from the lower or upper end, as the case may be, appears to be the most favourable position.

[RAYLEIGH.]

Friday, March 29, 1878.

Warren de La Rue, Esq. M.A. D.C.L. F.R.S. Vice-President,
in the Chair.

Professor James Dewar, M.A. F.R.S.

Experiments in Electro-Photometry.

(*Abstract.*)

Edmund Becquerel, in the year 1839, opened up a new field of chemical research through the discovery that electric currents may be developed during the production of chemical inter-actions excited by solar agency.

Hunt, in the year 1840, repeated, with many modifications, Becquerel's experiments, and confirmed his results.

Grove, in 1858, examined the influence of light on the polarized electrode, and concluded that the effect of light was simply an augmentation of the chemical action taking place at the surface of the electrodes.

Becquerel, in his well-known work, 'La Lumière,' published in 1868, gives details regarding the construction of an electro-chemical actinometer formed by coating plates of silver with a thin film of the sub-chloride, and subsequent heating for many hours to a temperature of 150° C.

Egeroff, in 1877, suggested the use of a double apparatus of Becquerel's form, acting as a differential combination, the plates of silver being coated with iodide instead of chloride.

The modifications of the halogen salts of silver when subjected to the action of light have up to the present time been used most successfully in the production of electric currents, and although mixtures of photographically sensitive salts have been shown by Smee to produce currents of a similar kind, yet no attempt has been made to examine the proper form of instrument required for the general investigation of the electrical actions induced by light on fluid substances.

This subject has occupied my attention for some time, and when the investigation is completed I shall deal with the results in greater detail. In the meantime the following description will give an idea of the method of investigation.

A little consideration shows that the amount of current produced by a definite intensity and quality of light acting during a short period of time on a given sensitive substance in solution, is primarily a function of the nature, form, and position of the poles in the cell relatively to the direction in which the light enters, and the selective absorption, concentration, and conductivity of the fluid.

The diffusive action taking place in such cells complicates the effects and is especially intricate when insoluble substances are formed. In order to simplify the investigation in the first instance, poles that are not chemically acted upon, and a sensitive substance yielding only soluble products on the action of light, were employed. For this purpose platinum poles and chlorous acid or peroxide of chlorine were selected.

The best form of cell had one of the poles made of fine platinum wire fixed as closely as possible to the inner surface where the light enters, the other pole being made of thicker wire placed deeper in the fluid.

As the action is confined to a very fine film where the light enters, the maximum amount of current is obtained when the composition of the fluid is modified deep enough to isolate temporarily the front pole in the modified medium. Under these conditions the formation of local currents is avoided, and the maximum electromotive force obtained.

In cells of this construction the amount of current is independent of the surface of the fluid acted upon by light, so that a mere slit, sufficient to expose the front poles, acts as efficiently as a larger surface. This prevents the unnecessary exhaustion of material and enables the cell to be made of very small dimensions. By means of such an apparatus the chemical actions of light and their electrical relations may be traced in many new directions.

The amount and direction of the current in the case of chlorous acid is readily modified by the addition of certain salts and acids, and thus electrical variations may be produced, resembling the effects observed during the action of light on the eye.

Certain modifications taking place in the chlorous acid by exposure to light increase its sensibility, and as a general result it is found that the fluid through these alterations increases in resistance. We have thus an anomalous kind of battery where the available electromotive force increases with the resistance. The addition of neutral substances which increase the resistance without producing new decompositions improves the action of the cell.

Care has to be taken in these experiments to use the same apparatus in a series of comparative experiments, as infinitesimal differences in the contact of the active pole render it difficult to make two instruments giving exactly the same results. Cells have been constructed with two, three, and four poles, and their individual and combined action examined. Quartz surfaces have also been employed instead of glass, thus enabling the chemical opacity of different substances to be determined.

The electrical currents derived through the action of light on definite salts are strong in the case of ferro- and ferri-cyanide of potassium, but remarkably so in the case of nitroprusside of sodium.

Of organic acids the tartrate of uranium is one of the most active. A mixture of selenious acid and sulphurous acid in presence of hydrochloric acid yields strong currents when subjected to light in the form of cell described. The list of substances that may be proved to undergo chemical decomposition by the action of light is very extensive; full details will be found in the completed paper.

[J. D.]

Friday, May 3, 1878.

Professor E. Frankland, D.C.L. F.R.S. &c. *M.R.I.* in the Chair.

W. Spottiswoode, Esq. LL.D. Tr.R.S. Sec.R.I.

A Nocturne in Black and Yellow.

It is well known that the coloured bands and rings shown by white light, when polarised and transmitted through crystals, fade, and cease to be visible when the retardation of the rays, due to the thickness of crystal traversed, is large. The feebleness of tint and confusion of definition arises from the overlapping of figures of different colours. But when monochromatic light is used no mixture of colour can take place, and the bands and rings remain perfectly defined, even when the thickness of the crystal is considerable. The more complicated figures produced by two plates of crystal are still more liable to obliteration; and the use of monochromatic light is in this case even more important for maintaining the integrity of the phenomena.

One essential requisite for bringing out the figures in question is purity of colour in the light employed. On this account the ordinary method of absorption by coloured media fails; and it is only by the use of a monochromatic source of light that a satisfactory result can be obtained. For eye observations, a spirit lamp, sometimes with the addition of a little salt, suffices; but the illumination from this source is insufficient for projection on a screen. My first attempt at supplying this requisite was made by replacing the lime in an oxy-hydrogen lamp by a cylinder of carbon through which a hole was bored and filled with chloride of sodium. This answered very well as long as there was enough sodium in the presence of the jet; but it was found difficult to maintain a constant supply at the particular point required. The carbon and sodium were next replaced by a kind of glass formed by melting borax, and to this small pieces of hard German glass were occasionally added. A bead of this substance was placed in a small platinum cup, so fixed that the jet of mixed gases could play upon it at a distance of about three-sixteenths of an inch. This arrangement serves perfectly for laboratory and experimental work; but for projection on a large scale a still more powerful source of light is required. For a burner adapted to lecture-purposes I am indebted to a suggestion of Professor Dewar. The burner consists of an oxy-hydrogen jet, with the addition to the hydrogen tube of a chamber containing metallic sodium. The metal is volatilized by a Bunsen's burner placed below it; so that the hydrogen emerges charged with sodium vapour. The result is a bright monochromatic light.

Let us begin our experiments with the simple case of a Babinet's compensator traversed by a beam of parallel rays. This instrument consists of two quartz wedges, having the axis of the crystal contained in one of the plane faces of each; but the axis in one wedge is parallel, in the other perpendicular to the refracting edge. The wedges are usually made with a refracting angle too small to show dispersion, but of such a thickness that the bands of colour due to their wedge form, shown by each singly, are pale. When the wedges are placed with their edges turned in opposite directions they form a plate substantially of uniform thickness; but owing to the fact that the axes are perpendicular to one another, the ray which has traversed one wedge as the ordinary or swifter, will traverse the other as the extraordinary or slower ray. Hence the total retardation of the rays will be that due to the difference of thickness of the wedges at each point; and therefore the compound plate will be optically of thickness zero along a central line parallel to the edges, and of increasing thickness towards each side.

If one of the wedges be turned through a right angle (say, to its second position), so that their axes are coincident, the two will form a wedge whose thickness, greatest at one angle, diminishes diagonally, at a rate double of that of each wedge singly.

If the same wedge be turned through a second right angle (say, to its third position), the refracting edges will be coincident, and the axes at right angles. The whole will then form substantially a wedge with a refracting angle double of that of each. But since the rays will at every point have traversed an equal thickness of each crystal, the one as an ordinary and the other as an extraordinary ray, the retardation will be everywhere neutralized.

In the first position, with white light, there will be a central band, dark or bright according as the polariser and analyser are crossed or coincident; and on each side first a white, or dark, band, and then coloured bands, whose tints are fainter in proportion to their distance from the centre. With monochromatic light, the bands are alternately dark and bright, and all equally well defined. In the second position, the bands are diagonal, and are perfectly distinct with monochromatic light, even when, as usual, the thickness of the wedges is too great to show them with white light. In the third position the field is uniformly dark or bright, whatever be the nature of light used.

If convergent or divergent light is used, uniaxal crystals (with which alone we are here concerned), when cut perpendicularly to their axes, show the well-known systems of isochromatic rings and dark brushes. When cut at other angles, they show portions of the same systems. But when two such plates are used in combination, theory indicates the presence of some secondary phenomena, which it is our business now to investigate. Of these only very small portions are visible with white light; but with monochromatic light the configuration may be traced throughout the entire field of view.

The effect of the two plates of crystal in producing secondary effects due to the crossing of two sets of isochromatic curves may be well illustrated by Tisley's harmonograph. In this instrument the figures due to vibrations having two rectangular components of (generally) different periods, are approximately produced. This is effected by a compound pendulum; but with the details of this ingenious piece of mechanism we are not here concerned. The curves due to various intervals between the components are well known; but the instrument does not accurately trace out circles or ellipses, or any other re-entering curves; for, owing to the friction, which gradually diminishes the amplitude of vibration, its traces are transcendental curves, or spirals. But on the other hand the friction, and the consequent deviation of the curves from their normal form, is so small that the various turns of the spirals may be taken as representing a series of rings with a sufficient degree of accuracy for our present purpose. Two features of the harmonograph curves should be noticed. First, as actually drawn, they are not geometrical lines, but bands of finite breadth; and on that account are the better suited to represent the interference rings of crystals. Secondly, the distance between the several convolutions of the harmonograph curves is greatest near the outer part of the figure and less towards the centre, while in the crystal rings the reverse is the case; and with reference to any secondary figure due to the crossing of the curves or rings, this difference must be borne in mind.

The curves here used as representing the rings of uniaxal crystals are those produced when the two vibrations of the pendulum are in unison, viz. they are as nearly circular as may be, but it is difficult to avoid a slightly elliptic form. A plate, originally drawn by the instrument, has been photographed twice; and the two facsimiles are now together projected on the screen. The secondary figures in question are then seen in the portion of the field comprised between the centres. I have selected three such pairs and fixed them with their centres at suitable distances; one of them shows ellipses; another parallel straight lines; the third hyperbolas, as secondary figures. If one plate be made to slide over the other, the following effects are usually observed. When the centres are near together, the crossing of the curves gives rise to secondary hyperbolas; as the centres recede, the hyperbolas, at first rectangular, become oblique; they then collapse into straight lines parallel to the plane passing through the axes; and finally they are converted into ellipses approximating more and more to circles as the centres recede still farther from one another. I have, however, found a pair of plates in which the order of the figures is reversed, and which consequently represent the phenomena as they actually occur with crystals.

Similar secondary figures are produced by two plates of crystal used together, as will presently be seen. In order to examine their mode of formation, it will be simplest to begin with some thick plates and moderately convergent light, by which means the details may be brought out on a sufficiently large scale. For this purpose I have three pairs of plates, cut respectively at 67° 30', 45°, 22° 30' to the axis. These, and more particularly the second, are used for producing the well-known Savart's bands. If convergent light be made to traverse any of these plates, some portion of the rings are produced. The proportion of the ring system contained in the field of view depends upon the convergence of the rays; and the distance of the centre of the rings from that of the field upon the inclination to the axis at which the plate is cut. In the specimens here used the rings, for reasons stated in connection with Babinet's compensator, are invisible with white, but visible with monochromatic light.

If a pair of these plates be used, and the principal plane of one, originally coincident with that of the other, is made to turn gradually through 180°, then with white light a series of coloured bands is produced. The distance between the bands increases, from a minimum when they are first visible, indefinitely as the angle of turning approaches 180°; while the brilliancy of their tints and accuracy of definition attains its maximum at 90°. With monochromatic light, when the angle between the principal planes of the plates is 0°, the vibrations will traverse the second plate in the same manner as they traversed the first; and the retardation between the ordinary and the extraordinary rays will be double of that due to each plate singly. The two plates will thus act as a plate of double thickness, and the number of bands (portions of rings) visible with one plate, will be doubled. When one plate is turned as before in front of the other, and the angle between their principal planes is gradually increased, the rings due to one plate cross those due to the other; and the intensity of illumination at the overlapping parts will depend upon the angle at which the rings cross one another. Bearing in mind the fact that at every point of the field the polarisation of one ray is parallel, and that of the other perpendicular to the direction of the ring, it is seen that when the rings cross at 90°, the ray which traversed the first plate as an ordinary, will traverse the second as an extraordinary ray; and consequently the retardation due to the two plates together will be the difference of that due to each alone. The result will therefore be similar to that produced by the two wedges when their refracting edges are at right angles; and the field will be crossed by diagonal straight lines.

For all other angles of crossing, including the case particularized above, the following will take place. Each ray which enters the first plate will emerge as two rays polarised, the one radially, the other tangentially with respect to the rings, and with a certain retardation. On entering the second plate, the vibrations of each of these rays will be again resolved radially and tangentially with respect to the rings due to the second plate. Each of the new components, whether radial or tangential, will therefore consist of two parts, generally of different intensities, one of them having been retarded behind the other in their passage through the first plate. The two parts of each tangentially vibrating ray will partially interfere; and so likewise will the two parts of each radially vibrating ray. In consequence of this interference, the two rays emerging from the second plate will in general be of different intensities, and one of them will be retarded behind the other in their passage through that plate. Finally all the vibrations will again be resolved into one plane by the analyser; and when so resolved they will in general partially interfere. This partial interference will cause the dark rings due to the first plate to be broken with patches of partial brightness, and the bright rings with patches of partial darkness, where the rings of one system cross those of the other. The general effect is in many cases that of a diaper-pattern over the field of view.

The distribution of these interruptions and the nature of the secondary figures which they form, will depend upon the curvature and angle of crossing, of the rings at each point of the field. The mathematical formulæ, which give the details of the illumination, present no difficulty beyond tediousness of calculation. And it will be sufficient here to say that, when carried only to a first approximation (with respect to the angle of incidence of the rays) they indicate, for the secondary figures, a series of straight lines alternately dark and bright, known as "Savart's bands." When carried to a second approximation the formulæ indicate that in the neighbourhood of the ring-centres, the secondary figures will be conic sections. When the principal planes of the crystals (planes containing the axis and the normal to the plate) are at 180° to one another, the conic sections are central. In that case, the expression for the square of one of the principal axes of the curve is a cubic in the line of the angle at which the crystal has been cut. This expression when equated to zero must, by the theory of equations, have one real root; in other words, it will vanish for one particular value of the angle, and be negative for greater values and positive for less values of the angle. If, therefore, the crystals be cut at an angle to the axis smaller than the angle given by the cubic equation, they will, when placed with the axes inclined to opposite sides of the field, show hyperbolas for the secondary figures; when cut at a certain angle (about 59° 50' in the case of quartz), the figures will be straight lines parallel to the line joining the ring-centres; and when cut at a greater angle, the figures will be ellipses, approximating to circles as the angle of section approaches to 90°.

But leaving aside the mathematical aspect of the question, the principal interest of the method of monochromatic light consists in the simplicity of the results, and in the opportunity which it affords of examining in detail all the effects due to two plates of crystal. It enables us in fact to follow the peculiarities of the secondary figures throughout the entire field of view, and to trace by a continuous process the modifications which these figures undergo when the relative positions of the crystals are changed.

Many of these effects are best seen with the optical arrangement usually employed for showing the crystal rings. I have here four pairs of quartz plates cut respectively at 45°, 59° 50′, 67° 30′, 90°, to the axis; the plates of each pair having their principal planes at 180° to one another. The first of these shows, in the region about the line joining the centres of the ring systems, a series of ellipses for the secondary figures; the second shows straight lines; the third oblique hyperbolas; the fourth rectangular hyperbolas. With a view, however, of exhibiting not only these the more important, but also all intermediate phases of the phenomena, I have prepared two curved sections of quartz, each forming nearly a quadrant, and cut at one end perpendicular, at the other nearly parallel to the axis. By placing these end to end, with the axes at opposite ends, and sliding one over the other, all the phases of the secondary figures can be shown in succession. And if the point midway between the axes be kept in the centre of the field of view, the figures will be symmetrical; otherwise unsymmetrical. Beginning with the ends, which are perpendicular to the axis, together in the field of view, and causing the quadrants to slide at the same rate in opposite directions, the secondary circles will be seen elongating into ellipses, the ellipses stretching out until they pass into parallel straight lines; and lastly, these lines contracting towards the centre of the field and diverging towards the sides, until they form hyperbolas, the obliquity of which gradually diminishes as we approach that part of the section which is parallel to the axis.

The case of two quartz plates cut at an angle of 67° 30′ to the axis, whereof one is free to revolve in front of the other, gives rise to some interesting transformations of the secondary figures. When the principal planes of the crystals are coincident, the field generally shows rings double in number to those due to one plate. But towards the side away from which the axes are directed the rays are more nearly parallel, while towards the opposite side the rays are more nearly perpendicular to the axis. Hence, towards the first side, the rings will approximate more nearly to circles, and towards the other they will show indications of becoming branches of hyperbolas. As the principal planes are turned round in opposite directions from their initial position, the secondary figures begin to appear. At 45° discontinuous bands with hyperbolic curvature towards the part of the field most distant from the ring centres are seen. At 90° these bands become continuous and sharply defined, while towards the ring-centres a portion of the ellipses may be observed entering the field. Beyond 90° the hyperbolic branches leave the field, and the rectilinear part of the bands is replaced by the diaper-pattern described in a former part of this lecture; while the ellipses are gradually elongated to parallel straight lines, and then are converted into hyperbolas. At 180° the hyperbolas occupy the centre of the field.

Many similar experiments may be made with biaxal crystals; but it would exceed our present limits to describe them here. I will, therefore, confine myself to a repetition with monochromatic light of the well-known experiment of showing the passage of mica from its proper biaxal form to an apparently uniaxal form, by crossing a number of films of the crystal.

It has perhaps been abundantly shown on more than one occasion in this theatre, by the spectra of polarised light, that colour is in fact a shadow; and that the varied tints, produced by crystals in light under this condition, are due to selective shadows thrown as it were over the various components of a colourless beam. And the present method of monochromatic light affords a striking illustration of the fact that suppression of light is a factor of all chromatic effects. But beside this, the method affords an opportunity of tracing, by a continuous process, the transformations of the results due to a continuous change of crystalline circumstance. The experiments supply a fresh instance of the fact that nature does nothing *per saltum;* and they may perhaps be regarded as adding one more link, however insignificant, to that great chain of continuity which is gradually binding more closely together the diversified phenomena of the material universe.

Lastly, they remind us that the most beautiful, the most delicate, the most instructive features of nature are dependent, not so much on abundance of material, on profusion of ornamentation, or variety of display, as upon simplicity of character, on fidelity to truth, and on strict but willing obedience to law.

[W. S.]

Friday, May 10, 1878.

C. WILLIAM SIEMENS, Esq. D.C.L. F.R.S. Vice-President,
in the Chair.

SIR WILLIAM THOMSON, LL.D. F.R.S.

The Effects of Stress on the Magnetization of Iron, Cobalt, and Nickel.

THOUGH, as discovered by Faraday, every substance has a susceptibility for inductive magnetization, the three metals, iron, nickel, cobalt, stand out so prominently from among the other known chemical elements, that they only are commonly regarded as *the* magnetic metals, and the magnetism of all other substances is so feeble as to be comparatively almost imperceptible.

The magnetization of each of the three magnetic metals is greatly affected by mechanical stress. From the beginnings of magnetic science it must have been known that the magnetism of iron and steel is disturbed, sometimes lost or much diminished, by blows, striking the metal with a hammer, or letting it fall on hard ground. Gilbert, nearly three hundred years ago, showed that bars of soft iron held in the direction of the dipping needle and struck violently by a hammer acquire much more magnetism, and again more reverse magnetism when inverted in that line, than when placed in those positions gently without shock of any kind. An ordinary fireside poker shows these effects splendidly, with no other apparatus than a little pocket-compass or a sewing needle, magnetized and slung horizontally, hanging by a fine silk thread. If habitually the poker rests upright the upper end will be found a true north pole, the lower a true south when first tested by the needle. Holding the poker with exceedingly gentleness, invert it :—The end that was down, though now up, is still a true south pole, and repels the north end (or true south pole) of the movable suspended needle. A gentle tap with the hand on the poker now produces a surprising result. Instantly it yields to the terrestrial influence, and its upper end, becoming a true north pole, attracts the northern end of the suspended needle. Even more surprising is the slightness of the agitation which suffices to shake the retained magnetism of a former position out of a soft iron wire, and let it take the magnetization due to the position in which it is held. A superstitious person would say ;—that is animal magnetism !—when he sees an iron wire becoming a notably effective magnet when held vertically and rubbed gently from end to end between finger and thumb.

Changes of magnetism produced by mechanical agitation are shown to a much greater degree in thin bars than in thick ones; and when the diameter exceeds a quarter of the length they are hardly sensible. Hence when the "Flinders bar" is applied to compensate the error produced in a ship's compass by change of magnetic latitude, its length ought not to be more than six or seven times its diameter; and for the same reason long iron rods or stanchions in the neighbourhood of a compass are very detrimental to its trustworthiness. Half a hundredweight of iron in the shape of rails or awning stanchions, too often to be found very near a compass, are more dangerous than tons or hundreds of tons in the shape of heavy steam steering gear, or of armoured turrets in an ironclad.

A piece of iron left in the Royal Institution by Faraday, with a label in his own handwriting to the effect that it had been between three hundred and four hundred years fixed in a vertical position in the stonework of the Oxford Cathedral, having been given to him by Dr. Buckland in May, 1835, was exhibited and tested. It was found to have its upper end a true north pole. It was inverted before the audience, and instantly that end became a true south pole and the other a true north pole. Thus nearly four hundred years in one position had done nothing to *fix* the magnetism. In its inverted position it was hammered violently on each end by a wooden mallet: this increased the magnetism somewhat, but did not *fix* it. The bar was inverted again, and then, in its first position, its original upper end, now up again, became again a true north pole.

The stoutness of the bar (that is to say, the greatness of the proportions of its breadth and thickness to its length) were such, that if of modern iron, it probably would not have behaved as it did; but probably also it may have been superior in "softness" to the ordinary run of modern bar iron.

Bars of nickel and cobalt, unique and splendid specimens, for which the speaker was indebted to the celebrated metallurgical chemist, Mr. Wharton, of Philadelphia, were exhibited, and found to show effects of concussion quite as do bars of iron of different qualities.

An altogether new effect of stress was discovered about ten years ago by Villari, according to which longitudinal pull augments the temporary induced magnetism of soft iron bars or wires when the magnetizing force is less than a certain critical value; and diminishes it when the magnetizing force exceeds that value; and augments the residual magnetism when the magnetizing force, whether it has been great or small, has been removed.

The speaker had measured approximately the Villari critical value, and found it to be about twenty-four times the vertical component of the terrestrial magnetic force (or about 10 C.G.S. units). The maximum effect in the way of augmentation by pull he had found with about six times the Glasgow vertical force. He had found for bars of nickel and cobalt opposite effects to those of Villari for soft iron, and had found a maximum value, with a certain degree of magnetizing force, and evidence making it probable that a critical magnetizing force would be found for each of these metals also, such that the magnetization would be *increased* by pull when the magnetizing force exceeds it.

The speaker had found corresponding effects of *transverse pull* in soft iron, and had found them to be correspondingly opposite to those discovered by Villari for longitudinal pull. The transverse pull was produced by water pressure in the interior of a gun-barrel applied by a piston and lever at one end. Thus a pressure of about 1000 lbs. per square inch, applied and removed at pleasure, gave effects on the magnetism induced in the vertical gun-barrel by the vertical component of the terrestrial magnetic force, and, again, by an electric current through a coil of insulated copper wire round the gun-barrel, which were witnessed by the audience. When the force magnetizing the gun-barrel was anything less than about sixty times the Glasgow value of the vertical component of the terrestrial force, the magnetization was found to be less with the pressure on than off. When the magnetizing force exceeded that critical value, the magnetization was *greater* with the pressure on than off. The residual (retained) magnetism was always less with the pressure on than off (after ten or a dozen "*ons*" and "*offs*" of the pressure to shake out as much of the magnetization as was so loosely held as to be shaken out by this agitation).

It is remarkable that the critical amount of the magnetizing force in respect to effect of transverse pull is more than double that of the Villari effect of longitudinal pull. Thus for intermediate amounts of force (say forces between 10 and 25 C.G.S. units), both longitudinal pull and transverse pull diminish the induced magnetization. Hence it is to be inferred that equal pull in all directions would diminish, and equal positive pressure in all directions would increase, the magnetization under the influence of force between these critical values, and through some range above and below them; and not improbably for all amounts, however large or small, of the magnetizing force; but further experiment is necessary to answer this question.

A most interesting further inquiry in connection with this subject is to find if æolotropic stress (pressure unequal in different directions), beyond the limits of elasticity, leaves in iron, nickel, or cobalt a permanent æolotropic difference of magnetic susceptibility in different directions analogous to that discovered thirty years ago by Tyndall in the diamagnetic quality of soft, imperfectly elastic material, such as fresh bread. Special difficulties prevented the speaker from obtaining any results thirty years ago, when he tried to discover corresponding effects in iron; but the investigation is not hopeless, and he intends to resume it.

[W. T.]

Friday, June 14, 1878.

GEORGE BUSK, Esq. F.R.S. Treasurer and Vice-President,
in the Chair.

PROFESSOR JAMES DEWAR, M.A. F.R.S.

The Liquefaction of Gases.

DALTON, twenty-two years before the discovery of the liquefaction of the gases, commences his essay 'On the Force of Steam or Vapour from Water and various other Liquids, both in a Vacuum and in Air,' with the following marvellous anticipation of subsequent research: "There can scarcely be a doubt entertained respecting the reducibility of all elastic fluids of whatever kind into liquids; and we ought not to despair of effecting it in low temperatures, and by strong pressure exerted upon the unmixed gases." * The same ideas are reiterated in Dalton's 'New System of Chemical Philosophy,' which appeared in 1808, yet no definite proof of the accuracy of his ideas regarding the relation of the gaseous and liquid states of matter were forthcoming until 1823.

The first information regarding the liquefaction of a gas is found in a letter sent by Faraday to Dr. Paris, the biographer of Sir Humphry Davy.

"DEAR SIR,
"The oil you noticed yesterday turns out to be liquid chlorine.
"Yours faithfully,
"MICHAEL FARADAY."

The letter is not dated, but we know from Dr. Paris that it must have been the 6th of March, 1823.

Faraday was then engaged in an investigation on the hydrate of chlorine. Davy suggested the examination of the action of heat in a closed vessel, as likely to yield interesting results, without stating, however, what he anticipated would occur. The letter addressed to Dr. Paris has reference to the result of this experiment. The solid hydrate of chlorine on being heated in a closed vessel, separated

---

* 'Literary and Philosophical Society of Manchester,' vol. v. 1802.

into water and a dense yellow oil, which Faraday proved to be pure liquid chlorine. Davy saw the importance of the discovery, and liquefied hydrochloric acid by the pressure of its own vapour, produced by generating it in a closed vessel.

Davy also succeeded in liquefying gases by a method which, at first view, appears very paradoxical—by the application of heat! The method consists in placing them in one leg of a bent sealed tube, confined by mercury, and applying heat to ether, or alcohol, or water, in the other end. In this manner, by the pressure of the vapour of ether, he liquefied prussic acid and sulphurous acid gas; which gases, on being reproduced, occasioned cold. There can be little doubt, he thinks, that these general facts of the condensation of the gases will have many practical applications. "They afford means of producing great diminutions of temperature, by the rapidity with which large quantities of liquids may be rendered aëriform; and as compression occasions similar effects to cold, in preventing the formation of elastic substances, there is great reason to believe that it may be successfully employed for the preservation of animal and vegetable substances for the purposes of food."

Faraday continued the experiments, and succeeded in liquefying many gases by the pressure caused by their continuous formation in a limited space, either through the ordinary reactions used for producing them, or by applying heat to some easily decomposed compound. By observing the volume of the liquid gas as compared with the same in the gaseous state the relative specific gravities could be determined, and by introducing small air manometers, the amount of pressure in atmospheres required to condense the various gases, and the influence of temperature on the pressure, could be observed. This series of experiments was completed in 1823.

Thilorier, in 1835, devised an apparatus to produce liquid carbonic acid on a large scale, and made the great discovery that the fluid ejected into air produced the solid in the form of snow. This substance while evaporating gave a temperature which he estimated about minus 100° C. The fluid gas was soluble in all proportions in ether, turpentine and bisulphide of carbon. It was four times more dilatable than air, and every additional degree of temperature added one atmosphere of pressure to the tension of the gas. The density of the vapour increased in a much greater proportion than the pressure, so that the law of Mariotte was no longer applicable to such a highly compressed condition of the gas.

Mitchell, in 1839, found the true temperature of the solid while evaporating in air was minus 78°, but mixed with ether and vaporized in vacuo it was reduced to minus 99°. With this powerful means of reducing temperature he solidified sulphurous acid.

In order to demonstrate the exceptional temperatures we have to deal with, a thermo-junction of iron-copper has the advantage over a fluid thermometer of having a very small mass and great sensibility, so that the length of a degree on our scale is easily observed. This junction placed in ice marks 0°. When placed in solid carbonic acid it maintains a fixed temperature of minus 80°, a point on the thermometer scale as far below the freezing point as the boiling point of spirits of wine is above it. If ether is added to the solid acid, a fluid which does not freeze, and which has the remarkable property of dissolving the carbonic acid, the same low temperature is kept up. The ether carbonic acid bath acts more rapidly, from the immersed body being wetted, and thus brought into close contact with the fluid. This thermometer is far more delicate than either an alcohol or bisulphide of carbon one, and yet it records a constant temperature, and why? The reason is, solid carbonic acid is actually boiling; this solution in ether is giving off continuously bubbles of gas like a boiling fluid. This is more readily seen if the snow is compressed into the form of ice by strong hydraulic pressure. This cylinder of clear carbonic acid ice weighs nearly half an ounce, and continues to give off gas when immersed in ether, as if it were a piece of marble dissolving in an acid. In this state it is one and a half times denser than water; and although the mass is at a temperature at least 20° below that of the Arctic regions, yet, strange to say, it is not covered with ice when dropped into a mass of water. It is in reality coated with a layer of gas constantly renewed, and thus the water never gets into actual contact. The solid is in a condition similar to that of the spheroidal state of liquids.

Another liquid gas that boils at a still lower temperature is nitrous oxide, or laughing gas. The temperature of boiling nitrous oxide freely exposed to the air is, as you observe, about 10° lower than that of carbonic acid. A temperature of minus 90° is thus the lowest we can reach through vaporization of a conveniently liquefiable gas, freely exposed to the atmosphere. As the boiling point, however, is dependent wholly upon the pressure of the gaseous atmosphere, if the pressure is removed from the surface of the liquid with a greater rapidity than it can accumulate through the generation of gas, much lower temperatures may be commanded. The lowest temperature recorded by Faraday was minus 110, and the experiment just made proves this is about the limit to be reached with his air-pump. The experiments of Cagniard de-la-Tour and Thilorier opened up new ideas and methods for further investigation, and Faraday attacked the subject for the second time in 1844. By the combined action of pressure, and the low temperature of the carbonic acid ether bath in vacuo, he now succeeded in liquefying all the gases with the exception of six, viz. oxygen, hydrogen, nitrogen, carbonic oxide, nitric oxide, and marsh gas, and obtained many of them in the form of solids. The most valuable and laborious part of the research was the determination of the tensions of the liquid gases. He was the first to show the great importance to be attached to these constants in defining the purity of gases. Thus he discovered that phosphuretted hydrogen, nitrous oxide, and olefiant gas, however carefully prepared, could by this method be proved to contain impurities. The experiments of Cagniard de-la-Tour induced him to conclude that no amount of pressure would liquefy the permanent gases unless the temperature was below some definite point of the thermometric scale. In fact he had a very clear idea of what is now called the critical point.

The following extracts from the investigation of 1844 are worthy of grave attention. They prove the wonderful accuracy of scientific prophecy and display the working of a mind, full of subtle powers of divination into nature's secrets:—

On the Liquefaction and Solidification of Bodies generally existing as Gases.[*]

"The experiments formerly made on the liquefaction of gases, and the results which from time to time have been added to this branch of knowledge, especially by M. Thilorier, have left a constant desire on my mind to renew the investigation. This, with considerations arising out of the apparent simplicity and unity of the molecular constitution of all bodies when in the gaseous or vaporous state, which may be expected, according to the indications given by the experiments of M. Cagniard de-la-Tour, to pass by some simple law into their liquid state, and also the hope of seeing nitrogen, oxygen, and hydrogen, either as liquid or solid bodies, and the latter probably as a metal, have lately induced me to make many experiments on the subject; and though my success has not been equal to my desire, still I hope some of the results obtained, and the means of obtaining them, may have an interest for the Royal Society; more especially as the applications of the latter may be carried much further than I as yet have had an opportunity of applying them.

"But as my hopes of any success beyond that heretofore obtained depended more upon depression of temperature than on the pressure which I could employ in these tubes, I endeavoured to obtain a still greater degree of cold. There are, in fact, some results producible by cold, which no pressure may be able to effect.

"Thus solidification has not as yet been conferred on the fluid by any degree of pressure.

"Again that beautiful condition which Cagniard de-la-Tour has made known, and which comes on with liquids at a certain heat, may have its point of temperature for some of the bodies to be experimented with, as oxygen, hydrogen, nitrogen, &c., below that belonging to the bath of carbonic acid and ether; and in that case, no pressure which any apparatus could bear would be able to bring them into the liquid or solid state.

"Thus though as yet I have not condensed oxygen, hydrogen, or nitrogen, the original objects of my pursuit, I have added six substances, usually gaseous, to the list of those that could previously be shown in the liquid state, and have reduced seven, including

---

[*] 'Phil. Trans.,' 1845.

ammonia, nitrous oxide, and sulphuretted hydrogen, into the solid form. And though the numbers expressing tension of vapour cannot (because of the difficulties respecting the use of thermometers and the apparatus generally) be considered as exact, I am in hopes they will assist in developing some general law governing the vaporization of all bodies, and also in illustrating the physical state of gaseous bodies as they are presented to us under ordinary temperature and pressure.

"Whether the same law may be expected to continue when the bodies approach near to the Cagniard de-la-Tour state is doubtful. That state comes on sooner in reference to the pressure required, according as the liquid is lighter and more expansible by heat and its vapour heavier; hence indeed the great reason for its facile assumption by ether.

"But though with ether, alcohol, and water, that substance which is most volatile takes up this state with the lowest pressure, it does not follow that it should always be so; and, in fact, we know that ether takes up this state at a pressure between thirty-seven and thirty-eight atmospheres, whereas muriatic acid, nitrous oxide, carbonic acid, and olefiant gas, which are far more volatile, sustain a higher pressure than this without assuming that peculiar state, and whilst their vapours and liquids are still considerably different from each other. Now whether the curve which expresses the elastic force of the vapour of a given fluid for increasing temperatures continues undisturbed after that fluid has passed the Cagniard de-la-Tour point or not is not known, and therefore it cannot well be anticipated whether the coming on of that state sooner or later with particular bodies, will influence them in relation to the more general law referred to above.

"The law already suggested gives great encouragement to the continuance of those efforts which are directed to the condensation of oxygen, hydrogen, and nitrogen, by the attainment and application of lower temperatures than those yet applied.

"If to reduce carbonic acid from the pressure of two atmospheres to that of one, we require to abstract only about half the number of degrees that is necessary to produce the same effect with sulphurous acid, it is to be expected that a far less abstraction will suffice to produce the same effect with nitrogen or hydrogen, so that further diminution of temperature and improved apparatus for pressure may very well be expected to give us these bodies in the liquid or solid state."

The classical researches of Regnault in 1847, on the compressibility of gases, proved that the permanent gases, with the exception of hydrogen, deviated from Mariotte's law in the same direction as those that were liquefiable, although to a much smaller amount. If Mariotte's law represents the perfect gaseous state, then hydrogen is a gas that shows greater resistance to alteration of volume than would result from this law. Natterer, in 1854, condensed hydrogen, oxygen, and nitrogen to a pressure of 2700 atmospheres without any change of state being observed. His experiments show that oxygen and nitrogen under great compression, behave like hydrogen, that is, become less and less compressible.

Andrews, in 1861, subjected the six gases that resisted the efforts of Faraday when cooled to the temperature of the carbonic acid ether bath to a pressure of at least 500 atmospheres without producing any change of state. During the course of this investigation Andrews observed that liquid carbonic acid raised to a temperature of 31° lost the sharp concave surface of demarcation between the liquid and gas and at last disappeared. The space was now occupied by a homogeneous fluid, which exhibited, when the pressure was suddenly diminished, or temperature slightly lowered, a peculiar appearance of moving or flickering striæ throughout its entire mass, owing to great alterations of density. At temperatures above 31° apparent liquefaction or separation into two distinct forms of matter could not be effected even when the pressure reached 400 atmospheres. This limiting temperature of the liquid state Andrews calls the "critical point." He has been engaged for the last twelve years in completing his researches on the gaseous state, which for accuracy and elegance can only be equalled by the work of Regnault.

Recent experiments resulting in the liquefaction of the permanent gases have been made simultaneously by M. L. Cailletet, of Paris, and M. R. Pictet, of Geneva. Each had large experimental resources and facility for conducting such investigations.

Both experimenters used the same ingenious method of reaching temperatures far below that of the carbonic acid bath in vacuo, by allowing the gases cooled to the above temperature and highly compressed, to expand suddenly. This expansion involves a rapid absorption of heat, and this is chiefly taken from the molecules of the gas forcing a portion to pass into the fluid or solid state.

Cailletet's apparatus is a modified form of that employed by Andrews in his great research on the continuity of the gaseous and liquid states of matter, and will be readily understood on seeing it in operation.*

Pictet's experiments were conducted on a manufacturing scale. A sulphurous acid ice machine cooled carbonic acid or nitrous oxide to a temperature of minus 65°, so that a pressure of from four to six atmospheres is all that is required to cause liquefaction. A system of compressing and exhausting pumps, worked at the rate of one hundred revolutions per minute by a steam engine, thus produce about 16 lbs. of liquid carbonic acid per hour. Vaporized in the very perfect vacua he can command, a temperature of minus 130° may be kept up for any length of time. The operations are so arranged in cycle that the

---

* Handsomely presented to the Royal Institution, by Dr. Warren De La Rue, for the purpose of illustrating this lecture.

same carbonic acid or nitrous oxide may be used again and again. [This photograph shows the general arrangements in the laboratory.] No pump is used by Pictet to compress the oxygen or hydrogen. The gases are produced by chemical reaction as in Faraday's tubes, only he replaces the glass by a strong tube of copper connected with a large iron bomb, in which by the application of heat the decomposition takes place. The narrow copper tube is cooled to minus 130° or 140° by the method explained, and here the liquefaction takes place. A screw stop-cock at the extremity of the tube allows the liquid gas to be ejected. The pressure of liquid oxygen at minus 138 is at least 273 atmospheres, and its density is a little less than that of water. Hydrogen in the liquid state at a temperature of minus 140° has a pressure of about 320 atmospheres and appears to solidify in the tube when the fluid jet is allowed to escape. The jet of liquid has a steel blue colour.

If oxygen, nitrogen, or air is compressed in Cailletet's apparatus to 250 atmospheres, something like three tons on the square inch, and the pressure suddenly released, there is an instantaneous cloud formed within the tube due to partial liquefaction.

This tube contains the first gas Cailletet succeeded in liquefying, the hydro-carbon called acetylene, which was discovered by an old Assistant of the Royal Institution, and is one of the most important bodies in the whole range of organic chemistry. After it is compressed to about 50 atmospheres a clear colourless fluid will result.

The work of Faraday has been completed; every gas may be forced to appear as a liquid.

[J. D.]

Friday, February 21, 1879.

WILLIAM SPOTTISWOODE, Esq. M.A. D.C.L. Pres. R.S. Vice-President, in the Chair.

PROFESSOR ROSCOE, LL.D. F.R.S.

A New Chemical Industry, established by M. Camille Vincent.

"AFTER I had made the discovery of the *marine acid air*, which the vapour of spirit of salt may properly enough be called, it occurred to me that, by a process similar to that by which this *acid* air is expelled from the spirit of salt, an *alkaline* air might be expelled from substances containing the volatile alkali. Accordingly I procured some volatile spirit of sal-ammoniac, and having put it into a thin phial and heated it with the flame of a candle, I presently found that a great quantity of vapour was discharged from it, and being received into a basin of quicksilver it continued in the form of a transparent and permanent air, not at all condensed by cold." These words, written by Joseph Priestley rather more than one hundred years ago, describe the experiment by which ammonia was first obtained in the gaseous state.

Unacquainted with the composition of this alkaline air, Priestley showed that it increased in volume when electric sparks are passed through it, or when the alkaline air (ammonia) is heated the residue consists of inflammable air (hydrogen).

Berthollet, in 1785, proved that this increase in bulk is due to the decomposition of ammonia into nitrogen and hydrogen, whilst Henry and Davy ascertained that two volumes of ammonia are resolved into one volume of nitrogen and three volumes of hydrogen.

The early history of sal-ammoniac and of ammonia is very obscure. The salt appears to have been brought into Europe from Asia in the seventh century, probably from volcanic sources. An artificial mode of producing the ammoniacal salts from decomposing animal matter was soon discovered, and the early alchemists were well acquainted with the carbonate under the name of *spiritus salis urinæ*. In later times sal-ammoniac was obtained from Egypt, where it was prepared by collecting the sublimate obtained by burning camels' dung.

Although we are constantly surrounded by an atmosphere of nitrogen, chemists have not yet succeeded in inducing this inert substance to combine readily, so that we are still dependant for our supply of combined nitrogen, whether as nitric acid or ammonia, upon the decomposition of the nitrogenous constituents of the bodies of plants and animals. This may be effected either by natural decay, giving rise to the ammonia which is always contained in the atmosphere, or by the dry distillation of the same bodies, that is, by heating them strongly out of contact with air; and it is from this source that the world derives the whole of its commercial ammonia and sal-ammoniac.

Coal, the remains of an ancient vegetable world, contains about 2 per cent. of nitrogen, the greater part of which is obtained in the form of ammonia when the coal undergoes the process of dry distillation. In round numbers two million tons of coal are annually distilled for the manufacture of coal gas in this country, and the ammoniacal water of the gasworks contains the salts of ammonium in solution.

According to the most reliable data 100 tons of coal were distilled so as to yield 10,000 cubic feet of gas of specific gravity $0 \cdot 6$, giving the following products, in tons :—

Gas.	Tar.	Ammonia Water.	Coke.
$22 \cdot 25$	$8 \cdot 5$	$9 \cdot 5$	$59 \cdot 75$ average.

This ammonia water contains about $1 \cdot 5$ per cent. of ammonia, hence the total quantity of the volatile alkali obtainable from the gasworks in England amounts to some 9000 tons per annum.

A singular difference is observed between the dry distillation of altered woody fibre as we have it in coal, and woody fibre itself. In the products of the first operation we chiefly find in the tar the aromatic hydrocarbons, such as benzene, whilst in the second we find acetic acid and methyl alcohol are predominant.

The year 1848 is a memorable one in the annals of revolutionary chemistry, for in that year Wurtz proved that ammonia is in reality only one member of a very large family. By acting with caustic potash on the nitriles of the alcohol radicals he obtained the first series of the large class of compound ammonias, the primary monamines. Of these methylamine is the first on our list:—

$$\left.\begin{matrix}CH_3\\CO\end{matrix}\right\}N + 2\,KOH = \left.\begin{matrix}CH_3\\H_2\end{matrix}\right\}N + CO\left\{\begin{matrix}OK\\OK\end{matrix}\right.$$

The years that followed, 1849–51, were prolific in ammoniacal discoveries. Hofmann pointed out that not only one atom of hydrogen in ammonia can be replaced by its equivalent of organic radical, but that two or all the three atoms of the hydrogen in ammonia can be likewise replaced, giving rise to the secondary and tertiary amines, by the following simple reactions :—

$$1.\ CH_3I + \left.\begin{matrix}H\\H\\H\end{matrix}\right\}N = HI + \left.\begin{matrix}CH_3\\H\\H\end{matrix}\right\}N$$

$$2.\ CH_3I + \left.\begin{matrix}CH_3\\H\\H\end{matrix}\right\}N = HI + \left.\begin{matrix}CH_3\\CH_3\\H\end{matrix}\right\}N$$

$$3.\ CH_3I + \left.\begin{matrix}CH_3\\CH_3\\H\end{matrix}\right\}N = HI + \left.\begin{matrix}CH_3\\CH_3\\CH_3\end{matrix}\right\}N$$

To these bodies the names of methylamine, di-methylamine, and tri-methylamine were given. They resemble ammonia in being volatile alkaline liquids or gases, which combine with acids to form crystalline and well-defined salts.

Hitherto these compound ammonias have been chemical curiosities; they have, however, recently become, as has so often been the case in other instances, of great commercial importance, and are now manufactured on a large scale.

We are all well aware that the French beet-root sugar industry is one of great magnitude, and that it has been largely extended in late years. In this industry, as in the manufacture of cane sugar, large quantities of molasses or treacle remain behind after the whole of the crystallizable sugar has been withdrawn. These molasses are invariably employed to yield alcohol by fermentation. The juice of the beet, as well as that of cane sugar, contains, in addition to the sugar, a large quantity of extractive and nitrogenous matters, together with considerable quantities of alkaline salts. In some sugar-producing districts the waste-liquors or spent-wash from the stills—called *vinasses* in French—are wastefully and ignorantly thrown away, instead of being returned to the land as a fertilizer, and thus the soil becomes impoverished. In France it has long been the custom of the distiller to evaporate these liquors (*vinasses*) to dryness, and to calcine the mass in a reverberatory furnace, thus destroying the whole of the organic matter, but recovering the alkaline salts of the beet-root. In this way 2000 tons of carbonate of potash are annually produced in the French distilleries. For more than thirty years the idea has been entertained of collecting the ammonia-water, tar, and oils which are given off when this organic matter is calcined, but the practical realization of the project has only quite recently been accomplished, and a most unexpected new field of chemical industry thus opened out, through the persevering and sagacious labours of M. Camille Vincent, of Paris.

The following is an outline of the process as carried out at the large distillery of Messrs. Tilloy, Delaune, and Co., at Courrières. The spent-wash having been evaporated until it has attained a specific gravity of $1\cdot31$, is allowed to run into cast-iron retorts, in which it is submitted to dry distillation. This process lasts four hours; the volatile products pass over, whilst a residue of porous charcoal and alkaline salts remains behind in the retort. The gaseous products given off during the distillation are passed through coolers, in order to condense all the portions which are liquid or solid at the ordinary temperature, and the combustible gases pass on uncondensed and serve as fuel for heating the retorts.

The liquid portion of the distillate is a very complex mixture of chemical compounds, resembling in this respect the corresponding product in the manufacture of coal gas. Like this latter, the liquid distillate from the spent-wash may be divided into

1. The ammonia-water.
2. The tar.

The ammonia-water of the vinasse resembles that of the coal-gas manufacture in so far as it contains carbonate, sulphydrate, and hydrocyanide of ammonia; but it differs from this (and approximates to the products of the dry distillation of wood) by containing in addition methyl alcohol, methyl sulphide, methyl cyanide, many of the members of the fatty acid series, and, most remarkable of all, *large quantities of the salts of trimethylamine.*

The tar, on re-distillation, yields more ammonia-water, a large number of oils, the alkaloids of the pyridene series, solid hydrocarbons, carbolic acid, and lastly, a pitch of fine quality.

The crude alkaline aqueous distillate is first neutralized by sulphuric acid, and the saline solution evaporated, when crystals of sulphate of ammonia are deposited; and these, after separating and draining off, leave a mother liquor, which contains the more soluble sulphate of trimethylamine. During the process of concentration, vapours of methyl alcohol, methyl cyanide, and other nitrils are given off, these being condensed, and the cyanide used for the preparation of ammonia and acetic acid by decomposing it with an alkali.

Trimethylamine itself is at present of no commercial value, though perhaps the time is not far distant when an important use for this substance will be found. The question arises as to how this material can be made to yield substances capable of ready employment in the arts. This problem has been solved by M. Vincent in a most ingenious way. He finds that the hydrochlorate of trimethylamine, when heated to a temperature of 260°, decomposes into (1) ammonia, (2) free trimethylamine, and (3) chloride of methyl.

$$3 \text{ NMe}_3\text{HCl} = 2 \text{ NMe}_3 + \text{NH}_3 + 3 \text{ MeCl}.$$

By bubbling the vapours through hydrochloric acid the alkaline gases are retained, and the gaseous chloride of methyl passes on to be purified by washing with dilute caustic soda and drying with strong sulphuric acid. This is then collected in a gas-holder, whence it is pumped into strong receivers and condensed.

The construction of these receivers is shown in Fig. 1. They consist of strong wrought-iron cylinders, tested to resist a pressure of 20 kilos. per square centimetre, and containing 50, 110, 220 kilos. chloride of methyl. The liquid is drawn from these receivers by opening the screw tap D, which is covered by a cap C, to prevent injury during transit.

Both ammonia and chloride of methyl are, however, substances possessing a considerable commercial value. The latter compound has up to this time, indeed, not been obtained in large quantities, but it can be employed for two distinct purposes: (1) it serves as a means of producing artificial cold; (2) it is most valuable for preparing methylated dyes, which are at present costly, inasmuch as they have hitherto been obtained by the use of methyl iodide, an expensive substance.

Methyl chloride was discovered in 1804 by MM. Dumas and

Péligot, who obtained it by heating a mixture of common salt, methyl alcohol, and sulphuric acid. It is a gas at the ordinary temperature, possesses an ethereal smell and a sweet taste, and its specific gravity

Fig. 1.

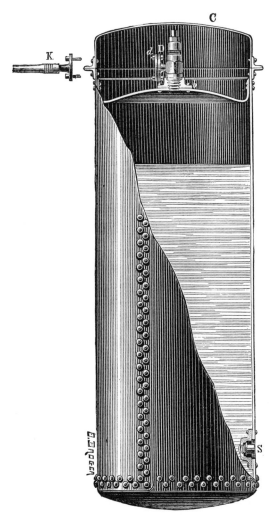

is 1·738. It is somewhat soluble in water (about 3 volumes), but much more in acetic acid (40 volumes), and in alcohol (35 volumes). It burns with a luminous flame, tinged at the edges with green, yielding carbonic and hydrochloric acids. Under pressure, methyl chloride can be readily condensed to a colourless, very mobile liquid, boiling at $-23°$ C. under a pressure of 760 mm. As the tension of the vapour is not high, and as it does not increase very rapidly with the temperature, the liquefaction can be readily effected, and the collection and transport of the liquefied chloride can be carried on without danger.

The following table shows the tension of chloride of methyl at varying temperatures:—

At $0°$ the tension of CH_3Cl is 2·48 atmospheres.
,, 15° ,, ,, 4·11 ,,
,, 20° ,, ,, 4·81 ,,
,, 25° ,, ,, 5·62 ,,
,, 30° ,, ,, 6·50 ,,
,, 35° ,, ,, 7·50 ,,

From these numbers we must of course subtract 1 to obtain the pressure which the vapour exerts on the containing vessel.

As a means of producing low temperatures chloride of methyl will prove of great service both in the laboratory and on a larger industrial scale. When the liquid is allowed to escape from the receiver into an open vessel, it begins to boil, and in a few moments the temperature of the liquid is lowered by the ebullition below $-23°$, the boiling point of the chloride. The liquid then remains for a length of time in a quiescent state, and may be used as a freezing agent. By increasing the rapidity of the evaporation by means of a current of air blown through the liquid, or better by placing the liquid in connection with a good air-pump, the temperature of the liquid can in a few moments be reduced to $-55°$, and large masses of mercury easily solidified. The construction of a small freezing machine employed by M. Camille Vincent is shown in Fig. 2. It consists of a double-cased copper vessel, between the two casings of which the methyl chloride (A) is introduced. The central space (M) is filled with some liquid such as alcohol, incapable of solidification. The chloride of methyl is allowed to enter from the cylindrical reservoir by the screw tap (B) and the screw (S) left open to permit of the escape of the gas. As soon as the whole mass of liquid has been reduced to a temperature of $-23°$, ebullition ceases, the screw (S) may be replaced, and if a temperature lower than $-23°$ be required, the tube (B) placed in connection with a good air-pump. By this simple means a litre of alcohol can be kept for several hours at temperatures either of $-23°$ or $-55°$, and thus a large number of experiments can be performed for which hitherto the expensive liquid nitrous oxide or solid carbonic acid was required.

M. Vincent has recently constructed a much larger and more perfect and continuous form of freezing machine, in which by means of an air-pump and a forcing pump the chloride of methyl is evaporated in the freezing machine and again condensed in the cylinders. This enlarged form of apparatus will probably compete favourably with the ether, and sulphurous acid, freezing machines now in use, as they can be simply constructed, and as the vapour and liquid do not attack metal and are non-poisonous, and as the frigorific effects which it is capable of producing are most energetic.

The second and perhaps more important application of methyl chloride is to the manufacture of methylated colours.

It is well known that rosaniline or aniline-red, $C_{20}H_{19}N_3$, yields compounds possessing a fine blue, violet, or green colour, when a

Fig. 2.

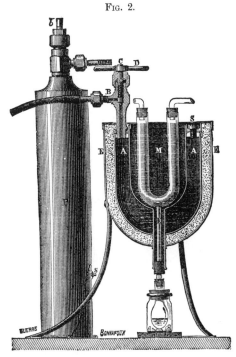

portion of the hydrogen has been replaced by the radicals methyl or ethyl, and the larger the proportion of hydrogen replaced the deeper is the shade of violet which is produced. Thus we have triethyl rosaniline or Hofmann's* violet, $C_{20}H_{16}(C_2H_5)_3N_3$.

By replacing one or two atoms of the hydrogen of aniline by methyl and by oxidizing the methyl anilines thus obtained, Charles Lauth obtained fine violet colours, whilst about the same time Hofmann observed the production of a bright green colouring matter,

---

* Hofmann, 'Proc. Roy. Soc.' xiii. 13 (1863).

now known as iodine green, formed during the manufacture of the violet, and produced from the latter colour by the action of methyl iodide.

In order to prepare aniline green from the pure chloride of methyl, a solution of methyl-aniline violet in methyl alcohol is placed in an iron digester and the liquid rendered alkaline by caustic soda. Having closed the digester, a given quantity of liquid chloride of methyl is introduced by opening a tap, and the digester thus charged is placed in a water bath and heated by a jet of steam, until the temperature reaches 95°, and the indicated pressure amounts to from 4 to 5 atmospheres. As soon as the reaction is complete, the hot water is replaced by cold, and the internal pressure reduced by opening the screw tap of the digester. The product of this reaction heated and filtered, yields the soluble and colourless base, whose salts are green. To the acidulated solution a zinc salt is added to form a double salt, and the green compound is then precipitated by the addition of common salt. By adding ammonia to a solution of the green salt, a colourless liquid is obtained, in which cloth mordanted with tannic acid and tartar emetic becomes dyed of a splendid green.

If rosaniline be substituted for methyl aniline in the preceding reaction Hofmann's violet is obtained. The application of methyl chloride to the preparation of violets and greens is, however, it must be remembered, not due to M. Vincent; it has been practised for some years by aniline-colour makers. M. Vincent's merit is in establishing a cheap method by which perfectly pure chloride of methyl can be obtained, and thus rendering the processes of the manufacture of colours much more certain than they have been hitherto.

The production of methyl violet from di-methyl aniline, may be easily shown by heating this body with a small quantity of chloral hydrate, and then introducing some copper turnings into the hot liquid. On pouring the mixture into alcohol, the violet colour is well seen.

In reviewing this new chemical industry of the beet-root vinasses, one cannot help being struck by the knowledge and ability which have been so successfully expended by M. Camille Vincent on the working out of the processes.

Here again we have another instance of the utilization of waste chemical products and of the preparation on a large scale of compounds hitherto known only as chemical rarities.

All those interested in scientific research must congratulate M. Camille Vincent on this most successful issue of his labours.

[H. E. R.]

Friday, Feb. 28, 1879.

C. WILLIAM SIEMENS, Esq. D.C.L. F.R.S. Vice-President, in the Chair.

SIR WILLIAM THOMSON, LL.D. F.R.S.

The Sorting Demon of Maxwell.

[Abstract.]

THE word "demon," which originally in Greek meant a supernatural being, has never been properly used to signify a real or ideal personification of malignity.

Clerk Maxwell's "demon" is a creature of imagination having certain perfectly well defined powers of action, purely mechanical in their character, invented to help us to understand the "Dissipation of Energy" in nature.

He is a being with no preternatural qualities, and differs from real living animals only in extreme smallness and agility. He can at pleasure stop, or strike, or push, or pull any single atom of matter, and so moderate its natural course of motion. Endowed ideally with arms and hands and fingers—two hands and ten fingers suffice—he can do as much for atoms as a pianoforte player can do for the keys of the piano—just a little more, he can push or pull each atom *in any direction*.

He cannot create or annul energy; but just as a living animal does, he can store up limited quantities of energy, and reproduce them at will. By operating selectively on individual atoms he can reverse the natural dissipation of energy, can cause one-half of a closed jar of air, or of a bar of iron, to become glowingly hot and the other ice cold; can direct the energy of the moving molecules of a basin of water to throw the water up to a height and leave it there proportionately cooled (1 deg. Fahrenheit for 772 ft. of ascent); can "sort" the molecules in a solution of salt or in a mixture of two gases, so as to reverse the natural process of diffusion, and produce concentration of the solution in one portion of the water, leaving pure water in the remainder of the space occupied; or, in the other case, separate the gases into different parts of the containing vessel.

"Dissipation of Energy" follows in nature from the fortuitous concourse of atoms. The lost motivity is essentially not restorable otherwise than by an agency dealing with individual atoms; and the mode of dealing with the atoms to restore motivity is essentially a process of assortment, sending this way all of one kind or class, that way all of another kind or class.

The classification, according to which the ideal demon is to sort them, may be according to the essential character of the atom; for instance, all atoms of hydrogen to be let go to the left, or stopped from crossing to the right, across an ideal boundary; or it may be according to the velocity each atom chances to have when it approaches the boundary: if greater than a certain stated amount, it is to go to the right; if less, to the left. This latter rule of assortment, carried into execution by the demon, disequalises temperature, and undoes the natural diffusion of heat; the former undoes the natural diffusion of matter.

By a combination of the two processes, the demon can decompose water or carbonic acid, first raising a portion of the compound to dissociational temperature (that is, temperature so high that collisions shatter the compound molecules to atoms), and then sending the oxygen atoms this way, and the hydrogen or carbon atoms that way; or he may effect decomposition against chemical affinity otherwise, thus:—Let him take in a small store of energy by resisting the mutual approach of two compound molecules, letting them press as it were on his two hands, and store up energy as in a bent spring, then let him apply the two hands between the oxygen and the double hydrogen constituents of a compound molecule of vapour of water, and tear them asunder. He may repeat this process until a considerable proportion of the whole number of compound molecules in a given quantity of vapour of water, given in a fixed closed vessel, are separated into oxygen and hydrogen at the expense of energy taken from translational motions. The motivity (or energy for motive power) in the explosive mixture of oxygen and hydrogen of the one case, and the separated mutual combustibles, carbon and oxygen, of the other case, thus obtained, is a transformation of the energy found in the substance in the form of kinetic energy of the thermal motions of the compound molecules. Essentially different is the decomposition of carbonic acid and water in the natural growth of plants, the resulting motivity of which is taken from the undulations of light or radiant heat, emanating from the intensely hot matter of the sun.

The conception of the "sorting demon," is purely mechanical, and is of great value in purely physical science. It was not invented to help us to deal with questions regarding the influence of life and of mind on the motions of matter, questions essentially beyond the range of mere dynamics.

[The discourse was illustrated by a series of experiments.]

[W. T.]

Friday, April 4, 1879.

WILLIAM SPOTTISWOODE, ESQ. D.C.L. LL.D. Pres. R.S.
Vice-President, in the Chair.

WILLIAM CROOKES, F.R.S.

Molecular Physics in High Vacua.

WHEN I was asked, a month or two ago, to illustrate in this theatre some of my recent researches on Molecular Physics in High Vacua, I exclaimed " How is it possible to bring such a subject worthily before a Royal Institution audience when none of the experiments can be seen more than three feet off?" If to-night I am fortunate enough to show all the experiments to those who are not far distant, and if I succeed in making most of them visible at the far end of the theatre, such a success will be entirely due to the great kindness of your late Secretary, Mr. Spottiswoode, who has placed at my disposal his magnificent induction-coil,—not only for this lecture, but for some weeks past in my own Laboratory,—thus enabling me to prepare apparatus and vacuum tubes on a scale so large as to relieve me of all anxiety so far as the experimental illustrations are concerned.

Before describing the special researches in molecular physics which I propose to illustrate this evening, it is necessary to give a brief outline of one small department of the modern theory of the constitution of gases. It is not easy to make clear the kinetic theory, but I will try to simplify it in this way:—Imagine that I have in a large box a swarm of bees, each bee independent of its fellow, flying about in all manner of directions and with very different velocities. The bees are so crowded that they can only fly a very short distance without coming into contact with one another or with the sides of the box. As they are constantly in collision, so they rebound from each other with altered velocities and in different directions, and when these collisions take place against the sides of the box pressure is produced. If I take some of the bees out of the box, the distance which each individual bee will be able to fly before it comes into contact with its neighbour will be greater than when the box was full of bees, and if I remove a great many of the bees I increase to a considerable extent the average distance that each can fly without a collision. This distance I will call the bee's *mean free path.* When the bees are numerous the mean free path is very short; when the bees are few the mean free path will be longer, the length being inversely proportional to the number of bees present. Let us now imagine a loose diaphragm to be introduced in the centre of the box, so as to divide the number of bees equally. The same number of bees being on each side, the impacts on the diaphragm will be equal; and the mean speed of the bees being the same, the pressure will be identical on each side of the diaphragm, and it will not move.

Let me now warm one side of this division so as to let it communicate extra energy to a bee when it touches it. As before, a bee will strike the diaphragm with its normal mean velocity, but will be driven back with extra velocity, the reaction producing an increase of pressure on the diaphragm. It will be found, however, that although the diaphragm is free to move, the extra strength of the recoil on the warm side does not produce any motion. This at first sight seems contrary to the law of action and reaction being equal. The explanation is not difficult to understand. The bees which fly away from the diaphragm have drawn energy from it, and therefore move quicker than those which are coming towards it; they beat back the crowd to a greater distance, and keep a greater number from striking the diaphragm. Near to the heated side of the diaphragm the density is less than the average, while beyond the free path the density is above the average, and this greater crowding extends to all other parts of the box. Thus it happens that the extra energy of the impacts against the warm side of the diaphragm is exactly compensated by the increased number of impacts on the cool side. In spite therefore of the increased activity communicated to a portion of the bees, the pressure on the two sides of the diaphragm will remain the same. This represents what occurs when the extent of the box containing the bees is so great, compared with the mean free path, that the abrupt change in the velocities of those bees which rebound from the walls of the box produces only an insensible influence on the motions of bees at so great a distance as the diaphragm.

I will next ask you to imagine that I am gradually removing bees from our box, still keeping the diaphragm warm on one side. The bees getting fewer the collisions will become less frequent, and the distance each bee can fly before striking its neighbour will get longer and longer, and the crowding in front of them will grow less and less. The compensation will also diminish, and the warmed side of the diaphragm will have a tendency to be beaten back. A point will at last be reached on the warm side, when the mean free path of the bees will be long enough to admit of their dashing right across from the diaphragm to the side of the box, without meeting more than a certain number of in-coming bees in their flight. In this case the bees will no longer fly quite in the same direction as before. They will now fly less sideways, and more forwards and backwards between the heated face of the diaphragm and the opposed wall of the box.

Because of this preponderating motion, and also because they will thereby less effectually keep back bees crowding in from the sides, there will now be a greater proportionate pressure both on the hot face of the diaphragm and on that part of the box which is in front of it. Hence the pressure on the hot side will now exceed that on the cool side of the diaphragm, which will consequently have a backward movement communicated to it.

I may diminish the size of the bees as much as I like, and by correspondingly increasing their number the mean free path will remain the same. Instead of bees let me call them molecules, and instead of having a few hundreds or thousands in the box let me have millions and billions and trillions; and if we also diminish the mean free path to a considerable extent, we get a rough outline of the kinetic theory of gases. (I may just mention that the mean free path of the molecules in air, at the ordinary pressure, is the ten-thousandth of a millimetre.)

Three years ago I had the honour of bringing before you the results of some researches on the Radiometer. Let me now take up the subject where I then left off. I have here two radiometers which have been rotating before you under the influence of a strong light shining upon them.

The explanation of the movement of the radiometer is this,—the light, or the total bundle of rays included in the term "light," falling upon the blackened side of the vanes, becomes absorbed, and thereby raises the temperature of the black side: this causes extra excitement of the air molecules which come in contact with it, and pressure is produced, causing the fly of the radiometer to turn round.

I have long believed that a well-known appearance observed in vacuum tubes is closely related to the phenomena of the mean free path of the molecules. When the negative pole is examined while the discharge from an induction-coil is passing through an exhausted tube, a dark space is seen to surround it. This dark space is found to increase and diminish as the vacuum is varied, in the same way that the ideal layer of molecular pressure in the radiometer increases and diminishes. As the one is perceived by the mind's eye to get greater, so the other is seen by the bodily eye to increase in size. If the vacuum is insufficient to permit the radiometer to turn, the passage of electricity shows that the "dark space" has shrunk to small dimensions. It is a natural inference that the dark space is the mean free path of the molecules of the residual gas.

The radiometer which has just been turning under the influence of the lime-light is not of the ordinary kind. Fig. 1 will explain its construction.

It is similar to an ordinary radiometer with aluminium disks for vanes, each disk coated on one side with a film of mica. The fly is supported by a hard steel instead of glass cup, and the needle point on which it works is connected by means of a wire with a platinum terminal sealed into the glass. At the top of the radiometer bulb a second terminal is sealed in. The radiometer can therefore be connected with an induction-coil, the movable fly being made the negative pole.

As soon as the pressure is reduced to a few millims. of mercury, a halo of velvety violet light forms on the metallic side of the vanes, the mica side remaining dark. As the pressure diminishes, a dark space is seen to separate the violet halo from the metal. At a pressure of half a millim. this dark space extends to the glass, and positive rotation commences. On continuing the exhaustion the dark space further widens out and appears to flatten itself against the glass, when the rotation becomes very rapid.

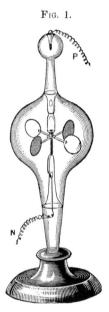

Fig. 1.

You perceive a dark space behind each vane and moving round with it. In the first experiment, radiation from the lime-light falling on the metallic sides of the vanes, produced a layer of molecular pressure which drove the fly round; so here the induction-current has produced molecular excitement at the surface of the vanes forming the negative pole, extending up to the side of the glass.

When the negative pole is in rapid rotation it is not easy to see this dark space, so I have arranged a tube in which the dark space will be visible to all present. The tube, as you will see by the diagram (Fig. 2), has a pole in the centre in the form of a metal disk, and other poles at each end. The centre pole is made negative, and the two end poles connected together are made the positive terminal. The dark space will be in the centre. When the exhaustion is not very great the dark space extends only a little distance on each side of the negative pole in the centre. When the exhaustion is very good, as it is in the tube before you, and I turn on the coil, the dark space is seen to extend for about 2 inches on each side of the pole.

Here, then, we see the induction spark actually illuminating the lines of molecular pressure caused by the excitement of the negative pole. The thickness of this dark space—nearly 2 inches—is the measure of the mean free path between successive collisions of the molecules of the residual gas. The extra velocity with which the negatively electrified molecules rebound from the excited pole keeps back the more slowly moving molecules which are advancing towards that pole. The conflict occurs at the boundary of the dark space, where the luminous margin bears witness to the energy of the discharge.

I will endeavour to throw on the screen an illustration of this dark space. A stream of water falls from a small jet on to a horizontal plate of glass. The water spreads over the plate and forms a thin film. The jet of water in the centre, from the velocity of its fall, drives the film of water before it on all sides, raising it into a ring-shaped heap. As I diminish the force of the jet the ring contracts: this is equivalent to the exhaustion getting less. When I increase the force of water the ring expands in size, the effect being analogous to an increase of exhaustion in my tubes. The extra velocity of the falling particles of water drives the in-coming particles of water before them, and raises a ridge round the side which exactly

FIG. 2.

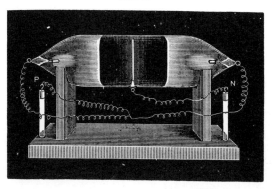

represents the luminous halo to the dark space to be seen in this tube.

If, instead of a flat disk, a metal cup is used for the negative pole, the successive appearances on exhausting the tube are somewhat different. The velvety violet halo forms over each side of the cup. On increasing the exhaustion the dark space widens out, retaining almost exactly the shape of the cup. The bright margin of the dark space becomes concentrated at the concave side of the cup to a luminous focus, and widens out at the convex side. When the dark space is very much larger than the cup, its outline forms an irregular ellipsoid drawn in towards the focal point. Inside the luminous boundary a dark violet light can be seen converging to a focus, and, as the rays diverge on the other side of the focus, spreading beyond the margin of the dark space; the whole appearance being strikingly similar to the rays of the sun reflected from a concave mirror through a foggy atmosphere. This proves a somewhat important point; it shows that the molecules thrown off the excited negative pole leave it in a direction almost normal to the surface.

I can illustrate this property of the molecular rays by an experiment. This diagram (Fig. 3) is a representation of the tube which is before you. It contains, as a negative pole, a hemi-cylinder (a) of polished aluminium. This is connected with a fine copper wire, b, ending at the platinum terminal, c. At the upper end of the tube is another terminal, d. The induction-coil is connected so that the hemi-cylinder is negative and the upper pole positive, and when exhausted to a sufficient extent, as is the case with this tube, the projection of the molecular rays to a focus is very beautifully shown. The rays are driven from the hemi-cylinder in a direction normal to its surface; they come to a focus and then diverge, tracing their path in brilliant green phosphorescence on the surface of the glass.

Fig. 3.

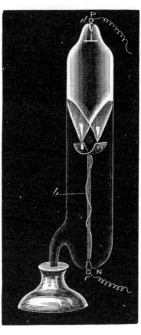

You will notice that the rays which project from the negative pole and cross in the centre have a bright green appearance; that colour is entirely due to the phosphorescence of the glass. At a very high exhaustion the phenomena noticed in ordinary vacuum tubes when the induction spark passes through them—an appearance of cloudy luminosity and of stratifications—disappears entirely. No cloud or fog whatever is seen in the body of the tube, and with such a vacuum as I am working with in these experiments—about a millionth part of an atmosphere—the inner surface of the glass glows with a rich green phosphorescence, the intensity of colour varying with the perfection of the vacuum. It scarcely begins to show much before the 800,000th of an atmosphere. At about a millionth of an atmosphere the phosphorescence is very strong, and after that it begins to diminish until there are not enough molecules left to allow the spark to pass.*

I have here a tube which will serve to illustrate the dependence of the green phosphorescence of the glass on the degree of perfection of the vacuum (Fig. 4). The two poles are at a and b, and at the end (c) is a small supplementary tube connected with the other by a narrow aperture, and containing solid caustic potash. The tube has been exhausted to a very high point, and the potash heated so as to drive off moisture and deteriorate the vacuum. Exhaustion has then

---

*   1·0 millionth of an atmosphere   =   0·00076 millim.
          1315·789 millionths of an atmosphere =   1·0 millim.
  1,000,000   ,,   ,,   ,,   = 760·0 millims.
      ,,   ,,   ,,   ,,   = 1 atmosphere.

been re-commenced, and the alternate heating and exhaustion have been repeated until the tube has been brought to the state in which it now appears before you. When the induction spark is first turned on nothing is visible—the vacuum is so high that the tube is non-conducting. I now warm the potash slightly, and liberate a trace of aqueous vapour. Instantly conduction commences, and the green phosphorescence flashes out along the length of the tube. I continue the heat, so as to drive off more gas from the potash. The green gets fainter, and now a wave of cloudy luminosity sweeps over the tube, and stratifications appear. These rapidly get narrower, until the spark passes along the tube in the form of a narrow purple line. I take the lamp away, and allow the potash to cool; as it cools, the aqueous vapour, which the heat had driven off, is re-

Fig. 4.

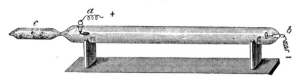

absorbed. The purple line broadens out, and breaks up into fine stratifications; these get wider, and travel towards the potash tube. Now a wave of green light appears on the glass at the other end, sweeping on and driving the last pale stratification into the potash; and now the tube glows over its whole length with the green phosphorescence. Would time allow I might keep it before you, and show the green growing fainter and the vacuum becoming non-conducting; but time is required for the absorption of the last traces of vapour by the potash, and I must pass on to the next subject.

This green phosphorescence is a subject that has much occupied my thoughts, and I have striven to ascertain some of the laws governing its occurrence. I soon perceived that the phosphorescence was not in the body of the tube itself, but was entirely on the surface of the glass. Another peculiarity of the rays producing this green phosphorescence is that they will not turn a corner in the slightest degree. Here is a V-shaped tube (Fig. 5), a pole being at each extremity. The pole at the right side (a) being negative, you see that the whole of the right arm is flooded with green light, but at the bottom it stops sharply, and will not turn the corner to get into the left side. When I reverse the current, and make the left pole negative, the green changes to the left side, always following the negative pole, leaving the positive side with scarcely any luminosity.

In the ordinary phenomena exhibited by vacuum tubes—phenomena with which we are all familiar—it is customary, for the more striking illustration of their contrasts of colour, to have the tubes bent into very elaborate designs. The positive luminosity caused by the phosphorescence of the residual gas follows all the convolutions and designs into which skilful glass-blowers can manage to twist the glass. The negative pole being at one end and the positive pole at the other, the luminous phenomena seem to depend more on the positive than on the negative at an ordinary exhaustion such as has hitherto been used to get the best phenomena of vacuum tubes. I have here two bulbs (Fig. 6), alike in shape and position of poles, the only difference being that one is at an exhaustion equal to a few millimetres of mercury—such a moderate exhaustion as will give stratifications or the ordinary luminous phenomena—whilst the other is exhausted to about the millionth of an atmosphere. I will

Fig. 5.

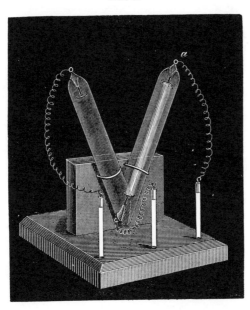

first connect the moderately exhausted bulb with the induction-coil, and, retaining the pole at one side (*a*) always negative, I will put the positive wire successively to the other three poles with which the bulb is furnished. You will see that as I change the position of the positive pole, the line of violet light joining the two poles changes. In this moderately exhausted bulb, therefore, the electric current always chooses the shortest path between the two poles, and moves about the bulb as I alter the position of the wires.

This, then, is the kind of phenomenon we get in ordinary exhaustions. I will now try the same experiment with a tube that is highly exhausted, and, as before, will make the side pole (a') the negative, the top pole (b) being positive. Notice how widely different is the appearance from that shown by the last bulb. The negative pole is in the form of a shallow cup. The bundle of rays from the cup crosses in the centre of the bulb, and thence diverging falls on the opposite side as a circular patch of green light. As I turn the bulb

Fig. 6.

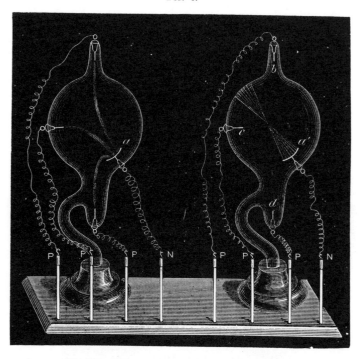

round you will all be able to see the faint blue focus and the green patch on the glass. Now observe, I remove the positive wire from the top, and connect it with the side pole (c). The green patch from the divergent negative focus is still there. I now make the lowest pole (d) positive, and the green patch still remains where it was at first, unchanged in position or intensity.

This, then, gives us another fact which brings us a little nearer to the cause of this green phosphorescence. It is this—that in the low vacuum the position of the positive pole is of every importance, whilst in a high vacuum it scarcely matters at all where the positive pole is; the phenomena seem to depend entirely on the negative pole. In very high vacua, such as we have been using, the phenomena follow altogether the negative pole. If the negative pole points in the direction of the positive all very well, but if the negative pole is entirely in the opposite direction it does not matter: the line of rays is projected all the same in a straight line from the negative.

I have hitherto spoken of and illustrated these phenomena in connection with *green* phosphorescence. It does not follow, however, that the phosphorescence is always of that colour. This colouration is a property of the particular kind of glass in use in my laboratory. I have here (Fig. 7) three bulbs composed of different glass: one is uranium glass (*a*), which phosphoresces of a dark green colour; another is English glass (*b*), which phosphoresces of a blue colour; and the third (*c*) is soft German glass—of which most of the apparatus before you is made—which phosphoresces of a bright apple-green

Fig. 7.

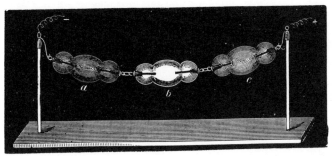

colour. It is therefore plain that this particular green phosphorescence is solely due to the glass which I am using. Were I to use English glass I should have to speak of blue phosphorescence, but I know of no glass which is equal to the German in brilliancy.

My earlier experiments were almost entirely carried on by the aid of the phosphorescence which glass takes up when it is under the influence of the electric discharge *in vacuo;* but many other substances possess this phosphorescent power, and some have it in a much higher degree than glass. For instance, here is some of the luminous sulphide of calcium prepared according to M. Ed. Becquerel's description. When it is exposed to light—even candlelight—it phosphoresces for hours with a rich blue colour. I have prepared a diagram with large letters written in this luminous sulphide; before it is exposed to the light the letters are invisible, but Mr. Gimingham has just exposed it in another room to burning magnesium, and now it is brought into the darkened theatre you will see the word " φως,"— *light,* a very suitable word for so beautiful a phosphorescence—shining brightly in luminous characters. The first letter, φ, shines with an orange light; it is a sulphide of calcium prepared from oyster-shells. The other letters, shining with a blue light, are sulphide of calcium prepared from precipitated carbonate of lime. Once the phosphorescence is excited the letters shine for several hours. I will put the diagram at the back, and we shall see how it lasts during the remainder of the lecture. This substance, then, is phosphorescent to light, but it is also much more strongly phosphorescent to the molecular discharge in a good vacuum, as you will see when I pass the discharge through this tube (Fig. 8). The white plate (a, b) in the centre of the tube is a sheet of mica painted over with the luminous sulphide of which the letter ϕ was composed in the diagram you have just seen. On connecting the poles with the coil the mica screen glows with a strong yellowish green light, bright enough to illuminate all the apparatus near it. But there is another phenomenon to which I now desire to draw attention: on the luminous screen is a kind of distorted star-shaped figure. A little in front of the negative pole I have fixed a star (c) cut out in aluminium, and it is the image of this star which you see on the screen. It is evident that the rays coming from the negative pole project an image of anything that happens to be in front of it. The discharge, therefore, must come from the pole in straight lines, and does not merely permeate all parts of the tube and fill it with light as it would were the exhaustion less good. Where there is nothing in the way the rays strike the screen and produce phosphorescence, and where there is an obstacle they are obstructed by it, and a shadow is thrown on the screen. I shall have more to say about this shadow presently; I merely now wish to establish the fact that these rays driven from the negative pole produce a shadow.

FIG. 8.

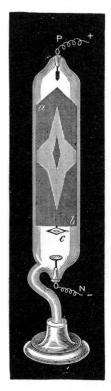

I must draw your attention to an important experiment connected with these molecular rays, but unfortunately it is a very delicate one, and very difficult to show to many at once; but I hope, if you know beforehand what to look for, you will all be able to see what I wish to show. In this pear-shaped bulb (Fig. 9 A) the negative pole (a) is at the pointed end. In the middle is a cross (b) cut out of sheet aluminium, so that the rays from the negative pole projected along the tube will be partly intercepted by the aluminium cross, and will project an image of it on the hemispherical end of the tube which is phosphorescent. I think you will all now see the shadow of the cross on the end of the bulb (c, d), and notice that the cross is black on a luminous ground. Now, the rays from the negative pole have been passing by the side of the aluminium cross to produce the shadow;

they have been hammering and bombarding the glass till it is appreciably warm, and at the same time they have been producing another effect on that glass—they have deadened its sensibility. The glass has got tired, if I may use the expression, by the enforced phosphorescence. Some change has been produced by this bombard-

Fig. 9 A.

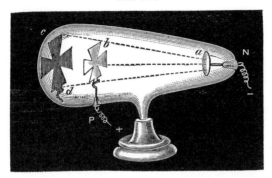

ment which will prevent the glass from responding easily to additional excitement; but the part that the shadow has fallen on is not tired—it has not been phosphorescing at all and is perfectly fresh; therefore if I throw this star down,—I can easily do so by giving the apparatus a slight jerk, for it has been most ingeniously constructed with a hinge by Mr. Gimingham,—and so allow the rays from the negative pole to fall uninterruptedly on to the end of the bulb, you will suddenly see the black cross (c, d, Fig. 9 B) change to a luminous

Fig. 9 B.

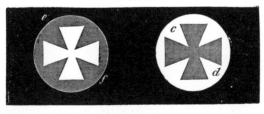

one (e, f), because the background is only faintly phosphorescing, whilst the part which had the black shadow on it retains its full phosphorescent power. The luminous cross is now dying out. This is a most delicate and venturous experiment, and I am fortunate in having succeeded so well, for it is one that cannot be rehearsed. After resting for a time the glass seems to partly recover its power of phosphorescing, but it is never so good as it was at first.

We have, therefore, found an important fact connected with this phosphorescence. Something is projected from the negative pole which has the power of hammering away at the glass in front of it, in such a way as to cause it not only to vibrate and become temporarily luminous while the discharge is going on, but to produce an impression upon the glass which is permanent. The explanation which has gradually evolved itself from this series of experiments is this :— The exhaustion in these tubes is so high that the dark space, as I showed you at the commencement of this Lecture, that extended around the negative pole, has widened out till it entirely fills the tube. By great rarefaction the mean free path has become so long that the hits in a given time may be disregarded in comparison to the misses, and the average molecule is now allowed to obey its own motions or laws without interference. The mean free path is in fact comparable to the dimensions of the vessel, and we have no longer to deal with a *continuous* portion of matter, as we should were the tubes less highly exhausted, but we must here contemplate the molecules *individually*. At first this was only a convenient working hypothesis. Long-continued experiment then raised this provisional hypothesis almost to the dignity of a theory, and now the general opinion is that this theory gives a fairly correct explanation of the facts. In these highly exhausted vessels the mean free path of the residual molecules of gas is so long that they are able to drive across from the pole to the other side of the tube with comparatively few collisions. The negatively electrified molecules of the gaseous residue in the tube therefore dash against anything that is in front, and cast shadows of obstacles just as if they were rays of light. Where they strike the glass they are stopped, and the production of light accompanies this sudden arrest of velocity.

Other substances besides English, German, and uranium glass, and Becquerel's luminous sulphides, are also phosphorescent. I think, without exception, the diamond is the most sensitive substance I have yet met for ready and brilliant phosphorescence. I have here a tube, similar to those already exhibited, containing a mica screen painted with powdered diamond, and when I turn on the coil, the brilliant blue phosphorescence of the diamond can be seen, quite overpowering the green phosphorescence of the glass. Here, again, is a very curious diamond, which I was fortunate enough to meet with a short time ago. By daylight it is green, produced, I fancy, by an internal fluorescence. The diamond is mounted in the centre of this exhausted bulb (Fig. 10), and the negative discharge will be directed on it from below upwards. On darkening the theatre you see the diamond shines with as much light as a candle, phosphorescing of a bright green.

In this other bulb is a remarkable collection of crystals of diamonds, which have been lent me by Professor Maskelyne. When I pass the discharge over them I am afraid you will only be able to see a few points of light, but if you will examine them after the Lecture, you will see them phosphoresce with a most brilliant series of colours—blue, apricot, red, yellowish green, orange, and pale green.

Next to the diamond the ruby is one of the most remarkable

Fig. 10.

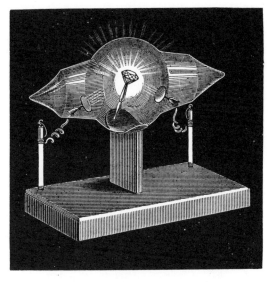

Fig. 11.

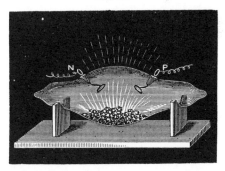

stones for phosphorescing. In this tube (Fig. 11) is a collection of ruby pebbles, for the loan of which I am indebted to my friend Mr. Blogg, of the firm of Blogg and Martin, who placed a small sackful at my disposal. As soon as I turn on the induction spark you will see these rubies shining with a brilliant rich red colour, as if they were glowing hot. Now the ruby is nothing but crystallised alumina with a little colouring-matter, and it became of great interest to ascertain whether the artificial ruby made by M. Feil, of Paris, would glow in the same manner. I had simply to make my wants known to M. Feil, and he immediately sent me a box containing artificial rubies and crystals of alumina of all sizes, and from those I have selected the mass in this tube which I now place under the discharge: they phosphoresce of the same rich red colour as the natural ruby. It scarcely matters what colour the ruby is, to begin with. In this tube of natural rubies there are stones of all colours— the deep red ruby and the pale pink ruby. There are some so pale as to be almost colourless, and some of the highly-prized tint of pigeon's blood; but in the vacuum under the negative discharge they all phosphoresce with about the same colour.

As I have just mentioned, the ruby is crystallised alumina. In a paper published twenty years ago by Ed. Becquerel[*] I find that he describes the appearance of alumina as glowing with a rich red colour in the phosphoroscope (an instrument by which the duration of phosphorescence in the sunlight can be examined). Here is some chemically pure precipitated alumina which I have prepared in the most careful manner. It has been heated to whiteness, and you see it glows with the rich red colour which is supposed to be characteristic of alumina. The mineral known as corundum is a colourless variety of crystallised alumina. Under the negative discharge in a vacuum, corundum phosphoresces of a rose-pink colour. There is another curious fact in which I think chemists will feel interested. The sapphire is also crystallised alumina, just the same as the ruby. The ruby has a little colouring-matter in it, giving it a red colour; the sapphire has a colouring-matter which gives it a blue colour, whilst corundum is white. I have here in a tube a very fine crystal of sapphire, and, when I pass the discharge over it, it gives alternate bands of red and green. The red we can easily identify with the glow of alumina; but what is the green? If alumina is precipitated and purified as carefully as in the case I have just mentioned, but in a somewhat different manner, it is found to glow with a rich green colour. Here are the two specimens of alumina in tubes, side by side. Chemists would say that there was no difference between one and the other; but I connect them with the induction-coil, and you see that one glows with a bright green colour, whilst the other glows with a rich red colour. Here is a fine specimen of chemically pure alumina, lent me by Messrs. Hopkin and Williams; by ordinary light it is a perfectly white powder. It is just possible that the rich fire of the ruby, which has caused it to be so prized, may be due, not entirely to the colouring-matter, but to its wonderful power of phosphorescing with a deep red colour, not only under the electric discharge in a vacuum, but whenever exposed to a strong light.

The spectrum of the red light emitted by all these varieties of

---

[*] Annales de Chimie et de Physique, 3rd series, vol. lvii. p. 50, 1859.

alumina—the ruby, corundum, or artificially precipitated alumina—is the same as described by Becquerel twenty years ago. There is one intense red line, a little below the fixed line B in the spectrum, having a wave-length of about 6895. There is a continuous spectrum beginning at about B, and a few fainter lines beyond it, but they are so faint in comparison with this red line that they may be neglected. This line may be called the characteristic line of alumina.

Fig. 12.

I now pass on to another fact connected with this negative discharge. Here is a tube (Fig. 12) with a negative pole (*a*, *b*) in the form of a hemi-cylinder, similar to the one you have already seen (Fig. 3), but in this case I receive the rays on a phosphorescent screen (*c*, *d*). See how brilliantly the lines of discharge shine out, and how intensely the focal point is illuminated; it lights the whole table. Now I bring a small magnet near, and move it to and fro; the rays obey the magnetic force, and the focus bends one way and the other as the magnet passes it. I can show this magnetic action a little more definitely. Here is a long glass tube (Fig. 13), very highly exhausted, with a negative pole at one end (*a*) and a long phosphorescent screen (*b*, *c*) down the centre of the tube. In front of the negative pole is a plate of mica (*b*, *d*) with a hole (*e*) in it, and the result is that when I turn on the current, a line of phosphorescent light (*e*, *f*) is projected along the whole length of the tube. I now place beneath the tube a powerful horse-shoe magnet: see how the line of light becomes curved under the magnetic influence (*e*, *g*), waving about like a flexible wand as I move the magnet up and down. The action of the magnet can be understood by reference to this diagram (Fig. 14).

Fig. 13.

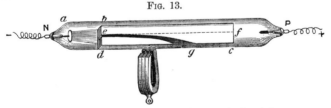

The north pole gives the ray of molecules a spiral twist one way, and the south pole twists it the other way; the two poles side by side compel the ray to move in a straight line up or down, along a plane at right angles to the plane of the magnet and a line joining its poles.

Now it is of great interest to ascertain whether the law governing the magnetic deflection of the trajectory of the molecules is the same as has been found to hold good at a lower vacuum. The former experiment was with a very high vacuum. This is a tube with a low vacuum (Fig. 15). On passing the induction spark it passes as a

Fig. 14.

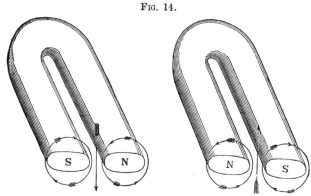

narrow line of violet light joining the two poles. Underneath I have a powerful electro-magnet. I make contact with the magnet, and the line of light dips in the centre towards the magnet. I reverse the poles, and the line is driven up to the top of the tube. Notice the difference between the two phenomena. Here the action is temporary. The dip takes place under the magnetic influence; the line of discharge then rises, and pursues its path to the positive pole. In the high exhaustion, however, after the ray of light had dipped to the magnet it did not recover itself, but continued its path in the altered direction.

Fig. 15.

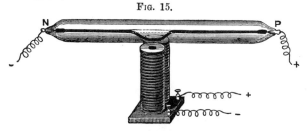

During these experiments another property of this molecular discharge has made itself very evident, although I have not yet drawn attention to it. The glass gets very warm where the green phosphorescence is strongest. The molecular focus on the tube which we have just seen (Fig. 12) would be intensely hot, and I have prepared an apparatus by which this heat at the focus can be intensified and rendered visible to all present. This small tube (*a*)

(Fig. 16) is furnished with a negative pole in the form of a cup (*b*). The rays will therefore be projected to a focus in the middle of the tube (Fig. 17, *a*). At the side of the tube is a small electro-magnet,

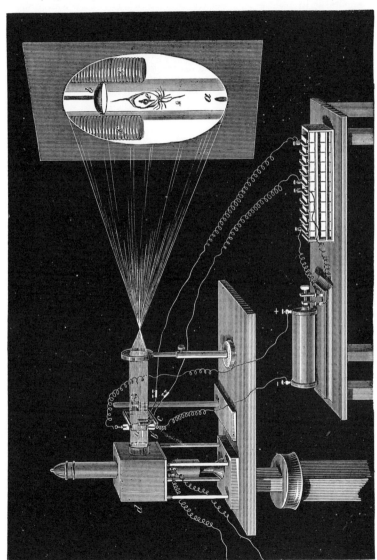

FIG. 16.

which I can set in action by touching a key, and the focus is then drawn to the side of the glass tube (Fig. 17, b). To show the first action of the heat I have coated the tube with wax. I will put the apparatus in front of the electric lantern (d), and throw a magnified image of the tube on the screen. The coil is now at work, and the focus of molecular rays is projected along the tube. I turn the magnetism on, and draw the focus on the side of the glass. The first thing you see is a small circular patch melted in the coating of wax. The glass soon begins to disintegrate, and cracks are shooting starwise from the centre of heat. The glass is softening. Now the atmospheric pressure forces it in, and now it melts. A hole (e) is perforated in the middle, the air rushes in, and the experiment is at an end.

Instead of drawing the focus to the side of the glass with a magnet, I will take another tube (Fig. 18), and allow the focus from

FIG. 17. FIG. 18.

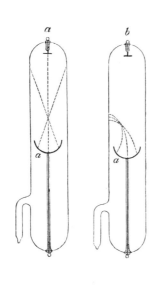

the cup-shaped negative pole (a) to play on a piece of platinum wire (b) which is supported in the centre of the bulb. The platinum wire not only gets white hot, but you can see sparks coming from it on all sides, showing that it is actually melting.

Here is another tube, but instead of platinum I have put in the focus that beautiful alloy of platinum and iridium which Mr. Matthey has brought to such perfection, and I think that I shall succeed in even melting that. I first turn on the induction-coil slightly, so as not to bring out its full power. The focus is now playing on the iridio-platinum, raising it to a white heat. I bring a small magnet near, and you see I can deflect the focus of heat just as I did the luminous focus in the other tube. By shifting the magnet I can drive the focus up or down, or draw it completely away from the metal, and render it non-luminous. I withdraw the magnet, and let the molecules have full play again; the metal is now white-hot. I increase the intensity of the spark. The metal glows with almost insupportable brilliancy, and at last melts.

There is still another property of this molecular discharge, and it is this:—You have seen that the molecules are driven violently from the negative pole. If I place something in front of these molecules, they show the force of impact by the heat which is produced. Can I make this mechanical action evident in a more direct way? Nothing is simpler. I have only to put some easily moving object in the line of discharge in order to get a powerful mechanical action. Mr. Gimingham,

Fig. 19.

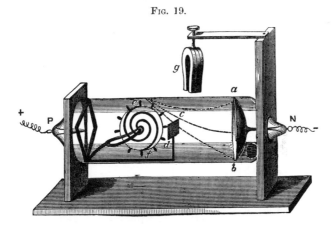

with great skill, has constructed a piece of apparatus which I will presently put in the electric lantern, so that all will be able to see its action. But first I will explain the construction by means of this diagram (Fig. 19). The negative pole (a, b) is in the form of a very shallow cup. In front of the cup is a mica screen (c, d), wide enough to intercept nearly all the molecular rays coming from the negative pole. Behind this screen is a mica wheel (e, f) with a series of vanes, making a sort of paddle-wheel of it. So arranged, the molecular stream from the pole $a\,b$ will nearly all be cut off from the wheel, and what escapes over and under the screen will hit the vanes equally, and will not produce any movement. I now put a magnet, *g*, over the tube, so as to deflect the stream over or under the obstacle *c d*, and the result will be rapid motion in one or the other direction, according to the way the magnet is turned. I now throw the image of the apparatus on the screen. The spiral lines painted on the wheel show which way it turns. I arrange the magnet to draw the molecular stream so as to beat against the upper vanes, and the wheel revolves rapidly, as if it were an over-shot water-wheel. I now turn the magnet so as to drive the molecular stream underneath; the wheel slackens speed, stops, and then begins to rotate the other way, as if it were an under-shot water-wheel. This can be repeated as often as I like to reverse the position of the magnet, the change of rotation of the wheel showing immediately the way the molecular stream is deflected.

This experiment illustrates the last of the phenomena which time allows me to bring before you, attending the passage of the induction spark through a highly exhausted atmosphere. It will now be naturally asked, What have we learned from the phenomena described and exhibited, and from the explanations that have been proposed? We find in these phenomena confirmation of the modern views of matter and energy. The facts elicited are in harmony with the theory that matter is not continuous but composed of a prodigious number of minute particles, not in mutual contact. The facts also are in full accordance with the kinetic theory of gases—to which I have already referred—and with the conception of heat as a particular kind of energy, expressing itself as a rapid vibratory motion of the particles of matter. This alone would be a lesson of no small value. In Science, every law, every generalisation, however well established, must constantly be submitted to the ordeal of a comparison with newly-discovered phenomena; and a theory may be pronounced triumphant when it is found to harmonise with and to account for facts which when it was propounded were still unrecognised or unexplained.

But the experiments have shown us more than this: we have been enabled to contemplate matter in a condition hitherto unknown,—in a fourth state,—as far removed from that of gas as gas is from liquid, where the well-known properties of gases and elastic fluids almost disappear, whilst in their stead are revealed attributes previously masked and unsuspected. In this ultra-gaseous state of matter phenomena are perceived which in the mere gaseous condition are as impossible as in liquids or solids.

I admit that between the gaseous and the ultra-gaseous state there can be traced no sharp boundary; the one merges imperceptibly into the other. It is true also that we cannot see or handle matter in this novel phase. Nor can human or any other kind of organic life conceivable to us penetrate into regions where such ultra-gaseous matter may be supposed to exist. Nevertheless, we are able to observe it and experiment on it, legitimately arguing from the seen to the unseen.

Of the practical applications that may arise out of these researches, it would now be premature to speak. It is rarely given to the discoverer of new facts and new laws to witness their immediate utilisation. The ancients showed a perhaps unconscious sagacity when they selected the olive, one of the slowest growing trees, as the symbol of Minerva, the goddess of Arts and Industry. Nevertheless, I hold that all careful honest research will ultimately, even though in an indirect manner, draw after it, as Bacon said, " whole troops of practical applications."

[W. C.]

Friday, May 16, 1879.

WILLIAM SPOTTISWOODE, Esq. M.A. D.C.L. LL.D. President R.S.
Vice-President, in the Chair.

PROFESSOR A. CORNU,

PROFESSOR AT THE POLYTECHMIC SCHOOL, MEMBER OF INSTITUTE OF FRANCE.

The Optical Study of the Elasticity of Solid Bodies.

PRELIMINARY REMARKS.

ALL solid bodies utilized in scientific and industrial applications are more or less elastic : and it is very important, in a practical as well as a theoretical point of view, to be able to predict the deformations due to given forces, or, conversely, to know the forces which correspond to given deformations.

Mathematical Calculation enables us to solve both problems for every description of form and of force in all their details, provided that it borrows from experience certain results obtained in very simple cases.

Homogeneous and Isotropic Bodies.

An elastic bar urged by external traction forces extends itself along its largest dimensions (longitudinal extension): at the same time, by the natural play of internal forces, its transversal dimensions diminish (transversal contraction).

[Illustration of this general fact with an indiarubber bar.]

In order to calculate all the circumstances of deformation of an elastic isotropic body, whatever may be its shape and acting forces, it is sufficient to know the rate of longitudinal extension (modulus of elasticity), and its ratio to the transversal contraction.

Various opinions amongst the Physicists upon the value of this Ratio.

A considerable number of physicists (Cagniard-Latour, Wertheim, Prof. Kirchhoff, Dr. Everett, &c.) have made a series of experiments on various supposed isotropic bodies.

The question, a very important one in a theoretical point of view, is to know if this ratio is a variable one according to the nature of the substance, or an invariable one, and equal to $\frac{1}{4}$ as given by the Navier's and Green's theories.

Double difficulty. 1. Is the body really homogeneous and isotropic ?

The metals are always annealed or crystallized: homogeneous glass is one of the bodies approaching nearest to theoretical isotropy.

2. The transversal contraction is extremely small.

Necessity of using any indirect mode of deformation to determine accurately the transversal contraction.

Mode of Experiment: Circular Flexion of a Rectangular Rod.

The upper surface, primitively plane, becoming curved with two opposite curvatures (and not cylindrical, as commonly supposed), like a *horse saddle*. The ratio of the main radii of curvature is, according to a theorem due to Mr. De St. Venant, precisely the ratio in question.

[Illustration of this general fact with an indiarubber plate.]

OPTICAL METHODS FOR TESTING THE DEFORMATION OF THE SURFACE OF ELASTIC BODIES.

1. *Variation of Focus of a Beam of Light Reflected from the Polished Surface of the Elastic Body.*

2. *Use of the Newton's Coloured Rings.*

Newton's rings are produced by illuminating with white or with monochromatic light the thin film of air comprised between a fixed surface and the exterior surface of the elastic body.

Extreme sensitiveness of this method, according to the small difference of thickness, corresponding to the successive rings.

The lines of equal intensity of the rings correspond to the lines of equal thickness of the film of air. The successive rings correspond to a difference of thickness of about $\frac{1}{4}$ of a thousandth of a millimeter (a hundred thousandth of an inch).

If the fixed surface is a plane one, the appearance of the rings is exactly the *topographic map* of the deformed surface, of which the *scale of elevation* is the small length above defined.

In a small part of the field, the rings coincide, in form, with conic sections, concentric with the *indicatrix curve* of Ch. Dupin.

Illustration of various forms of Newton's rings—circular, elliptic, hyperbolic—with monochromatic light (sodium vapour in electrical arc).

Optical Method of Testing the Circular Flexion.

A piece of plate-glass is used. The Newton's rings, before flexion, more or less regular according to the perfection of polish, become, by increasing forces, more and more regular, and take the form of conjugate hyperbolas, the axes of which being parallel and perpendicular to the main dimension of the rod.

The trigonometrical tangent of the semi angle of the common asymptote converges towards the value $\frac{1}{2}$; therefore, the ratio of the curvatures, and consequently the ratio in question, is $\frac{1}{4}$ with the best isotropic body.

The theoretical solution of the problem seems to be in favour of Navier's and Green's theories.

Generality of the Optical Method.

Application to the torsion of a rectangular plate.

The shape of the deformed surface becomes a hyperbolic paraboloid; but the asymptotes of the hyperbolæ (and not the axis, as before) are parallel and perpendicular to the main dimension of the rod.

Fixing of the Newton's Rings by Photography.

In order to study at leisure, and with accuracy, the *topographic surfaces* of elastic deformation, it is very convenient to keep an exact and fixed image of the field.

The induction spark between two poles of magnesium supplies a source of light which fulfils the three necessary conditions—to be intense, photographic, and monochromatic.

Amongst the bright lines of the magnesium spectrum none is useful for that purpose; the radiation utilized as a source of photographic light is invisible, but becomes visible when projected on Prof. Stokes's fluorescent screen.

Though the photography of the Newton's rings be a delicate operation, the experiment will be tried before the audience.

Newton's rings were photographed for the first time by the illustrious Dr. Young at the Royal Institution in the year 1803.

Friday, June 6, 1879.

GEORGE BUSK, Esq. F.R.S. Treasurer and Vice-President,
in the Chair.

JAMES DEWAR, M.A. F.R.S.
FULLERIAN PROFESSOR OF CHEMISTRY, ROYAL INSTITUTION, ETC.

Spectroscopic Investigation.

IN Kirchhoff's celebrated paper "On the Relation between the Radiating and Absorbing Powers of different Bodies for Light and Heat," the remarkable experiments of reversing the bright lines of lithium and sodium by causing sunlight to pass through the vapours of those metals, volatilized in the flame of a Bunsen's burner, are described. Bunsen and Kirchhoff reversed the stronger lines of potassium, calcium, strontium, and barium by deflagrating their chlorates with milk-sugar, before the slit of the solar spectroscope. Recent researches on the artificial formation of Fraunhofer lines have been made by Cornu, Lockyer, and Roberts.

Cornu improved upon a method previously used by Foucault. It depends upon so arranging the electric arc that the continuous spectrum of the intensely heated poles is examined through an atmosphere of the metallic vapours volatilized around them. By this means Cornu succeeded in reversing several lines in the spectra of the following metals, in addition to those above mentioned, viz. thallium, lead, silver, aluminium, magnesium, cadmium, zinc, and copper. He observed that, in general, the reversal began with the least refrangible of a group of lines, and gradually extended to the more refrangible lines of the group, and drew the conclusion that a very thin layer of vapour was sufficient for the reversal. In almost every case the lines reversed are the more highly refrangible of the lines characteristic of each metal.

Lockyer's plan was to view the electric arc through the vapours of the metals volatilized in a horizontal iron tube. The iron tube had its ends covered with glass plates, and was heated in a furnace, a current of hydrogen passing during the experiment. He did not succeed in observing any new reversal of bright lines, with the exception of an unknown absorption line which sometimes appeared when zinc was experimented on. He confirmed, however, the channelled-space absorption spectra observed by Roscoe and Schuster in the cases of potassium and sodium, and recorded channelled-space spectra in the case of antimony, phosphorus (?), sulphur, and arsenic (probably). "As the temperature employed for the volatilization of the metals did not exceed bright redness, or that at which cast iron readily melts, the range of metals examined was necessarily limited; and in order to extend these observations to the less fusible metals, as well as to ascertain whether the spectra of those volatilized at the lower temperature would be modified by the application of a greater degree of heat," a new series of experiments were undertaken by Lockyer and Roberts, in which the combined action of a charcoal furnace and the oxy-hydrogen blowpipe was employed. A lime crucible after the form of Stas was used to replace the iron tube. By this means they obtained still no new reversal of a metallic line, but they observed channelled-space spectra in the cases of silver, manganese, chromium, and bismuth. They observed, however, that the metal thallium gave the characteristic *bright* green line, the light of the arc not being reversed.

In the above-mentioned experiments, the coolness of the ends of the tube, which acted as condensers of the metallic vapours, and the continual change of density and temperature necessarily produced by the maintenance of a current of hydrogen through the tube, appear to account for the failure in observing reversals.

The following facts have been acquired during the course of a long series of conjoint experiments with my distinguished colleague, Professor Liveing, of Cambridge.*

In order to examine the reversal of the spectra of metallic vapours, it is more satisfactory to observe the absorptive effect produced on the continuous spectrum emitted by the sides and end of the tube in which the volatilization takes place. For this purpose it is convenient to use iron tubes about half an inch in internal diameter, and about 27 inches long, closed at one end, thoroughly cleaned inside, and coated on the outside with borax, or with a mixture of plumbago and fireclay. These tubes are inserted in a nearly vertical position in a furnace fed with Welsh coal, which will heat about 10 inches of the tube to about a welding heat, and observations are made through the upper open end of the tube, either with or without a cover of glass or mica. To exclude oxygen, and avoid as much as possible variations of temperature, hydrogen is introduced in a gentle stream by a narrow tube into the upper part only of the iron tube, so that the hydrogen floats on the surface of the metallic vapour without producing convection currents in it. By varying the length of the small tube conveying the hydrogen, the height in the tube to which the metallic vapour reaches may be regulated. Thus different depths of metallic vapour may be maintained at a comparatively constant temperature for considerable periods of time. The general plan of the apparatus is given in Plate I. (at the end of the paper).

"On the Reversal of the Lines of Metallic Vapours," Nos. I., II., III., IV., V., VI., 'Proc. Roy. Soc.,' 1878–1879.

By this means the characteristic lines of the volatile metals thallium and indium may be easily reversed.

Metallic lithium, alone or mixed with sodium, gave no results. Similarly, chloride of lithium and metallic sodium, introduced together, gave no better results. To a tube containing mixed potassium and sodium vapour, lithium chloride was added. Now the bright-red lithium line was sharply reversed, and remained well defined for a long time. The lithium line was only reversed in a mixture of the vapours of potassium and sodium, and it seems highly probable that a very slightly volatile vapour may be diffused in an atmosphere of a more volatile metal, so as to secure a sufficient depth of vapour to produce a sensible absorption. This would be analogous to well-known actions which take place in the attempt to separate organic bodies of very different boiling points by distillation, where a substance of high boiling point is always carried over, in considerable quantity, with the vapour of a body boiling at a much lower temperature.

Sodium and potassium, when observed in such tubes, give none of the appearances noted by Lockyer, "On a New Class of Absorption Phenomena," in the 'Proceedings of the Royal Society,' vol. xxii., but the channelled-space spectrum of sodium described by Roscoe and Schuster in the same volume of the 'Proceedings' was often seen. Potassium gives no channelled-space absorption, but continuous absorption in the red, and one narrow absorption band, with a wave-length of about 5,730, not corresponding with any bright line of that metal.

The absorption spectrum of sodium vapour is by no means so simple as has been generally represented. The fact that the vapour of sodium in a flame shows only the reversal of the D lines, while the vapour, volatilized in tubes, shows a channelled-space absorption corresponding to no known emission spectrum, appears to be a part of a gradational variation of the absorption spectrum, which may be induced with perfect regularity. Experiments with sodium exhibit the following succession of appearances, as the amount of vapour is gradually diminished, commencing from the appearance when the tube is full of the vapour of sodium, part of it condensing in the cooler portion of the tube, and some being carried out by the slow current of hydrogen. During this stage, although the lower part of the tube is at a white heat, as long as the cool current of hydrogen displaced metallic vapour, on looking down the tube it appeared perfectly dark. The first appearance of luminosity is of a purple tint, and, with the spectroscope, appears as a faint blue band, commencing with a wave-length of about 4,500, and fading away into the violet. Next appears a narrow band in the green, with a maximum of light, with a wave-length of about 5,420, diminishing in brightness so rapidly on either side as to appear like a bright line. This green band gradually widens, and is then seen to be divided by a dark band with a wave-length of about 5,510. Red light next appears, and between the red and green light is an enormous extension of the D absorption line, while a still broader dark space intervenes between the green and the blue light. The dark line in the green (wavelength about 5,510) now becomes more sharply defined. This line appears to have been observed by Roscoe and Schuster, and regarded by them as coinciding with the double sodium line next in strength to the D lines, but it is considerably more refrangible than that double line. In the next stage, the channelled-space spectrum comes out in the dark space between the green and blue, and finally in the red. Gradually the light extends, the channels disappear, the D line absorption narrows, but still the dark line in the green is plainly discernible. Lastly, there is only D lines absorption.

The method of observation described may be used to observe emission as well as absorption spectra; for if the closed end of the tube be placed against the bars of the furnace so as to be relatively cooler than the middle of the tube, the light emitted by the vapours in the hottest part is more intense than that emitted by the bottom of the tube. This succeeds admirably with sodium.

The volatility of rubidium and cæsium rendered it advisable to try the effects first in glass tubes. For this purpose a piece of combustion tubing had one end drawn out and the end turned up sharply, and sealed off (like an ill-made combustion tube of the usual form), so as to produce an approximately plane face at the end of the tube; a small bulb was then blown at about an inch from the end, and the tube drawn out at about an inch from the bulb on the other side, so as to form a long narrower tube. Some dry rubidium or cæsium chloride was introduced into the bulb, and a fragment of fresh cut sodium, and the narrow part of the tube turned up, so as to allow the tube and bulb to be seen through in the direction of the axis of the tube. The shape of tube is given in Plate I. The open end was then attached to a Sprengel pump, and the air exhausted; the sodium was then melted, and afterwards either dry hydrogen or dry nitrogen admitted, and the end of the tube sealed off at nearly the atmospheric pressure. It is necessary to have this pressure of gas inside the tube, otherwise the metal distilled so fast on heating that the ends were speedily obscured by condensed drops of metal. Through these tubes placed lengthways in front of a spectroscope, a lime light was viewed. On warming the bulb of a tube in which rubidium chloride had been sealed up with sodium, the D lines were of course very soon seen; and very soon there appeared two dark lines near the extremity of the violet light, which, on measurement, were found to be identical in position with the well-known violet lines of rubidium. Next appeared faintly the channelled spectrum of sodium in the green, and then a dark line in the blue, very sharp and decided, in the place of the more refrangible of the characteristic lines of cæsium in the flame spectrum. As the temperature rose, these dark lines, especially those in the violet, became sensibly broader; and then another fine dark line appeared in the blue, in the

PLATE I.

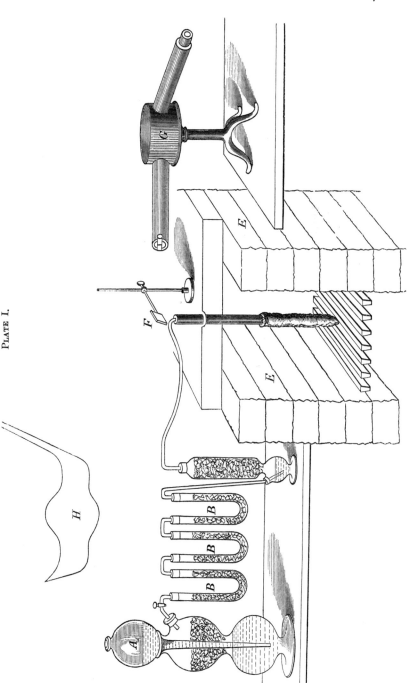

place of the less refrangible of the cæsium blue lines. During this time no dark line could be observed in the red; but as the temperature rose, a broad absorption band appeared in the red, with its centre about midway between B and C, ill defined at the edges, and though plainly visible not very dark. The lines in the violet had now become so broad as to touch each other and form one dark band. On cooling, the absorption band in the red became gradually lighter without becoming defined, and was finally overpowered by the channelled spectrum of sodium in that region. The double dark line in the violet became sharply defined again as the temperature fell. There are two blue lines in the spectrum of rubidium taken with an induction-coil very near the two blue lines of cæsium; but they are comparatively feeble, and the two dark lines in the blue observed in the places of the characteristic blue lines of cæsium must have been due to a small quantity of cæsium chloride in the sample of rubidium chloride. These blue lines were not, however, visible when some of the rubidium chloride was held in the flame of a Bunsen's burner, nor when a spark was taken from a solution of the chloride; but the more refrangible of them ($Cs\alpha$) was visible in the spark of an induction-coil, without a Leyden jar, taken between beads of the rubidium chloride fused on platinum wires.

When a tube containing cæsium chloride and sodium was observed, in the same way, the two dark lines in the blue were seen very soon after the heating began, and the more refrangible of them broadened out very sensibly as the temperature increased. The usual channelled spectrum of sodium was seen in the green, and an additional channelling appeared in the yellow, which may be due to cæsium or to the mixture of the two metals. Indeed the cæsium chloride was not free from rubidium, and the dark lines of rubidium were distinctly seen in the violet. Metallic lithium acts on the chlorides of cæsium and rubidium, giving the same results as sodium.

It is remarkable that these absorption lines of cæsium coincide with the blue lines of cæsium as seen in the flame, not with the green line which that metal shows when heated in an electric spark of high density. It is to be observed, however, that when sparks from an induction-coil without a jar are taken between beads of cæsium chloride fused on platinum wires, a spectrum similar to the flame spectrum is seen, and it is only when a Leyden jar is used that the spectrum is reduced to a green line. In like manner both the violet lines of rubidium are reversed, and both these violet lines are seen when the spark of an induction-coil, without jar, is passed between beads of rubidium chloride fused on platinum wire, though only one of them appears when a Leyden jar is used.

Mixtures of carbonate of cæsium with carbon, and of carbonate of rubidium with carbon, prepared by charring the tartrates, heated in narrow porcelain tubes, placed vertically in a furnace, gave sharp results. A small quantity of the cæsium mixture, introduced into a tube at a bright red heat, showed instantly the two blue lines reversed and so much expanded as to be almost in contact. The width of the dark lines decreased as the cæsium evaporated, but they remained quite distinct for a long time. A similar effect was produced by the rubidium mixture, only it was necessary to have the tube very much hotter, in order to get enough of violet light to see the reversal of the rubidium lines. In this case the two lines were so much expanded as to form one broad dark band, which gradually resolved itself into two as the rubidium evaporated. The reversal of these lines of cæsium and rubidium seems to take place almost or quite as readily as that of the D lines by sodium, and the vapours of those metals must be extremely opaque to the light of the refrangibility absorbed, for the absorption was conspicuous when only very minute quantities of the metals were present. The red, yellow, and green parts of the spectrum were carefully searched for absorption lines, but none due to cæsium or rubidium could be detected in any case. It is perhaps worthy of remark that the liberation of such extremely electro-positive elements as cæsium and rubidium from their chlorides by sodium and by lithium, though it is probably only partial, is a proof, if proof were wanting, that so-called chemical affinity only takes a part in determining the grouping of the elements in such mixtures; and it is probable that the equilibrium arrived at in any such case is a dynamical or mobile equilibrium, continually varying with change of temperature.

It is difficult to prevent the oxidation of magnesium in the iron tubes, and tubes wider than half an inch did not give satisfactory results. With half-inch tubes, the lines in the green were clearly and sharply reversed, also some dark lines, not measured, were seen in the blue. The sharpness of these lines depended on the regulation of the hydrogen current, by which the upper stratum of vapour could be cooled at will.

(1) The absorption spectrum of magnesium consists of two sharp lines in the green, of which one, which is broader than the other, and appears to broaden as the temperature increases, coincides in position with the least refrangible of the b group, while the other is less refrangible, and has a wave-length very nearly 5,210. These lines are the first and the last to be seen, and were first taken for the extreme lines of the b group.

(2) A dark line in the blue, always more or less broad, difficult to measure exactly, but very near the place of the brightest blue line of magnesium. This line was not always visible, indeed rarely when magnesium alone was placed in the tube. It was better seen when a small quantity of potassium or sodium was added. The measure of the less refrangible edge of this band gave a wave-length of very nearly 4,615.

(3) A third line or band in the green rather more refrangible than the b group. This is best seen when potassium and magnesium are introduced into the tube, but it may also be seen with sodium and magnesium. The less refrangible edge of this band is sharply

defined, and has a wave-length about 5,140, and it fades away towards the blue.

These absorptions are all seen both when potassium and sodium are used along with mgnesium, and may be fairly ascribed to magnesium, or to magnesium together with hydrogen.

But besides these, other absorptions are seen which appear to be due to mixed vapours.

(4) Thus when sodium and magnesium are used together a dark line, with ill-defined edges, is seen in the green, with a wave-length about 5,300. This is the characteristic absorption of the mixed vapours of sodium and magnesium; it is not seen with either vapour separately, nor is it seen when potassium is used instead of sodium.

(5) When potassium and magnesium are used together, a pair of dark lines are seen in the red. The less refrangible of these sometimes broadens into a band with ill-defined edges, and has a mean wave-length of about 6,580. The other is always a fine sharp line, with a wave-length about 6,475. These lines are as regularly seen with the mixture of potassium and magnesium as the above-mentioned line (5,300) is seen with the mixture of sodium and magnesium, but are not seen except with that mixture.

There is a certain resemblance between the absorptions above ascribed to magnesium, and the emission spectrum seen when the sparks of a small induction-coil, without Leyden jar, are taken between electrodes of magnesium.

The coincidences of the series of the solar spectrum hitherto observed have, for the most part, been with lines given by dense electric sparks; while it is not improbable that the conditions of temperature, and the admixtures of vapours in the upper part of the solar atmosphere, may resemble much more nearly those in our tubes.

It became a question of interest to find the conditions under which the same mixtures would give luminous spectra, consisting of the lines which had been seen reversed. On observing sparks from an induction-coil taken between magnesium points in an atmosphere of hydrogen, a bright line regularly appeared, with a wave-length about 5,210, in the same position as one of the most conspicuous of the dark lines observed to be produced by vapour of magnesium with hydrogen in our iron tubes. This line is best seen, i.e. is most steady, when no Leyden jar is used, and the rheotome is screwed back, so that it will but just work. It may, however, be seen when the coil is in its ordinary state, and when a small Leyden jar is interposed; but it disappears (except in flashes) when a larger Leyden jar is used, if the hydrogen be at the atmospheric pressure. This line does not usually extend across the whole interval between the electrodes, and is sometimes only seen near the negative electrode. Its presence seems to depend on the temperature, as it is not seen continuously when a large Leyden jar is employed, until the pressure of the hydrogen and its resistance is very much reduced. When well-dried nitrogen or carbonic oxide is substituted for hydrogen, this line disappears entirely; but if any hydrogen or traces of moisture be present it comes out when the pressure is much reduced. In such cases the hydrogen lines C and F are always visible as well. Sometimes several fine lines appear on the more refrangible side of this line, between it and the *b* group, which give it the appearance of being a narrow band, shaded on that side. Various samples of magnesium used as electrodes, and hydrogens prepared and purified in different ways, gave the same results.

In addition to the above-mentioned line, there is also produced a series of fine lines, commencing close to the most refrangible line of the *b* group, and extending with gradually diminishing intensity towards the blue. These lines are so close to one another, that in a small spectroscope they appear like a broad shaded band. We have little doubt that the dark absorption line, with wave-length about 5,140, shading towards the blue, observed in our iron tubes, was a reversal of part of these lines, though the latter extend much further towards the blue than the observed absorption extends.

Charred cream of tartar in iron tubes, arranged as before, gave a broad absorption band extending over the space from about wave-length 5,700 to 5,775, and in some cases still wider, with edges ill-defined, especially the more refrangible edge. By placing the charred cream of tartar in the tube before it was introduced into the furnace, and watching the increase of light as the tube got hot, this band was at first seen bright on a less bright background, it gradually faded, and then came out again reversed, and remained so. No very high temperature was required for this, but a rise of temperature had the effect of widening the band. Besides this absorption, there appeared a very indefinite faint absorption in the red, with the centre at a wave-length of about 6,100, and a dark band, with a tolerably well-defined edge on the less refrangible side, at about a wave-length of 4,850, shading away towards the violet. A fainter dark band was sometimes seen beyond, with a wave-length of about 4,645; but sometimes the light seemed abruptly terminated at about wave-length 4,850. It will be noticed that these absorptions are not the same as those seen when potassium is heated in hydrogen, nor do they correspond with known emission lines of potassium, although the first, which is also the most conspicuous and regularly visible of these absorptions, is very near a group of three bright lines of potassium. It seemed probable that they might be due to a combination of potassium with carbonic oxide. Potassium heated in carbonic oxide in glass tubes, united readily with the gas, but the compound did not appear to volatilize at a dull red heat, and no absorption, not even that which potassium gives when heated in nitrogen under similar circumstances, could be seen. Induction sparks between an electrode of potassium and one of platinum in an atmosphere of carbonic oxide, gave the usual bright lines of potassium, and also a bright band, identical in position with the above-

---

* With greater dispersion this line is seen as the sharp edge of a series of very fine lines shading off towards the blue like the ordinary hydrocarbon spectrum.

mentioned band, between wave-lengths about 5,700 and 5,775. This band could not be seen when hydrogen was substituted for carbonic oxide. A mixture of sodium carbonate and charred sugar, heated in an iron tube, gave only the same absorption as sodium in hydrogen. There were also no indications of any absorption due to a compound of rubidium or of cæsium with carbonic oxide.

A mixture of barium carbonate, aluminium filings, and lamp-black, heated in a porcelain tube, gave two absorption lines in the green, corresponding in position to bright lines seen when sparks are taken from a solution of barium chloride, at wave-lengths 5,242 and 5,136, marked α and β by Lecoq de Boisbaudran. These two absorptions were very persistent, and were produced on several occasions. A third absorption line, corresponding to line δ of Boisbaudran, was sometimes seen; and on one occasion, when the temperature was as high as could be obtained in the furnace fed with Welsh coal, and a mixture of charred barium tartrate with aluminium was used, a fourth dark line was seen with wave-length 5,535. This line was very fine and sharply defined, whereas the other three lines were ill-defined at the edges; it is, moreover, the only one of the four which corresponds to a bright line of metallic barium.

Repeated experiments with charred tartrates of calcium and of strontium mixed with aluminium gave no results; but on one occasion, when sodium carbonate was used along with the charred tartrate of strontium and aluminium, the blue line of strontium was seen reversed: and on another occasion, when a mixture of charred potassium, calcium, and strontium tartrates, and aluminium was used, the calcium line, with wave-length 4,226, was seen reversed.

In order to command higher temperatures, experiments were made with the electric arc enclosed in lime, magnesia, or carbon crucibles. The different forms used are represented in Plate II. Figs. 1, 2, 3, 4, and 5; and the plan for projecting reversals in Plate III.

In the first experiments thirty cells of Grove were employed; in the later ones the Siemens arc from the powerful dynamo-machine belonging to the Royal Institution.

The electric arc in lime crucibles gives a very brilliant spectrum of bright lines, a copious stream of vapours ascending the tube. On drawing apart the poles, which could be done for nearly an inch without stopping the current, the calcium line with wave-length 4,226 almost always appears more or less expanded with a dark line in the middle, both in the lime crucibles and in carbon crucibles into which some lime has been introduced; the remaining bright lines of calcium are also frequently seen in the like condition, but sometimes the dark line appears in the middle of K (the more refrangible of Fraunhofer's lines H), when there is none in the middle of H. On throwing some aluminium filings into the crucible, the line 4,226 appears as a broad dark band, and both H and K, as well as the two aluminium lines between them, appear for a second as dark bands on a continuous background. Soon they appear as bright bands with dark middles;

PLATE II.

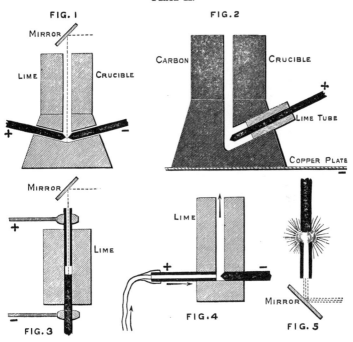

PLATE III.

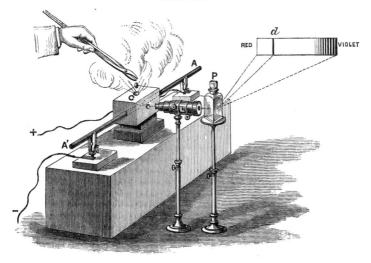

gradually the dark line disappears from H, and afterwards from K, while the aluminium lines remain with dark middles for a long time. When a mixture of lime and potassium carbonate was introduced into a carbon crucible, the group of three lines with wave-lengths 4,425, 4,434, and 4,454 were all reversed, the least refrangible being the most strongly reversed, and remaining so longest, while the most refrangible was least strongly reversed, and for the shortest time.

When aluminium was put into the crucible, only the two lines of that metal between H and K were seen reversed. The lines at the red end remained steadily bright.

When magnesium was put into a lime crucible, the *b* group expanded a little without appearing reversed, but when some aluminium was added, the least refrangible of the three lines appeared with a dark middle, and on adding more magnesium the second line put on the same appearance; and lastly, the most refrangible was reversed in like manner. The least refrangible of the three remained reversed for some time; and the order of reversibility of the group is that of refrangibility. Of the other magnesium lines, that in the yellowish-green (wave-length 5,527) was much expanded, while the blue line (wave-length 4,703), and a line still more refrangible than the hitherto recorded lines, with wave-length 4,354, were still more expanded each time that magnesium was added.

The following experiments were made in carbon crucibles:—

With strontia the lines with wave-lengths 4,607, 4,215 and 4,079 were all seen with dark lines in the middle, but no reversal of any strontium line less refrangible could be seen.

A mixture of barium and potassium carbonates produced the reversal of the lines with wave-lengths 5,535 and 4,933. When barium chlorate was dropped into a crucible, the four lines with wave-lengths 4,553, 4,933, 5,545, and 5,518 were reversed.

To observe particularly the effects of potassium a mixture of lime and potassium carbonate previously ignited was thrown in. The violet lines of potassium, wave-length 4,044, came out immediately as a broad black band, which soon resolved into *two* narrower dark bands having wave-lengths nearly 4,042 and 4,045. On turning to the red end the two extreme red lines were both seen reversed. No lines of potassium between the two extremes could be seen reversed, but the group of three yellow lines were all expanded, though not nebulous, and other lines in the green were seen much expanded.

Sodium carbonate gave only the D lines reversed, though the other lines were expanded, and the pairs in the green had each become a very broad nebulous band, and D almost as broad a black band. When sodium chlorate was dropped into a crucible, the pair of lines with wave-lengths 5,681, 5,687, were both momentarily reversed, the latter much more strongly than the former.

When a very little charred rubidium tartrate was put in, the two violet lines were sharply reversed, appearing only as black lines on a continuous light background. Turning to the red end, the more refrangible of the two lines in the extreme red (wave-length 7,800) was seen to have a decided dark line in the middle, and it continued so for some time. The addition of more rubidium failed to cause any reversal of the extreme red line, or of any but the three lines already mentioned.

On putting lithium carbonate into the crucible, the violet line of lithium appeared as a nebulous band, and on adding some aluminium this violet band became enormously expanded, but showed no reversal. The blue lithium line (wave-length 4,604) was well reversed, as was also the red line, while a fine dark line passed through the middle of the orange line. On adding a mixture of aluminium filings and the carbonates of lithium and potassium, the red line became a broad black band, and the orange line was well reversed. The green line was exceedingly bright, but not nebulous or reversed, and the violet line still remained much expanded, but unreversed.

Metallic indium placed in the crucible gave the lines with wave-lengths 4,101 and 4,509, and both were seen strongly reversed. No other absorption line of indium could be detected.

In some cases a current of hydrogen or of coal-gas was introduced into the crucibles by means of a small lateral opening, or by a perforation through one of the carbon electrodes, as is shown in Plate II. Fig. 4; sometimes the perforated carbon was placed vertically, and we examined the light through the perforations. When no such current of gas is introduced, there is frequently a flame of carbonic oxide burning at the mouth of the tube. The current of hydrogen produces very marked effects. As a rule, it increases the brilliance of the continuous spectrum, and diminishes relatively the apparent intensity of the bright lines, or makes them altogether disappear with the exception of the carbon lines. When this last is the case, the reversed lines are seen simply as black lines on a continuous background. The calcium line with wave-length 4,226 is always seen under these circumstances as a more or less broad black band on a continuous background, and when the temperature of the crucible has risen sufficiently, the lines with wave-lengths 4,434 and 4,454, and next that with wave-length 4,425, appear as simple black lines. So, too, do the blue and red lines of lithium, and the barium line of wave-length 5,535, appear steadily as sharp black lines, when no trace of the other lines of these metals, either dark or bright, can be detected. Dark bands also frequently appear, with ill-defined edges, in the positions of the well-known bright green and orange bands of lime.

With sodium chloride, the pair of lines (5,687, 5,681) next more refrangible than the D group were repeatedly reversed. In every case the less refrangible of the two was the first to be seen reversed, and was the more strongly reversed, as has also been observed by Mr. Lockyer. But our observations on this pair of lines differ from his in so far as he says that "the double green line of sodium shows scarcely any trace of absorption when the lines are visible," while we have repeatedly seen the reversal as dark lines appearing on the expanded bright lines; a second pair of faint bright lines, like ghosts of the first, usually coming out at the same time on the more refrangible side.

Potassium carbonate gave, besides the violet and red lines which had been reversed before, the group, wave-lengths 5,831, 5,802, and 5,872, all reversed, the middle line of the three being the first to show reversal. Also the lines wave-lengths 6,913, 6,946, well reversed, the less refrangible remaining reversed the longer. Also the group, wave-lengths 5,353, 5,338, 5,319 reversed, the most refrangible not being reversed until after the others. Also the line wave-length 5,112 reversed, while two other lines of this group, wave-lengths 5,095 and 5,081, were not seen reversed.

Using lithium chloride, not only were the red and blue lines, as usual, easily reversed, and the orange line well reversed for a long time, but also the green line was distinctly reversed; the violet line still unreversed, though broad and expanded. Had this green line belonged to cæsium, the two blue lines of that metal which are so easily reversed could not have failed to appear; but there was no trace of them.

In the case of rubidium, the less refrangible of the red lines was well reversed as a black line on a continuous background, but it is not easy to get, even from the arc in one of our crucibles, sufficient light in the low red to show the reversal of the extreme ray of this metal.

With charred barium tartrate, and also with baryta and aluminium together, the reversal of the line with wave-length 6,496 was observed, in addition to the reversals previously described. The less refrangible line, wave-length 6,677, was not reversed.

With charred strontium tartrate, the lines with wave-lengths 4,812, 4,831, and 4,873, were reversed, and by the addition of aluminium, the line wave-length 4,962 was reversed for a long time, and also the lines wave-lengths 4,895, 4,868.

On putting calcium chloride into the crucible, the line wave-length 4,302 was reversed, this being the only one of the well-marked group to which it belongs which appeared reversed. On another occasion, when charred strontium tartrate was used, the line wave-length 4,877 was seen reversed, as well as the strontium line near it. The lines wave-lengths 6,161, 6,121, have been seen momentarily reversed.

With magnesium, when a stream of hydrogen or of coal-gas was led into the crucible, the line wave-length 5,210, previously seen in iron tubes, and ascribed to a combination of magnesium with hydrogen, was regularly seen, usually as a dark line, sometimes with a tail of fine dark lines on the more refrangible side similar to the tail of bright lines seen in the sparks taken in hydrogen between magnesium points. Sometimes, however, this line (5,210) was seen bright. It always disappeared when the gas was discontinued, and appeared again sharply on readmitting hydrogen. These effects were however, only well defined in crucibles having a height of at least 3 inches above the arc.

On putting a fragment of metallic gallium into a crucible, the less refrangible line, wave-length 4,170, came out bright, and soon a dark line appeared in the middle of it. The other line, wave-length 4,031, showed the same effect, but less strongly.

Reviewing the series of reversals which have been observed, in many cases the least refrangible of binary groups is the most easily reversed, as has been previously remarked by Cornu.

Making a general summation of the results respecting the alkaline earth metals, potassium and sodium, having regard only to the most characteristic rays, which for barium may be taken as 21, for strontium 34, for calcium 37, for potassium 31, and for sodium 12, the reversals number respectively 6, 10, 11, 13, and 4. That is in the case of the alkaline earth metals about one-third, and these chiefly in the more refrangible third of the visible spectrum, the characteristic rays remaining unreversed in the more refrangible part of the spectrum being respectively 2, 5, and 4.

The curious behaviour of the lines of different spectra with regard to reversal induced a comparison with the bright lines of the chromosphere of the sun, as observed by Young. It is well known that some of the principal lines of the metals giving comparatively simple spectra, such as lithium, aluminium, strontium, and potassium, are not represented amongst the dark lines of Fraunhofer, while other lines of those metals are seen: and an examination of the bright chromospheric lines shows that special rays highly characteristic of bodies which appear from other rays to be present in the chromosphere are absent, or are less frequent in their occurrence than others.

In the following table the relation between the observations on reversals and Young's on the chromospheric lines is shown.

Lines in Wave-Lengths.	Frequency in Chromosphere.	Behaviour. Reversal in our Tubes.	Remarks.
Sodium .. 6,160 } 6,154 }	0	Expanded.	
D	50	Most easy	Principal ray.
5,687 } 5,681 }	2	Difficultly reversed.	
5,155 } 5,152 }	2	Very diffused.	
4,983 } 4,982 }	0	,, ,,	
Lithium .. 6,705	0	Readily reversed ..	Most characteristic, at low temperature and low density.
6,101	3	Difficultly reversed.	
4,972	0	,, ,,	
4,603	0	Readily reversed.	
4,130	0	Very diffused ..	Described by Boisbaudran.

Lines in Wave-Lengths.	Frequency in Chromosphere.	Behaviour. Reversal in our Tubes.	Remarks.
Magnesium 5,527	40	Expanded.	
b_1 5,183	50	Reversed ..	
b_2 5,172	50	,,	Most characteristic.
b_4 5,167	30	Difficultly reversed	
4,703	0	Much expanded.	
? 4,586	0	,, ,,	Doubtful whether due to magnesium.
4,481	0	Not seen either bright or reversed.	Characteristic of spark absent in arc.
Barium .. 6,677	25	0	May be either Ba or Sr.
6,496	18	Reversed.	,, ,, ,,
6,140	25	0	
5,534	50	Readily reversed ..	Most persistent.
5,518	15	Reversed.	
4,933	30	,,	Well-marked ray.
4,899	30	0	
4,553	10	Pretty readily reversed.	
Strontium 6,677	25	0	May be Sr or Ba.
6,496	18	0	,, ,, ,,
4,902			
4,895			
4,873			
4,868	..	Reversed.	
4,812			
4,831			
4,607	0	Readily and strongly reversed.	Most characteristic.
4,215	40	Readily reversed ..	Well marked.
4,077	25	,, ,, ..	,, ,,
Calcium .. 6,161	8	Reversed difficultly	Very bright.
6,121	5	,, ,,	
5,587	2	Doubtful reversal.	
5,188	10	Reversed.	
4,877			
4,587	2	0	
4,576	4	0	
4,453	0	Readily reversed.	
4,435	1	,, ,,	
4,425	2	,, ,,	
4,302			
4,226	3	Most easily reversed	Very characteristic.
4,095 (?)	0	Strongly reversed.	
3,968	75	Well reversed.	
3,933	50	Rather more readily than the last.	
Aluminium 6,245	8	0	Strong lines
6,237	8	0	
3,961 / 3,943	0	Strongly reversed ..	Very marked.

Lines in Wave-lengths.		Frequency in Chromosphere.	Behaviour. Reversal in our Tubes.	Remarks.
Potassium	7,670 7,700	0	Strongly reversed ..	Chief rays.
	6,946 6,913	..	Reversed.	
	5,872 5,831 5,802	..	,,	
	5,353 5,338 5,319	..	,,	
	5,112	..	,,	
	4,044 4,042	3	..	Well marked.
Cæsium	.. 5,990	10	0	
	4,555	10	Strongly reversed ..	Most marked.

The group calcium, barium, and strontium on the one hand, and sodium, lithium, magnesium, and hydrogen, on the other, seem to behave in a similar way in the chromosphere of the sun; but before definite conclusions can be reached regarding the sequence of the reversals, a further series of long and laborious experiments must be executed.

[J. D.]

Friday, January 16, 1880.

GEORGE BUSK, Esq. F.R.S. Treasurer and Vice-President,
in the Chair.

PROFESSOR JAMES DEWAR, M.A. F.R.S.
FULLERIAN PROFESSOR OF CHEMISTRY R.I.

(Abstract.)

Investigations at High Temperatures.

I INTEND to discuss on the present occasion the results of a preliminary study of the chemical interactions taking place at the temperature of the electric arc, and the inferences which can be deduced from a series of radiation experiments as to the probable temperature of this source of heat.

On the Formation of Hydrocyanic Acid in the Electric Arc.

The conclusion that the so-called carbon spectrum is invariably associated with the formation of acetylene,* induced me to try and ascertain whether this substance can be extracted from the electric arc, which invariably shows this peculiar spectrum at the positive pole, when it is powerful and occasionally intermittent. For this purpose the carbons were used in the form of tubes, as shown in the following figure, so that a current of air could be drawn by means of an aspirator through either pole, and the products thus extracted from the arc, collected in water, alkalies, and other absorbents. Gases may be led through one of the poles, and suction induced through the other, in order to examine their effect on the arc and the products obtained from it.

The following results were obtained by means of the Siemens and De Méritens magneto-machines, recently presented to the Royal Institution through the munificence of the Duke of Northumberland and Mr. Siemens :—

Air drawn by an aspirator from the arc through a drilled negative carbon, and the gases passed through potash, iodide of potassium, and

---

* As suggested by Plücker, Ångström, and Thalén.

starch paste, gave no reaction for the presence of nitrites. The potash contained sulphides.

Hydrogen led in through the positive pole, and the gases extracted as above gave the well-known acetylene compound with ammoniacal sub-chloride of copper; while, at the same time, a wash-bottle containing water gave distinct evidence of the presence of hydrocyanic acid.

Fig. 1.

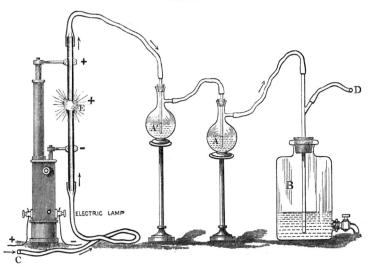

A hydrogen flame burning between the carbon poles gave no sulphides or hydrocyanic acid, when treated in the above manner. The condensed water from the combustion gave the reaction for nitrites.

Air drawn through the negative carbon gave considerable quantities of hydrocyanic acid, which was greatly increased by extracting the gases through the positive carbon. Air was aspirated at the rate of about one litre per minute.

The same carbons used with the long arc of the De Méritens magneto-machine gave no hydrocyanic acid.

Carbons purified in chlorine and hydrogen gave with De Méritens' arc nothing; with Siemens' and a draught of air through the negative pole, a small quantity of hydrocyanic acid, but a larger yield when the positive pole was used. The gases extracted from the arc after the absorption of the hydrocyanic acid contained acetylene. If the carbons are not purified, sulphuretted hydrogen is always found along with the other gases.

The inference drawn from the above experiments is that the high temperature of the positive pole is required to produce the reaction, which is in all probability the result of acetylene reacting with free nitrogen, as when induction sparks are passed through the mixed gases, viz.:—

$$C_2H_2 + N_2 = 2HCN,$$

and that the hydrogen is obtained from the decomposition of aqueous vapour, and the combined hydrogen in the carbons. It is possible that traces of alkaline salts in the carbon poles may favour the formation of hydrocyanic acid, but, as all attempts to purify the poles so as to stop the reaction failed, I am inclined to believe it is a direct synthesis. The acetylene reaction is one of the many remarkable syntheses discovered by Professor Berthelot, of Paris. The presence of sulphuretted hydrogen is doubtless due to the reduction of the sulphates, invariably present in the ash of the carbon.

The discovery of the formation of hydrocyanic acid in the electric arc necessitated a more complete examination of the various reactions taking place in the arc with poles of various kinds, and in presence of different gaseous media.

Various difficulties have impeded the satisfactory progress of the investigation. During the course, however, of numerous experiments, facts of interest have been recorded which are worthy of appearing as preliminary results in a very extensive and difficult research.

Formation of Cyanogen Compounds.

The influence of impurities in the carbon on the production of hydrocyanic acid had first to be ascertained. For this purpose, drilled Siemens' carbons were placed in a porcelain tube, and treated for several days at a white heat with a rapid stream of chlorine, until the greater part of the silica, oxide of iron, alumina, &c., were volatilized in the form of chlorides. Sometimes the carbons had a subsequent treatment with hydrogen, or were directly treated with a current of chlorine while the arc was in operation.

Carbons treated in this way continued to yield hydrocyanic acid, when a steady current of air was drawn through the positive pole as formerly described, even when the same pole had several successive treatments with chlorine during the electric discharge. Natural graphite poles gave the same result.

As it was evident that the elimination of a large portion of the impurities had little influence on the production of the hydrocyanic acid, the only other explanation of its formation appeared to be the presence of aqueous vapour, and organic impurities in the air, or a direct formation of cyanogen from carbon and nitrogen through the acetylene reaction formerly described. To obtain a pure and dry atmosphere in which such experiments could be carried out, the following apparatus was employed:—

A tin vessel, Fig. 2, about 2 feet high and 1 foot in diameter, had an annular space, through which a constant stream of water was kept flowing. This cylinder was placed upon a porcelain stand, having a narrow groove filled with mercury, so as to make an air-tight joint.

The lamp was placed inside this vessel, the wires connecting it with the machine being brought through the bottom of the stand. A tube passed through the porcelain base, which allowed a current of dry air to be forced through the vessel. A small aperture in the top of the tin vessel allowed the glass tube coming from the positive pole to pass with little friction, through which the products from the arc were drawn. This annular vessel was very convenient, not only for examining the products formed in the arc, but also those formed outside of it, and the water flowing round it served the double purpose of keeping it cool and enabling a determination of the amount of total radiation in heat units to be made.

Fig. 2.

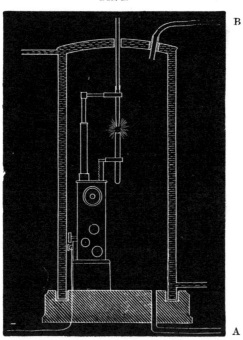

Vessels containing pumice moistened with sulphuric acid and phosphoric anhydride were placed inside the cylinder in order to dry the interior as completely as possible.

Numerous experiments made by forcing perfectly dry air into the vessel through the tube A, and drawing it out by the tube B through a weighed sulphuric acid bulb, gave after an hour a few milligrams of increase, owing, no doubt, to some slight defect in the soldering of the tin, which allowed a capillary film of water to cover part of the surface and diffuse into the interior.

When the ordinary Siemens' carbons were used as poles in this almost dry atmosphere, the yield of hydrocyanic acid was still very marked, purified carbons yielding the same results.

As the yield of cyanogen compounds did not appear to be diminished, and it seemed almost impossible to get the large volume of air in the tin vessel perfectly dry, another plan was adopted. The poles were enclosed in an egg-shaped glass globe about 8 inches long and 6 inches in diameter, in order to diminish the volume of air to be dried and dispense with the water covering. The globe, balanced through a system of pulleys, was firmly attached to the lower or negative pole, with which it moved without impeding the automatic action of the lamp.

Dry air was sometimes forced through the negative carbon itself, at other times through a glass tube passing up the side of it into the globe, the products from the arc being drawn through the positive pole as before.

As the glass globe soon became very hot, and as a far larger supply of dry air was forced through the globe than was drawn out from the arc, it is inconceivable that any moisture could remain near the arc after it had been in operation for a few minutes.

Seven consecutive experiments, each of ten minutes' duration, made with the same purified carbon poles, did not show any diminution in the quantity of hydrocyanic acid, unless in one of the experiments, when the arc would not be drawn into the interior of the carbon tube, but persisted in rotating round it.*

These experiments show that drilled carbons even after prolonged treatment with chlorine, still contained a quantity of combined hydrogen, and organic analyses showed that the amount of ash and combined hydrogen in the various samples was never less than about 0.75 of the former, and as much as 0.1 of the latter. Poles made with especially purified carbon by Messrs. Siemens for these experiments proved to be no better in respect to the quantity of hydrogen and ash they contained.

The well-nigh impossible problem of eliminating hydrogen from masses of carbon such as can be employed in experiments of this kind, proves conclusively that the inference drawn by Mr. Lockyer,† as to the elementary character of the so-called carbon spectrum from an examination of the arc in dry chlorine, cannot be regarded as satisfactory, seeing that undoubtedly hydrogen was present in the carbon, and in all probability nitrogen in the chlorine.

---

* Cyanogen is difficult to recognize in presence of prussic acid when in small quantity, especially when impurities from the carbons complicate the tests. In speaking generally of the formation of this acid in the arc, I do not mean to exclude the possibility of cyanogen being formed as well.

† "Note on the Existence of Carbon in the Coronal Atmosphere of the Sun," Proc. Roy. Soc.,' vol. xxvii. p. 308.

Experiments with Carbon Tubes.

In order to ascertain whether the formation of hydrocyanic acid and acetylene in the arc was really due to transformations induced by some occult power located in the arc, or was simply the result of the high temperature attained by the carbons, experiments were made in carbon tubes, the arc being merely used as a means of heating. The method of arranging the arc for this experiment is represented in Fig. 3.

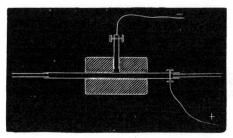

A block of lime about 5 inches long by 3 inches thick was drilled horizontally, as shown in the drawing, another hole being drilled so as to meet it in the centre of the mass.

The new bricks used in the Bessemer converters do very well for all the experiments of this description.

A drilled purified carbon was placed in the horizontal channel and made the positive pole, the negative pole being a solid rod of carbon passing through the vertical aperture. Gases were passed through the positive carbon, and were thus subjected to the intense heat of the walls of the tube, the arc passing outside.

The walls of the positive carbon were pierced by the arc with great rapidity, not lasting, as a rule, more than fifteen minutes. This action could only be retarded by using thicker carbons, or by rotating the tube.

The porosity of the carbons, which allowed a constant diffusion of gases through their walls, was a great source of difficulty.

In order to prove that the temperature in the interior of the carbon tube is higher than that of the oxyhydrogen flame, it is sufficient to place in it a few small crystals of diamond, and to maintain a current of hydrogen to prevent oxidation. In a few minutes the diamond is transformed into coke.

On passing a mixture of three volumes hydrogen and one volume nitrogen thoroughly dried through the positive pole, a large yield of hydrocyanic acid was always obtained, and on using equal volumes of hydrogen and nitrogen the quantity was, if anything, increased.

Pure dry hydrogen by itself gave a trace of hydrocyanic acid, and a considerable quantity of acetylene.

Pure dry air gave no hydrocyanic acid or acetylene; moist air, on the contrary, giving abundance of the former, but only a trace of the latter.

The yield in all these experiments altered considerably with the rate at which the gases were passed, a quick stream always producing more than a slow one, unless when oxygen was present.

Formation of Nitrites in the Arc.

In these experiments the annular vessel was made use of, in which the lamp was allowed to work automatically, often for an hour or two. A continuous stream of dry air was kept circulating through the interior, being afterwards passed through a series of wash bottles containing dilute caustic soda, or directly through strong sulphuric acid, to absorb the oxides of nitrogen. The nitrous acid was estimated in the former case by titration with permanganate of potash, and the total combined nitrogen by the mercury process.

In this way many experiments were made with a Siemens lamp, both with a long and short arc; Jablochkoff's candles without any insulating material between the poles were also employed with the highest intensity current of a De Méritens machine, in order to have the greatest variety in the character of the discharge.

The stream of dry air was forced through the vessel at varying degrees of speed, and was found to have a decided effect on the quantity of nitrites produced, the more rapid stream giving the largest yield of nitrites.

The following table gives the amount of the nitrous acid produced in a number of different experiments.

The nitrites are calculated as nitrous acid.

1. Siemens' machine and lamp.   2. Jablochkoff's candles.

	Nitrites produced in 1 hour.		De Méritens' Highest Intensity Current.
	Siemens'		
	Long Arc.	Short Arc.	
	milligrams.	milligrams.	milligrams.
1st experiment =	193	28	769
2nd ,, =	804	97	723
3rd ,, =	618	73	1225
4th ,, =	500	121	548
5th ,, =	622	90	955
6th ,, =	474	85	1006
7th ,, =	380	..	1257
8th ,, =	459	..	964
9th ,, =	..	..	664
10th ,, =	..	..	489
11th ,, =	..	..	693
=	509 mean	..	930 mean

In these experiments, the total nitrogen estimated by the mercury process was almost identically the same as the amount of nitrogen obtained by a very careful dilution of the acid in a large quantity of water and titration with permanganate, proving that the main product was nitrous anhydride, which may be explained by the fact that the quantity of oxygen in presence of nitrogen in the immediate neighbourhood of the poles is greatly diminished by the combustion of the carbons, or that the nitric peroxide formed is subsequently reduced by contact with the red hot carbon, or other reducing products.

It is thus proved the carbon, nitrogen, oxygen, and hydrogen being present at the temperature of the electric arc, the compound substances hydrocyanic acid, acetylene, and nitrous acid are invariably formed.

Radiation Experiments.

In a report to the British Association * on the determination of high temperatures in the year 1873, it was experimentally proved that the law of Dulong and Petit could not be used as a basis for the estimation of high temperatures, seeing that it " gives a far too rapid increase for the total radiation." It was further observed that the value of the radiation emitted by the same substance at different temperatures expressed in terms of the thermo-electric current increase of intensity, plotted in terms of the temperature, represented a "parabolic curve." Assuming the general accuracy of this law for high temperatures, the total radiation may be taken as nearly proportional to the square of the temperature. From this law the hypothetical temperature of the sun was "estimated as at least 11,000 C." Rosetti has recently made a more elaborate investigation on the subject, and has arrived independently at a formula of a parabolic order. Rosetti † represents his results by the equation—

$$\mu = a\,T^2(T - \theta) - b(T - \theta),$$

where μ is the total radiation measured by intensity of thermo-electric current, $T°$ the absolute temperature of the source, $\theta°$ that of the medium surrounding the pile, and a and b constants. However well this formula may represent the complete series of the experiments, it is certain that his results for temperatures above 150° may be expressed within the limits of probable error as proportional to the square of the temperature. To be convinced of this, it is sufficient to plot the logarithm of the respective values of the radiation and temperature, when it will be found the results arrange themselves in a straight line, the tangent of which may be 1·9 or 2 for the observations above 150°. Experiments made with the thermopile, surrounded with an annular vessel, through which a continuous current of water

---

* Report of the Committee for determining High Temperatures by means of the Refrangibility of the Light evolved by Fluid or Solid Substances. Bradford, 1873. Page 461.

† " Recherches Expérimentales sur la Température du Soleil " (Accad. R. dei Lincei. 1877–78).

at constant temperature is caused to circulate, as represented in Fig. 4, where EF represents the section of the vessel, and CD a large water screen, on the same plan, each having a narrow opening, about half an inch in diameter, through which the radiant heat passed to the pile, have confirmed the earlier results. The vessel holding the

Fig. 4.

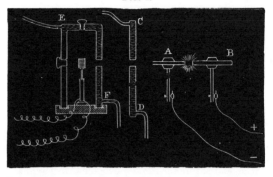

mercury or other substance to be heated to different temperatures has a radiating face, which was made of the sheet iron used in the construction of telephone plates, and the thermometer must be placed close to the back of the front surface, and the face guarded with a screen, FG. The tube, CE, is connected with a condenser, when substances at their boiling point are employed for giving fixed points. The form of the apparatus is shown in Fig. 5.

Fig. 5.

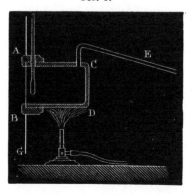

This arrangement of the apparatus is necessary in order to get anything like comparable results. The two following tables give the records of two series of experiments, without any correction being made in the numbers representing the deviations of the Thomson galvanometer:—

TABLE I.

Temperature.	Deviation.	Difference.	Temperature.	Deviation.	Difference.
°80	32	6·5	°160	95·5	10·5
90	38·5	6·0	170	106	11·0
100	44·5	6·5	180	117	13·5
110	52	7·5	190	130·5	13·0
120	59·5	7·0	200	143·5	14·5
130	66·5	9·0	210	158	14·5
140	75·5	9·5	220	172·5	15·5
150	85	10·5	230	188	

TABLE II.

Temperature.	Deviation.	Temperature.	Deviation.
°100	21	°200	71
120	29	220	86
150	41	355	240
160	46	448	370
180	57		

If the differences in the galvanometer readings for every ten degrees in the first table be tabulated, it will be observed the second difference may be regarded as constant, considering the errors of this kind of observation. A parabolic formula can therefore represent the results with sufficient accuracy. These second differences are far more constant than similar numbers deduced from Rosetti's observations, and his more complete formula in terms of the absolute temperature is too extensive, considering the range of the experiments where temperature was accurately known. The results of Table II. extend to the boiling points of mercury and sulphur, and the numbers are in near accord with the simple square of the temperature. The alteration in the condition of the radiating surface at high temperatures causes great complications, and until this difficulty is overcome, experiments at high temperatures must remain uncertain. All the experiments show that for an approach to a knowledge of temperatures beyond the range of our actual thermometric scale, the law given in 1873 is a sufficiently correct reproduction of the facts, considering the limited data at our disposal.

The intensity of the radiation of the positive pole of the Siemens' arc, as compared with the same surface heated with a large oxyhydrogen blowpipe, was determined by employing a hollow negative carbon which allowed the intensely heated surface to radiate directly on to the pile, as shown in Fig. 4. A large number of observations have been made by this method at different times, and with slight modifications in the order of the experiments, leading to the average result that the intensity of the total radiation of the positive pole of the Siemens' arc is ten times that of the same substance at the temperature of the oxyhydrogen flame. If we take an average result of nine to one, then we may infer that the temperature of the limiting positive pole is about 6000° C., seeing that the mean temperature of the oxyhydrogen may be taken as 2000° C. The mean value of the total radiation of the Siemens' arc was determined by observing the rate of flow of the water through the annular vessel, represented in Fig. 1, together with the mean increment of temperature. This gave on the average 34,000 gram-units per minute, or a little more than three horse-power.

[J. D.]

Friday, January 21, 1881.

WILLIAM BOWMAN, Esq. LL.D. F.R.S. Vice-President, in the Chair.

WARREN DE LA RUE, Esq. M.A. D.C.L. F.R.S. Sec. R.I.
Cor. Mem. Inst. France, Hon. Mem. Impl. Academy of St. Petersburg, &c.

The Phenomena of the Electric Discharge with 14,400 *Chloride of Silver Cells.*

FOR the last six years I have, in conjunction with my friend Dr. Hugo Müller, been engaged with experiments on the electric discharge, using as the source of electricity a constant voltaic battery which we devised.* It is in principle the same as that invented by Daniell, but in our battery a solid electrolyte, insoluble in water or a weak saline solution, namely, chloride of silver, replaces the soluble sulphate of copper, so that no porous cell is needed in the chloride of silver battery. The results of our experiments my colleagues think of sufficient interest to be brought under the notice of the Members of the Royal Institution, and I will endeavour to make them as clear as possible in the limited time at our disposal. I must, however, ask your kind indulgence if I fail, as I have not the practice of lecturing. It is true that it is not the first time that I have had the honour to occupy this chair, which I did upwards of forty years ago.†

I may as well commence by describing the tool which I am about to use in the experiments: the diagram will help you to understand it. The chloride of silver battery is made up as follows: A glass tube $1\frac{1}{8}$ inch in diameter, $5\frac{1}{2}$ inches long, and containing about 2 fluid ounces of liquid; into this is fitted a paraffin stopper with two holes perforated through it; through one of these a zinc rod $\frac{3}{16}$ inch diameter and $5\frac{1}{2}$ inches long is inserted, and fastened by melting a little of the paraffin around it; the other element is formed of a flattened silver wire, which passes between the stopper and the glass; so that the metallic elements are zinc and silver. On the flattened silver wire is cast the electrolyte—namely, a rod of chloride of silver $2\frac{1}{8}$ inches long and $\frac{5}{16}$ diameter, and the cell is charged through the second perforation in the stopper with a solution of chloride of ammonium containing $2\frac{1}{2}$ per cent. of salt (Fig. 1). When the circuit is not closed—that is, when the silver element is not connected by means of a conductor to the zinc, no action whatever takes place; and in proof of this I may state that I have a battery which was made

---

* 'Phil. Trans.' Part I. vol. clxix. pp. 55–121, pp. 155–241; vol. clxxi. pp. 65–116.
† May 19, 1837.

up more than six years ago, and is still in action, loss of the fluid by evaporation having been from time to time made up. But as soon as connection is established, then the chloride of silver parts with its chlorine and the zinc dissolves, and metallic silver is separated, in a

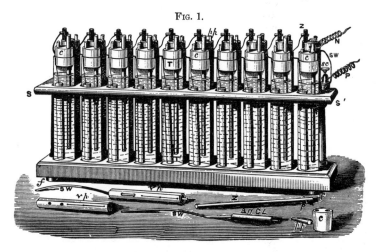

FIG. 1.

spongy state, from the chloride, and remains attached to the silver wire, retaining still the form of a rod. Such an element has the electromotive force of a volt,* nearly (1·03 volt).

A Volt is that electromotive force which, working through a resistance of one Ohm, would deposit 0·0011363 gramme of silver from a salt of silver; or decompose 0·0000947 gramme (0·00146 grain) of water in one second.

A column of mercury at 0° Cent., one square millimetre in section and 1·05 metre high, offers a resistance of 1 ohm; a pure copper wire $\frac{1}{16}$ inch diameter and 129 yards long offers a resistance of an ohm.

These cells are grouped together in trays containing twenty or more, and the trays are placed in cabinets containing in some instances 1200 cells, in others 2160 cells; a cabinet of 1200 cells is shown in Fig. 2. The total number of elements I am about to use is 14,400, and these possess a potential of 14,832 volts, which is con-

---

* The units adopted for electrical measurements are those of the Centimetre Gramme Second (C. G. S.); where the length is 1 centimetre, the mass 1 gramme, and the interval of time 1 second.

The expression for the volt in this system is 10^8 C. G. S.
,, ,, ohm ,, 10^9 ,,
,, ,, microfarad ,, 10^{-15} ,,

For a complete account of these and other units see 'Everett's Units and Physical Constants,' 1879. Macmillan.

siderably greater than that of any battery hitherto united in series. The illustrious Sir Humphry Davy used in 1808, in this theatre, a battery of 2000 plates 4 inches square, with double plates of copper, the battery being charged with a dilute mixture of sulphuric and nitric acids. With this magnificent instrument, placed at his disposal by the subscriptions of a few patrons of science, he obtained a spark $\frac{1}{40}$ to $\frac{1}{30}$ of an inch, when the terminals were made to approach each

FIG. 2.

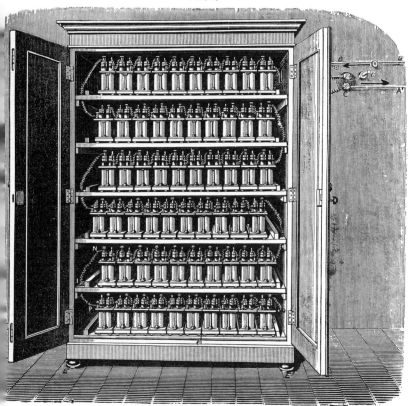

other (a striking distance of $\frac{1}{40}$ of an inch would accord with our experiments with the chloride of silver battery, if the difference of potential of the two batteries is taken into account). When the discharge had once taken place, then the terminals might be separated 4 inches without causing its discontinuance.

My friend the late Mr. Gassiot constructed several batteries of high potential, and at the time of his death there were 3000 Leclanché cells in action at his laboratory; on January 26th, 1875, I measured the length of the spark between points and found it to be 0·025 inch; 3000 of our cells produced a spark of more than twice this length, namely 0·0564 inch, on account of its better insulation.

I propose, in order to show the power we have at command, in the first instance to accumulate, by means of a condenser, the electricity from 3240 cells, and to send its charge through a platinum wire $\frac{1}{80}$ of an inch thick. In charging the condenser I will pass the current through a voltameter, in order that you may judge of the very small chemical force concerned in the production of the enormous mechanical effect of the electric discharge. I may as well at once tell you that the current necessary to charge the condenser I am employing would decompose merely $\frac{1}{5000}$ of a grain of water. I will first of all pass the current from twenty cells through the voltameter; you will see that there is a rapid evolution of mixed gases (oxygen and hydrogen) into which the water is resolved. The evolution of gas, you will at once perceive, is very much slower when the current is charging the condenser; also it is more rapid at first and then gradually lessens, and would entirely cease if there were no leakage of the charge.

When I send the charge of the condenser, which has the enormous capacity of 42·8 microfarads * (or equal to 6485 Leyden jars, like that I have before me, which has coatings of 442 square inches), through $2\frac{1}{2}$ inches of gold wire $\frac{1}{80}$ inch diam. strained on a glass plate, it will be violently deflagrated with a loud report, and the metal will be scattered into dust, which the microscope shows to be composed of minute metallic globules, and not an oxide resulting from combustion. Faraday proved that the quantity of electricity necessary to produce a powerful flash of lightning would result from the decomposition of a single grain of water. This can be realised when it is remembered that it would be 5000 times as great as the charge of the 42·8 m.f. condenser just shown you. If we place the glass plate on which the wire was strained before the microscope, then it will be perceived that the distribution of particles of gold is not uniform along the space which the wire occupied, but on the contrary, they present a stratified appearance, indicating a series of pulsations during the apparently instantaneous discharge. I hope to show you shortly that the most steady discharge through a vacuum tube is in reality intermittent.

As I shall for this purpose cause the current passing through the tube to pass at the same time through an induction coil, so as to induce a secondary current, I will render evident to you, in a striking manner, that when electricity is caused to pass through a wire, it induces another or secondary current in an adjacent wire. I have

---

* A condenser, which holds the charge of a current produced by 1 volt working for 1 second through a resistance of 1 ohm to a potential of 1 volt, has the capacity of 1 farad. The farad is too large a quantity for practical purposes, therefore the millionth part of it, or the microfarad, is employed as the unit of capacity.

here two insulated wires, each 350 yards long, coiled side by side on a reel; to the extremities of one coil is attached a platinum wire six inches long and $\frac{1}{500}$ inch diameter; through the other coil I will send the charge of electricity from a condenser of seven microfarads capacity (about the sixth of that just used) charged with 10,800 cells. You perceive that the platinum wire is violently deflagrated with a loud report by the induced current.

The mechanical effects produced by the charge of a condenser are as the square of the number of cells used to charge it, and although the condenser which I have just used has only one-sixth of the capacity of that I first showed you, yet its mechanical effects are nearly twice as great; for the square of 10,800 is to the square of 3240 as 11 to 1. In order to show the enormous power of its charge I will send it through 29 inches of platinum $\frac{1}{100}$ of an inch in diameter; this is immediately deflagrated. And if I allow the charge to pass between the terminals of a discharger the loud report of the spark renders evident the enormous power stored up by the condenser. I had hoped to show you the condenser charged with 14,400 cells, but it is not capable of withstanding this potential, for one after the other of the coated glass plates, of which it is made up, has broken down with the charge shortly before the lecture.

In order to afford you an opportunity of forming a pictorial conception of that which it is wished to convey, respecting the stratified discharge, I will recall to your recollection an experiment often shown to you by Dr. Tyndall (Fig. 3). With a reservoir of water, placed at a height of a few feet, when the tap at the lower portion is turned on the water flows out, apparently in a continuous stream; but when the thread of water is examined by means of an intermittent beam of light, it is at once seen that the flow is not continuous, but (in consequence of the tendency of water to assume a globular form) the stream as it descends breaks up into a series of drops, one following the other in rapid succession. It is not my purpose here to refer to the cause of the phenomenon, which has been explained to you by Dr. Tyndall in his lectures on Sound, but only to recall this elegant experiment in order to present a mental picture of what may occur in the aggregation of the molecules of gases conveying electricity.

Now I will cause a discharge of electricity to pass through a vacuum tube containing residual carbonic acid at a pressure of 0·5 millim. (Fig. 4), and you will at once perceive that the residual gas groups itself into a series of luminous strata, the molecules which compose them being held together by the balance of electric forces, whereas in the case of the water stream the particles composing the globules are held together by cohesive attraction.

The strata do not flow on like the drops of water, but remain stationary; they are, as it were, so many Leyden jars charged on one side with positive and the other with negative electricity; each imparts say its positive charge to the next negative end of the succeeding stratum, and receives a charge from that behind it; and thus the flow of electricity goes on from one terminal to the other without any movement of the strata necessarily taking place.

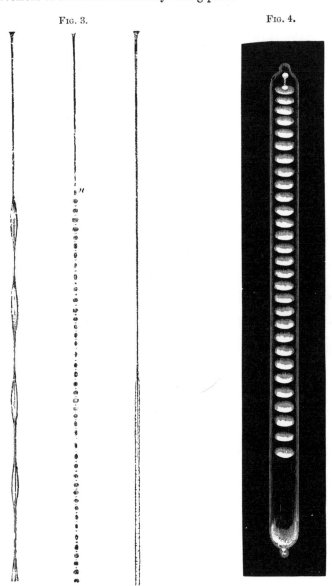

Fig. 3.   Fig. 4.

If we examine the electric arc passing between the terminals of a battery, either at ordinary atmospheric pressure or at other less pressures, it is seen that there is a resemblance to the discharge in vacuum tubes, the light emitted by different parts of it not having the same intensity throughout, and that under most circumstances there is a tendency to break up into distinct entities of the nature of strata and ultimately to take a stratified appearance like the discharge in vacuum tubes; from this we may infer that the discharge in a vacuum tube is in reality a magnified arc.

FIG. 5.

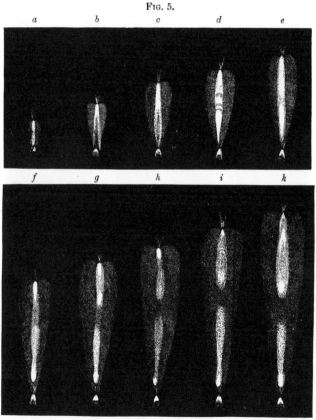

I cannot show these phenomena in a way that you could make them out at the distance you are from me, but I will, with the assistance of Mr. Cottrell, exhibit to you copies of photographs of the arc in atmospheric air (*a* to *n*, Figs. 5 and 5A) taken in my laboratory under various conditions as to distance between the terminals and pressure, as set forth in the following table:—

Fig.	Distance.	Pressure.		Cells.	Current.
		mm.	M.		Weber.
	inch.				
a	0·58	748·6	985,000	10,940	
b	× 2 1·16	294·9	388,026	,,	0·02881
c	× 3 1·74	191·3	251,711	,,	0·04060
d	× 4 2·32	142·6	187,631	,,	0·04474
e	× 5 2·90	112·6	148,157	,,	0·03459
f	× 6 3·48	99·4	130,789	,,	0·03071
g	× 7 4·06	85·9	113,026	,,	0·03259
h	× 8 4·64	71·6	94,210	,,	0·02693
i	× 9 5·22	65·5	86,184	,,	0·02693
k	×10 5·80	64·4	84,737	,,	0·03071
l	× 6	67	88,158	11,000	Too small to measure.
m	× 6	8	10,526	11,000	0·01771
n	× 6·3	2	2,632	2,400	Not measured.

Fig. 5a.

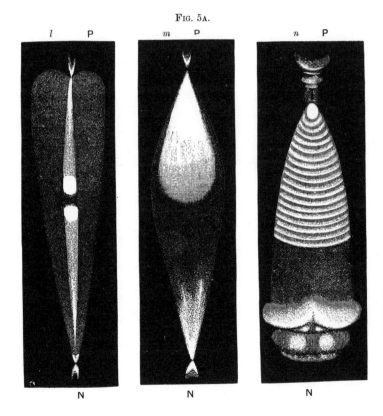

I can let you see the arc, although I am unable to show the details of its structure; thus, when I move the discharging key the arc passes between the two points 0·7 inch apart fixed in the micrometer-discharger (Fig. 6), in which, however, the terminals shown consist of a point and disc, instead of two points, which I am now using. And I may mention that before the discharge takes place there is neither condensation nor dilatation of a gaseous medium in contiguity with the charged terminals, as has been suggested,

Fig. 6.

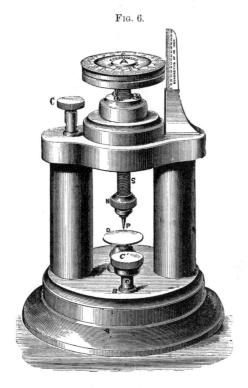

whatever may be their form. The length of the arc varies with the potential of the battery, and with the form of the terminals; between points, the length of the striking distance increases as the square of the number of cells employed. Thus, with 1000 cells the striking distance is 0·0051 inch; with 11,000 cells it is 0·62 inch, as shown in the diagram (Fig. 7). The potential of 11,000 cells put our means of insulation to a severe test, and 14,400 cells overcomes it to such an extent as to interfere so seriously with the striking distance that I only obtain a spark 0·7 inch long.

On the supposition that a cloud would act very much as a mere point at the great distance at which a lightning discharge occurs between clouds or a cloud and the earth, we may from these data

Fig. 7.

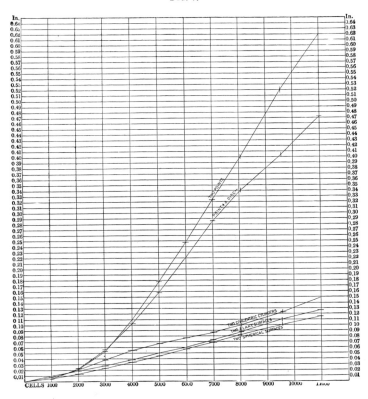

calculate the potential necessary to produce a lightning flash a mile, or 63,360 inches, long. It would require nearly 243 units of 14,400 cells united in series, or say 3,500,000 cells about.

The striking distance may be increased by an arrangement of condensers to form what is called a cascade,* and in this way I shall be able to produce a spark an inch long with only 1200 cells. Such a battery I now use to charge twenty-five plates of a small condenser, and by means of a rotating commutator, connect, so to speak, the outside of one plate to the inside of the next, and thus multiply the potential

---

* The battery itself is a cascade.

twenty-five times; 1200 cells have, as you see, a very short striking distance; it is only 0·00608 inch, so that the spark obtained with the cascade is 164 times as long as with the battery alone. If there were no loss in converting quantity into potential, it would be 625 times or the square of 25. The apparatus I am using is the so-called Rheostat of Gaston-Planté. Franklin, it will be remembered, was the inventor of the cascade. It is not impossible that the effects of lightning may at times be increased by a kind of cascade arrangement formed by the charged layers of cloud floating one over the other.

Between discs the law of the electric discharge is not the same as between points; its length does not increase nearly so rapidly, as

FIG. 8.

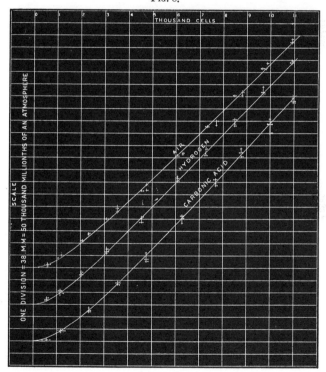

will be seen by a reference to the diagram (Fig. 7), which represents the discharge between two points, a point and disc, between spherical surfaces, and between concentric cylinders. But the increment of potential necessary to produce a discharge for a given distance between discs, say a centimetre, becomes less as the distance between discs, and consequently the potential, are increased. Thus the electrostatic force per centimetre with 1000 volts and a striking distance of 0·0205 centimetre, is 163 electrostatic units, while it is only 113 units with 11,000 volts and a striking distance of 0·3245 centimetre.

We have found, moreover, that the discharge between discs in air, hydrogen, carbonic acid, and probably also in other gases, may be represented by a hyperbolic curve, and this is the case whether we send the discharge through the gas at a constant pressure and increase

FIG. 9.

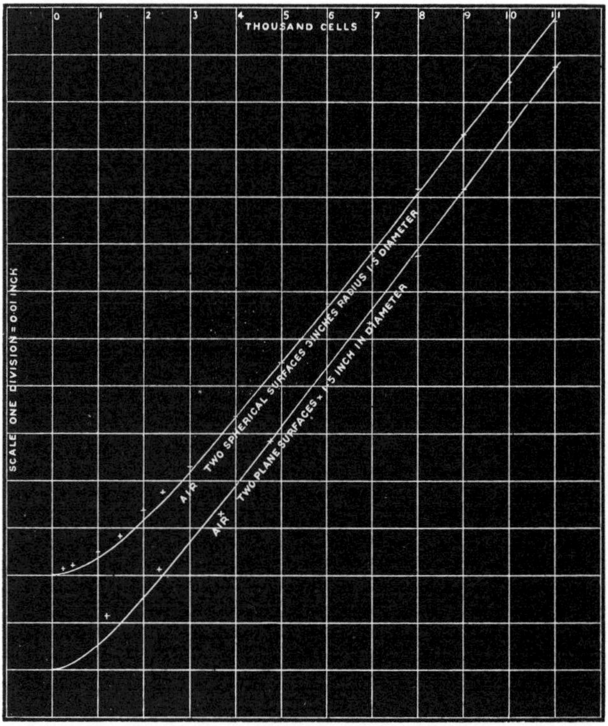

the distances between the terminals and also the number of cells, or send the discharge at a constant distance and vary the pressure and number of cells; the obstacle in the way of a discharge being as the number of molecules between the terminals *up to a certain point*, as will be seen in the diagrams (Figs. 8 and 9), in which the cross marks represent the actual observations. For although the potential necessary to produce a discharge diminishes as the pressure decreases, yet

LIBRARY OF SCIENCE

Fig. 10.

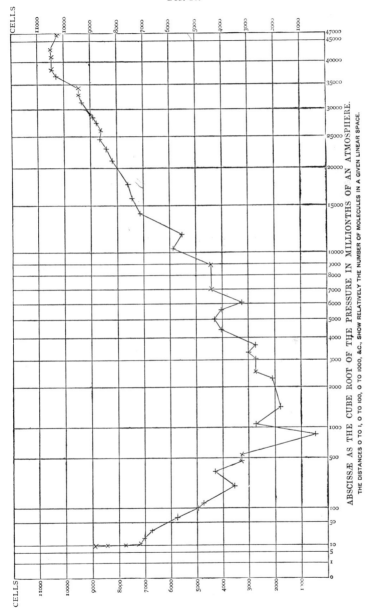

this is true only up to a certain limit; after this has been reached it rapidly increases, and ultimately the resistance becomes so great, as the exhaustion is carried further, that it is easier for a spark to pass between terminals placed at the same distance outside the tube in air at atmospheric pressure. The diagram, Fig. 10, in which the abscissæ are the cube roots of the pressures in millionths of an atmosphere, and the ordinates the number of cells necessary to produce a discharge in a hydrogen tube 30 inches between the terminals, shows the results of experiment. The pressure of minimum resistance varies for different gases. We have determined it to be 0·642 mm. = 845 millionths of an atmosphere for hydrogen, at which pressure the potential necessary to produce a discharge through a tube with terminals 30 inches apart was found to be only 430 cells. At a pressure of 0·0065 mm., 8·6 millionths, it requires as high a potential, 8937 cells, as at a pressure of 21·7 mm., 28,553 millionths, to cause the discharge to take place. At 0·00137 mm., 1·8 millionth, 11,000 cells will not pass. The greatest exhaust we have obtained in a hydrogen vacuum and an absorption by spongy palladium was 0·000055 mm., 0·07 millionth, which offered so great a resistance that a 1-inch spark from an induction coil could not traverse the tube. I will now exemplify what I have said by showing you a tube with an absorption chamber (Fig. 11). I expect that the vacuum will prove to be so good that the whole of the battery, 14,400 cells,

Fig. 11.

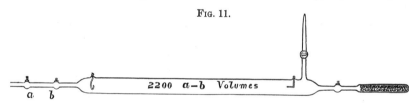

will not cause a current to pass: you see there is no illumination of the tube; if I now heat the absorbing material so as to cause gas to enter the tube, then the discharge of a much smaller number of cells, namely 3600, illuminates the tube, and if I allow it to cool again the discharge ceases.

It has been suggested that there is a polarisation of the terminals of a vacuum tube during the passage of electricity just like that which occurs in a voltameter, and that this increases the obstacle to the discharge; but by an elaborate series of experiments we have proved that such is not the case under the conditions of the experiment. It is quite true that, after the connection between the battery and the terminals of the tube has been broken, there is a deflection of the needle when they are connected with a galvanometer, but we have shown that this is entirely due to a minute static charge proportionate to the capacity of the terminals.*

---

* 'Roy. Soc. Proc.' No. 205, 1880.

Our experiments * enable us to throw some light on another atmospheric electrical phenomenon—namely, the probable height of the aurora borealis, which the accompanying figure (Fig. 12) of a discharge roughly resembles. I will now pass the current of the whole 14,400 cells through the large tube 199, containing a residual charge of atmospheric air at a pressure of 1 millimetre, and you will perceive a carmine luminosity touching the positive pole and reaching half-way down the tube. This reminds one of those ruddy glows frequently seen in auroral displays. Fig. 12 in the plate is copied from a photograph since taken in my laboratory of this appearance. Around the bright luminosity is a dark band which shuts off a portion of the fluorescence of the glass tube, a blue fluorescence produced by the ruddy light of the luminosity, showing that around the luminosity there is an absorbent zone of less elevated temperature. Many estimates have been made from time to time of the height of aurorae, founded upon observations made by persons at a distance from each other, and supposed to be observing the same feature in the display; but it must be remarked that there is always much uncertainty in these estimates, from the difficulty of knowing whether the different observers have noticed the self-same streamer.

FIG. 12.

Frequently very considerable altitudes have been assigned to these displays; for example, as much as 281 miles. We shall presently see that it is very improbable that any electrical discharge could occur at such a height. We have calculated from experiment that the pressure of least resistance for air is 0·397 millimetre, 498·6 millionths, and therefore in air it results that a maximum electric discharge, and consequent brilliancy, of the aurora, would occur at an elevation where the atmosphere has that pressure—namely, 37·67 miles. The greatest exhaust we have produced—and this has not been surpassed—is 0·000055 millimetre, 0·07 millionth, which is the pressure the atmosphere would have at 81·47 miles; and as 11,000 cells failed to produce a discharge even in hydrogen at this low pressure, it may be assumed that at this height the discharge would be considerably less brilliant than at 37·67 miles, should such occur.†

At a height of 281 miles the atmosphere would only

---

* 'Roy. Soc. Proc.' No. 203, 1880.
† It is conceivable that the aurora may occur at times at an altitude of a few thousand feet.

have a pressure of 0·00000000000000000000018 millimetre, or 0·00000000000000000024 millionth, and even at 124·15 miles in height the atmospheric pressure would be only 0·00000001 millimetre, or 0·00001 millionth. It is highly improbable that a display of aurora would occur at a height even of 124 miles, and it is difficult to conceive that an electrical potential could possibly exist necessary to overcome the enormous resistance that would be offered at 281 miles, the tenuity of the air being 54,000,000,000,000 (54 million million) times as great as at 124 miles.

Pressure mm.	Pressure M.	Height, miles.	Visible at miles.	Remarks.
0·00000001	0·000013	124·15	1061	No discharge could occur.
0·000055	0·07	81·47	860	Pale and faint.
0·379	499·0	37·67	585	Maximum brilliancy.
0·800	1053·0	33·96	555	Pale salmon.
1·000	1316·0	32·87	546	Salmon coloured.
1·500	1974·0	30·86	529	,, ,,
3·000	3947·0	27·42	499	Carmine.
20·660	27184·0	17·86	403	,,
62·000	81579·0	12·42	336	,,
118·700	156184·0	9·20	289	Full red.

There are some phenomena connected with the discharge from the voltaic battery which I will bring under notice before we proceed to the study of the discharge in vacuum tubes. I have already spoken of the difference in the length of the discharge between points and discs; and I have now to call your attention to the influence of the form of the point on the length of the spark. At first it would naturally be supposed that the longest discharge would occur with the sharpest point, but this is not the case; a great number of experiments with various forms of points have shown that a point in the form of a paraboloid gives the longest spark; and longer in the proportion 1·29 to 1 than one in the form of a cone of the same length and diameter at the base. It is difficult to account for this difference in the length of the spark, but it is evident the potential must be greater at the extremity of a paraboloidal point than it is at the extremity of a conical one (Fig. 13).

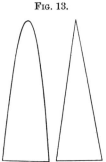

Fig. 13.

If a point and a disc be used together as terminals and the point be made alternately positive and negative, the spark is longest when the point is negative for low tensions up to 3000 cells, and longest when the point is positive beyond that number.

FIG. 14.

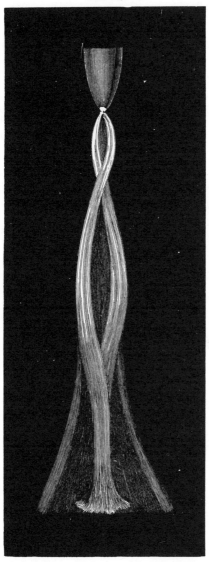

The nature of the metal makes not the slightest difference on the length of the spark, with one exception. Brass, copper, silver, steel, platinum, magnesium, and graphite, all give, under similar circumstances, precisely the same length of spark: aluminium, however, gives a spark longer in the proportion of five to four.

Before the spark jumps and the arc forms it is preceded by what we have called the streamer discharge. This is different in appearance at the positive and negative terminals. You are not able to see the characteristics now that I produce the streamer discharge before you, but they are represented enlarged on the diagram; the terminals being supposed to be a point and a disc, and the point being made alternately positive and negative. When the point is positive the discharge takes the form of a series of twisted streamers (Fig. 14), when negative it is in the form of a brush (Fig. 15). The current which

Fig. 15.

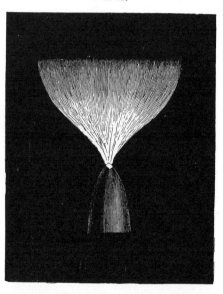

takes place in the form of streamers at a distance but a very little beyond the true striking distance, namely, that at which the arc is formed, is only the $\frac{1}{2500}$ part of the current which passes with the arc, and this is only one-half of that of the battery when short-circuited. When the streamer discharge is examined in a rotating mirror with a microscope suitably constructed, it is seen that the negative current is much the more continuous, for with the same velocity of the rotating mirror the positive discharge breaks up into a series of distinct images, whilst the light of the negative is spread out into a sheet, as you will see on the diagram (Fig. 16). The above effects were produced with

8000 cells, but with 11,000 cells we obtained a further confirmation of them, the effects being shown in another figure on the diagram (Fig. 17), which represents a streamer discharge between two points. The negative discharge is a brush which is seen continuously on the lower terminal, while the positive consists of a series of intermittent, ever-

FIG. 16.

changing spiral streamers which envelop the negative brush discharge without in the least disturbing its form. They go past the negative point and then curl upwards towards it.

If I insert a very high resistance between the battery and the terminals the streamer discharge ceases, and a static spark passes from time to time which is exactly like that from an ordinary frictional machine; it pierces a thin strip of paper just as a static charge would do. The battery gathers up at intervals a charge at the terminals, and the discharge occurs as soon as the potential is sufficient to force its way across the obstacle opposed by the intervening air.

The same thing occurs if I attach a condenser to the terminals of the battery: it takes a longer time for the battery to charge it, and consequently the discharge occurs at longer intervals, shorter or longer according as the terminals are adjusted to a less or greater distance. The condenser I am now using has a capacity of $1 \cdot 5$ microfarad, and hence the accumulated charge is very considerable, and the discharge is like that of a powerful electric battery. But whether the capacity of the accumulator be large or by comparison infinitesimally small like that of the points in the discharging micrometer, there is always an interval of time which elapses between successive discharges; the interval may be so extremely minute that thousands of millions of discharges may occur in a second, but the flow is nevertheless discontinuous, like the drops constituting the stream of water before referred to.

I will endeavour to prove to you that an apparently perfectly steady discharge through a vacuum tube, in which there is no apparent motion of the strata, and in which even the rotating mirror would fail to detect any intermittence, is nevertheless discontinuous. It is true that the period of pulsation must be of a very high order, millions in a second, and it is necessary, therefore, to have recourse to special means in order to detect it, which I will describe.

The current from the battery of 2400 cells is made to pass through the primary of an induction coil, and then through a vacuum tube (No. 142, Fig. 11 in the plate—the figure is copied from a photograph obtained in 2½ seconds) containing a residue of carbonic acid at a pressure of 0·4 millimetre. The arrangement is shown in Fig. 18, where T T′ is the vacuum tube connected in circuit with the primary of coil 819; Z A the battery, on the A side connected with fluid resistances F R, F R′, and wire resistances amounting in all to a megohm. On the Z side is shown a condenser C connected through a fluid resistance F R″ with the Z pole. The terminals of the secondary wire of the induction coil are connected with a sensitive Thomson galvanometer. If there is any intermittence in the current through the tube, an effect will be produced on the galvanometer *under certain circumstances*, that is, provided the rise and fall of the current occur in unequal periods. I will now make connection with the battery; you perceive that there is a deflection of the galvanometer to the left on making contact; I will now allow the spot of light to come to rest, and then break contact, and there is a deflection in the reverse direction, that is, to the right. The first is called the inverse current, the last the direct current. I will now send the current again through the primary of the induction coil and thence through the tube, but in the first instance I will short-circuit the secondary current so that the galvanometer may not be disturbed when the connection with the battery is made; now that the strata are perfectly steady, I will allow the induced current to go through the galvanometer by removing the short-circuit plug, and you see that there is a slight *permanent* deflection

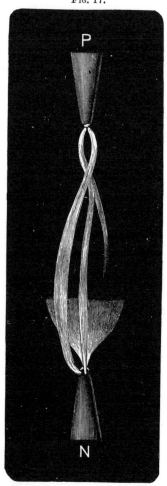

Fig. 17.

of the galvanometer. This shows that the discharge in the vacuum tube, although apparently quite steady, is a pulsating one; as the swing is to the right we know that the current is a direct or break contact one, thus indicating that the discharge through the tube increases compara-

Fig. 18.

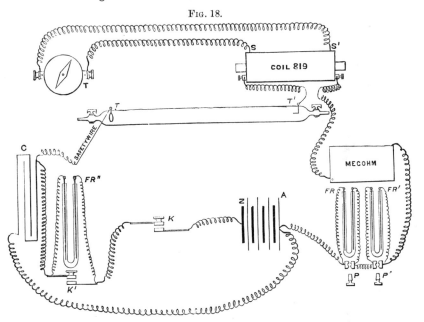

tively slowly, then drops more suddenly. If the rise and fall were in equal times, there would be no deflection of the galvanometer.

If the terminals of a telephone are placed in the circuit between the battery and a vacuum tube, the pulsations are sometimes sufficiently slow to produce audible sounds when the telephone is placed to the ear. But the telephone is not adapted to render evident intermittences of a very high order.

There is a remarkable phenomenon which occurs when a charge is sent through a closed vessel containing air or gas, within certain limits of pressure, which I will endeavour to show you. As soon as the connection is made between the battery and the terminals a sudden expansion of gas takes place, as you will see (Fig. 19) by the depression of the mercury in the gauge connected with the bell jar, and as soon as the connection with the battery is broken the gas returns suddenly almost exactly to its original volume, showing only a small increase due to a slight elevation of temperature; the mercury in the gauge rises, therefore, nearly to its original position. The effect is similar to that which would be produced if an empty bladder suspended between the terminals had been suddenly inflated and as suddenly emptied. The ratio of the increased volume (pres-

FIG. 19.

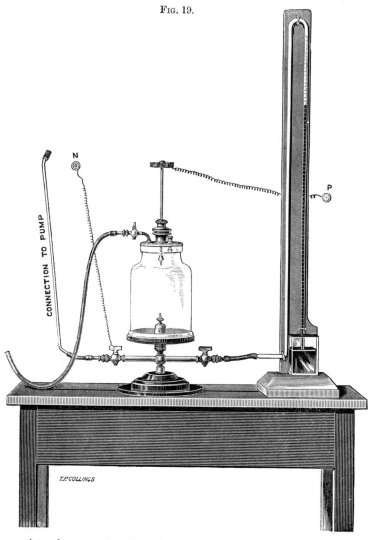

sure) to the normal volume in our experiments rose sometimes as 1·71 to 1, and in others there was scarcely any appreciable increase; in the present instance it is as 1·4 to 1.

The discharge, in the bell jar, was photographed on one occasion and the central spindle or arc proper was measured on the photograph, and a calculation made of what its temperature must have been on the supposition that the sudden dilatation might be due to it, and the result was 16000° Cent. Experiments were also made to ascertain the temperature of different parts of the arc, and it was found that platinum wire $\frac{1}{1000}$ of an inch in diameter was immediately fused, but there was no vaporisation of the platinum, which certainly would have occurred had such a temperature as 16000° Cent. existed. It was ultimately concluded, from a number of experiments and considerations, that the enormous and sudden dilatation could not be attributed to a sudden increase of temperature, but must be caused by the scattering of the gas molecules away from the terminals, and their projection by electrification against the walls of the containing vessel.

We have proved experimentally that the discharge in a vacuum tube does not differ essentially from that in air and other gases at ordinary atmospheric pressures; it cannot be considered as a current in the ordinary acceptation of the term, nor as at all analogous to conduction through metals, and must consequently be of the nature of a disruptive discharge, the particles acting as carriers of electrification. For example, a wire having a given difference of potential between its ends, can permit one, and only one current to pass; whereas, we have found by accurate measurements that with a given difference of potential between the terminals of a vacuum tube, currents of strength varying from 1 to 135 can flow.

We have found, moreover, that the resistance of a vacuum tube, unlike that of a wire, does not increase in the ratio of the distance between the terminals. As an example may be cited that, in a Spottiswoode tube (Fig. 20) with one shifting terminal, which can be placed at any required distance from the other, for seven times the distance between the terminals the resistance was found to be only twice as great. Moreover the fall of potential is not uniform for equal increments of distance between the terminals of a vacuum tube as it is for equal increments of the length of a wire. In order to determine this we used a tube with seventeen rings inserted in it at equal distances (Fig. 20); to these were attached wires which projected through the tube, and were soldered to it. One pole of the battery was connected to No. 1 ring of the tube, and the last ring as well as the second pole of the battery were connected to earth and stood at zero. By means of an electrometer, shown in Fig. 21, the induction plate of which could be made to communicate with each ring successively, it was found that the difference of potential between the first pair of rings, reckoning from the terminal connected with the battery, was five times as great as that between the eighth and ninth; again, that between the sixteenth and seventeenth it was twice that between the eighth and ninth. If I, by way of illustration, suspend a number of pith balls to a wetted thread, one end of which I connect with one pole of the battery and the other to earth, the other pole of the battery being also connected to earth, you will notice a uniform decrease of divergence of the balls, because

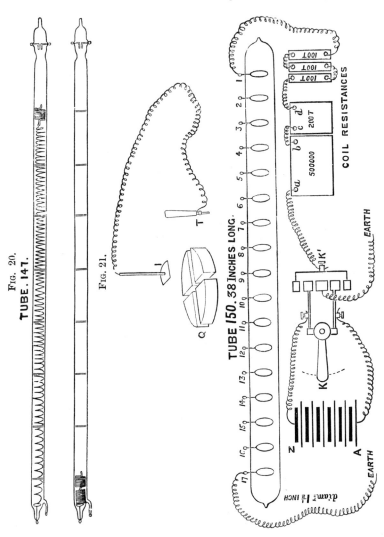

the potential decreases uniformly for equal distances, whereas this does not occur when pith balls are suspended to the rings of a vacuum tube, as you will see when I connect the first ring to one pole of the battery and the other to earth.

We will now take up the phenomena exhibited by vacuum tubes. It will be seen that the strata have their origin at the positive pole. Thus in a given tube, with a certain gas, there is produced at a certain pressure, in the first instance, only one luminosity,* which forms at the positive terminal; then, as the exhaustion is gradually carried further, it detaches itself, moving towards the negative, and being followed by other luminosities, which gradually increase in number up to a certain point. This I will show you, with Mr. Cottrell's aid, by projecting copies of photographs, made in my laboratory, from tubes containing hydrogen at gradually decreasing pressures.

If I now connect the fixed terminal of the Spottiswoode tube, containing residual carbonic acid at a pressure of 1 millimetre, with the positive pole of a battery of 2400 cells, having first caused the movable terminal (which I have connected previously to the negative pole) to approach quite close to the positive wire, you will see only one stratum. I incline the tube and allow the negative terminal to recede. Now there are three strata (Fig. 10 in the plate), and as the negative recedes further and further fresh strata pour in one by one from the positive until the whole tube is filled to within a constant distance from the negative with our electric drops (Fig. 9 in the plate).

I may here pause to draw attention to the resemblance of the strata produced by an electrical discharge in a vacuum tube to the lycopodium records of sound-pulsations in air which are given in Tyndall's work on 'Sound' (Fig. 22).

FIG. 22.

With the same potential the phenomena vary with the amount of current which is allowed to pass through the tube, and this amount we can easily regulate by inserting resistances between the battery and the tube. As the current is increased the number of strata in some tubes increases, but with other tubes the number decreases.

A change in the amount of current frequently produces an entire change in the colour of the strata. For example, in a hydrogen tube from a cobalt blue to a pink (Figs. 4 and 3 in the plate). It also changes the spectra of the strata. Moreover, the spectra of the illuminated terminals and those of the strata differ; usually the most brilliant spectra are obtained from the negative terminal.

---

* It is not improbable that ball-lightning may be of the nature of this single luminosity or stratum, charged like it as a Leyden jar, and projected by an electric discharge taking place behind it; in the same way that a mechanical impulse sends forth a vortex ring.

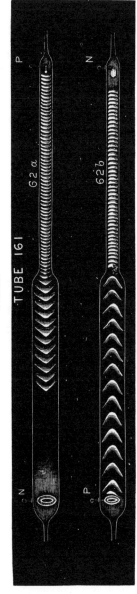

Fig. 23.

If the discharge is irregular and the strata indistinct, an alteration of the amount of current makes the strata distinct and steady; most frequently a point of steadiness is produced by the careful introduction of external resistance; subsequently, with the introduction of more resistance, a new phase of unsteadiness, and still more resistance, another phase of steady and distinct stratification.

At the same pressure, and with the same current, the diameter of the tube affects the character and closeness of the stratification (Fig. 23), as will be seen when I cause the current to pass in tube No. 161, which contains residual hydrogen. It consists of two portions, one 18 inches long and 1·15 internal diameter, the other 17·5 inches long and 0·975 inch diameter. The battery I am using consists of 4800 cells, and you perceive that whether the terminal in the small tube is positive or negative there is a marked difference in the form and closeness of the strata in the two tubes.

The greatest heat is developed in the vicinity of the strata. This fact we established most easily when the tube contained only one stratum, or a small number separated by a broad interval. There is reason to believe that even in the dark discharge like that in the neighbourhood of the negative terminal there may be a kind of stratified formation, for we have found a development of heat in part of a tube in which there was no illumination except on the terminals.

There are vacuum tubes made which are not open from end to end, and which consist of a number of separate chambers, some inserted into the others. The induction coil illuminates these very beautifully, but the battery will not do so, as no discharge can take place through them. On the other hand, the alternating currents of an induction coil charge up such a tube first with positive, then with negative; electricity, and produce an illumination in consequence of the alternating charge and discharge of the walls of it. When I turn on the battery current you will perceive a flash, then if I reverse the current another flash, and if I do this quickly I make the illumination a little more persistent. But I have a rapidly reversed commutator by which I can reverse the current 350 times in a second, and you see that with its help I can illuminate the tube very beautifully.

In almost every case there is a dark interval near the negative terminal, but occasionally we have met with tubes in which the strata completely fill the tube, the last ones threading themselves on the wire used for the negative terminal (Fig. 24). Unfortunately I have not one which I can show you, as these tubes which have shown this phase completely change after the current had passed a very short time.

FIG. 24.

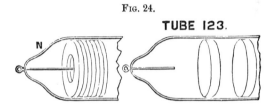

TUBE 123.

I now propose to show, with the aid of my two assistants, Messrs. James Fram and Ernest Davis, some tubes with various gases at different degrees of exhaust, in order that you may see the strata in all their beauty and witness the changes of which I have spoken.

I will in the first place show a tube in which there are produced a series of luminosities like those in one of the photographs which were projected on the screen. It is No. 148, with residual hydrogen at a pressure of 4 millimetres, and connected with 7920 cells. Fig. 8 in the plate shows the phenomenon; it is from a photograph obtained in four seconds.

Tube 201, shown by Fig. 7 in the plate, is a hydrogen vacuum at a pressure of 0·8 millimetre; with 3600 cells and an interposed resistance of 1,500,000 ohms a perfectly steady close stratification is produced. The figure is copied from a photograph obtained in three seconds.

Tube 139, shown in the plate by Fig. 4, is a hydrogen vacuum, pressure 0·8 millimetre, with 3600 cells and an interposed resistance of 200,000 ohms. A series of beautiful blue double strata are produced, with a carmine line between the double strata. The figure is copied from a photograph obtained in one and a half second.

Tube 139.—On interposing 500,000 ohms resistance in the circuit,

118 LIBRARY OF SCIENCE

instead of 200,000, the strata are reduced in number and turn pink. The phenomenon, except as regards colour, is shown in the plate (Fig. 6), copied from a photograph obtained in three seconds.

Tube 130, a hydrogen vacuum 0·8 millimetre, with 2400 cells and an interposed resistance of 60,000 ohms. A series of tongue-shaped strata are produced, which cross each other like the components of the letter X and remain perfectly steady. The phenomenon is precisely like that shown by another tube, the photograph of which you saw projected on the screen. Tube 130 is represented in the plate (Fig. 6); it is copied from a photograph obtained in two and a half seconds.

Tube 333 is a hydrogen vacuum, pressure 0·8 millimetre, with 3600 cells and an interposed resistance of 1,500,000 ohms. There are produced a series of double tongue-shaped strata, united at their narrowest parts. This phenomenon is represented in Fig. 5 in the plate, which is copied from a photograph obtained in three seconds.

Before showing the next tube, I will exhibit one (No. 51) with a carbonic acid vacuum, in which the negative terminal consists of a wire nineteen inches long formed into a helix, the positive being a ring. On passing the current from a battery of 1200 cells through the tube, first interposing a resistance of 500,000 ohms, about two inches only of the negative is illuminated; on gradually, however, removing the resistance, more and more of the spiral negative glows until at last the whole of it is brilliantly illuminated. It will be seen by this that the negative discharge requires a greater outlet than the positive.

FIG. 25.

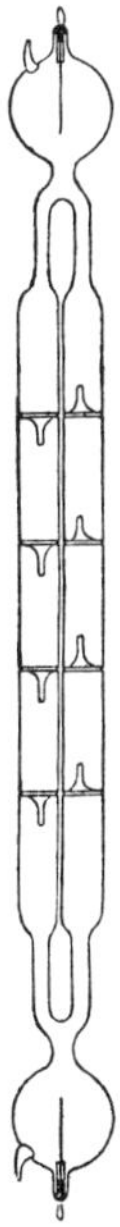

I will now exhibit a tube (No. 163), Fig. 25, consisting of two branches united at the top and bottom. In each of these is a series of funnels, the broad end of which fills the branch; in one branch the mouths of the funnels are placed in a contrary direction to that in the other. On connecting the terminals with the battery of 3600 cells, the current is free to pass either in both branches, or through one or the other, but it invariably passes down that branch in which the wide mouth of the funnel is towards the negative. It traverses alternately the right or left hand branch, according as I make the top or bottom terminal negative; thus again exemplifying the necessity for a greater space for the negative discharge to pass than is required for the positive. The phenomenon is shown in Figs. 11 and 12 in the plate.

The photographs from which the figures in the plate are copied were taken in my laboratory, by Mr. H. Reynolds, on dry plates.

Very frequently, when the exhaust is very great, the discharge becomes most sensitive to the approach of the finger or any conductor in connection with the earth, or charged by a separate source of electricity; the same thing occurs if the current is made intermittent by an interval of air in one of the connections (an air spark). See Fig. 26.

The preparations which were necessary in order that this lecture could be given have occupied a considerable time. In the first place, most of the tubes I have shown you had to be specially prepared, during the last three months, and reserved for this occasion. For all tubes completely alter their character if a current is repeatedly passed through them, and then they no longer present the beautiful phases of stratification you have witnessed. Moreover, it was not possible to

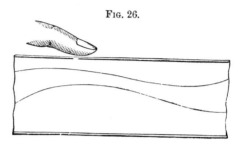

FIG. 26.

remove the battery from my laboratory, as its construction would not permit it. I therefore have had built up by Messrs. Tisley and Co. an entirely new series, in such a way that the battery can be carried away, when requisite, without injury. The construction of the battery was commenced in June 1879, and was finished in August 1880. The charging of the battery occupied a fortnight, and was finished in the second week in December.

It only remains for me to thank you for your flattering attendance under such adverse meteorological conditions.*

---

* Occasioned by a very heavy snowstorm on the 18th of January.

Friday, January 28, 1881.

THOMAS BOYCOTT, Esq. M.D. F.L.S. Vice-President, in the Chair.

ARTHUR SCHUSTER, Esq. Ph D. F.R.S.

The Teachings of Modern Spectroscopy.

A SCIENCE, like a child, grows quickest in the first few years of its existence; and it is therefore not astonishing that, though twenty years only have elapsed since Spectrum Analysis first entered the world, we are able to speak to-day of a modern spectroscopy, with higher and more ambitious aims, striving to obtain results which shall surpass in importance any of those achieved by the old spectroscopy, to the astonishment of the scientific world.

A few years ago the spectroscope was a chemical instrument. It was the sole object of the spectroscopist to find out the nature of a body by the examination of the light which that body sends out when it is hot. The interest which the new discovery created in scientific and unscientific circles was due to the apparent victory over space which it implied. No matter whether a body is placed in our laboratory or a thousand miles away—at the distance of the sun or of the furthest star—as long as it is luminous and sufficiently hot, it gives us a safe and certain indication of the elements it is composed of.

To-day, we are no longer satisfied to know the chemical nature of sun and stars; we want to know their temperature, the pressure on their surface; we want to know whether they are moving away from us or towards us; and still further, we want to find out, if possible, what changes, in their physical and chemical properties, the elements with which we are acquainted have undergone under the influence of the altered conditions which must exist in the celestial bodies. Every sunspot, every solar prominence, is a study in which the unknown quantities include not only the physical conditions of the solar surface, but also the possibly changed properties, under these conditions, of our terrestrial elements. The spectroscope is rapidly becoming our thermometer and pressure gauge; it has become a physical instrument.

The application of the spectroscope to the investigation of the nature of celestial bodies has always had a great fascination to the scientific man as well as to the amateur; for in stars and nebulæ one may hope to read the past and future of our own solar system. But it is not of this application that I wish to speak to-day.

As there is no other instrument which can touch the conditions of the most distant bodies of our universe, bodies so large that their size surpasses our imagination, so is there no other instrument which equals it in the information it can yield on the minute particles at the other end of the scale, particles which in their turn are so small, that we can form no conception of their size or number. The range of the spectroscope includes both stars and atoms, and it is about these latter that I wish to speak.

The idea that all matter is built up of atoms, which we cannot further divide by physical or chemical means, is an old one. As a scientific hypothesis, however—that is, an hypothesis which shall not only qualitatively, but also quantitatively, account for actual phenomena—it has only been worked out in the last thirty years. The development of molecular physics was contemporaneous with that of spectroscopy, but the two sciences grew up independently. Those who strove to advance the one paid little attention to the other, and did not trouble to know which of their conclusions were in harmony, which in discordance, with the results of the sister science. It is time, I think, now that the bearing of one branch of inquiry on the other should be pointed out: where they are in agreement, their conclusions will be strengthened, while new investigations will lead to more perfect truths where disagreement throws doubt on apparently well-established principles.

What I have ventured to call modern spectroscopy, is the union of the old science with the modern ideas of the dynamical theory of gases, and includes the application of the spectroscope to the experimental investigation of molecular phenomena, which without it might for ever remain matters of speculation or of calculation.

A body, then, is made up of a number of atoms. These are hardly ever, perhaps never, found in isolation. Two or more of them are bound together, and do not part company as long as the physical state of the body remains the same. Such an association of atoms is called a molecule. When a body is in the state of a gas or vapour, each molecule for the greater part of the time is unaffected by the other molecules in its neighbourhood, and therefore behaves as if these were not present. The gaseous state, then, is the one in which we can best study these molecules. They move about amongst each other, and within each molecule the atoms are in motion. Each atom, again, has its own internal movement. But if the world was made up of atoms and molecules alone, we should never know of their existence; and to explain the phenomena of the universe, we must recognise the presence of a continuous universal medium penetrating all space and all bodies. This medium, which we call the luminiferous ether or simply the ether, serves to keep up the connection between atoms or molecules. All communications from one atom to another and from one molecule to another are made through this ether. The internal motions of one atom are communicated to this medium, propagated through space, until they reach another atom; attraction, repulsion, or some other manifestation takes place; and if you examine any of the changes which you see constantly going on around you, and follow it backwards through its various stages, you will always find the motion of atoms or molecules at the end of the chain.

The importance of studying the motion of molecules is therefore clear; and it is the special domain of the modern spectroscopy to investigatate one kind of these motions.

When a tuning-fork or a bell is set into vibration, its motion is taken up by the surrounding air, waves are set up, they spread and produce the sensation of sound in our ears. Similarly when an atom vibrates, its motion is taken up by the ether, waves are set up, they spread, and if of sufficient intensity produce the sensation of light in our eyes. Both sound and light are wave motions. A cursory glance at a wave in water will lead you to distinguish its two most prominent attributes. You notice at once that waves differ in height. So the waves both of light and sound may differ in height, and to a difference in height corresponds a difference in the intensity of the sound you hear or of the light you see. The higher the wave the greater its energy, the louder is the sound or the brighter is the light. But in addition to a difference in height you have noticed that in different waves the distance from crest to crest may vary. The distance from crest to crest is the length of the wave, and waves not only differ in height but also in length. A difference in the length of a wave of sound corresponds to a difference in the pitch of the sound; the longer a sound-wave is, the lower is the tune you hear. In the case of light a difference in the length of the wave corresponds to a difference in the colour you see. The longest waves which affect our eyes produce the sensation of red, then follows orange, yellow, green, blue, and the shortest waves which we ordinarily see seem violet. If a molecule vibrates, it generally sends out a great number of waves which vary in length. These fall together on our retina, and produce a compound sensation which does not allow us to distinguish the elementary vibrations, which we want to examine. A spectroscope is an instrument which separates the waves of different lengths before they reach our retina; the elementary vibrations after having passed through a spectroscope no longer overlap, but produce their impressions side by side of each other, and their examination and investigation is therefore rendered possible.

The elements of spectroscopy will be familiar to most of you, but you will forgive me if I briefly allude to some points, which, though well known, are of special importance in the considerations which I wish to bring before you to-night.

When a body is sufficiently hot it becomes luminous, or to speak in scientific language, the vibrations which are capable of producing a luminous sensation on our retina, are increased in intensity as the temperature is raised, until they produce such a sensation. By means of a strong electric current I can in the electric lamp raise a piece of carbon to a high temperature. When looked at with the unaided eye it seems white hot, but when I send the rays through a prism and project them, as I do now, on a screen, you see a continuous band of light. This fact we express by saying that the spectrum of the carbon poles in the electric lamp is a continuous one. You see side by side the different colours known to you by the familiar but incorrect name of "the rainbow colours"; and the experiment teaches you that the carbon pole of the electric lamp sends out rays in which all wave lengths which produce a luminous sensation are represented.

But if now I introduce into the electric arc a small piece of a volatile metal you see no longer a continuous band of light. The band is broken up into different parts. Narrow bands or lines of different colours are separated by a space sometimes black, sometimes slightly luminous. The metal has been converted into vapour by the great heat of the electric current, and the vibrations of its molecules take place in distinct periods, so that the waves emanating from it have certain definite lengths. If the molecule could only send out one particular kind of waves, I should in its spectrum only see one single line. We know of no body which does so, though we know of several in which the possible periods of vibration are comparatively few; the spectrum of these will therefore contain a few lines only. Thus we have two different kinds of spectra, continuous spectra and line spectra. But there is a certain kind intermediate in appearance between these two. The spectra of "fluted bands," as they are called, appear, when seen in spectroscopes of small dispersive powers, as made up of bands, which have a sharp boundary on one side and gradually fade away on the other. When seen with more powerful instruments each band seems to be made up of a number of lines of nearly equal intensity which gradually come nearer and nearer together as the sharp edge is approached. This sharp edge is generally only the place where the lines are ruled so closely that we can no longer distinguish the individual components. The edge is sometimes towards the red, sometimes towards the violet end of the spectrum. Occasionally, however, the fluted bands do not show any sharp edge whatever, but are simply made up of a series of lines which are, roughly speaking, equidistant. No one who has seen a spectrum of fluted bands can ever fail to distinguish it from the other types of spectra which I have described.

What, then, is the cause for the existence of these different types? The first editions of text-books in which our science was discussed stated that a solid or liquid body gave a continuous spectrum, while a gaseous body had a spectrum of lines; the spectra of bands were not mentioned. The more recent editions give a few exceptions to this rule, and the editions which have not appeared yet, will—so I hope, at least—tell you that the state of aggregation of a body does not directly affect the nature of the spectrum. The important point is not whether a body is solid, liquid, and gaseous, but how many atoms are bound together in a molecule, and how they are bound together. This is one of the teachings of modern spectroscopy. A molecule containing a few atoms only gives a spectrum of lines. Increase the number of atoms and you will obtain a spectrum of fluted bands; increase it once more, and you will obtain a continuous spectrum. The scientific evidence for the statements I have made is unimpeachable. In the first place, I may examine spectra of bodies which I know to be compound. Special precautions often are necessary to accomplish this purpose, for too high a temperature would invariably break up the compound molecule into its more elementary constituents. For some bodies I may employ the low temperature of an ordinary Bunsen burner. With others, a weak electric spark taken from their liquid solutions will supply a sufficient quantity of luminous undecomposed matter to allow the light to be analysed by a spectroscope of good power. The spectrum of a compound body is never a line spectrum. It is either a spectrum of bands or a continuous spectrum. The spectra of the oxides, chlorides, bromides, or iodides of the alkaline earths, for instance, are spectra of fluted bands. All these bodies are known to contain atoms of different kinds, the metallic atoms of calcium, barium, or strontium, and the atoms of chlorine, bromine, iodine, or oxygen.

But to obtain these spectra of bands we need not have necessarily recourse to molecules containing different kinds of atoms. Elementary bodies show these spectra, and we must conclude therefore that the dissimilarity of the atoms in the molecule has nothing to do with the appearance of the fluted bands. Similarity in the spectrum must necessarily be due to a similarity in the forces which bind the atoms together, and this at once suggests that it is the compound nature of the molecule which is the true cause of the bands, but that the molecule need not be necessarily a compound of an atom with an atom of different kind, for it may be a compound of an element with itself. We have ample proof that this is the true explanation of the different types of spectra. I shall presently give you a few examples in support of the view which is now nearly unanimously adopted by spectroscopists.

I have hitherto left unmentioned one important method of investigating the periods of molecular vibrations, a method which is applicable to low temperatures. If I have a transparent body and allow light sent out by a body giving a continuous spectrum to fall through it, I often observe that the transparent body sifts out of the light falling through it certain kind of rays. Spectra are thus produced which are called absorption spectra, because the body which is under examination does not send out any light, but absorbs some vibrations which are made to pass through it. It is an important fact that a molecule absorbs just the rays which it is capable itself of sending out. I can therefore investigate the spectrum of a body just as well by means of the absorption it produces as by means of the light which it sends out.

Vapours like bromine or iodine examined in this way give us a spectrum of fluted bands. A powerful spark in these gases gives, however, a line-spectrum. Here, then, a change of spectrum has taken place. The same body at different temperatures gives us a different spectrum, and the change which takes place is the same as that observed in the spectrum of a compound body the moment the temperature has risen sufficiently to decompose that body. I conclude from spectroscopic observations, therefore, that the molecules of bromine and iodine just above their boiling-point are complex molecules, which are broken up at the temperature of the electric spark. At high temperatures the molecules of these bodies contain a smaller number of atoms, and it follows from this that the gases must be lighter or that their density must be smaller. These conclusions, which on spectroscopic grounds have been definite and clear for some years, have recently, by independent methods, been confirmed by Victor Meyer and others. It has been directly proved that at high temperatures the molecules of iodine and bromine contain a smaller number of atoms than they do just above their boiling-point. In other cases the change of density has not been directly proved, only because these necessary measurements are difficult or even impossible at very high temperatures, but we may be perfectly sure that chlorine, as well as the metallic vapours of silver, sodium, potassium, &c., which show an analogous change of their spectra, will ultimately be proved to undergo a change of density at high temperatures.

As we can trace the change from a line-spectrum to a band-spectrum taking place simultaneously with an increase of density, so may we follow the change from a band-spectrum to a continuous spectrum indicating the formation of a molecule still more complex.

Sulphur vapour, at a temperature just above its boiling-point, contains three times the number of atoms in one molecule that it does at a temperature of a thousand degrees. The spectrum of sulphur vapour observed by absorption is continuous when the heavier molecule only is present. At the higher temperatures, when each molecule is decomposed into three, the spectrum belongs to the type of fluted band-spectra. From the cases in which we can thus prove the change in the spectra and in the densities to go on simultaneously, we are justified in concluding that also in other cases, where no such change of density has yet been observed, it yet takes place; and it is not a very daring generalisation to believe that a change in spectra is always due to a change in molecular arrangement, and generally, perhaps always, accompanied by a change in the number of atoms which are bound together into one molecule.

With regard to the well-known statement, that solids and liquids give continuous spectra, while gases give line-spectra, it must be remarked that metallic vapours show in nearly all cases a continuous spectrum before they condense. Oxygen gives a continuous spectrum at the lowest temperature at which it is luminous. Examining liquids and solids by the method of absorption, we find that many of them show discontinuous spectra, presenting fairly narrow bands. It is not denied that the nearness of molecules does not affect the spectrum. It may render the bands more wide and indistinct at their edges, but its influence is more of a nature which in gas spectra is sometimes observed at high pressures when the lines widen, and does not consist of an alteration in type. Though in a solid or liquid body the molecules are much nearer together, they are less mobile; and hence the number of actual collisions need not be necessarily much increased. The fact that a crystal may show a difference in the absorption spectrum according as the vibrations of the transmitted light take place along or across the axis, shows, I think, that mutual impacts cannot much affect the vibrations, but that each molecule, at least in a crystal, must be kept pretty well in its place.

We have divided spectra into three types, but in all attempts at classification we are met by the same difficulty. The boundaries between the different types are not in all cases very well marked. Every one will be able to distinguish a well-defined band-spectrum from a line-spectrum, but there are spectra taking up intermediate positions both between the line- and band-spectra and between band-spectra and continuous spectra. With regard to these it may be difficult to tell to which type the spectrum really belongs. It may happen that a change of spectrum takes place, the spectrum retaining its type; but in these cases, as a rule, the more complex molecule will have a spectrum approaching the lower type, although it may not actually belong to that lower type. To be perfectly general, we may say that a combination of atoms always produces an alteration in the spectrum in the direction of the change from the line-spectrum, through the band-spectrum to the discontinuous spectrum.

If we accept the now generally received opinion as to the cause of the different types of spectra, we may obtain information on molecular arrangement and complexity where our ordinary methods fail. At high temperatures, or under much diminished pressure, measures of density become difficult or impossible; and it is just in these cases that the spectroscope furnishes us with the most valuable information. If we find three spectra of nitrogen and the same number for oxygen, we must accept the verdict, and conclude that these gases can exist in three different allotropic states.

Amongst the remarkable phenomena observed in vacuum tubes, perhaps not the least curious is the spectrum observed at the negative pole, which in several cases is only observed there, and under ordinary circumstances in no other part of the tube. Both oxygen and nitrogen have a spectrum which is generally confined to the negative glow. Some years ago I tried to prove that also in these cases we have only to deal with a special modification of the gases which, curiously enough, only exists near the negative pole, and is broken up and decomposed in every other part of the tube. The experiments I then made seem to me to prove the point conclusively. After a current of electricity had passed through the tube for some time in one direction, the current was suddenly reversed; the negative pole now became positive, but the spectrum still was visible for some time in its neighbourhood, and only gradually disappeared. This experiment shows that the spectrum may exist in other parts of the tube, and that it is therefore due to a peculiar kind of molecule, and not to anything specially related to electric phenomena taking place in the neighbourhood of the negative pole. Other experiments supported this view.

The classification of spectra, according to the complexity of the vibrating molecule, is of great theoretical importance; for by its means we may hope to obtain some information on the nature of the forces which bind together the atoms into one molecule. Our whole life is a chemical process, and a great part of the mysteries of nature would be cleared up if we could gain a deeper insight into the nature of chemical forces. I believe no other line of investigation to be as hopeful in this respect as the one which examines directly the vibrations of the molecules which take place under the influence of these chemical forces. If we could find a connection between the vibrations of a compound molecule and the vibrations of the simpler elements which it contains, we should have made a very decided step in the desired direction. I need not say that various attempts have been made to clear up so important a point; but we have to deal with complicated forces, and the attempts have as a rule not been crowned with much success.

There are, however, a few exceptions, a few cases of greater simplicity than the rest, where we are able to trace to their mechanical causes, the spectroscopic changes which take place on chemical combination. These few and simple cases may serve as the fingerposts which show us the way to further research, and we may hope, to further success. To make the spectroscopic changes of which I am speaking clear to you, I must have recourse to the analogy between sound and light, and remind you of the fact that when the prongs of a tuning-fork are weighted its tone is lowered, which means that the period of vibration is increased, and consequently that the length of the wave of sound sent out is lengthened. Now, suppose a molecule or atom, the spectrum of which I am acquainted with, enters into combination with another. And suppose that the vibrations of the second molecule are weak or lie outside the visible range of the spectrum, then the most simple assumption which I could make would be that the addition of the new molecule is equivalent to an increase of the mass of the other. An increase of mass without alteration of the force of the molecule, will, as in the case of the tuning-fork, lengthen the period of vibration, and increase the wave length. If a case of that kind were actually to happen, I should observe the whole spectrum shifting towards the red; and this is what is observed in the few simple cases to which I have referred. The first observation to that effect is due to Professor Bunsen, of Heidelberg. Examining the absorption spectra of different didymium salts, he found such a close resemblance between them, that no difference could be detected with instruments of small powers; but with larger instruments it was found that the bands varied slightly in position, that in the chloride they were placed most towards the blue end of the spectrum, that when the sulphate was substituted for the chloride, a slight shift towards the less refrangible end took place, and that a greater shift in the same direction occurred on examining the acetate. Professor Bunsen remarks that the molecular weight of the acetate is larger than that of the sulphate, and that the molecule of the sulphate again is heavier than that of the chloride. He adds: " These differences in the absorption spectra of different didymium compounds cannot in our present complete state of ignorance of any general theory for the absorption of light in absorptive media, be connected with other phenomena. They remind one of the slight gradual alterations in pitch which the notes from a vibrating elastic rod undergo when the rod is weighted, or of the change of tone which an organ pipe exhibits when the tube is lengthened." The accompanying woodcut, copied from Professor Bunsen's paper, may serve to illustrate the shift observed in one of the absorption bands.

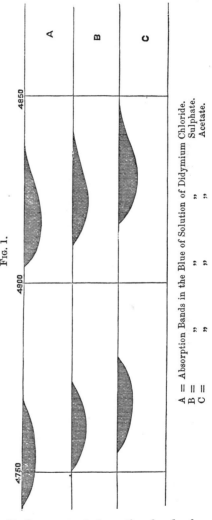

Fig. 1.

A = Absorption Bands in the Blue of Solution of Didymium Chloride.
B = ,, ,, ,, ,, ,, ,, Sulphate.
C = ,, ,, ,, ,, ,, ,, Acetate.

Similar changes take place when some substances like cyanin and chlorophyll are dissolved in different liquids. Absorption bands characteristic of these various substances appear, but they slightly vary in position. Professor Kundt, who has carefully examined this displacement of absorption bands, has come to the conclusion that as a rule the liquids of high dispersive powers were those which shifted the bands most towards the red end of the spectrum. But though there is an apparent tendency in this direction, no rule can be given which shall be absolutely true whatever the substance which is dissolved. Fig. 2 shows the absorption spectrum of cyanin when dissolved in different liquids. The measurements made by Claes* are employed. We have here an interesting

FIG. 2.

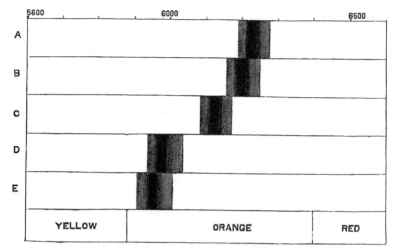

A = Absorption of Cyanin in Bisulphide of Carbon.
B =   ,,   ,,   Nitrobenzene.
C =   ,,   ,,   Benzene.
D =   ,,   ,,   Ether.
E =   ,,   ,,   Alcohol.

proof that a solution is sometimes much more of a chemical compound than is generally supposed. The solvent and the substance must, indeed, be closely connected in order to produce a shifting of the absorption band. On the other hand, it is not astonishing that no general law can be given which connects the displacement with the physical properties of the solvent, for the closeness of connection depending on the special chemical affinity for each solvent has as much to do with the amount of shifting observed, as the molecular weight or the dispersion or refractive power may have. The shifting of the absorption bands in different solutions of the same substances is only one of many applications of spectroscopes to the examination of molecular phenomena in liquids. Into the interesting researches of Professor Russell, who has greatly extended this field of inquiry, we have no time to enter.

---

* Wied. Ann., iii. p. 388, 1878.

The changes of spectra due to molecular combinations and rearrangements have in addition to their theoretical importance a great practical interest, for they will afford us some day a means of answering approximately a great many questions relating to the temperature of sun and stars. The gases and vapours in the solar atmosphere are for the greater part in the molecular condition in which they give a line-spectrum, and we know of stars the spectra of which resemble our solar spectrum very nearly. We shall not be far wrong in ascribing to such stars a temperature similar to that of our sun. Other stars have absorbing envelopes showing spectra of fluted bands. We know that fluted bands belong to a more complex molecular condition, which only can exist at lower temperatures. These stars, therefore, must have a lower temperature than our sun. Dr. Huggins, who has succeeded in obtaining most valuable photographs of star-spectra, has been able to classify and arrange star-spectra; and it is more than likely that in the series of stars arranged in order by means of their spectra, we have at one end those of the highest, at the other those of the lowest temperature. We are as yet far from being able to assign any particular temperature to a star, but the question by means of the spectroscope has been reduced to one which can be decided in our laboratories, and however difficult it may be, we may rest assured that it will ultimately be solved. As to our sun, its temperature has been the subject of many investigations. Attempts have been made to deduce it (at least approximately) from the amount of heat it sends out. Different experimental laws have been proposed to connect together the heat radiation of a body, and the temperature of that body. The first law which was thus proposed gives ten million degrees Centigrade as a lower limit; the second law reduces that lower limit to a little over 1500 degrees. Both these laws we now know to be wrong. More accurate laws give something like ten or twenty thousand degrees, but the whole method employed is one which is open to a great many objections.

We measure the combined heat radiation of different layers on the solar surfaces, all of which are at different temperatures, and we observe only an average effect which is much influenced by the absorption in the outer layers of the solar atmosphere and in the corona, and does not admit of easy interpretation. The spectroscopic method, which is yet in its infancy, has the advantage that we can observe separately each layer of the sun, and we thus examine the temperature not as an average, but for every part of the solar body. Our way to proceed would consist in carefully observing the spectra in different layers of the sun. Supposing we observe a change at one point, we may investigate at what temperature that change takes place, and we may then ascribe the same temperature to that particular place at the solar surface, if no other cause has interfered which may have affected our result. This last conditional limitation leads us to the discussion of the important but difficult question, whether we can determine any such interfering cause, which, not being temperature, yet produces the same change in a spectrum which we have hitherto only ascribed to changes of temperature.

I must here remark that a change in type is not the only spectroscopic change in the spectrum which is observed to take place on varying the temperature. Line spectra especially are subject to curious variations in the relative intensities of their lines. These variations follow no general rule, and must be investigated separately for each element. The cause of this variation is a subject on which there exists a great difference of opinion; but, whatever this cause may be, if the changes always take place at one fixed temperature, we can turn them into account in measuring that temperature. However strong our wish that such a spectroscopic measurement of temperature may ultimately be obtained, a remarkable complication of facts has delayed the realisation of this hope for at least a considerable period of time.

We have to enter partly into a theoretical question, and I must necessarily allude to some of the facts recognised by all who believe in the molecular theory of gases. Each molecule, which, as we have seen, sends out rays of light and heat on account of its internal motion, is surrounded by other molecules. These are, indeed, very closely packed, and continually moving about with enormous velocities. Generally they move in straight lines, but it must necessarily happen that often they come very near, and then affect and deflect each other. Perhaps they come into actual contact, perhaps they repel each other so strongly when near, that contact never takes place. The time elapsing between two such collisions is very small. If you can imagine one second of time to be magnified to the length of a hundred years, it would only take about a second, on the average, from the time a molecule has encountered one other molecule until it encounters the second. During the greatest part of this very short time, it moves in a straight line, for the forces between molecules are so small that they do not affect each other, unless their distance is exceedingly small. It is, therefore, only during a very small fraction of time that one molecule is under the influence of another, and it is one of the greatest problems of molecular physics to find out what happens during that short element of time. I should like to explain to you how I believe the spectroscope may contribute its share to the settlement of that question. In his first great paper on the molecular theory of gases, the late Professor Clerk Maxwell assumed that two molecules may actually come into contact, that they may strike each other, as two billiard balls do, and then separate, according to the laws of elastic bodies. This theory is difficult of application when a molecule contains more than one atom, and especially as it did not in the case of conduction of heat give results ratified by the experimental test, Maxwell abandoned it in favour of the idea that molecules repel each other according to the inverse fifth power of the distance. This second theory not only gave what at the time was believed to be the correct law for the dependence of the coefficient of conduction on temperature, but it also helped its author over a considerable mathematical difficulty. Further experiments have shaken our faith in the first of these two reasons, and the second is not sufficient to induce us to adopt without further inquiry the new law of action between two molecules.

It is exceedingly likely that the forces acting between two molecules when they are in close proximity to each other are partly due to, or at least modified by the vibrations of the molecules themselves. Such vibrations must, as in the case of sound, produce attractive and repulsive forces, and vibrating molecules will affect each other in a similar way as two tuning-forks would. Now, if the forces due to vibrations play any important part in a molecular encounter, the spectroscope will, I fancy, give us some information. If two molecules of the same kind encounter, the periods of vibration are the same, and the forces due to vibration will remain the same during, perhaps, the whole encounter. If two dissimilar molecules encounter, the relative phase of the vibrations, and hence the forces, will constantly change. Attraction will rapidly follow repulsion, and the whole average effect will be much smaller than in the case of two atoms of the same kind. We have no clear notion how such differences may act, and we must have recourse to experiment to decide whether any change in the effect of an encounter is observed when a molecule of a different kind is substituted for a molecule having the same periods of vibration.

When a body loses energy by radiation, that energy is restored during an encounter; the way in which this energy is restored will profoundly affect the vibrations of the molecule, and hence the observed spectrum. I have endeavoured by means of theoretical considerations, or speculations, as you may perhaps feel inclined to call them, to lead you on to an experimental law which I believe to be of very great importance. The spectrum of a molecule is in fact variable at any given temperature, and changes if the molecule is surrounded by others of different nature.

Placing a molecule in an atmosphere of different nature without change of temperature produces the same effect as would be observed on lowering of temperature.

Let me give you one example. Lithium at the temperature of the Bunsen flame has almost exclusively one red line in its spectrum. At the high temperature of the arc or spark the red line becomes weak, and almost entirely disappears. It is replaced by a strong orange line, which is already slightly visible, though weak, at low temperatures, and by additional green and blue lines.

But even at the high temperature of the spark we may obtain again a spectrum containing the red line only if we mix a small quantity of lithium with a large quantity of other material. The same spark, for instance, will give us the low temperature spectrum of lithium when taken from a dilute solution of a lithium salt, and the high temperature spectrum when that solution is concentrated.

The spectra of zinc and tin furnish us other examples in the same direction, but the spectra of nearly all bodies show the same law in a more or less striking way.

If this law which I have given you is a true one,* and I believe it will stand any test to which no doubt it will be subjected, we shall be able to draw some important conclusions from it. In the first place, it will be proved that the forces between atoms do depend on their vibrations. If this is true, any change in the vibrations of the spectrum, however small, will entail a corresponding change in all the other properties of the body. On the other hand, any change in the affinities of the element observed by other means will be represented by a change in the spectrum.

It is also possible that the introduction of forces due to vibratory motion will help us over a considerable difficulty in the molecular theory of gases. Some of the conclusions of that theory are at present absolutely contrary to fact. A spectroscopist, for instance, who is acquainted with the mercury spectrum and all the changes in that spectrum which can take place, feels more than sceptical when he is told that the molecule of mercury contains only one atom, which neither rotates nor vibrates.

Nor can it be of advantage to science to pass silently over this difficulty, or to neglect it as unessential, as is often done by modern writers. The late Professor Maxwell, at least, was well aware of its importance, and has often expressed in private conversation how serious a check he considered the molecular theory of gases to have received. This is not the place to enter more fully into this point and to consider how the vibratory forces may affect some of the suppositions on which the theoretical consequences are founded.

However important the effects of concentration or dilution on the spectra may be, they render the spectroscope less trustworthy as a thermometric instrument; for if the company in which a molecule is placed changes the spectrum in the same way as temperature would, it will be difficult to interpret our results. But although the discussion of our observations may be rendered more arduous and complicated, we need not on that account despair. It is one of the problems of spectroscopy to find out the composition of bodies, not only qualitatively, but also quantitatively, and when we shall know in what proportion different bodies are distributed in the sun, we may

---

* Lockyer, 'Studies in Spectrum Analysis,' p. 140, draws attention to the fact that an admixture of a second element dims the spectrum of the first, and he expresses this fact by saying: "In encounters of dissimilar molecules the vibrations of each are damped." Later he has shown that the lines of oxygen and nitrogen, which are wide at atmospheric pressure, thin out when the gases are only present in small quantities. Lecoq de Boisbaudran in his 'Atlas' gives several examples of the differences in the relative brilliancy of lines produced by concentrating or diluting the solution from which the spark is taken. The complete parallelism of this change to the changes produced by increased temperature has, however, never received sufficient attention.

reduce the problem of finding out this temperature to the much simpler one of finding out the temperature of a given electric spark.

I hope that the few facts which I have been able to bring before you to-night have given you some idea of the important questions which have been brought under the range of spectroscopic research. Many of these questions still await an answer, some have only been brought into the preliminary stage of speculative discussion, but the questions have been raised, and the student of the history of science knows that this is an important step in its development and progress. The spectrum of a molecule is the language which that molecule speaks to us. This language we are endeavouring to understand. The unexperienced in a new tongue which he is trying to learn does not distinguish small differences of intonation or expression. The power over these is only gradually and slowly acquired. So it is in our science. We have passed by, and no doubt still are passing by, unnoticed differences which appear slight and unimportant, but which when properly understood will give us more information than the rough and crude distinctions which have struck us at first. We have extended our methods of research; we have extended our power over the physical agents; we can work with the temperature of sun and stars almost as we can with those in our laboratories. No one can foretell the result, and perhaps in twenty years' time another lecturer will speak to you of a spectroscopy still more modern in which some questions will have received their definite answer, and by which new roads will have been opened to a further extension of science.

[A. S.]

Friday, March 4, 1881.

WILLIAM BOWMAN, Esq. F.R.S. Vice-President, in the Chair.

Sir WILLIAM THOMSON, LL.D. F.R.S. ETC.

Elasticity viewed as possibly a Mode of Motion.

WITH reference to the title of his discourse the speaker said: "The mere title of Dr. Tyndall's beautiful book, 'Heat, a Mode of Motion,' is a lesson of truth which has manifested far and wide through the world one of the greatest discoveries of modern philosophy. I have always admired it; I have long coveted it for Elasticity; and now, by kind permission of its inventor, I have borrowed it for this evening's discourse.

"A century and a half ago Daniel Bernouilli shadowed forth the kinetic theory of the elasticity of gases, which has been accepted as truth by Joule, splendidly developed by Clausius and Maxwell, raised from statistics of the swayings of a crowd to observation and measurement of the free path of an individual atom in Tait and Dewar's explanation of Crookes' grand discovery of the radiometer, and in the vivid realisation of the old Lucretian torrents with which Crookes himself has followed up their explanation of his own earlier experiments; by which, less than two hundred years after its first discovery by Robert Boyle, 'the Spring of Air' is ascertained to be a mere statistical resultant of myriads of molecular collisions.

"But the molecules or atoms must have elasticity, and *this* elasticity must be explained by motion before the uncertain sound given forth in the title of the discourse, 'Elasticity viewed as possibly a Mode of Motion,' can be raised to the glorious certainty of 'Heat, a Mode of Motion.'"

The speaker referred to spinning-tops, the child's rolling hoop, and the bicycle in rapid motion as cases of stiff, elastic-like firmness produced by motion; and showed experiments with gyrostats in which upright positions, utterly unstable without rotation, were maintained with a firmness and strength and elasticity as might be by bands of steel. A flexible endless chain seemed rigid when caused to run rapidly round a pulley, and when caused to jump off the pulley, and let fall to the floor, stood stiffly upright for a time till its motion was lost by impact and friction of its links on the floor. A limp disc of indiarubber caused to rotate rapidly seemed to acquire the stiffness of a gigantic Rubens' hat-brim. A little wooden ball which when thrust down under still water jumped up again in a moment, remained down as if embedded in jelly when the water was caused to rotate rapidly, and sprung back as if the water had elasticity like that of jelly, when it was struck by a stiff wire pushed down through the centre of the cork by which the glass vessel containing the water was filled. Lastly, large smoke rings discharged from a circular or elliptic aperture in a box were shown, by aid of the electric light, in their progress through the air of the theatre when undisturbed. Each ring was circular, and its motion was steady when the aperture from which it proceeded was circular, and when it was not disturbed by another ring. When one ring was sent obliquely after another the collision or approach to collision sent the two away in greatly changed directions, and each vibrating seemingly like an indiarubber band. When the aperture was elliptic each undisturbed ring was seen to be in a state of regular vibration from the beginning, and to continue so throughout its course across the lecture-room. Here, then, in water and air was elasticity as of an elastic solid, developed by mere motion. May not the elasticity of every ultimate atom of matter be thus explained? But this kinetic theory of matter is a dream, and can be nothing else, until it can explain chemical affinity, electricity, magnetism, gravitation, and the inertia of masses (that is, crowds of vortices).

Le Sage's theory might easily give an explanation of gravity and of its relation *to inertia of masses,* on the vortex theory, were it not for the essential aeolotropy of crystals, and the seemingly perfect isotropy of gravity. No finger-post pointing towards a way that can possibly lead to a surmounting of this difficulty, or a turning of its flank, has been discovered, or imagined as discoverable. Belief that no other theory of matter is possible is the only ground for anticipating that there is in store for the world another beautiful book to be called "Elasticity, a Mode of Motion."

[W.T.]

Friday, March 11, 1881.

WILLIAM SPOTTISWOODE, Esq. M.A. D.C.L. Pres. R.S. &c.
in the Chair.

SHELFORD BIDWELL, M.A. LL.B. *M.R.I.*

Selenium and its applications to the Photophone and Telephotography.

BEFORE entering upon my subject, I must claim your indulgence upon two grounds. A week ago I had not the remotest idea that I was to have the honour of addressing you here this evening; the time which I have had for preparation has, therefore, been exceedingly limited. In the second place, it is my desire (in accordance with the traditions of this Institution) not merely to give a description of the experiments in which I have for the last few months been engaged, but, as far as possible, to reproduce them before you. Now these experiments are mostly of a very delicate nature. In the quiet of a laboratory—where time is practically unlimited, and where an operation, if it should fail at first, may be repeated an indefinite number of times—success is tolerably certain to be finally obtained; but in exhibiting delicate experiments before an audience, one is working under the most unfavourable conditions, and, in case of failure in the first instance, the attempt cannot generally be repeated. Moreover, the substance with which we are chiefly concerned, selenium, is apparently extremely capricious in its behaviour. This appearance is, of course, really due to our present ignorance of its properties; but the fact remains that, on account of the great uncertainty of its action, it is a very difficult substance to deal with.

Selenium is a rare chemical element which was discovered in the beginning of the present century. In many of its properties it closely resembles sulphur, and, like sulphur and some other substances, it is capable of existing in more than one form.

The ordinary form is that called vitreous. Selenium in this condition is as absolutely structureless as glass, and in appearance resembles nothing so much as bright black sealing-wax, with, perhaps, somewhat of a metallic lustre; its real colour, however, when seen in thin films, is ruby red. Its melting-point is a little higher than 100° C. In its second modification selenium is crystalline. When in this form its surface is dull, its fracture is metallic (not unlike that of cast iron), its colour is grey or leaden, and it is quite opaque to light; its melting-point also is considerably higher, being 217° C.

Vitreous selenium, if melted and kept for a certain length of time at a temperature between its own fusing-point and that of crystalline selenium, will crystallise; and I think I am right in saying, from casual observation, though I have made no experiments to verify the point, that the length of time necessary for crystallisation depends upon the degree of temperature, being proportionately shorter as the temperature approaches 217° C.

Vitreous selenium is an exceedingly bad conductor of electricity; it is, indeed, an almost perfect insulator. Crystalline selenium is a moderately good conductor, and it possesses this very remarkable property, which has been utilised in the photophone and other inventions, that it conducts better in the light than in the dark, the change in its resistance to the passage of a current of electricity through it varying, according to Professor W. G. Adams, as the square root of the illuminating power.

Let a galvanometer be connected to the two poles of a battery by means of two copper wires. The passage of a current of electricity will at once be denoted by the deflection of the magnetic needle; or, if a little mirror is attached to the needle, and a beam of light be reflected from it upon a scale, the movement of the spot of light will indicate the movement of the needle. Let now one of the wires be cut, and the two ends be joined together by a piece of crystalline selenium. The spot of light will again move, but its deflection will be very much less than it was before, showing that the resistance of the selenium is very much greater than that of the wire. Moreover, if the piece of selenium be alternately exposed to and screened from a beam of light, the deflection will be greater when it is in the light than when it is in the dark, showing a corresponding variation in its resistance. This remarkable property of selenium was first announced and exhibited by Mr. Willoughby Smith in 1873. But the effects produced by the simple arrangement which I have just described are small, and very delicate instruments are required for their observation.

Since that date several devices have been proposed for exaggerating the effect, but they all depend upon the fact that the amount of the variation increases with the extent of the selenium surface acted upon. It has lately been the fashion to call these arrangements "cells," which, in most cases at all events, seems to be a very inappropriate name. It has been suggested that they should be termed "rheostats," a name which well expresses the purposes for which they are generally used, and is less likely to lead to confusion than the other. In deference to custom, however, I shall to-night call them by the usual name.

The simplest selenium cell which could be devised, would be made by placing two short pieces of copper wire parallel to each other, and very near together, and connecting them by a narrow strip of selenium. The effect produced by light is increased by lengthening this arrangement. We might go on increasing the wires with advantage until they were 10 or 12 feet long or more, every additional inch of length producing an increase of "sensitiveness" as it is called. But a cell of this length would be cumbersome and unwieldy to use, and in fact could hardly be lifted without being destroyed. Dr. Werner Siemens therefore adopted the device (among others) of coiling up the wires so as to form a double spiral, and thus made a convenient and portable cell of great sensitiveness. But it is very difficult indeed, as I know by experience, to produce these double wire spirals of any considerable size without the two wires touching one another at some point. After many attempts I succeeded in producing spiral cells about $\frac{3}{4}$ inch in diameter, but I found it impossible to exceed this size, and as it was not large enough I adopted a simple and very effective variation of Siemens's method, which I believe has not been previously suggested. A copper wire is wound after the fashion of a flat screw around a narrow slip of mica, the threads of the screw being about $\frac{1}{16}$ inch apart. Beside this wire, and at a distance of $\frac{1}{32}$ inch from it, a second wire is wound in exactly the same manner, each of its turns coming midway between two consecutive turns of the first. Care is taken that the two wires do not touch each other at any point. Over one surface of the mica a film of melted selenium is spread, and after being worked smooth and uniform it is crystallised. By this means the two wires are connected with each other through half their entire length, by a series of very narrow strips of crystalline selenium.

I have here a tiny selenium cell which has been constructed in this manner. Each wire makes about six turns, and the area of the selenium upon its surface is about half that of a threepenny piece, its thickness not much exceeding that of a sheet of ordinary note-paper. Its resistance, though it is very high relatively to that of good conductors, is, compared with anything of the kind that I had ever seen before, remarkably low, and its sensitiveness to light is great. When a batswing gas-flame is held at a distance of three inches from it, its resistance is less than one-third of that which it measures in the dark. Larger and more carefully constructed cells, of course, show better results. No form, however, that I have tried (and I have made several dozens) has in my hands been superior to that of a double flat screw.

When cells are to be made of any considerable size, the labour of winding on the wires with sufficient regularity is very great. By the help of a lathe this difficulty may be reduced to a minimum. The method of proceeding is this:—A cylinder is turned of hard wood, of length and diameter slightly greater than that of the proposed cell. This cylinder is cut longitudinally into two equal parts, and between the two semi-cylinders thus formed a slip of mica is placed sandwich-like. The ends are secured with screws, and the whole is smoothed down in the lathe. When the edges of the mica are quite flush with the surface of the wood, a screw of from thirty to forty threads to the inch is cut upon the cylinder. On removing the mica from the wood its two edges are found to be beautifully and regularly notched. The first wire is then wound into alternate notches, and the second into the others.

I will throw upon the screen the image of a slip of mica with the two wires wound upon it, and ready for the reception of the selenium coating. It will be seen that the turns are perfectly regular, and close as the wires are to each other, they do not touch at any point. (Fig. 1.)

FIG. 1.

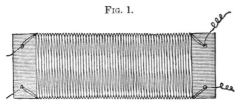

Mica Plate, wound with Two Copper Wires ready for Selenium Coating.

The next step is to apply the selenium, and to do this properly is an operation which requires a certain amount of practice and patience. The mica is heated to a temperature slightly above 217° C., and melted selenium is spread over its surface as evenly as possible with a metal spatula. The cell is then cooled, and its surface should be smooth and lustrous. Before you is an embryo cell which has reached this stage of its preparation. The selenium being still in the vitreous condition, is a perfect insulator, and when the cell is connected in circuit with a battery and a reflecting galvanometer, the spot of light is found to be absolutely motionless. I now propose to crystallise the selenium in your presence. The mere crystallisation occupies a very short time. It is only necessary to place the cell upon a brass plate, and raise it by means of a Bunsen burner to a temperature somewhat below the fusing point of crystalline selenium. The method described by Professor Adams in his classical paper published in the 'Philosophical Transactions,' is entirely different. He heated a bucket of sand by placing in it a red-hot iron ball. At the expiration of an hour he removed the ball, and placed in the heated sand his pieces of vitreous selenium, wrapped up in paper. After remaining for twenty-four hours, the selenium was generally found to have attained the crystalline form, and the resistance of some of his specimens thus prepared was far lower than that of any which have been made by myself. Their sensitiveness, however, does not appear to have been great. The method of crystallisation which I generally adopt, and which is due to Professor Graham Bell, has at all events the merits of simplicity and rapidity. In two or three minutes the whole surface of the selenium film becomes dull and slate-coloured, and if, when the cell has attained this condition, it be removed from the hot metal plate, I have little doubt that on testing it with a galvanometer, it will be found to conduct electricity and to be sensitive to light. (Exp.) According to Professor Graham Bell, nothing more is necessary for obtaining the greatest degree of sensitiveness. The old-fashioned process of long heating and slow cooling may, he says, be altogether dispensed with. In this matter my experience differs entirely from his, for I find that cells which have been kept for some hours at a temperature just below the point of fusion, and then allowed to cool very gradually, are vastly more sensitive to light than those which have not been thus annealed.

The following table shows the resistances in the dark, and under different degrees of illumination, of a few cells taken from my stock. The resistance of No. 6, when exposed to a lime-light at 10 inches, is less than one-fiftieth of its resistance in the dark.

RESISTANCES IN OHMS.

Cell No.	In Dark.	Gas Jet at			
		12 Inches.	6 Inches.	3 Inches.	
4	400,000	190,000	150,000	80,000	Lime-light at 10 Inches, 5,700
6	290,000	80,000	54,000	29,000	
7	430,000	160,000	110,000	64,000	
9	87,000	52,000	42,000	33,000	With Alum Cell, 14,000
10	62,000	32,000	26,000	17,000	
11	100,000	63,000	50,000	33,000	24,000
12	22,700	9,100	6,600	4,500	..

It is an interesting question, which of the coloured components of white light has the greatest power in effecting these changes in the resistance of selenium; or, again, whether the effect is produced by light at all, or is due simply to heat. Captain Sale came to the conclusion, on moving a piece of selenium through the solar spectrum, that the maximum effect was produced at or just outside the extreme end of the red, at a point nearly coinciding with the maximum of the heat rays. Professor Adams performed the same experiment with the spectrum both of the sun and of the electric light, and found that the action on the selenium was greatest "in the greenish-yellow and in the red portions of the spectrum." The greenish-yellow is the point of maximum illumination, which is a remarkable fact; but his words seem to imply that there was a second maximum in the red. The violet and the ultra-violet rays, he says, produced very little, if any, effect. In consequence of the discrepancies in these results, I determined to repeat the experiment for myself. The source of light which I used in the first instance was an oxy-hydrogen lime-light, and the spectrum was formed with a bisulphide of carbon prism. The experiment was repeated six times, and three different selenium cells were used. The results were precisely the same in every case, and proved in the most marked manner that the greatest influence occurred in the boundary-line between the red and the orange; thus differing completely from the results obtained both by Captain Sale and by Professor Adams. Moreover, the resistance when the selenium cells were placed in the ultra-red, two inches beyond the limits of the visible spectrum, was in every case lower than when it was in the blue, indigo, and violet But even in the ultra-violet the resistance of the cells was lower than when they were quite removed from the spectrum.

My friends Mr. Preece and Mr. W. H. Coffin, who were present during these experiments, suggested that it would be desirable to vary them by making use of different sources of light, and different methods of dispersion; and a few days afterwards, by the great kindness of Mr. Norman Lockyer, they were repeated in Mr. Lockyer's laboratory by Mr. Preece and myself with the electric light and a magnificent diffraction grating. Nine experiments were made with three cells, and the results were as absolutely concordant as those which we had previously obtained; but they all concurred in placing the maximum at the extreme edge of the red, thus agreeing with Captain Sale's observations. One other remarkable effect must be noticed. In the case of a single cell—that which I distinguish as No. 6—with which three experiments were made, a second maximum was observed in every case in the greenish yellow, though the effect was about 20 per cent. smaller than at the extreme red. The electric light, however, is from its great unsteadiness most unsuitable for experiments of this nature, and since no such exceptional phenomenon was ever observed before or since, I am inclined to believe that, by a coincidence which however remarkable is by no means impossible, the light happened to be unusually intense just on the three occasions when this particular cell was in the greenish yellow. A third series of experiments made with a gas flame and a bisulphide of carbon prism, agreed with the first in placing the maximum at the orange end of the red. Many more combinations of sources of light and dispersion remain to be tried, but time for these and for innumerable other experiments which have suggested themselves has hitherto been wanting: for an operation which may be described in a dozen words not unfrequently requires as many hours for its performance.

By the help of a reflecting galvanometer I now propose to show you the various effects produced by different parts of the spectrum of the electric light formed by a bisulphide prism upon the resistance of a selenium cell. The maximum deflection is seen to occur when the selenium is at the extreme outer edge of the red.

The effect of interposing various coloured glasses between a gas flame and the selenium cell was also tried. The greatest effect was produced by orange glass, the smallest by green. It was, too, observed as a remarkable fact that the light transmitted by a dark-blue glass produced a greater effect than that which had been passed through a blue glass of much lighter tint. But on a spectroscopic examination the darker one was found to transmit a certain portion of red light. I also tried the effect of radiation from a black-hot poker held at a distance of about 6 inches from the selenium, and upon the first trial found that the resistance, instead of being diminished, was increased by several thousand ohms. I imagined this to be due to a rise of temperature in the selenium, and was thus led to experiment upon the effect of temperature. In this matter, too, there is a remarkable discrepancy between the authorities. Professor Adams says that an increase of temperature *increases* the resistance of selenium, and even suggests that a selenium bar should be used for the construction of a very delicate thermometer. Dr. Guthrie, Messrs. Draper and Moss, and others, make the directly opposite assertion that the resistance of selenium *diminishes* with heat. I repeated my poker experiment, which had in the former case apparently corroborated Professor Adams, and now to my utter astonishment I found that the resistance was greatly diminished. This second experiment, therefore, seemed to support Dr. Guthrie's statement. A great number of experiments were now undertaken for the purpose of arriving at the truth of the matter, with the details of which I will not weary you. Solutions of alum in water, of iodine in bisulphide of carbon, plates of glass and of ebonite were interposed between the selenium and the sources of light and heat. The selenium was now fried, now frozen; and the most contradictory results were obtained. At one moment I felt convinced that Professor Adams was right, at the next there appeared to be no shadow of doubt that Dr. Guthrie's was the true theory. In fact, it seemed as if the selenium was possessed by a demon which produced the variations in accordance with the caprices of its own unaccountable will. At length, when the confusion was at its height and the demon most bewildering, the true explanation was suddenly revealed, and so exceedingly simple is it that now the only marvel is that it should have so long eluded discovery. The secret of the matter is this: and it discloses one of the most remarkable properties of this most remarkable substance. There is a certain degree of temperature at which a piece of crystalline selenium has a maximum resistance. If a piece of selenium at this temperature is exposed to either heat or cold—it matters not which—its resistance will at once be diminished; and extremes of either produce a far greater variation than is ever effected by the action of light. A selenium cell which at the ordinary temperature measured in a dim light 110,000 ohms, was reduced by immersing it in oil at 115° C. to 18,000 ohms. The resistance of the same cell was reduced by immersing it in turpentine at $-6°$ C. to 49,000 ohms. In the case of the single cell with which I have hitherto made the experiment, the temperature on each side of which the resistance is diminished is

24° C.* Let this piece of selenium be gradually raised from a temperature of zero to a temperature of 100° C. While passing from zero to 24°, its resistance will rapidly increase. Passing from 24° to 100° its resistance will again rapidly diminish. (This experiment was successfully shown.)

Until Professor Bell directed his attention to selenium, all observations concerning the effect of light upon its conductivity had been made by means of the galvanometer. But it occurred to him that the marvellously sensitive telephone which he has invented might with advantage be used for the purpose, and on the 17th May, 1878, he announced in this theatre "the possibility," to use his own words, "of hearing a shadow by interrupting the action of light upon selenium." A few days afterwards Mr. Willoughby Smith informed the Society of Telegraph Engineers that he had carried this idea into effect, and had heard the action of a ray of light upon a piece of crystalline selenium.

When a selenium cell, a telephone, and a battery are connected in circuit, a uniform current of electricity will, under ordinary circumstances, flow through the telephone, and a person listening would hear nothing. Suppose now that a series of flashes of light were allowed to fall upon the selenium. In the intervals of darkness the selenium cell would offer a greater resistance to the passage of the electric current than during the intervals of light. The strength of the current would be constantly varying; and if the flashes succeeded one another quickly enough and with sufficient regularity, a musical note would now be heard by a person listening at the telephone. The exact pitch of this note would of course depend upon the rate at which the flashes succeeded one another, being high when the succession is rapid, low when it is slow. The nature of this sound is very peculiar, reminding one of the moaning of a syren or the rising and falling of the wind. With a sufficiently sensitive cell, powerful battery, and delicate telephone, the sound may be heard at a distance of many feet.

I shall interrupt the steady beam of light which is now falling upon the cell by causing a zinc disk with radial slits cut in it to rotate in the path of the beam, and the sound produced by the rapid succession of light and shade upon the selenium cell will be heard in the telephone. When the cell is screened from the light, the sound at once ceases. When the screen is removed, the sound is again heard as before. By using a system analogous to that of dots and dashes, an intermittent beam of light might be employed to convey photophonic messages to a distance.

But Professor Bell has gone further than this. He was not satisfied with merely *interrupting* a steady beam of light, producing alternately strong light and total darkness, but he aimed at graduating its

---

* The experiment has since been repeated with five other cells, and their temperatures of maximum resistance were found to be 23°, 14°, 30°, 25°, and 22° respectively.

intensity in correspondence with the varying phases of the complex sound-waves produced by the human voice. It is evident that if a beam so regulated were allowed to fall upon the selenium cell, the exact words spoken, with their articulation unimpaired, would be reproduced in the telephone. Professor Bell adopted a device which is equally marvellous for its extraordinary simplicity and for its perfect efficiency. The beam of light is made to fall upon the face of a small flexible mirror, whence it is reflected to the distant selenium cell, lenses being used for the purpose of rendering the rays parallel and condensing them where required. The speaker directs his voice upon the back of the mirror, which takes up the sound-waves and is thrown into a state of vibration, thus becoming alternately concave and convex. Now, when it is concave the light reflected by it is more concentrated, and the selenium cell more brightly illuminated. On the other hand, when it is convex, the opposite effect is produced: the rays are more dispersed and the illumination of the cell less intense. And since the movements of the mirror are in exact correspondence with the sound-waves of the voice, so also will be the intensity of the illumination of the selenium cell. The strength of the current passing through it will vary in the same proportion, and will cause the telephone plate to vibrate in consonance with the mirror, and thus to reproduce the exact sounds by which the mirror was set in motion.

In the small experimental photophone which is before you, the receiving station is within 20 feet of the transmitter, and any sounds heard in the telephone would of course be utterly drowned by the actual voice of the speaker at the mirror. It is necessary, therefore, to prolong the telephone wires, and carry them to a distant room, where the sounds that have travelled along the beam of light can be heard without interruption. Professor Bell, using instead of a lens a large reflector for receiving the beam of light, has heard words which were spoken when the mirror was 700 feet away from the selenium cell.

It is impossible to exhibit the photophone in action to an audience, because the effects can only be heard by a single person at a time. But I may mention, that in the course of some experiments with this little instrument which Professor Tyndall very kindly permitted me to make here on the 7th of December last, every word transmitted by it was perfectly understood.

I propose now to say a few words upon another and very different application of selenium. In point of interest and importance it cannot be compared with the photophone: but since it is a child of my own I naturally regard it with a certain amount of affection. It occurred to me a few months ago that the wonderful property of selenium, which we have been discussing this evening, might be applied in the construction of an instrument for transmitting pictures of natural objects to a distance along a telegraph wire. I have constructed a rough experimental apparatus in order to ascertain whether my ideas could be carried out in practice, and it is so far successful, that although the pictures hitherto transmitted are of a very rudimentary character, I think there can be little doubt that further elaboration of the instrument would render it far more effective.

Iodide of potassium is very easily decomposed by a current of electricity. If a piece of paper which has been soaked in a solution of this substance be laid on a piece of metal M, Fig. 2, which is connected to the negative pole of a battery B, and a piece of platinum

FIG. 2.

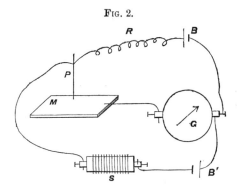

wire P, which is connected with the positive pole, be drawn over its surface, the path of the point will be marked by a brown line, due to the liberation of iodine. Let the platinum wire and the metal plate be connected to a second battery B' in such a manner that a current of electricity may pass through the paper in the opposite direction; and let a variable resistance R be inserted between the platinum wire and the first battery B, and a selenium cell S between the platinum wire and the second battery. And let the resistance be so adjusted that when the selenium cell is exposed to a strong light, the two opposite currents through the paper and the galvanometer G neutralise each other; then the point when drawn over the paper will make no mark. But if the selenium cell is shaded, its resistance will be immediately increased, and the current from the first battery will predominate. The point, if moved over the paper, will now trace a strong line, which, if the selenium is again exposed, will be broken off or enfeebled according to the intensity of the light. (Exp.)

If a series of these brown lines were drawn parallel to one another and very close together, it is evident that by regulating their intensity and introducing gaps in the proper places any design or picture might be represented. This is the principle of Bakewell's copying telegraph, which will transmit writing or pictures drawn upon tinfoil with a non-conducting ink. My instrument differs from his in that the current is varied simply by the action of light. The transmitting instrument Y, Fig. 3, consists of a small cylindrical box 2 inches deep, mounted upon a horizontal spindle, upon which is cut a screw having sixty-four threads to the inch. This works in two bearings 4 inches apart, one of which has an inside screw corresponding to that upon the spindle. At a point midway between the two ends of the cylinder a pin-hole H is drilled, and behind the hole a selenium

FIG. 3.

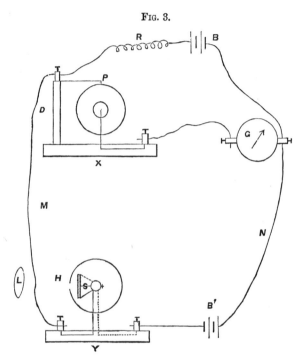

cell S is fixed. One terminal of the selenium cell is connected (through the spindle and stand of the instrument) with the negative pole of a battery B', the other with the line wire M to the distant station. The receiving instrument X contains a similar brass cylinder, similarly mounted. A platinum point P presses gently upon its surface, and is connected both to the line wire and, through a variable resistance R, with the positive pole of a local battery B, the negative pole of which is connected through the galvanometer G with the cylinder. A wire or earth connection N, between the negative pole of the local battery and the positive pole of the other, completes the arrangement.

To prepare the instruments for work the cylinder of the transmitting instrument is brought to its middle position and a picture not more than 2 inches square is focussed upon its surface by means of a photographic lens L. The hole H in the cylinder is then brought to the brightest point of the focussed picture, and a scrap of sensitised paper being placed under the platinum point of the receiver, the variable resistance is adjusted so that the two opposite currents through the paper neutralise each other. When this is accomplished the two cylinders are screwed back as far as they will go, the cylinder of the receiver is covered with sensitised paper, and all is ready to commence operations.

The two cylinders are caused to rotate slowly and synchronously. The little hole in the transmitting cylinder will in the course of its spiral path cover successively every point of the focussed picture, and the amount of light falling at any moment upon the selenium cell will be proportional to the illumination of that particular spot of the picture which, for the time being, is occupied by the pin-hole. During the greater part of each revolution the platinum point will trace a uniform brown line upon the prepared paper, but when the pin-hole happens to be passing over a bright part of the picture, this line is enfeebled or broken. The spiral traced by the point is so close as to produce, at a little distance, the appearance of a uniformly coloured surface, and the breaks in the continuity of the line constitute a picture which, if the instrument were perfect, would be a counterpart of that projected upon the transmitter.

The pictures upon which I have hitherto operated have been mostly simple designs, such as diamonds and squares cut out of thin metal, and projected by a magic lantern (see Fig. 4). But the instrument is in its earliest stage of infancy. It is at present hardly a

Fig. 4.

Image Focussed upon Transmitter.   Image as Reproduced by Receiver.

single month old, and I regret to say that since its birth it has been shamefully neglected, circumstances having prevented me from giving it even the ordinary care and attention which all young creatures ought to receive. Nevertheless, I cannot but think that it is capable of indefinite development; and should there ever be a demand for telephotography, it may in time turn out to be a useful member of society.

[S. B.]

Friday, April 8, 1881.

GEORGE BUSK, Esq. F.R.S. Treasurer and Vice-President,
in the Chair.

PROFESSOR TYNDALL, D.C.L. F.R.S. *M.R.I.*

The Conversion of Radiant Heat into Sound.

THE Royal Society has done me the honour of publishing a long series of memoirs on the interaction of radiant heat and gaseous matter. These memoirs did not escape criticism. Distinguished men, among whom the late Professor Magnus and the late Professor Buff may be more specially mentioned, examined my experiments, and arrived at results different from mine. Living workers of merit have also taken up the question, the latest of whom,[*] while justly recognising the extreme difficulty of the subject, and while verifying, so far as their experiments reach, what I had published regarding dry gases, find that I have fallen into what they consider grave errors in my treatment of vapours.

None of these investigators appear to me to have realised the true strength of my position in its relation to the objects I had in view. Occupied for the most part with details, they have failed to recognise the stringency of my work as a whole, and have not taken into account the independent support rendered by the various parts of the investigation to each other. They thus ignore verifications, both general and special, which are to me of conclusive force. Nevertheless, thinking it due to them and me to submit the questions at issue to a fresh examination, I resumed some time ago the threads of the inquiry. The results shall in due time be communicated to the Royal Society; but meanwhile I would ask permission to bring to the notice of the Fellows a novel mode of testing the relations of radiant heat to gaseous matter, whereby singularly instructive effects have been obtained.

Last year I became acquainted with the ingenious and original experiments of Mr. Graham Bell, wherein musical sounds are obtained through the action of an intermittent beam of light upon solid bodies. From the first I entertained the opinion that these singular sounds were caused by rapid changes of temperature, producing corresponding changes of shape and volume in the bodies impinged

---

[*] Lecher and Pernter, 'Philosophical Magazine,' January, 1881; 'Sitzb. der k. Akad. der Wissensch. in Wien,' July, 1880.

upon by the beam. But if this be the case, and if gases and vapours really absorb radiant heat, they ought to produce sounds more intense than those obtainable from solids. I pictured every stroke of the beam responded to by a sudden expansion of the absorbent gas, and concluded that when the pulses thus excited followed each other with sufficient rapidity, a musical note must be the result. It seemed plain, moreover, that by this new method many of my previous results might be brought to an independent test. Highly diathermanous bodies, I reasoned, would produce faint sounds, while highly adiathermanous bodies would produce loud sounds; the strength of the sound being, in a sense, a measure of the absorption. The first experiment made with a view of testing this idea, was executed in the presence of Mr. Graham Bell;* and the result was in exact accordance with what I had foreseen.

The inquiry has been recently extended so as to embrace most of the gases and vapours employed in my former researches. My first source of rays was a Siemens' lamp connected with a dynamo-machine, worked by a gas-engine. A glass lens was used to concentrate the rays, and afterwards two lenses. By the first the rays were rendered parallel, while the second caused them to converge to a point about seven inches distant from the lens. A circle of sheet zinc provided first with radial slits and afterwards with teeth and interspaces cut through it, was mounted vertically on a whirling table, and caused to rotate rapidly across the beam near the focus. The passage of the slits produced the desired intermittence,† while a flask containing the gas or vapour to be examined received the shocks of the beam immediately behind the rotating disk. From the flask a tube of indiarubber, ending in a tapering one of ivory or boxwood, led to the ear, which was thus rendered keenly sensitive to any sound generated within the flask. Compared with the beautiful apparatus of Mr. Graham Bell, the arrangement here described is rude: it is, however, very effective.

With this arrangement the number of sounding gases and vapours was rapidly increased. But I was soon made aware that the glass lenses withdrew from the beam its most effectual rays. The silvered mirrors employed in my previous researches were therefore invoked; and with them, acting sometimes singly and sometimes as conjugate

---

* On November 29: see 'Journal of the Society of Telegraph Engineers,' December 8, 1880.

† When the disk rotates the individual slits disappear, forming a hazy zone through which objects are visible. Throwing by the clean hand, or better still by white paper, the beam back upon the disk, it appears to stand still, the slits forming so many dark rectangles. The reason is obvious, but the experiment is a very beautiful one.

I may add that when I stand with open eyes in the flashing beam, at a definite velocity of recurrence, subjective colours of extraordinary gorgeousness are produced. With slower or quicker rates of rotation the colours disappear. The flashes also produce a giddiness sometimes intense enough to cause me to grasp the table to keep myself erect.

mirrors, the curious and striking results which I have now the honour to submit to the Members were obtained.

Sulphuric ether, formic ether, and acetic ether being placed in bulbous flasks, their vapours were soon diffused in the air above the liquid. On placing these flasks, whose bottoms only were covered by the liquid, behind the rotating disk, so that the intermittent beam passed through the vapour, loud musical tones were in each case obtained. These are known to be the most highly absorbent vapours which my experiments revealed. Chloroform and bisulphide of carbon, on the other hand, are known to be the least absorbent, the latter standing near the head of diathermanous vapours. The sounds extracted from these two substances were usually weak and sometimes barely audible, being more feeble with the bisulphide than with the chloroform. With regard to the vapours of amylene, iodide of ethyl, iodide of methyl and benzol, other things being equal, their power to produce musical tones appeared to be accurately expressed by their ability to absorb radiant heat.

It is the vapour, and not the liquid, that is effective in producing the sounds. Taking, for example, the bottles in which my volatile substances are habitually kept, I permitted the intermittent beam to impinge upon the liquid in each of them. No sound was in any case produced, while the moment the vapour-laden space above an active liquid was traversed by the beam, musical tones made themselves audible.

A rock-salt cell filled entirely with a volatile liquid and subjected to the intermittent beam produced no sound. This cell was circular and closed at the top. Once, while operating with a highly adiathermanous substance, a distinct musical note was heard. On examining the cell, however, a small bubble was found at its top. The bubble was less than a quarter of an inch in diameter, but still sufficient to produce audible sounds. When the cell was completely filled the sounds disappeared.

It is hardly necessary to state that the pitch of the note obtained in each case is determined by the velocity of rotation. It is the same as that produced by blowing against the rotating disk and allowing its slits to act like the perforations of a syren.

Thus, as regards vapours, prevision has been justified by experiment. I now turn to gases. A small flask, after having been heated in the spirit-lamp so as to detach all moisture from its sides, was carefully filled with dried air. Placed in the intermittent beam it yielded a note so feeble as to be heard only with attention. Dry oxygen and hydrogen behaved like dry air. This agrees with my former experiments, which assigned a hardly sensible absorption to these gases. When the dry air was displaced by carbonic acid, the sound was far louder than that obtained from any of the elementary gases. When the carbonic acid was displaced by nitrous oxide the sound was much more forcible still, and when the nitrous oxide was displaced by olefiant gas it gave birth to a musical note which, when the beam was in good condition and the bulb well chosen, seemed as loud as that of an ordinary organ-pipe. We have here the exact order in which my former experiments proved these gases to stand as absorbers of radiant heat. The amount of the absorption and the intensity of the sound go hand in hand.

In 1859 I proved gaseous ammonia to be extremely impervious to radiant heat. My interest in its deportment when subjected to this novel test was therefore great. Placing a small quantity of liquid ammonia in one of the flasks, and warming the liquid slightly, the intermittent beam was sent through the space above the liquid. A loud musical note was immediately produced.

In this relation the vapour of water interested me most, and as I could not hope that at ordinary temperatures it existed in sufficient amount to produce audible tones, I heated a small quantity of water in a flask almost up to its boiling-point. Placed in the intermittent beam, I heard—I avow with delight—a powerful musical sound produced by the aqueous vapour.

Small wreaths of haze, produced by the partial condensation of the vapour in the upper and cooler air of the flask, were however visible in this experiment; and it was necessary to prove that this haze was not the cause of the sound. The flask was therefore heated by a spirit-flame beyond the temperature of boiling water. The closest scrutiny by a condensed beam then revealed no trace of cloudiness above the liquid. From the perfectly invisible vapour however the musical sound issued, if anything, more forcible than before. I placed the flask in cold water until its temperature was reduced from about $90°$ to $10°$ C., fully expecting that the sound would vanish at this temperature; but notwithstanding the tenuity of the vapour, the sound extracted from it was not only distinct but loud.

Three empty flasks filled with ordinary air were placed in a freezing mixture for a quarter of an hour. On being rapidly transferred to the intermittent beam, sounds much louder than those obtainable from dry air were produced.

Warming these flasks in the flame of a spirit-lamp until all visible humidity had been removed, and afterwards urging dried air through them, on being placed in the intermittent beam the sound in each case was found to have fallen almost to silence.

Sending, by means of a glass tube, a puff of breath from the lungs into a dried flask, the power of emitting sound was immediately restored.

When, instead of breathing into a dry flask, the common air of the laboratory was urged through it, the sounds became immediately intensified. I was by no means prepared for the extraordinary delicacy of this new method of testing the adiathermancy and diathermancy of gases and vapours, and it cannot be otherwise than satisfactory to me to find that particular vapour, whose alleged deportment towards radiant heat has been most strenuously denied, affirming thus audibly its true character.

After what has been stated regarding aqueous vapour we are prepared for the fact that an exceedingly small percentage of any highly adiathermanous gas diffused in air suffices to exalt the sounds. An accidental observation will illustrate this point. A flask was filled with coal gas and held bottom upwards in the intermittent beam. The sounds produced were of a force corresponding to the known absorptive energy of coal-gas. The flask was then placed upright, with its mouth open upon a table, and permitted to remain there for nearly an hour. On being restored to the beam, the sounds produced were far louder than those which could be obtained from common air.*

Transferring a small flask or a test tube from a cold place to the intermittent beam it is sometimes found to be practically silent for a moment, after which the sounds become distinctly audible. This I take to be due to the vaporisation by the calorific beam of the thin film of moisture adherent to the glass.

My previous experiments having satisfied me of the generality of the rule that volatile liquids and their vapours absorb the same rays, I thought it probable that the introduction of a thin layer of its liquid, even in the case of a most energetic vapour, would detach the effective rays, and thus quench the sounds. The experiment was made and the conclusion verified. A layer of water, liquid formic ether, sulphuric ether, or acetic ether one-eighth of an inch in thickness rendered the transmitted beam powerless to produce any musical sound in the vapours. These liquids being transparent to light, the efficient rays which they intercepted must have been those of obscure heat.

A layer of bisulphide of carbon about ten times the thickness of the transparent layers just referred to, and rendered opaque to light by dissolved iodine, was interposed in the path of the intermittent beam. It produced hardly any diminution of the sounds of the more active vapours—a further proof that it is the invisible heat rays, to which the solution of iodine is so eminently transparent, that are here effectual.

Converting one of the small flasks used in the foregoing experiments into a thermometer bulb, and filling it with various gases in succession, it was found that with those gases which yielded a feeble sound, the displacement of a thermometric column associated with the bulb was slow and feeble, while with those gases which yielded loud sounds the displacement was prompt and forcible.

On January 4 I chose for my source of rays a powerful lime-light, which, when sufficient care is taken to prevent the pitting of the cylinder, works with admirable steadiness and without any noise. I also changed my mirror for one of shorter focus, which permitted a

---

* The method here described is, I doubt not, applicable to the detection of extremely small quantities of fire-damp in mines.

nearer approach to the source of rays. Tested with this new reflector the stronger vapours rose remarkably in sounding power.

Improved manipulation was, I considered, sure to extract sounds from rays of much more moderate intensity than those of the lime-light. For this light, therefore, a common candle flame was substituted. Received and thrown back by the mirror, the radiant heat of the candle produced audible tones in all the stronger vapours.

Abandoning the mirror and bringing the candle close to the rotating disk, its direct rays produced audible sounds.

A red-hot coal, taken from the fire and held close to the rotating disk, produced forcible sounds in a flask at the other side.

A red-hot poker, placed in the position previously occupied by the coal, produced strong sounds.

The temperature of the iron was then lowered till its heat just ceased to be visible. The intermittent invisible rays produced audible sounds.

The temperature was gradually lowered, being accompanied by a gradual and continuous diminution of the sound. When it ceased to be audible the temperature of the poker was found to be below that of boiling water.

As might be expected from the foregoing experiments an incandescent platinum spiral, with or without the mirror, produced musical sounds. When the battery power was reduced from ten cells to three the sounds, though enfeebled, were distinct.

My neglect of aqueous vapour had led me for a time astray in 1859, but before publishing my results I had discovered the error. On the present occasion this omnipresent substance had also to be reckoned with. Fourteen flasks of various sizes, with their bottoms covered with a little sulphuric acid, were closed with ordinary corks and permitted to remain in the laboratory from December 23 to January 4. Tested on the latter day with the intermittent beam, half of them emitted feeble sounds, but half were silent. The sounds were undoubtedly due, not to dry air, but to traces of aqueous vapour.

An ordinary bottle containing sulphuric acid for laboratory purposes, being connected with the ear and placed in the intermittent beam, emitted a faint, but distinct, musical sound. This bottle had been opened two or three times during the day, its dryness being thus vitiated by the mixture of a small quantity of common air. A second similar bottle, in which sulphuric acid had stood undisturbed for some days, was placed in the beam: the dry air above the liquid proved absolutely silent.

On the evening of January 7, Professor Dewar handed me four flasks treated in the following manner:—Into one was poured a small quantity of strong sulphuric acid; into another a small quantity of Nordhausen sulphuric acid; in a third were placed some fragments of fused chloride of calcium; while the fourth contained a small quantity of phosphoric anhydride. They were closed with well-fitting indiarubber stoppers, and permitted to remain undisturbed throughout the night. Tested after twelve hours, each of them emitted a feeble sound, the flask last mentioned being the strongest. Tested again six hours later, the sound had disappeared from three of the flasks, that containing the phosphoric anhydride alone remaining musical.

Breathing into a flask partially filled with sulphuric acid instantly restores the sounding power, which continues for a considerable time. The wetting of the interior surface of the flask with the sulphuric acid always enfeebles, and sometimes destroys, the sound.

A bulb less than a cubic inch in volume, and containing a little water lowered to the temperature of melting ice, produces very distinct sounds. Warming the water in the flame of a spirit-lamp, the sound becomes greatly augmented in strength. At the boiling temperature the sound emitted by this small bulb * is of extraordinary intensity.

These results are in accord with those obtained by me nearly nineteen years ago, both in reference to air and to aqueous vapour. They are in utter disaccord with those obtained by other experimenters, who have ascribed a high absorption to air and none to aqueous vapour.

The action of aqueous vapour being thus revealed, the necessity of thoroughly drying the flasks when testing other substances becomes obvious. The following plan has been found effective:— Each flask is first heated in the flame of a spirit-lamp till every visible trace of internal moisture has disappeared, and it is afterwards raised to a temperature of about 400° C. While the glass is still hot a glass tube is introduced into it, and air freed from carbonic acid by caustic potash, and from aqueous vapour by sulphuric acid, is urged through the flask until it is cool. Connected with the ear-tube, and exposed immediately to the intermittent beam, the attention of the ear, if I may use the term, is converged upon the flask. When the experiment is carefully made, dry air proves as incompetent to produce sound as to absorb radiant heat.

I also tried to extract sounds from perfumes, which I had proved in 1861 to be absorbers of radiant heat. I limit myself here to the vapours of pachouli and cassia, the former exercising a measured absorption of 30, and the latter an absorption of 109. Placed in dried flasks, and slightly warmed, sounds were obtained from both these substances, but the sound of cassia was much louder than that of pachouli.

Many years ago I had proved tetrachloride of carbon to be highly diathermanous. Its sounding power is as feeble as its absorbent power.

In relation to colliery explosions, the deportment of marsh-gas

---

* In such bulbs even bisulphide of carbon vapour may be so nursed as to produce sounds of considerable strength.

was of special interest. Professor Dewar was good enough to furnish me with a pure sample of this gas. The sounds produced by it, when exposed to the intermittent beam, were very powerful.

Chloride of methyl, a liquid which boils at the ordinary temperature of the air, was poured into a small flask, and permitted to displace the air within it. Exposed to the intermittent beam, its sound exceeded in power that of marsh-gas.

The specific gravity of marsh-gas being about half that of air, it might be expected that the flask containing it, when left open and erect, would soon get rid of its contents. This, however, is not the case. After a considerable interval the film of this gas clinging to the interior surface of the flask was able to produce sounds of great intensity.

A small quantity of liquid bromine being poured into a well-dried flask, the brown vapour rapidly diffused itself in the air above the liquid. Placed in the intermittent beam, a somewhat forcible sound was produced. This might seem to militate against my former experiments, which assigned a very low absorptive power to bromine vapour. But my former experiments were conducted with obscure heat; whereas in the present instance I had to deal with the radiation from incandescent lime, whose heat is in part luminous. Now the colour of the bromine vapour proves it to be an energetic absorber of the luminous rays; and to them, when suddenly converted into thermometric heat in the body of the vapour, I thought the sounds might be due.

Between the flask containing the bromine and the rotating disk I therefore placed an empty glass cell: the sounds continued. I then filled the cell with transparent bisulphide of carbon: the sounds still continued. For the transparent bisulphide I then substituted the same liquid saturated with dissolved iodine. This solution cut off the light, while allowing the rays of heat free transmission: the sounds were immediately stilled.

Iodine vaporised by heat in a small flask yielded a forcible sound, which was not sensibly affected by the interposition of transparent bisulphide of carbon, but which was completely quelled by the iodine solution. It might indeed have been foreseen that the rays transmitted by the iodine as a liquid would also be transmitted by its vapour, and thus fail to be converted into sound.*

To complete the argument:—While the flask containing the bromine vapour was sounding in the intermittent beam, a strong solution of alum was interposed between it and the rotating disk. There was no sensible abatement of the sounds with either bromine or iodine vapour.

In these experiments the rays from the lime-light were converged to a point a little beyond the rotating disk. In the next experiment they were rendered parallel by the mirror, and afterwards rendered

---

* I intentionally use this phraseology.

convergent by a lens of ice. At the focus of the ice-lens the sounds were extracted from both bromine and iodine vapour. Sounds were also produced after the beam had been sent through the alum solution and the ice-lens conjointly.

Several vapours other than those mentioned in this abstract have been examined, and sounds obtained from all of them. The vapours of all compound liquids will, I doubt not, be found sonorous in the intermittent beam. And, as I question whether there is an absolutely diathermanous substance in nature, I think it probable that even the vapours of elementary bodies, including the elementary gases, when more strictly examined, will be found capable of producing sounds.

[J. T.]

Friday, May 27, 1881.

WILLIAM BOWMAN, Esq. LL.D. F.R.S. Vice-President, in the Chair.

H. E. ROSCOE, Esq. LL.D. F.R.S. &c.
PRESIDENT OF THE CHEMICAL SOCIETY.

Indigo, and its Artificial Production.

MORE than eleven years ago the speaker had the pleasure of bringing before this audience a discovery in synthetic chemistry of great interest and importance, viz. that of the artificial production of alizarin, the colouring substance of madder. To-day it is his privilege to point out the attainment of another equally striking case of synthesis, viz. the artificial formation of indigo. In this last instance, as in the former case, the world is indebted to German science, although to different individuals, for these interesting results, the synthesis of indigo having been achieved by Professor Adolf Baeyer, the worthy successor of the illustrious Liebig in the University of Munich. Here then we have another proof of the fact that the study of the most intricate problems of organic chemistry, and those which appear to many to be furthest removed from any practical application, are in reality capable of yielding results having an absolute value measured by hundreds of thousands of pounds.

In proof of this assertion, it is only necessary to mention that the value of the indigo imported into this country in the year 1879 reached the enormous sum of close on two millions sterling, whilst the total production of the world is assessed at twice that amount; so that if, as is certainly not impossible, artificial indigo can be prepared at a price which will compete with the native product, a wide field is indeed open to its manufacturers.

Indigo, as is well known, is a colouring matter which has attracted attention from very early times. Cloth dyed with indigo has been found in the old Egyptian tombs. The method of preparing and using this colour is accurately described by both Pliny and Dioscorides, and the early inhabitants of these islands were well acquainted with indigo, which they obtained from the European indigo plant, *Isatis tinctoria*, the woad plant, or pastel. With this they dyed their garments and painted their skins. After the discovery of the passage to India by the Cape of Good Hope, the eastern indigo, derived from various species of *Indigofera*, gradually displaced woad, as containing more of the colouring matter. But this was not accomplished without great opposition from the European growers of woad; and severe enactments were promulgated against the introduction of the foreign colouring matter, an edict condemning to death persons "who used that pernicious drug called devil's food" being issued by Henry the Fourth of France. The chief source of Indian indigo is the *Indigofera tinctoria*, an herbaceous plant raised from seed which is sown in either spring or autumn. The plant grows with a single stalk to a height of about three feet six inches, and about the thickness of a finger. It is usually cut for the first time in June or July, and a second or even a third cutting obtained later in the year. The value of the crop depends on the number of leaves which the plant puts forth, as it is in the leaves that the colouring principle is chiefly contained. Both the preparation of the colouring matter from the plant, and its employment as a dye or as a paint, are carried on at the present day exactly as they have been for ages past. The description of the processes given by Dioscorides and Pliny tally exactly with the crude mode of manufacture carried on in Bengal at the present day as follows:—

"The Bengal indigo factories usually contain two rows of vats, the bottom of one row being level with the top of the other. Each series numbers from fifteen to twenty, and each vat is about 7 yards square and 3 feet deep; they are built of brickwork lined with stone or cement. About a hundred bundles of the cut indigo plants are placed in each vat in rows, and pressed down with heavy pieces of wood; this is essential to the success of the operation. Water is then run in so as to completely submerge the plants, when a fermentation quickly ensues, which lasts from nine to fourteen hours, according to the temperature of the atmosphere. From time to time a small quantity of the liquor is taken from the bottom of the vat to see how the operation is proceeding. If the liquor has a pale-yellow hue the product obtained from it will be far richer in quality, but not so abundant as if it had a golden-yellow appearance. The liquor is then run off into the lower vats, into which men enter and agitate it by means of bats or oars, or else mechanically by means of a dash-wheel, each vat requiring seventeen or eighteen workpeople, who are kept employed for three or four hours. During the operation, the yellow liquor assumes a greenish hue, and the indigo separates in flakes. The liquor is then allowed to stand for an hour, and the blue pulpy indigo is run into a separate vessel, after which it is pumped up into a pan and boiled, in order to prevent a second fermentation, which would spoil the product by giving rise to a brown matter. The whole is then left to stand for twenty hours, when it is again boiled for three or four hours, after which it is run on to large filters, which are placed over vats of stonework about 7 yards long, 2 yards wide, and 1 yard deep. The filters are made by placing bamboo canes across the vats, covering these with bass mats, and over all stretching strong canvas. The greater part of the indigo remains under the form of a dark blue or nearly black paste, which is introduced into small wooden frames having holes at the bottom and lined with strong canvas. A piece of canvas is then placed on the top of the frame, a perforated wooden cover, which fits into the box, put over it, and the whole submitted to a gradual pressure. When as much of the water as possible has been squeezed out the covers are removed, and the indigo allowed to dry slowly in large drying sheds, from which light is carefully excluded. When dry, it is ready for the market. Each vat yields from 36 to 50 lbs. of indigo." *

The same process carried out in the times of the Greeks is thus described by Dioscorides: "Indigo used in dyeing is a purple coloured froth formed at the top of the boiler; this is collected and dried by the manufacturer; that possessing a blue tint and being brittle is esteemed the most."

The identity of the blue colouring matter of woad and that of the Bengal plant was proved by Hellot, and by Planer and Trommsdorff at the end of the last century. These latter chemists showed that the blue colour of the woad can be sublimed, and thus obtained in the pure state, a fact which was first mentioned in the case of indigo by O'Brien, in 1789, in his treatise on calico printing. Indigo thus purified is termed indigotin. It has been analysed by various chemists, who ascertained that its composition may be most simply expressed by the formula C_8H_5NO.

Indigo is a blue powder, insoluble in water, alkalis, alcohol, and most common liquids. In order to employ it as a dyeing agent it must be obtained in a form in which it can be fixed or firmly held on to the fibres of the cloth. This is always effected by virtue of a property possessed by indigo-blue of combining with hydrogen to form a colourless body, soluble in alkalis, known as indigo-white, or reduced indigo, of which the simplest formula is C_8H_6NO. This substance rapidly absorbs oxygen from the air and passes into the blue insoluble indigo, which, being held in the fibre of the cloth, imparts to it a permanent blue dye. This reduction to white indigo may be effected in various ways. The old cold vat, or blue-dip vats as they are termed, consist of a mixture of indigo, slacked lime, and green vitriol. The latter salt reduces the indigo, and the white indigo dissolves in the lime water. This process of indigo dyeing is both expensive and troublesome, owing to loss of indigo and formation of gypsum, so that many plans have been proposed to remedy these evils.

Concerning the origin of indigo in the leaves of the *Indigofera*, various and contradictory views have been held. Some have supposed that blue indigo exists ready formed in the plant; others, that white indigo is present, which on exposure to air is converted into indigo-blue. Schunck has, however, proved beyond doubt that the woad plant (*Isatis tinctoria*), the *Indigofera tinctoria* of India, and the Chinese and Japanese indigo plant (*Polygonum tinctorium*) contain neither indigo-blue or white indigo ready formed. It is now known

---

* Crace-Calvert.

that by careful treatment the leaves of all these indigo-yielding plants can be shown to contain a colourless principle termed indican, and that this easily decomposes, yielding a sugar-like body and indigo-blue. That white indigo is not present in the leaves is proved by the fact that this compound requires an alkali to be present in order to bring it into solution, whereas the sap of plants is always acid. The decomposition is represented by Schunck as follows:—

$$\underset{\text{Indican.}}{C_{26}H_{31}NO_{17}} + 2H_2O = \underset{\text{Indigotin.}}{C_8H_5NO} + \underset{\text{Indiglucin.}}{3C_6H_{10}O_6}.$$

So readily does this change from indican to indigo take place, that bruising the leaf or exposing it to great cold is sufficient to produce a blue stain. Even after mere immersion in cold alcohol or ether, when the chlorophyll has been removed, the leaves appear blue, and this has been taken to show the pre-existence of indigo in the plant. But these appearances are deceptive, for Schunck has proved that if boiling alcohol or ether be used, the whole of the colour-producing body as well as the chlorophyll is removed, the leaves retaining only a faint yellow tinge, whilst the alcoholic extract contains no indigo-blue, but on adding an acid to this liquid the indican is decomposed and indigo-blue is formed.

Passing now to the more immediate subject of his discourse, the speaker again reminded his hearers that indigo was the second natural colouring matter which has been artificialy prepared; alizarin the colouring matter of the madder-root being the first. As a rule, the simpler problems of synthetic chemistry are those to which solutions are the soonest found, and these instances form no exception to the rule. The synthetic production of indigo is a more difficult matter than the artificial formation of alizarin, and hence the speaker did not apologise for leading up to the complex through the more simple phenomenon.

When the ingenious Japanese workman who had never seen a watch had one given to him with an order to make a duplicate, he took the only sensible course open to him, and carefully pulled the watch to pieces, to see how the various parts were connected together. Having once ascertained this, his task was a comparatively easy one, for he then had only to make the separate parts, and fit them together, and he thus succeeded so well in imitating the real article that no one could tell the difference. So it is with the chemist, until he knows how the compound is built up, that is, until he has ascertained its constitution, any attempt at synthesis is more like groping in the dark than like shaping the course by well-known landmarks into harbour.

In the case of alizarin it was comparatively easy to reduce it to its simplest terms, and to show that the backbone of this colouring matter is anthracene $C_{14}H_{10}$, a hydro-carbon found in coal-tar. This fact being ascertained, the next step was the further process of clothing the hydro-carbon by adding four atoms of oxygen and subtracting the two atoms of hydrogen present in excess, and this was soon successfully accomplished, so that now, as we know, artificial alizarin has excluded the natural colouring matter altogether.

$$C_{14}H_{10} \qquad C_{14}H_6O_2(OH)_2.$$
Anthracene. Alizarin.

What now was the first step gained in our knowledge concerning the constitution of indigo, of which the simplest formula is C_8H_5NO?

STEP No. 1.—This was made so long ago as 1840, when Fritsche proved that aniline, $C_6H_5NH_2$, can be obtained from indigo. The name for this now well-known substance is indeed derived from the Portuguese "anil," a word used to designate the blue colour from indigo. This result of Fritsche's is of great importance, as showing that indigo is built up from the well-known benzene ring C_6H_6, the skeleton of all the aromatic compounds, and moreover that it contains an amido-group.

STEP No. 2 was also made by Fritsche in the following year, when, by boiling indigo with soda and manganese dioxide, he obtained ortho-amido-benzoic acid, or, as he then termed it, anthranilic acid. The following is the reaction which here occurs:—

$$C_8H_5NO + O + 2H_2O = C_7H_5NH_2O_2 + CH_2O_2.^*$$
Indigo. Ortho-amido-benzoic acid.

What light does this fact shed upon the constitution of indigo? It shows (1) that one of the eight atoms of carbon in indigo can be readily separated from the rest; (2) that the carboxyl and the amido-

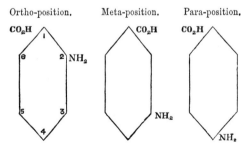

group are in neighbouring positions in the benzene ring, viz. 1 and 2. For we have three isomeric acids of the above composition.

---

* Bottinger, Deut. Chem. Ges. 1877, i. 269.

Step No. 3.—The next advance of importance in this somewhat complicated matter is the discovery by Erdmann and Laurent independently, that indigo on oxidation yields a crystalline body, which, however, possesses no colouring power, to which they gave the name of isatin.

$$C_8H_5NO + O = C_8H_5NO_2.$$
Indigo.   Isatin.

Step No. 4.—The reverse of this action, viz. the reduction of isatin to indigo, was accomplished by Baeyer and Emmerling in 1870 and 1878, by acting with phosphorus pentachloride on isatin, and by the reducing action of ammonium sulphide on the chloride thus formed.

Understanding now something of the structure and of the relationships of the body which we wish to build up, let us see how this edifice has, in fact, been reared. Three processes have been successfully employed for carrying out this object. But of these three only one is of practical importance. A synthetic process may yield the wished-for result, but the labour incurred may be too great and the losses during the campaign may be too severe to render it possible to repeat the operation with advantage on a large scale; just as it costs, at the usual rate of wages, more than twenty shillings to wash a sovereign's worth of gold out of the Rhine sands, so that this employment is only carried on when all other trades fail.

For the sake of completeness, let us, however, consider all three processes, although Nos. 1 and 2 are at present beyond the pale of practical schemes.

These three processes have certain points in common. (1) They all proceed from some compound containing the benzene nucleus. (2) They all start from compounds containing a nitrogen atom. (3) They all commence with an ortho-compound.

They differ from one another; inasmuch as process No. 1 starts from a compound containing seven atoms of carbon (instead of eight), and to this, therefore, one more atom must be added; process No. 2, on the other hand, starts from a body which contains exactly the right number (eight) of carbon atoms; whilst No. 3 commences with a compound in which nine atoms of carbon are contained, and from which, therefore, one atom has to be abstracted before indigo can be reached.

Process No. 1 (Kekulé—Claissen and Shadwell).—So long ago as 1869 Kekulé predicted the constitution of isatin, and gave to it the formula which we now know that it possesses, viz.

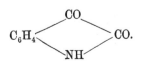

Following up this view, Claissen and Shadwell, two of Kekulé's pupils, succeeded in preparing isatin, and, therefore, now indigo, from ortho-nitro-benzoic acid.

The following are the steps in the ascent:—

1. Ortho-nitro-benzoic acid acted on by phosphorus pentachloride yields the chloride $C_6H_4(NO_2)COCl$.
2. This latter heated with silver cyanide yields the nitril $C_6H_4(NO_2)CO.CN$.
3. On heating this with caustic potash it yields ortho-nitro-phenyl-glyoxylic acid, $C_6H_4(NO_2)CO.CO_2H$.
4. This is converted by nascent hydrogen into the amido-compound $C_6H_4(NH_2)CO.CO_2H$.
5. And this loses water and yields isatin, $C_6H_4NH.CO.CO$.

(Q. E. D.)

The reasons why this process will not work on a large scale are patent to all those who have had even bowing acquaintance with such unpleasant and costly bodies as phosphorus pentachloride or cyanogen.

Process No. 2.—Baeyer's (1878) synthesis from ortho-nitro-phenyl acetic acid.

This acid can be obtained synthetically from toluol, and it is first converted into the amido-acid, and which, like several ortho-compounds, loses water, and is converted into a body called oxindol, from which isatin, and therefore indigo, can be obtained. The precise steps to be followed are:—

1. Ortho-amido-phenylacetic yields oxindol:

2. This on treatment with nitrous acid yields nitrosoxindol:

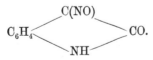

3. This again with nascent hydrogen gives amidoxindol:

$$C_6H_4 \diagdown^{CH(NH_2)}_{NH} \diagdown CO.$$

4. Which on oxidation gives isatin,

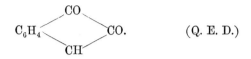 (Q. E. D.)

This process, the feasibility of which had also been foreseen by Kekulé, is however not available as a practical scheme for various reasons.

Process No. 3.—This may be called the manufacturing process, and was also proposed by Baeyer. It starts from cinnamic acid, a substance contained in gum benzoin, balsam of Peru, and some few other aromatic bodies. These sources are, however, far too expensive to render this acid thus obtained available for manufacturing purposes. But Bertagnini, in 1856, had obtained cinnamic acid artificially from oil of bitter almonds, and other processes for the same purpose have since been carried out. Of these, that most likely to be widely adopted is the following practical modification by Dr. Caro of Mr. Perkin's beautiful synthesis of cinnamic acid:—

1. $C_6H_5CH_3 + 4Cl = C_6H_5CHCl_2 + 2HCl.$
   Toluene.　　　　　　Benzylene dichloride.

2. $C_6H_5CHCl_2 + 2CH_3.CO.O.Na = C_6H_5CH\!=\!CH.CO.OH. + 2Nacl.$
   Benzylene　　　Sodium acetate.　　　　Cinnamic acid.
   dichloride.

But why did Baeyer select this nine-carbon acid from which to prepare indigo? For this he had several reasons. In the first place, it had long been known that all indigo compounds when heated with zinc-dust yield indol, C_8H_7N, a body which stands therefore to indigo in the same relation as anthracene to alizarin, and Baeyer and Emmerling had so long ago as 1869 prepared this indol from ortho-nitro-cinnamic acid thus:—

$$C_8H_6(NO_2)CO_2H = C_8H_7N + O_2 + CO_2.$$

Secondly, the ortho-nitro-cinnamic acid required (for we must remember that indigo is an ortho-compound and also contains nitrogen) can be readily prepared from cinnamic acid, and this itself again can be obtained on a large scale. Thirdly, this acid readily parts with one atom of carbon, and thus renders possible its conversion into eight-carbon indigo.

The next steps in the process are (3) the formation of ortho-nitro-cinnamic acid, (4) the conversion of this into its dibromide, (5) the separation from this of the two molecules of hydrobromic acid, giving rise to ortho-nitro-phenyl-propiolic acid, and (6), and lastly, the conversion of this latter into indigo by heating its alkaline solution with grape sugar, xanthate of soda, or other reducing agent. These reactions are thus represented :—

3. $C_6H_5CH{=}CHCOOH \qquad C_6H_4(NO_2)CH{=}CH.COOH.$
  Cinnamic acid   yields   Ortho-nitro-cinnamic acid.

In this process the para-acid is also obtained, and as this is useless for the manufacture of indigo, it has to be removed. This is effected by converting the acids into their ethyl ethers, which, possessing different degrees of solubility, can be readily separated from one another.

4. This is next converted into the dibromide
$$C_6H_4(NO_2)CHBr.CHBrCOOH.$$

5. And by careful treatment with caustic soda this yields ortho-nitro-phenyl-propiolic acid, thus :—
$$C_6H_4(NO_2)CHBrCHBrCOOH + 2NaOH =$$
$$C_6H_4(NO_2)C_2.COOH + 2NaBr + 2H_2O.$$

6. $n[C_6H_4(NO_2)C_2.COOH + H_2 = C_8H_5NO + CO_2 + H_2O].$
 Ortho-nitro-phenyl-propiolic acid.   Indigotin.

(Q. E. D.)

The last of these reactions is in reality not so simple as the equation indicates. For only about 40 per cent. of indigo is obtained, whereas according to theory 68 per cent. should result. Indeed, although, as we have seen, indigo can be prepared by these three methods, chemists are as yet in doubt as to its molecular weight, the probability being that the molecule of indigo contains twice 16 atoms of carbon, or has the formula $4(C_8H_5NO)$ or $C_{32}H_{20}N_4O_4$. Still it must be remembered that according to Sommaruga the vapour density of indigo is 9·45, a number corresponding to the simpler formula $C_{16}H_{10}N_2O_2$.

The artificial production of indigo may even now be said to be within measurable distance of commercial success, for the ortho-nitro-phenyl-propiolic acid, the colourless substance which on treatment with a reducing agent yields indigo-blue, is already in the hands of the Manchester calico printers, and is furnished by the Baden Company for alkali and aniline colours at the price of 6s. per lb. for a paste containing 25 per cent. of the dry acid.

With regard to the nature of the competition between the artificial and the natural colouring matters it is necessary to say a few words. In the first place, the present price at which the manufacturers are able to sell their propiolic acid is 50s. per kilo. But 100 parts of this can only yield, according to theory, 68·58 parts of indigo-blue, so that the price of the artificial (being 73s. per kilo.) is more than twice that of the pure natural colour. Hence competition with the natural dye-stuff is not to be thought of until the makers can reduce the price of dry propiolic acid to 20s. per kilo., and also obtain a theoretical yield from their acid. This may, or it may not, be some day accomplished, but at present it will not pay to produce indigo from nitro-phenyl-propiolic acid. Nevertheless a large field lies open in the immediate future for turning Baeyer's discovery to practical account. It is well known that a great loss of colouring matter occurs in all the processes now in use for either dyeing or printing with indigo. It has already been stated that a large percentage of indigo is lost in the "cold vats" in the sediment. Another portion is washed off and wasted after the numerous dippings, whilst in order to produce a pattern much indigo must be destroyed before it has entered into the fibre of the cloth. Moreover, the back of the piece is uselessly loaded with colour. In the processes of printing with indigo the losses are as great, or even greater, and, in addition, such considerable difficulties are met with that only a few firms (Potter, Grafton in Manchester, and Schlieper in Elberfeld) have been successful in this process. But a still more important fact remains, that no printing process exists in which indigo can be used in combination with other colours in the ordinary way, or without requiring some special mode of fixing after printing. Hence it is clear that the weak points of natural indigo lie in the absence of any good process for utilising the whole of its colouring matter, and in the impossibility, or at any rate, great difficulty of employing it in the ordinary madder styles of calico printing. Such were the reasons which induced the patentees to believe that although the artificial dye cannot be made at a price to compete with natural indigo for use in the ordinary dye-beck, it can even now be very largely used for styles to which the ordinary dye-stuff is inapplicable.

To begin with, Baeyer employed (Patent 1177) grape sugar as a reducing agent. The reduction in this case does not take place in the cold, and even on long standing only small traces of indigo are formed, but if heated to 70° or upwards the change takes place. Unfortunately this production of indigo-blue is rapidly followed by its reduction to indigo-white, and it is somewhat difficult in practice to stop the reaction at the right moment. But "necessity is the mother of invention," and Dr. Caro of Mannheim, to whom the speaker is greatly indebted for much of the above information, found that sodium xanthate is free from many of the objections inherent to the glucose reduction process, inasmuch as the reaction then goes on in the cold. Moreover, he finds that the red isomeride of indigo-blue, Indirubin, which possesses a splendid red colour, also occurring in natural indigo, but whose tinctorial power is less than that of the blue, is produced in less quantity in this case than when glucose is employed. On this cloth, alumina and iron mordants may be printed, and this afterwards dyed in alizarin, &c., or this colouring matter may also be printed on the cloth and the colour fixed by moderate steaming without damage to the indigo-blue. This process is now in actual use by printers both in England and on the Continent, so that, thanks especially to the talent and energy of Dr. Caro, Baeyer's discovery has been practically applied within the short space of twelve months of its conception. Operations on a manufacturing scale have been successfully carried on in the Baden Soda and Aniline Works at Ludwigshafen for the last two months, and the directors see no reason why they should not be able to supply any demand, however great, which may be made for ortho-nitro-phenyl-propiolic acid.

The proper way of looking at this question at present is, therefore, to consider ortho-nitro-phenyl-propiolic acid and indigo as two distinct products not comparable with each other, inasmuch as the one can be put to uses for which the other is unfitted, and there is surely scope enough for both. Still, looking at the improvements which will every day be made in the manufacturing details, he must be a bold man who would assert the impossibility of competition with indigo in all its applications. For we must remember that we are only at the beginning of these researches in the indigo field. Baeyer and other workers will not stay their hands, and possibly other colouring matters of equal intensity and of equal stability to indigo may be obtained from other as yet unknown or unrecognised sources, and it is not improbable that these may turn out to be more formidable competitors in the race with natural indigo than ortho-nitro-phenyl-propiolic acid.

Looking at this question of the possible competition of artificial with the natural indigo from another point of view, it must, on the other hand, be borne in mind that the present mode of manufacturing indigo from the plant is extremely rude and imperfect, and that by an improved and more careful carrying out of the process, great saving in colouring matter may be effected, so that it may prove possible to produce a purer article at a lower price, and thus to counterbalance the production of the artificial material.

The following are the directions issued by the patentees to calico printers for using the new colour:—

PRINTING WITH ARTIFICIAL INDIGO.

No. I.—On Unprepared Cloth.

Standard.

Take 4 lb. propiolic acid paste (equal to 1 lb. dry acid), and 1 lb. borax finely powdered; mix well. The mixture first becomes fluid and at last turns stiff. Then add 3 quarts white starch thickening (wheat starch), mix well, and strain.

Printing Colour.

Take the above standard and dissolve in it immediately before printing 1½ lb. xanthate of soda, stir well, and ready for use.

For lighter shades reduce the above printing colour with the following: In 1 gallon white starch paste dissolve 1 lb. xanthate of soda.

Directions for use.—Print and dry as usual. The pieces ought not to be placed in immediate contact with drying cylinders, or otherwise be subjected to heat above 100° C. The indigo-blue is best developed by allowing the printed goods to remain in a dry atmosphere and at an ordinary temperature for about 48 hours. Damp air ought to be excluded as much as possible until the colour is fully developed. Then the pieces may be passed through the ageing machine, or steamed at low pressure if such treatment should be required for fixing any other colour or mordant printed along with the indigo-blue.

After the blue is ready formed, the pieces are first thoroughly washed in the washing machine and *then boiled* in the clean water, or better, in a weak solution of hyposulphite of soda (1 lb. to 10 gallons), *and at a full boil* for half an hour in order to volatilise the smell which would otherwise adhere to the goods.

Clean in a soap-bath, at a temperature not above 40° C.; wash, dry, and finish.

Observations.—Wheat starch gives the best results in the colour, then follows gum tragacanth. The colour is considerably reduced by using gum senegal, dark British gum, or calcined farina as thickening materials.

So far borax has answered best as an alkaline solvent of propiolic acid, it may however be replaced in the above standard by acetate of soda (from 1 to 1½ lb.) or by 6 oz. pearlash or soda. Any excess of caustic-potash, or soda, destroys propiolic acid.

The above standard keeps unchanged for any length of time, it is likewise not sensibly altered by a small amount of xanthate of soda, but when mixed with its full proportion of xanthate, as in the above printing colour, it gradually loses strength after several hours.

The xanthate ought therefore to be mixed with the standard immediately before printing, and any colour remaining unused may then be saved by mixing with the same a large proportion of starch paste.

Propiolic acid may be printed along with aniline black, catechu brown and drabs, and with alumina and iron mordants for madder colours.

After the indigo-blue is fully developed, the mordants are fixed in the ordinary manner, dyed with alizarin, padded with Turkey-red oil, steamed, and otherwise treated as usual.

Indigo-blue, whether natural or artificial, suffers by prolonged steaming at high pressure. For this reason, only such steam colours can be associated with propiolic acid as may be fixed by short steaming at low pressure.

No. II.—ON PREPARED CLOTH (FOR FULL SHADES).

Dissolve 2 lb. of xanthate of soda in 1 gallon of cold water. Pad the goods with the above; dry, print with standard, and after printing follow the above treatment. The pieces may also be first printed with xanthate and then covered with standard. Alumina and iron mordants for madder colours may be likewise printed on cloth thus prepared, or printed with xanthate of soda.

The potential importance, from a purely commercial point of view, of the manufacture, may be judged of by reference to the following statistics, showing that the annual value of the world's growth of indigo is no less than four millions sterling.

ESTIMATED YEARLY AVERAGE OF THE PRODUCTION OF INDIGO IN THE WORLD, TAKEN FROM THE TOTAL CROP FOR A PERIOD OF TEN YEARS.

	Pounds Weight.	Pounds Sterling.
Bengal, Tirhoot, Benares, and N.-W. India	8,000,000	2,000,000
Madras and Kurpah	2,200,000	400,000
Manilla, Java, Bombay, &c.	..	500,000
Central America	2,250,000	600,000
China and elsewhere, consumed in the country	..	Say 500,000
		4,000,000

How far the artificial will drive out the natural colouring matter from the market cannot, as has been said, be foreseen. It is interesting, as the only instance of the kind on record, to cast a glance at the history of the production of the first of the artificial vegetable colouring matters, alizarin. In this case the increase in the quantity produced since its discovery in 1869 has been enormous, such indeed that the artificial colour has now entirely superseded the natural one, to the almost complete annihilation of the growth of madder-root. It appears that whilst for the ten years immediately preceding 1869 the average value of the annual imports of madder-root was over one million sterling, the imports of the same material during last year (1880) amounted only to 24,000l. The whole difference being made up by the introduction of artificial alizarin. In 1868, no less a quantity than 60,000 tons of madder-root were sent into the market, this containing 600,000 kilos. of pure natural alizarin. But in ten years later a quantity of artificial alizarin more than equal to the above amount was sent out from the various chemical factories. So that in ten years the artificial production had overtaken the natural growth, and the 300,000 or 400,000 acres of land which had hitherto been used for the growth of madder, can henceforward be better employed in growing corn or other articles of food. According to returns, for which the speaker had to thank Mr. Perkin, the estimated growth of madder in the world previous to 1869 was 90,000 tons, of the average value of 45l. per ton, representing a total of 4,050,000l.

Last year (1880) the estimated production of the artificial colouring matter was 14,000 tons, but this contains only 10 per cent. of pure alizarin. Reckoning 1 ton of the artificial colouring matter as equal to 9 tons of madder, the whole artificial product is equivalent to 126,000 tons of madder. The present value of these 14,000 tons of alizarin paste, at 122l. per ton, is 1,568,000l. That of 126,000 tons of madder at 45l. is 5,670,000l., or a saving is effected by the use of alizarin of considerably over four millions sterling. In other words, we get our alizarin dyeing done now for less than one-third of the price which we had to pay to have it done with madder.

Our knowledge concerning the chemistry of alizarin has also proportionately increased since the above date. For whilst at that time only one distinct body having the above composition was known, we are now acquainted with no less than nine out of the ten di-oxyanthraquinones whose existence is theoretically possible, according as the positions of the two semi-molecules of hydroxyl are changed.

Of the nine known di-oxyanthraquinones, only one, viz. alizarin, or that in which the hydroxyls are contained in the position 1, 2, is actually used as a colouring agent. Then again, three tri-oxyanthraquinones, $C_4H_5O_2(OH)_3$, are known. One of these is contained in madder-root, and has long been known as purpurin. The other tri-oxyanthroquinones can be artificially prepared. One termed anthra-purpurin is an important colouring matter, especially valuable to Turkey-red dyers, as giving a full or fiery red. The other, called flavo-purpurin, gives an orange dye with alumina mordants. All these various colouring matters can now be artificially produced, and by mixing these in varying proportions a far greater variety of tints can be obtained than was possible with madder alone, and thus the power of diversifying the colour at will is placed in the hands of the dyer and calico printer.

It is quite possible that in an analogous way a variety of shades of blue may be ultimately obtained from substituted indigos, and thus our catalogue of coal-tar colours may be still further increased.

To Englishmen it is a somewhat mortifying reflection, that whilst the raw materials from which all these coal-tar colours are made are produced in our country, the finished and valuable colours are nearly all manufactured in Germany. The crude and inexpensive materials are, therefore, exported by us abroad, to be converted into colours having many hundred times the value, and these expensive colours have again to be bought by English dyers and calico printers for use in our staple industries. The total annual value of manufactured coal-tar colours amounts to about three and a half millions; and as England herself, though furnishing all the raw material, makes only a small fraction of this quantity, but uses a large fraction, it is clear that she loses the profit on the manufacture. The causes of this fact, which we must acknowledge, viz. that Germany has driven England out of the field in this important branch of chemical manufacture, are probably various. In the first place, there is no doubt that much of the German success is due to the long-continued attention which their numerous Universities have paid to the cultivation of Organic Chemistry as a pure science. For this is carried out with a degree of completeness, and to an extent to which we in England are as yet strangers. Secondly, much again is to be attributed to the far more general recognition amongst German than amongst English men of business of the value, from a merely mercantile point of view, of high scientific training. In proof of this it may be mentioned, that each of two of the largest German colour-works employs no less a number than from twenty-five to thirty highly educated scientific chemists, at salaries varying from 250l. to 500l. or 600l. per annum. A third cause which doubtless exerts a great influence in this matter is the English law of patents. This, in the special case of colouring matters at least, offers no protection to English patentees against foreign infringement, for when these colours are once on the goods they cannot be identified. Foreign infringers can thus lower the price so that only the patentee, if skilful, can compete against them, and no English licencees of the patent can exist. This may to some extent account for the reluctance which English capitalists feel in embarking in the manufacture of artificial colouring matters. That England possesses both in the scientific and in the practical direction ability equal to the occasion none can doubt. But be that as it may, the whole honour of the discovery of artificial indigo belongs to Germany and to the distinguished chemist Professor Adolf Baeyer, whilst towards the solution of the difficult problem of its economic manufacture the first successful steps have been taken by Dr. Caro and the Baden Aniline and Soda Works of Mannheim.

[H. E. R.]

Friday, June 10, 1881.

WILLIAM BOWMAN, Esq. LL.D. F.R.S. Vice-President, in the Chair.

JAMES DEWAR, Esq. M.A. F.R.S.

FULLERIAN PROFESSOR OF CHEMISTRY AT THE ROYAL INSTITUTION, AND JACKSONIAN IN THE UNIVERSITY OF CAMBRIDGE.

Origin and Identity of Spectra.

ON a former occasion I detailed the results of a joint research made in concert with my esteemed colleague Professor Liveing, on the "Reversibility of the Rays of Metallic Vapours."* The present lecture will be devoted to a record of the results of our work in relation to three disputed questions in spectroscopic investigation, viz. (1) the Carbon Spectrum, (2) the Magnesium Spectrum, and (3) the Identity of the Spectral Lines of different Elements.

Spectrum of Carbon Compounds.

The spectrum of the flame of hydrocarbons burning in air has been repeatedly described,-first by Swan in 1856, and afterwards by Attfield, Watts, Morren, Plücker, Huggins, Boisbaudran, Piazzi Smyth, and others. The characteristic part of this spectrum consists of four groups of bands of fine lines in the orange, yellow, green, and blue respectively, which are hereafter referred to as the hydrocarbon bands. These four groups, according to Plücker and Hittorf, also constitute the spectrum of the discharge of an induction coil in an atmosphere of hydrogen between carbon electrodes. They are also conspicuous in the electric discharge in olefiant gas at the atmospheric and at reduced pressures.

Plücker and Hittorf notice the entire absence in the flame of olefiant gas of the two bright groups of lines (blue and violet as described below) characteristic of the flame of cyanogen.

Several observers have described the spectrum of the flame of burning cyanogen. Faraday, as long ago as 1829, called the attention of Herschel and Fox Talbot to it, and the latter, writing of his observations,† points out as a peculiarity that the violet end of the spectrum is divided into three portions with broad dark intervals, and that one of the bright portions is ultra-violet. More recently Dibbits, Morren, Plücker, and Hittorf have particularly described this spectrum. Dibbits ‡ mentions in the cyanogen flame fed with oxygen a series of

* 'Proceedings of the Royal Institution,' vol. ix. p. 204.
† 'Phil. Mag.' ser. iii. vol. iv. p. 114.   ‡ 'Pogg. Ann.' 1864.

orange and red bands shaded on the less refrangible side (i. e. in the opposite way to the hydrocarbon bands), the four hydrocarbon bands more or less developed, a group of seven blue lines, a group of two or three faint blue (indigo) lines, then a group of six violet lines, and, lastly, a group of four ultra-violet lines. When the cyanogen is burnt in air, the hydrocarbon bands are less developed, and the three faint indigo lines are scarcely visible, but the rest of the spectrum is the same, only less brilliant.

Plücker and Hittorf[*] state that in the flame of cyanogen burning in air under favourable circumstances, the orange and yellow groups of lines characteristic of burning hydrocarbons are not seen, the brightest line of the green group appears faintly, the blue group is scarcely indicated; but a group of seven fluted bands in the blue, three in the indigo, and seven more in the violet, are well developed, especially the last. When the flame was fed with oxygen instead of air, they state that an ultra-violet group of three fluted bands appeared. They notice also certain red bands with shading in the reverse direction, which are better seen when the flame is fed with air than with oxygen. Other observers give similar accounts, noticing the brilliance of the two series of bands in the blue and violet above mentioned, and that they are seen equally well in the electric discharge through cyanogen.

Ångström and Thalén, in a memoir "On the Spectra of Metalloids,"[†] contend that the channelled spectra of the hydrocarbon and cyanogen flames are the spectra respectively of acetylene and cyanogen, and not of carbon itself, and that in the flame of burning cyanogen we sometimes see the spectrum of the hydrocarbon superposed on that of the cyanogen, the latter being the brighter; and that in vacuum tubes containing hydrocarbons the cyanogen spectrum observed is due to traces of nitrogen.

No chemist who remembers the extreme sensibility of spectroscopic tests, and the difficulty, reaching almost to impossibility, of removing the last traces of air and moisture from gases, will feel any surprise at the presence of small quantities of either hydrogen or nitrogen in any of the gases experimented on.

Mr. Lockyer [‡] obtained a photograph of the spectrum of the electric arc in an atmosphere of chlorine, which shows the series of fluted bands in the ultra-violet, on the strength of which he throws over the conclusion of Ångström and Thalén, and draws inferences regarding the existence of carbon vapour above the chromosphere in the coronal atmosphere of the sun, which, if true, would be contrary to all we know of the properties of carbon.

The conclusions of Ångström and Thalén have been much strengthened by the results of a series of observations carried out by Professor Liveing and myself.

---

[*] 'Phil. Trans.' 1865.   [†] 'Nova Acta, Roy. Soc. Upsala,' vol. ix.
[‡] 'Proc. Roy. Soc.' vol. xxvii. p. 308.

Electric Arc in different Gases.

The experiments were made with a De Meritens dynamo-electric machine, arranged for high tension, giving an alternating current capable of producing an arc between carbon poles in air of from 8 to 10 millims. in length. The carbon poles used were 3 millims. in diameter, and had been previously purified by prolonged heating in a current of chlorine. This treatment, though it removes a large part of the metallic impurities present in the commercial carbons, will not remove the whole, so that lines of calcium, iron, magnesium, and sodium may still be recognised in the arc. Besides the traces of metallic impurities, a notable quantity of hydrogen always remains unremovable by this treatment with chlorine.

The arc was taken in different gases inside a small glass globe (aa in Pl. I. Fig. 1) about 60 millims. in diameter, blown in the middle of a tube. The two ends of the tube (bb) were closed with dry corks, through which were passed (1) the carbons (cc), inserted through two pieces of narrow glass tubing; (2) two other glass tubes (dd) through which currents of the different gases experimented on were passed.

The arc in the globe filled with air gave a tolerably bright continuous spectrum, on which the green and blue hydrocarbon bands were seen, also the seven bands in the indigo (wave-lengths 4600 to 4502, Watts) as in the flame of cyanogen, and much more brightly the six bands in the violet (wave-lengths 4220 to 4158, Watts) and five ultra-violet. Besides these bands, lines of iron, calcium, and sodium were visible. The arc in this case was practically taken in a mixture of nitrogen and carbonic oxide, for in a short time the oxygen of the air is converted into carbonic oxide.

On passing through the globe a current of carbonic acid gas, the bands in the indigo, violet, and ultra-violet gradually died out until they ceased to be visible continuously, and when momentarily seen were only just discernible. On the other hand, the hydrocarbon bands, yellow, green, and blue, came out strong, and were even brilliant. Lines of iron and calcium were still visible. On stopping the current of carbonic acid gas and allowing air to diffuse into the globe, the violet and ultra-violet bands soon began to appear, and presently became permanent and bright, the hydrocarbon bands remaining bright.

When a continuous current of dry hydrogen was passed through the globe, the arc, contrary to what would be expected from the behaviour of the spark discharge in hydrogen, would not pass through more than a very short space, very much less than in air or carbonic acid gas. There was a tolerably bright continuous spectrum, with no trace of bands in the indigo, violet, or ultra-violet, and no metallic lines, with the exception of a fairly bright line in the red, which we identified, by comparison with the spark in a vacuum tube, with the C line of hydrogen. The F line, identified in like manner, was also seen as a faint diffuse band. This last line was in general overpowered by the continuous spectrum, but was regularly seen when, from some variation in the discharge, the continuous spectrum became less brilliant. This was the first occasion on which we had seen the hydrogen lines in the arc, though Secchi * states that he had seen them by the use of moist carbon poles. The hydrocarbon bands in the green and blue were at intervals well seen. Those in the yellow and orange were, owing doubtless to the smaller dispersion of the light in that region, overpowered by the continuous spectrum. Whereas when air and carbonic acid gas were used, the inside of the globe was quickly covered with dust from the disintegrated poles, scarcely any dust was thrown off when the arc was passed in hydrogen.

In nitrogen a longer arc could be formed, and the indigo, violet, and ultra-violet bands of cyanogen all came out at intervals brilliantly. The green and blue hydrocarbon bands were also well developed.

On filling the globe with chlorine, keeping a current of that gas passing through it, the arc would not pass through a greater distance than about 2 millims. No metallic lines were visible. At first the violet bands, as well as the green and blue hydrocarbon bands, were visible; but gradually, when the current of chlorine had been passing for some minutes, there was nothing to be seen but a continuous spectrum with the green and blue hydrocarbon bands. Neither of these bands were strong, and at intervals the blue bands disappeared altogether.

The arc would not pass in a current of carbonic oxide through any greater space than in chlorine. There was much continuous spectrum; the yellow, green, and blue hydrocarbon bands were well seen, some of the indigo bands were just discernible, the violet had nearly, and the ultra-violet quite, gone from sight. No trace of the carbonic oxide bands, as seen in the spark discharge in that gas, was visible. This is the more remarkable since under similar circumstances two of the characteristic lines of hydrogen were seen.

In nitric oxide a very long arc could be obtained. The violet and ultra-violet cyanogen bands were well seen, the indigo bands were seen, but weaker. The blue and green hydrocarbon bands were also seen well when the arc was short, not so well when the arc was long. Many metallic lines of iron, calcium, and magnesium were seen.

In ammonia only a short arc could be obtained. All the bands were faint, but the indigo and violet and ultra-violet cyanogen bands were always visible.

These experiments with different gases eliminate to a large extent the influence of electric conductivity on the character of the spectrum.

Apart from the relative electric conductivity of gases, it is clear, from the experiments, that the length and character of alternating electric discharges between carbon poles in different gases do not follow the law which we should expect. It will require a pro-

---

* 'Compt. Rend.' 1873.

longed series of experiments to arrive at definite conclusions on this matter; but, in the meantime, it is highly probable that one of the main factors in producing these remarkable variations in the arc will be found to be the relative facility with which the carbon of the poles combines with the gaseous medium.

On a review of the above series of observations, certain points stand out plainly. In the first place, the indigo, violet, and ultra-violet bands, characteristic of the flame of cyanogen, are conspicuous in the arc taken in an atmosphere of nitrogen, air, nitric oxide, or ammonia, and they disappear almost, if not quite, when the arc is taken in a non-nitrogenous atmosphere of hydrogen, carbonic oxide, carbonic acid, or chlorine. These same bands are seen brightly in the flames of cyanogen and hydrocyanic acid, but are not seen in those of hydrocarbons, carbonic oxide, or carbon disulphide. The conclusion seems irresistible that they belong to cyanogen; and this conclusion does not seem to us at all invalidated by the fact that they are seen weakly, or by flashes, in the arc or spark taken in gases supposed free from nitrogen on account of the extreme difficulty of removing the last traces of air. They are never, in such a case, the principal or prominent part of the spectrum, and in a continuous experiment they are seen to fade out in proportion as the nitrogen is removed. This conclusion is strengthened by the recent discovery that cyanogen is always generated in the electric arc in atmospheric air.

The green and blue bands, characteristic of hydrocarbon flames, are well seen when the arc is taken in hydrogen; but, though less strong when the arc is taken in nitrogen or in chlorine, they seem to be always present in the arc, whatever the atmosphere. This is what we should expect, if they be due, as Ångström and Thalén suppose, to acetylene; for we have found that the carbon electrodes always contain, even when they have been long treated with chlorine at a white heat, a notable quantity of hydrogen.

The hydrocarbon bands are well developed in the blowpipe flame, that is, under conditions which appear, at first sight, unfavourable to the existence of acetylene. We have, however, satisfied ourselves, by the use of Deville's aspirator, that acetylene may be withdrawn from the interior of such a flame, and from that part of it which shows the hydrocarbon bands brightly.

The question as to whether these bands are due to carbon itself or to a compound of carbon with hydrogen, has been somewhat simplified by the observations of Watts and others on the spectrum of carbonic oxide. There is, we suppose, no doubt now that that compound has its own spectrum quite distinct from the hydrocarbon flame spectrum. The mere presence of the latter spectrum feebly developed in the electric discharge in compounds of carbon supposed to contain no hydrogen, appears to us to weigh very little against the series of observations which connect this spectrum directly with hydrocarbons.

In the next place, it appears, from experiments, that the development of the violet bands of cyanogen, or the less refrangible hydrocarbon bands, is not a matter of temperature only. For the appearance of the hydrogen lines C and F in the arc taken in hydrogen indicates a temperature far higher than that of any flame. Yet the violet bands disappear at that temperature, and the green bands are well developed. The violet bands are, nevertheless, seen equally well at the different temperatures of the flame, arc, and spark, provided cyanogen be the compound under observation in the flame, and nitrogen and carbon are present together at the higher temperatures of the arc and spark.

The question of the constitution of comets, since the discovery by Huggins[*] that the spectra of various comets are identical with the hydrocarbon spectrum, naturally leads to some speculation in connection with the conclusions to which our experiments point. Provided we admit that the materials of the comet contain ready-formed hydrocarbons and that chemical or electrical actions may take place, generating a high temperature, then the acetylene spectrum might be produced at temperatures no higher than that of ordinary flames without any trace of the cyanogen spectrum, or of metallic lines. Such actions might be brought about by the tidal disturbances involving collisions and projections of the constituents of the swarms of small masses circulating in orbits round the sun, which we have every reason to believe constitute the cometic structure. If, on the other hand, we assume only the presence of uncombined carbon and hydrogen, we know that the acetylene spectrum can only be produced at a very high temperature; and if nitrogen were also present, that we should at such a temperature have the cyanogen spectrum as well. Either then the first supposition is the true one, not disproving the presence of nitrogen; or else the atmosphere which the comet meets is hydrogen only and contains no nitrogen.

The Flame of Cyanogen.

The accompanying diagram (Pl. I. Fig. 2) shows the relative position of the bands in that part of the spectrum of the flame of cyanogen fed with a jet of oxygen which is more refrangible than the Fraunhöfer line F. Only those bands which are less refrangible than the solar line L have been before described, but photographs show another set of two shaded bands slightly less refrangible than the solar line N accompanied by a very broad diffuse band of less intensity on the more refrangible side of N; also a strong shaded band, which appears to be absolutely coincident with the remarkable shaded band in the solar spectrum, which has been designated by the letter P; and near this, on the less refrangible side, a much fainter diffuse band, which also seems to coincide with a part of the solar spectrum sensibly less luminous than the parts on either side of it. Watts found that the spectrum cyanogen of the flame did not disappear when the

---

[*] 'Proc. Roy. Soc.' vol. xvi. p. 386; vol. xxiii. p. 154; ' Phil. Trans.' 1868, p. 555.

flame was cooled by diluting the cyanogen with carbonic acid ; and we have found that it retains its characters when the cyanogen is burnt in nitric oxide. The flame in the last case must be one of the hottest known, from the large amount of heat evolved in the decomposition of cyanogen and nitric oxide, namely, 41,000 and 43,300 units respectively. There is in the case of cyanogen, as in the case of so many other substances, a difference in the relative intensities of the different parts of the spectrum at different temperatures, but no other change of character.

On the theory that these groups of lines are the product of an exceptional temperature in the case of the cyanogen flame, it is inconceivable that they could disappear by combustion in oxygen, instead of in ordinary air. Our observations accord with the statement of Morren, Plücker, Hittorf, and Thalén, that a cyanogen flame, fed with oxygen, when it is intensely luminous, still yields these peculiar groups. We have found these peculiar groupings in the flame when it had a current of oxygen in the middle, and was likewise surrounded outside with oxygen. There is nothing remarkable in the fact that only a continuous spectrum is seen to proceed from any hydrocarbon or nitrocarbon burning in excess of oxygen, as we know from Frankland's experiments that carbonic acid and water vapour at the high temperature of flame under compression give in the visible portion a continuous spectrum. In fact, this is what we should anticipate, provided intermediate, and not the final, compounds are the active sources of the banded spectrum.

Each of the five sets of bands shown in the diagram is attended on its more refrangible side by a series of rhythmical lines extending to a considerable distance, not shown in the diagram, but easily seen in the photographs.

Coal gas burning in oxygen gives no bands above that near G within the range of the diagram, Fig. 2 ; but beyond this our photographs show a spectrum of a character quite different from that at the less refrangible end. The most remarkable part of this spectrum is a long series of closely set strong lines, filling the region between the solar lines R and S, and ending abruptly with two strong lines a little beyond S. These are lines of various intensities, not regularly arranged so as to give shaded bands like those in the less refrangible part of the spectrum. Beyond these lines there is another large group of lines, not so strong or so closely set, but sharp and well defined. This peculiar part of the spectrum is really due to the vapour of water, and shall be discussed in the sequel.

Spark Discharge in various Gases.

Mr. Lockyer's experiments on the spectrum of carbon compounds are directly opposed to the results given above, as will be understood by the following extract from one of his papers on the subject : *—

---

* 'Proc. Roy. Soc.' vol. xxx. p. 336.

"I beg permission, therefore, in the meantime, to submit to the notice of the Society an experiment with a tube containing CCl_4,* which, I think, establishes the conclusions arrived at by prior investigators. And I may add that it is the more important to settle the question, as Messrs. Liveing and Dewar have already based upon their conclusions theoretical views of a kind which appear to me calculated to mislead, and which I consider to have long been shown to be erroneous." The following experiments have been made to test the accuracy of our previous work, and to confirm or disprove Mr. Lockyer's views.

The form of sparking tube employed was similar to that used by Salet. This was attached by thick rubber tubing to a straight glass tube of which one half, about 6 inches long, was filled with phosphoric anhydride, and the other half with small fragments of soda-lime to prevent any chlorine from the decomposition of the tetrachloride by the spark from reaching the Sprengel pump. The tetrachloride used had been prepared in our own laboratory, and fractionated until it had a constant boiling point of 77° C. Sufficient of it was introduced into the sparking tube to fill nearly one quarter of the bulb at the end, and the whole interior of the tube thoroughly wetted with it in order to facilitate the removal of the last traces of air.

When the tube containing the tetrachloride had been so far exhausted that little but condensible vapours were pumped out, the bulb was heated so as to fill the apparatus with vapour of tetrachloride, the pump still going, and this was repeated as long as any incondensible gas was extracted. Sparks were then passed through the tube for a short time, the pump still being kept going. After a short time it was unnecessary to keep the pump going, as all the chlorine produced by decomposition of the tetrachloride was absorbed by the soda-lime. On now examining the spectrum, no trace of any of the bands we ascribe to nitrocarbons could be detected, either by the eye or by photography, however the spark might be varied. The violet lines of chlorine described by Salet were more or less visible, coming out brightly when a condenser was used. Several tubes were treated in this way, and many photographs taken, but always with the same result; no trace appeared of either the seven blue, the six violet, the five ultra-violet, or of the still more refrangible bands of the cyanogen flame. All the photographs showed three lines in the ultra-violet, but these do not at all resemble the nitrocarbon bands, as they are not shaded. The least refrangible of the three is nearly coincident with the middle maximum in the ultra-violet set of five bands, but the other two do not coincide with any of these maxima. When a condenser is used, these three lines come out with much greater intensity, and two other triplets appear on the more refrangible side, as well as other lines. In order to compare the positions of these lines with the cyanogen bands, we have taken several

---

* Carbon tetrachloride.

photographs of the spark in tetrachloride simultaneously with a cyanogen flame, the latter being thrown in by reflection in the usual way.

Not one of many photographs so taken showed any traces of the cyanogen bands. The general character of the violet part of the spectrum of the spark in carbon tetrachloride taken without a condenser (but not the exact position to scale of wave-lengths of all the lines) is shown at B in Fig. 2. At C of the same diagram are shown the brightest of the additional lines which come out with the use of a condenser. Photographs of sparks taken in hydrochloric acid showed a precisely similar group of ultra-violet lines, so that the three lines which our photographs show amongst the five ultra-violet nitrocarbon bands are due to chlorine.

Having satisfied ourselves by repeated trials that pure carbon tetrachloride or trichloride, if free from nitrogen, does not give any of the bands we ascribe to nitrocarbon compounds, our next step was to determine whether the addition of nitrogen would bring them out, and if so, what quantity of nitrogen would make them visible.

For this purpose we introduced a minute fragment of bichromate of ammonia, carefully weighed, and wrapped in platinum foil, into the neck of one of the sparking tubes containing carbon tetrachloride, connected the tube to the Sprengel pump, and removed the air as before. The spark examined in the tube showed no trace of any nitrocarbon band. A pinch-cock was now put on the rubber tube, and the bichromate heated by a spirit-lamp to decomposition (whereby it is resolved into nitrogen, water, and oxide of chromium). On now passing the spark the six violet bands were well seen. There was no change in the condition of the coil or rheotome, so that the spark was of the same character as it had been before when no nitrocarbon bands were visible, and the change in the spectrum cannot be attributed to any change in the spark. The weight of the bichromate was between ·0005 and ·0006 grm.; and the nitrogen this would evolve would fill just about $\frac{1}{20}$ of a cubic centimetre at atmospheric pressure. The tube held 30 cub. centims., so that vapour of carbon tetrachloride when mixed with $\frac{1}{600}$ part of its volume of nitrogen, gives under the action of the electric spark the nitrocarbon bands distinctly. Other similar experiments confirmed this result. It is worthy of remark that the nitrocarbon bands were not seen instantaneously on the admission of nitrogen into the tube, but were gradually developed, as if it was necessary that a certain quantity of nitrocarbon compound should be formed under the influence of the electric discharge and accumulated before its spectrum became visible.

A tube, containing naphthaline, previously well washed with dilute sulphuric acid, dried and resublimed, was attached to the Sprengel pump, and treated as the tubes with tetrachloride had been. The spark in this tube likewise showed no nitrocarbon bands. After a time the tube cracked, and then the nitrocarbon bands made their appearance, and on setting the pump going a good deal of gas was pumped out. When the air had again been pretty completely exhausted, the nitrocarbon bands were no longer visible, but gradually reappeared again as air leaked through the crack. Another tube, containing a mixture of naphthaline and benzol, showed no trace of the nitrocarbon bands.

The observation of the nitrocarbon bands in the spectrum of the spark in naphthaline was one of the reasons which led Watts at one time to ascribe these bands to free carbon.

In our first experiments with carbonic oxide the gas was made by the action of sulphuric acid on dried formiate of sodium.

At first the six violet cyanogen bands were well seen, and the seven blue bands faintly; but gradually, as the air became more completely expelled, the blue bands disappeared entirely, and then the violet bands so far died out that it was only by manipulating the coil that they could be made visible, and then only very faintly. A bubble of air about $\frac{1}{400}$ part of the volume of gas in the generating flask and tube, was now introduced, when almost immediately the bands reappeared brightly. As the stream of gas continued, they again gradually died away until they were represented only by a faint haze. It was subsequently found that each introduction of fresh acid into the flask was attended with a marked increase in the brightness of the nitrocarbon bands, which died away again when the current of gas was continued without fresh introduction of acid. On testing the acid it was found to contain, as is frequently the case with sulphuric acid, a very small quantity of the oxides of nitrogen. The difficulty of getting all the air expelled from the apparatus and reagents led us to adopt another method of making carbonic oxide. Carbonic oxide was generated by heating in a tube of hard glass in a combustion furnace a mixture of pure dry potassium oxalate with one quarter of its weight of quicklime, the mixture having been previously heated for some time to expel traces of ammonia. No trace whatever of the nitrocarbon bands could be detected in this carbonic oxide, however the spark might be varied. The pressure of the gas was reduced to 1 inch of mercury, while the spectrum was observed from time to time. Still no trace of the nitrocarbon bands could be detected. More of the oxalate was heated, and the observations repeated again and again, always with the same result. Carbonic oxide, therefore, if quite free from nitrogen, does not give, at the atmospheric or any less pressure, the nitrocarbon spectrum.

On passing the spark between carbon poles in nitrogen, the nitrocarbon bands are plainly seen; and remain visible through great variations in the character of the spark. Photographs taken, with and without the use of the condenser, showed the violet and ultra-violet nitrocarbon bands, including those near N and P. If the nitrogen was swept out by a current of carbonic acid gas, on passing the spark the nitrocarbon bands could no longer be detected, and photographs showed no trace of any of the ultra-violet bands.

In all the foregoing experiments the bands which Ångström and Thalén ascribe to hydrocarbons were always more or less plainly seen. Much more care than has generally been thought necessary is needed if the last traces of hydrogen and its compounds are to be removed from spectral tubes. Indeed, water cannot be completely removed from apparatus and reagents which do not admit of being heated to redness.

Thus a mixture of carbonate of sodium and boric anhydride, previously to admixture heated red hot, was introduced into one end of a piece of combustion tube, near the other end of which wires had been sealed, and the open end drawn out; the mixture was then heated, and when it was judged that all the air was expelled, the tube was sealed off at atmospheric pressure. On passing sparks through it carbonic oxide bands and oxygen lines could be seen, but no hydrogen, hydrocarbon, or nitrocarbon bands could be detected. It appears, therefore, that the application of a red heat is likely to prove a more effectual means of getting rid of moisture than the use of any desiccating agent.

Are the groups of shaded bands seen in the more refrangible part of the spectrum of a cyanogen flame, of which the three which can be detected by the eye are defined by their wave-lengths (4600 to 4502, 4220 to 4158, and 3883 to 3850), due to the vapour of uncombined carbon, or, as we conclude, to a compound of carbon with nitrogen?

The evidence that carbon can take the state of vapour at the temperature of the electric arc is at present very imperfect. Carbon shows at such temperatures only incipient fusion, and that uncombined carbon should be vaporised at the far lower temperature of the flame of cyanogen is so incredible an hypothesis, that it ought not to be accepted if the phenomena admit of any other probable explanation. On the other hand, cyanogen or hydrocyanic acid is generated in large quantity in the electric arc taken in nitrogen, and Berthelot has shown that hydrocyanic acid is produced by the spark discharge in a mixture of acetylene and nitrogen, so that in the cases in which these bands shine out with the greatest brilliance, namely, the arc in nitrogen and the cyanogen flame, we know that nitrocarbon compounds are present. Further, we have shown that these bands fade and disappear in proportion as nitrogen is removed from the arc. Ångström and Thalén had previously shown the same thing with regard to the discharge between carbon electrodes; and the conclusion to which they and we have come would probably have commanded universal assent if it had not been for the fact that these bands had been seen in circumstances where nitrogen was supposed to be absent, but where, in reality, the difficulty of completely eliminating nitrogen, and the extreme sensibility of the spectroscopic test, had been inadequately apprehended.

Our argument is an induction from a very long series of observations which lead to one conclusion, and hardly admit of any other explanation. Mr. Lockyer, however, attempts to explain the disappearance of the bands when nitrogen is absent by the statement, "that the tension of the current used now brings one set of flutings into prominence, and now another." This is no new observation. It is well known that variations in the discharge produce variations in the relative intensities of different parts of a spectrum. Certain lines of magnesium, cadmium, zinc, and other metals, very brilliant in the spark, are not seen, or are barely seen, at all in the arc. His remark might be applied to the spectra of compounds as well as to those of elements. Variation in the discharge accounts very well for some of the variations of intensity in the bands if they be due to a nitrocarbon; it will not, however, account for the fact observed by us, that the bands, or those of them which have the greatest emissive power, and are best developed by the particular current used, come out on the addition of a minute quantity of nitrogen, when there is every reason to think that no variation of the current occurs.

Much the same may be said with regard to the changes of the spectrum produced by changes of temperature. We cannot infer from any of these changes that the spectrum is not due to a compound.

Again, Mr. Lockyer attempts to get over the difficulties of his case by the supposition that "the sets of carbon flutings represent different molecular groupings of carbon, in addition to that or those which give us the line spectrum."

Now, until independent evidence can be adduced that carbon can exist in the state of uncombined vapour at the temperature of a cyanogen flame, and that different groupings in such vapour exist, the hypothesis here enunciated is a gratuitous one, so long as the existence of nitrocarbon compounds in the flame, arc, and spark will sufficiently explain the facts.

The observation above recorded, that there is in the spectrum of cyanogen a strong shaded band coincident with the very characteristic dark shaded band P of the solar spectrum, strengthens materially the evidence in favour of the existence of these bands in the solar spectrum; the more so, as the series of lines at P has far more of the distinctive character of the cyanogen spectrum than any other series in the ultra-violet part of the solar spectrum.

The hypothesis that if present they are due to vapour of carbon uncombined in the upper cooler region of the chromosphere seems absurd. One object of our investigations has been to determine the permanence of compounds of the non-metallic elements, and the sensitiveness of the spectroscopic test in regard to them. It appeared probable that if such compounds exist in the solar atmosphere their presence would be most distinctly revealed in the more refrangible part of the spectrum. In the meantime it is sufficiently clear that the presence of nitrogen in the solar atmosphere may be recognised through cyanogen when free nitrogen might escape detection.

The series of experiments, unless proved to be wrong, are almost conclusive proof that Mr. Lockyer's views regarding the origin and variation of the carbon spectrum have no real experimental basis.

Spectrum of Magnesium.

The absorption spectrum of magnesium and of magnesium with potassium and sodium, as seen in iron tubes in a hydrogen atmosphere, described in the former lecture, correspond to no known emission lines of magnesium. We could only ascribe their origin to the mixtures employed as distinct from the separate elements, and therefore were led to investigate the conditions under which corresponding emission lines could be produced.

In 'Proc. Roy. Soc.' vol. xxvii. p. 494, the emission spectrum of sparks from an induction coil taken between magnesium points in an atmosphere of hydrogen is described as follows:—

" A bright line regularly appeared with a wave-length about 5210. .... This line does not usually extend across the whole interval between the electrodes, and is sometimes seen only at the negative electrode. Its presence seems to depend on the temperature, as it is not seen continuously when a large Leyden jar is employed, until the pressure of the hydrogen, and its resistance, is very much reduced. When well-dried nitrogen or carbonic oxide is substituted for hydrogen, this line disappears entirely; but if any hydrogen or traces of moisture be present, it comes out when the pressure is much reduced. In such cases the hydrogen lines C and F are always visible as well. Sometimes several fine lines appear on the more refrangible side of this line between it and the b group, which give it the appearance of being a narrow band shaded on that side." " In addition to the above-mentioned line, we observed that there is also produced a series of fine lines, commencing close to the most refrangible line of the b group, and extending, with gradually diminishing intensity, towards the blue .... from forty-five to fifty being visible, and placed at nearly equal distances from each other."

In a paper entitled "A New Method of Spectrum Observation," * Mr. Lockyer regards this spectrum as illustrative and confirmatory of his views regarding the possibility of elemental dissociation at different heat-levels. The view taken by Mr. Lockyer may be expressed in his own words:—

" The flame spectrum of magnesium perhaps presents us best with the beautiful effects produced by the passage from the lower to the higher heat-level, and shows the important bearing upon solar physics of the results obtained by this new method of work. ... In the flame the two least refrangible of the components of b are seen associated with a line less refrangible so as to form a triplet. A

---

* 'Proc. Roy. Soc.' vol. xxx. p. 22.

series of flutings and a line in the blue are also seen. . . . On passing the spark all these but the two components of b are abolished. We get the wide triplet replaced by a narrow one of the same form, the two lines of b being common to both.

"May we consider the existence of these molecular states as forming a true basis for Dalton's law of multiple proportions? If so, then the metals in different chemical combinations will exist in different molecular groupings, and we shall be able, by spectrum observations, to determine the particular heat-level to which the molecular complexity of the solid metal, induced by chemical affinity, corresponds. . . . Examples.—None of the lines of magnesium special to the flame spectrum are visible in the spectrum of the chloride either when a flame or a spark is employed."

In order to ascertain the true cause of the variations in the magnesium spectrum, the following experiments and observations were made, and they demonstrate that the views of Mr. Lockyer on this question must also be regarded as resting on faulty experimenting:—

1. *Observations on the Spark between Magnesium Points in Nitrogen and Carbonic Oxide at various Pressures.*

The points were pieces of magnesium wire. Round one end of each a platinum wire was tightly coiled and fused into the side of a glass tube. This tube was attached by fusion at one end to another tube filled with phosphoric anhydride, which in turn was connected with a Sprengel pump. The other end of the tube was connected by a thick rubber tube, capable of being closed by a pinchcock, with a gasholder containing nitrogen over strong sulphuric acid. The tube having been exhausted and filled with nitrogen two or three times, it was found that no line at 5210 was visible in the spark. The tube was now gradually exhausted, and the spark watched as the exhaustion proceeded. No line at 5210 was seen, although the exhaustion was carried nearly as far as the pump would carry it; nor was any hydrogen line (C or F) visible, either with or without the use of a jar. The communication with the gasholder was now opened, and the tube refilled with nitrogen at the atmospheric pressure; a communication was then made with another vessel containing hydrogen, which was allowed to diffuse into the tube for a very short time. On now passing the spark, the line at 5210 at once appeared, although the quantity of hydrogen diffused into the nitrogen must have been very small. The experiments with nitrogen at reduced pressure were repeated several times, with the same result. It was found necessary to have the phosphoric anhydride, as without it traces of moisture were left or found their way through the pump into the tube, and then, when the exhaustion was carried far enough, both the line at 5210 and the hydrogen lines, C and F, made their appearance. We have never, however, been able to detect the line at 5210, in nitrogen, without being able to detect C or F either at the same time or by merely varying the discharge by means of a Leyden jar.

Experiments made in the same way with carbonic oxide instead of nitrogen led to precisely the same results.

2. *Observations on the Spark between Magnesium Points in Hydrogen at reduced Pressures.*

A tube, similar to those employed with nitrogen and carbonic oxide, was attached at one end to a Sprengel pump and mercury gauge, and at the other end to an apparatus for generating hydrogen. Dry hydrogen was passed through for some time, and the connection with the hydrogen apparatus closed. On sparking with the hydrogen at the atmospheric pressure, the line at 5210 and its attendant series were visible, and were still visible when a small Leyden jar was used with the induction coil, but disappeared almost entirely when a large Leyden jar was used. When the pressure of the hydrogen was reduced to half an atmosphere, the line at 5210 was seen faintly when a large Leyden jar was used, but not the series of fine lines. When the pressure was reduced to 180 millims., the series of fine lines began to show when the large jar was used. By still further reducing the pressure the whole series was permanently visible when the large jar was used; but when the exhaustion was carried still further they grew fainter, and almost disappeared. On gradually readmitting hydrogen, the same phenomena recurred in the reverse order.

3. *Observations on the Arc with Magnesium and Hydrogen.*

The line at 5210 is not seen in the arc in a lime or carbon crucible when magnesium is dropped in without the introduction of hydrogen. If, however, a gentle stream of hydrogen or of coal gas be led in through a perforation in one of the electrodes, the line at 5210 immediately makes its appearance, and, by varying the current of gas, it may be made to appear either bright or reversed. However small the current of hydrogen be made, the line can be detected as long as the current and the supply of magnesium continue, but disappears very quickly when the current of gas ceases.

4. *Observations on the Flame of Burning Magnesium.*

The line at 5210 may often be seen in the flame of magnesium burning in air, but both it and the series of fine lines which accompany it come out with greatly increased brilliance if the burning magnesium be held in a jet of hydrogen, of coal gas, or of steam.

The experiments above described, with nitrogen and carbonic oxide at reduced pressures, are almost if not quite conclusive against the supposition that the line at 5210 is due merely to the lower temperature of the spark in hydrogen. From De La Rue and Müller's observations it would appear that nitrogen at a pressure of 400 millims. should produce much the same effect on the spark as hydrogen at 760 millims. Now the pressures of the nitrogen and carbonic oxide were reduced far below this without any trace of the line in question being visible. Moreover, the magnesium line at 4481, which is not seen in the arc, and may be reasonably ascribed to the higher temperature of the spark, may be seen in the spark at the same time as the line at 5210 when hydrogen is present. Nevertheless temperature does seem to affect the result in some degree, for when a large Leyden jar is used, and the gas is at the atmospheric pressure, the line almost disappears from the spark, to reappear when the pressure is reduced; but by no variation of temperature have we been able to see the line when hydrogen was carefully excluded.

A line of the same wave-length has been seen by Young in the chromosphere once. Its absence from the Fraunhöfer lines leads to the inference that the temperature of the sun is too high (unless at special times and places) for its production. If it be not due to a compound of magnesium with hydrogen, at any rate it occurs with special facility in the presence of hydrogen, and ought to occur in the sun if the temperature were not too high.

We have thus far been careful to ascribe this line and its attendant series to a mixture of magnesium and hydrogen rather than to a chemical compound, because this sufficiently expresses the facts, and we have not yet obtained any independent evidence of the existence of any chemical compound of those elements. We have independent evidence that mixtures which are not probably chemical compounds favour the production of certain vibrations which are not so strong or are not seen at all when the elements of those mixtures are taken separately. The remarkable absorptions produced by mixtures of magnesium with potassium and sodium above-mentioned belong to this class. We have not been able to obtain the emission spectra corresponding to these absorptions, but in the course of our observations on the arc we have frequently noticed that certain lines of metals present in the crucible are only seen, or come out with especial brilliance, when some other metal is introduced. This is the case with some groups of calcium lines which are not seen, or are barely visible, in the arc in a lime crucible, and come out with great brilliance on the introduction of a fragment of iron, but are not developed by other metals such as tin.

Spark Spectrum of Magnesium in Hydrogen under increased Pressures.

In order to ascertain if this peculiar spectrum could be produced at a high temperature in the presence of hydrogen, which we have already shown to be essential to its production at the atmospheric and at reduced pressures, experiments were made with hydrogen at pressures increasing up to twenty atmospheres.

On the supposition that this spectrum originates from the formation of some chemical compound, probably formed within certain limits of temperature when vapour of magnesium is in presence of hydrogen, the stability of the body ought to depend largely on the pressure of the gaseous medium. Like Graham's hydrogenium, this body might be formed in hydrogen of high pressure at a temperature at which it would under less pressure be decomposed. In fact, it has been shown by Troost that the hydrides of palladium, sodium, and potassium all follow strictly the laws of chemical dissociation enunciated by Deville; and increased pressure, by rendering the compound more stable, ought, if the secondary effect of such pressure in causing a higher temperature in the electric discharge were not overpowering, to conduce to a more continuous and brilliant spectrum of the compound. Conversely, if such a more continuous and brilliant spectrum be found to result, in spite of the higher temperature, from increased pressure, it can only be explained by the stability of the substance being increased with the pressure.

Now, what are the facts? When the spark of an induction coil, without a Leyden jar, is passed between magnesium electrodes in hydrogen at atmospheric pressure, the flutings in the green are, as before described, always seen, but they are much stronger at the poles and do not always extend quite across the field. As the pressure is increased, however, they increase in brilliance and soon extend persistently from pole to pole, and go on increasing in intensity, until, at fifteen and twenty atmospheres, they are fully equal in brilliance to the b group, notwithstanding the increased brightness these have acquired by the higher temperature, due to the increased pressure. The second set of flutings, those in the yellowish green, come out as the pressure is increased, and, in fact, at twenty atmospheres only the b group and the flutings are noticeable; if the yellow magnesium line be visible at all it is quite lost in the brilliance of the yellow flutings. The tail of fine lines of these flutings extend at the high pressure quite up to the green, and those of the green flutings quite up to the blue. On again letting down the pressure the like phenomena occur in the reverse order, but the brilliance of the flutings does not diminish so rapidly as it had increased. If, now, when the pressure has again reached that of the atmosphere, a large Leyden jar be interposed in the circuit, on passing the spark the flutings are still seen quite bright, and they continue to be seen with gradually diminishing intensity until the sparks have been continued for a considerable time. It appears that the compound, which had been formed in large quantity by the spark without jar at the higher pressures, is only gradually decomposed, and not re-formed, by the high temperature of the spark with jar. This experiment, which was several times repeated, is conclusive against the supposition that the flutings are merely due to a lower temperature. When the pressure was increased at the same time that the jar was employed, the flutings did not immediately disappear, but the expansion of the magnesium lines and the increase of the continuous spectrum seemed to overpower them.

When nitrogen was substituted for hydrogen, the strongest lines of the green flutings were seen when the spark without jar was first passed at atmospheric pressure, probably from hydrogen occluded, as it usually is, in the magnesium electrodes. As the pressure was increased they speedily disappeared entirely, and were not again seen either at high or low pressures.

With carbonic oxide the same thing occurred as with nitrogen; but in this gas the flutings due to the oxide of magnesium (wave-length 4930 to 5000) were, for a time, very well seen.

Fig. 4, Plate III., shows more completely than we have given it before the general character of the magnesium-hydrogen spectrum, which consists of two sets of flutings closely resembling in character the hydrocarbon flutings, each fluting consisting of a multitude of fine lines closely set on the less refrangible side, and becoming wider apart and weaker towards the more refrangible side, but extending under favourable circumstances much farther than is shown in the figure. The set in the green is the stronger, and it was to this that our former observations were confined. It has two flutings, one beginning at about wave-length 5210 and the other close to b_1 on its more refrangible side. The other set consists of three principal flutings, of which the first begins at about wave-length 5618, the next at about wave-length 5566, and the third begins with three strong lines at about the wave-lengths 5513, 5512, 5511. Both sets are very well seen when a magnesium wire is burnt in the edge of a hydrogen flame, and in the arc in a crucible of magnesia when a gentle current of hydrogen is led into it. There is also a pair of bands in the blue beginning at about the wave-lengths 4850, 4802.

Mr. Lockyer states (*loc. cit.*) that none of the lines of magnesium, special to the flame spectrum, are visible in the spectrum of the chloride, either when flame or spark is employed. But we find that when the spark is taken between platinum points, from a solution of the chloride of magnesium, in a tube such as those used by Delachanal and Mermet, the line at wave-length 5210 can frequently be seen in it when the tube is filled with air, and that if the tube be filled with hydrogen the green flutings of magnesium-hydrogen are persistent and strong.

Repeated observations have confirmed our previous statements as to the facility with which the magnesium-hydrogen spectrum can be produced in the arc by the help of a current of the gas. In a magnesia crucible, by regulating the current of hydrogen, the flutings can be easily obtained either bright or reversed.

The variations in the spectrum of magnesium, and the conditions under which it is observed, throw additional light on the question of the emissive power for radiation of short wave-lengths of substances at the temperature of flames to which we alluded in our paper on the spectrum of water.*

Ultra-Violet Spectrum of the Flame of Burning Magnesium.

When magnesium wire or ribbon is burnt in air, we see the three lines of the b group, the blue line about wave-length 4570, first noticed by us in the spark spectrum;† and photographs show, besides, the well-known triplet in the ultra-violet between the solar lines K and L sharply defined, and the line for which Cornu has found the wave-length 2850 very much expanded and strongly reversed. These lines are all common to the flame, arc, and spark spectra; and the last of them (2850) seems to be by far the strongest line both in the flame and arc, and is one of the strongest in the spark. But, in addition to these lines, the photographs of the flame show a very strong, somewhat diffuse, triplet, generally resembling the other magnesium triplets in the relative position of its components, close to the solar line M; and a group of bands below it extending beyond the triplet near L. These bands have, for the most part, each one sharply defined edge, but fade away on the other side; but the diffuse edges are not all turned towards the same side of the spectrum. The positions of the sharp edges of these bands, and of the strong triplet near M, are shown in Pl. III., Fig. 1. It is remarkable that the triplets near P and S are absent from the flame spectrum, and that the strong triplet near M is not represented at all either in the arc or spark. The hydrogen-magnesium series of lines, beginning at a wave-length about 5210, are also seen sometimes, as already described by us,‡ in the spectrum of the flame; but we have never observed that the appearance of these lines, or of the strong line with which they begin, is connected with the non-appearance of b_4. Indeed, we can almost always see all three lines of the b group in the flame, though as b_4 is the least strong of the three, it is likely to be most easily overpowered by the continuous spectrum of the flame.

Burning magnesium in oxygen instead of atmospheric air does not bring out any additional lines; on the contrary, the continuous spectrum from the magnesia overpowers the line spectrum, and makes it more difficult of observation.

Magnesia heated in the oxyhydrogen jet does not appear to give the lines seen in the flame, except that at 2850.

Spectrum of the Arc.

The spectrum of magnesium, as seen in the arc, contains several lines besides those heretofore described. These lines come out brightly, generally considerably expanded, when a fragment of magnesium is dropped into the crucible through which the arc is passing,

---

* 'Proc. Roy. Soc.' 1880, No. 201, p. 152.
† Ibid. vol. xxvii. p. 350.   ‡ Ibid. vol. xxx. p. 96.

but rapidly contract and gradually become very faint or disappear entirely.

By examining the arc of a Siemens machine, taken in a crucible of dense magnesia under the dispersion of the spectrum of the fourth order, given by a Rutherford grating of 17,296 lines to the inch, we are able to separate the iron and magnesium lines which form the very close pair b_4 of the solar spectrum. Either of the two lines can be rendered the more prominent of the pair at will, by introducing iron or magnesium into the crucible. The less refrangible line of the pair is thus seen to be due to iron, the more refrangible to magnesium. Comparison of the solar line and the spark between magnesium points confirms this conclusion, that the magnesium line is the more refrangible of the two.

In the ultra-violet part of the spectrum photographs show several new lines. First, a triplet of lines above U at wave-lengths about 2942, 2938·5, 2937. These lines are a little below a pair of lines given by the spark for which Cornu has found the wave-lengths 2934·9, 2926·7. The latter pair are not seen at all in photographs of the arc, nor the former three in those of the spark. The strong line, wave-length about 2850, is always seen, very frequently reversed. Of the quadruple group in the spark to which Cornu has assigned the wave-lengths 2801·3, 2797·1, 2794·5, and 2789·9, the first and third are strongly developed in the arc, the other two hardly at all. Next follows a set of five nearly equidistant lines, well-defined and strong, but much less strong than the two previously mentioned, wave-lengths about 2782·2, 2780·7, 2779·5, 2778·2, 2776·9. The middle line is a little stronger than the others. The same lines come out in the spark.

Beyond these follow a series of pairs and triplets; probably they are triplets in every case; but the third, most refrangible, line of the triplets is the weakest, and has not in every case been noticed as yet. These succeed one another at decreasing intervals with diminishing strength, and are alternately sharp and diffuse, the diffuse triplets being the strongest. The positions are shown in Pl. III., Fig. 2. The series resembles in general character the sodium and the potassium series described by us in a former communication, and we cannot resist the inference that they must be harmonically related, though they do not follow a simple harmonic law. The most refrangible line in the figure at wave-length 2605 represents a faint diffuse band which is not resolvable into lines; it belongs, no doubt, to the diffuse members of the series, and, to complete the series, there should be another sharp group between it and the line at wave-length 2630. This belonging to the weaker members of the series is too weak to be seen.

It is worthy of remark that the line at wave-length 5710, described by us in a previous communication,[*] is very nearly the octave of the strong line at 2850. Moreover, the measures we have taken of the wave-length of this last line, with a Rutherford grating of 17,296

---

[*] 'Proc. Roy. Soc.' No. 200, p. 98.

lines to the inch, indicate a wave-length 2852 nearly, which is still closer to the half of 5710.

When metallic magnesium is dropped into a crucible of magnesia or lime through which the arc is passing, the electric current seems sometimes to be conducted chiefly or entirely by the vaporised metal, so that the lines of other metals almost or wholly disappear; but the line at wave-length 3278 does not in such cases appear, though the other magnesium lines are very strongly developed. The line at wave-length 2850 is often, under such circumstances, enormously expanded and reversed, those at wave-lengths 2801, 2794, and the alternate diffuse triplets, including those near L and near S, much expanded and reversed, and the group of five lines (2776–2782) sometimes reversed.

When the arc of a Siemens machine is taken in a magnesia crucible, the strong line of the flame spectrum, wave-length 4570, is well seen sharply defined; it comes out strongly and a little expanded on dropping in a fragment of magnesium. When a gentle stream of hydrogen is led in through a hollow pole, this line is frequently reversed as a sharp black line on a continuous background. From comparing the position of this line with those of the titanium lines in its neighbourhood, produced by putting some titanic oxide into the crucible, we have little doubt that it is identical with the solar line $4570 \cdot 9$ of Angström.

When the arc is taken in a crucible into which the air has access, it may be assumed that the atmosphere about the arc is a mixture of nitrogen and carbonic oxide. When a stream of hydrogen is passed, either through a perforated pole or by a separate opening, into the crucible, the general effect is to shorten the length to which the arc can be drawn out, increase the relative intensity of the continuous spectrum, and diminish the intensity of the metallic lines. Thus, with a very gentle stream of hydrogen in the magnesia crucible, most of the metallic lines, except the strongest and those of magnesium, disappear. Those lines which remain are sometimes reversed; those at wave-length 2850 and the triplet near L being always so. With a stronger stream the lines of magnesium also disappear, the b triplet being the last in that neighbourhood to go, and b_1 and b_2 remaining after b_4 had disappeared.

Chlorine seems to have an opposite effect to hydrogen, generally intensifying the metallic lines, at least those of the less volatile metals, but it does not sensibly affect the spectrum of magnesium. Nitrous oxide produces no marked effect; coal-gas acts much as hydrogen.

Spectrum of the Magnesium Spark, in Gases under High Pressures.

In the spark of an induction coil taken between magnesium points in air we get all the lines seen in the arc except two blue lines at wave-lengths 4350 and 4166, three lines above U, and the series of triplets more refrangible than the quintuple group about wave-length 2780. The blue line wave-length about 4570 is best seen in the spark without a jar when the magnesium electrodes are close together, and the rheotome made to work slowly; and this and the other faint lines of the spark at about 4586 and 4808 require for their detection a spectroscope in which the loss of light is small.

On the other hand, some additional lines are seen. Of these, the strong line at wave-length 4481 and the weaker line at 4586 are well known. Another faint line in the blue at wave-length 4808 has been observed by us in the spark, and two diffuse pairs between H and the triplet near L. Two ultra-violet lines at wave-lengths $2934 \cdot 9$, $2926 \cdot 7$ (Cornu) are near, but not identical, with two lines of the arc above mentioned; and two more lines at wave-lengths $2797 \cdot 1$, $2789 \cdot 9$ (Cornu) make a quadruple group with the very strong pair conspicuous in the arc in this region. The spectrum of the spark ends, so far as we have observed, with the quintuple group (2782-2776) already described in the arc.

When a Leyden jar is used with the coil, some of the lines are reversed. This is notably the case with the triplet near L, the line at wave-length 2850, and those at 2801 and 2794. Cornu* noticed the reversal of the two less refrangible lines of the triplet near L under these circumstances. This effect is very much increased by increasing the pressure of the gas in which the spark is taken. The Cailletet pump is well suited for such experiments. The gases used were hydrogen, nitrogen, and carbonic oxide; and the image of the spark was thrown on to the slit of the spectroscope by a lens. In hydrogen, when no Leyden jar was used, the brightness of the yellow and of the blue lines of magnesium, except at first that at wave-length 4570, diminished as the pressure increased; while, on the other hand, the b group was decidedly stronger at the higher pressure. The pressure was carried up to 20 atmospheres, and then the magnesium lines in the blue and below almost or entirely disappear, leaving only the b group very bright, and the magnesium-hydrogen bands which are described below; even the hydrogen lines F and C were not visible. When a jar was used, the magnesium lines expanded as the pressure was increased; all three lines of the b group were expanded and reversed at a pressure of 5 atmospheres; the yellow line, wave-length 5528, was also expanded but not reversed; and the line at 4481 became a broad, very diffuse band, but the line at wave-length 4570 was but very little expanded. The expansion both of the b group and of the yellow line seemed to be greater on the less refrangible than on the more refrangible side of each line, so that the black line in those which were reversed was not in the middle. When the jar was used, the pressure could not be carried beyond 10 or $12\frac{1}{2}$ atmospheres, as the resistance became then so great that the spark would not pass across the small distance of about 1 millim. between the electrodes.

* 'Compt. Rend.' 1871.

At a pressure of 2½ atmospheres, with a jar, the ultra-violet magnesium triplet near L was very well reversed, and the two pairs of lines on its less refrangible side (shown in Plate III., Fig. 3) were expanded into two diffuse bands.

In nitrogen and in carbonic oxide the general effects of increased pressure on the magnesium lines (not the magnesium-hydrogen bands) seemed to be much the same as in hydrogen. Without a jar the blue and yellow lines were enfeebled, and at the higher pressures disappeared, while the b group was very brilliant but not much expanded. With the jar all the lines were expanded, and all three lines of the b group strongly reversed. The bands of the oxide (wavelength 4930–5000) were not seen at all in hydrogen or nitrogen; they were seen at first in carbonic oxide, but not after the sparking had been continued for some time.

The disappearance of certain lines at increased pressure is in harmony with the observations of Cazin,* who noticed that the banded spectrum of nitrogen, and also the lines, grew fainter as the pressure was increased, and finally disappeared. When a Leyden jar is employed there is a very great increase in the amount of matter volatilised by the spark from the electrodes, as is shown by the very rapid blackening of the sides of the tube with the deposited metal, and this increase in the amount of metallic vapour may reasonably be supposed to affect the character of the discharge, and conduce to the widening of the lines and the reversal of some of them. Without a jar the amount of matter carried off the electrode also doubtless increases with the pressure and consequent resistance, and may be the cause of the weakening, as Cazin suggests, of the lines of the gas in which the discharge is passed. It is to be noted, moreover, that the disappearance of the hydrogen lines depends, in some degree, on the nearness of the electrodes. The lines C and F which were, as above stated, sometimes invisible in the spark when the electrodes were near, became visible, under circumstances otherwise similar, when the magnesium points had become worn away by the discharge.

Comparison of the Spectra.

When we compare the spectra of magnesium in the flame, arc, and spark, we observe that the most persistent line is that at wave-length 2850, which is also the strongest in the flame and arc, and one of the strongest in the spark. The intensity of the radiation of magnesium at this wave-length is witnessed by the fact that this line is always reversed in the flame as well as in the arc when metallic magnesium is introduced into it, and in the spark between magnesium electrodes when a Leyden jar is used. It is equally remarkable for its power of expansion. In the flame it is a broad band, and equally so in the arc when magnesium is freshly introduced, but fines down to a narrow line as the metal evaporates.

---

* 'Phil. Mag.' 1877, vol. iv. p. 154.

Almost equal in persistence are the series of triplets. Only the least refrangible pair of these triplets are seen in the flame, another pair are seen in the spark, but the complete series is only seen in the arc. We regard the triplets as a series of harmonics, and to account for the whole series being seen only in the arc we must look to some other cause than the temperature. This will probably be found in the greater mass of the incandescent matter contained in the crucible in which the arc was observed.

The blue line of the flame at wave-length 4570 is well seen in the arc, and is easily reversed, but is always a sharp line, increased in brightness but not sensibly expanded by putting magnesium into the crucible. In the spark, at atmospheric pressure, it is only seen close to the pole or crossing the field in occasional flashes; but seems to come out more decidedly at rather higher pressures, at least in hydrogen.

The series of bands near L, well developed in the flame, but not seen at all in the arc or spark, look very much like the spectrum of a compound, but we have not been able to trace them to any particular combination. Sparks in air, nitrogen, and hydrogen have alike failed to produce them. The very strong, rather diffuse triplet at M, with which they end, so closely resembles in general character the other magnesium triplets, that it may well be connected with that constitution of the magnesian particle which gives rise to the triple sets of vibrations in other cases, but, if so, its presence in the flame alone is not easily explained.

The occurrence of this triplet in the ultra-violet, and of the remarkable series of bands associated with it, as well as the extraordinary intensity of the still more refrangible line at wave-length 2850, which is strongly reversed in the spectrum of the flame, corroborates what the discovery of the ultra-violet spectrum of water had revealed, that at the temperature of flame substances while giving in the less refrangible part of the spectrum more or less continuous radiation, may still give, in the regions of shorter wave-length, highly discontinuous spectra, such as have formerly been deemed characteristic of the highest temperatures. This subject we will not discuss further at present, but simply remark that "it opens up questions as to the emissive power for radiation of short wave-lengths of gaseous bodies at the comparatively low temperature of flame with regard to which we are accumulating facts."

In the arc and spark, but not in the flame, we have next a very striking group of two very strong lines at wave-lengths about 2801 and 2794, and a quintuple group of strong but sharp lines above them. The former are usually reversed in the spark with jar, and all are reversed in the arc when much magnesium is present. There are also several single lines in the visible part of the spectrum common to the arc and spark. All of these may be lines developed by the high temperature of the arc and spark. Two blue lines in the arc have not been traced in the spark, but their non-appearance may be due to the same cause as that above suggested for the non-appearance of the higher triplets, the smallness of the incandescent mass in the spark.

A triplet of lines in the arc near U appear to be represented in the spark by an equally strong, or stronger, pair near but not identical in position. The possibility of such a shift, affecting these two lines only in the whole spectrum and affecting them unequally, must in the present state of our knowledge be very much a matter of speculation. Perhaps sufficient attention has not hitherto been directed to the probability of vibrations being set up directly by the electric discharge independently of the secondary action of elevation of temperature. Some of the observations above described, and many others well known, indicate a selective action by which an electric discharge lights up certain kinds of matter in its path to the exclusion of others; and it is possible that in the case of vibrations which are not those most easily assumed by the particles of magnesium, the character of the impulse may slightly affect the period of vibration. The fact that, so far as observations go, the shift in the case of this pair of magnesium lines is definite and constant, militates against the supposition suggested. On the other hand, the ghost-like pairs of lines observed in the spark below the triplet near L, suggest the idea that some of the particles have their tones flattened by some such cause.

The strong pair at wave-lengths 2801, 2794, are accompanied in the spark, but not usually in the arc, by a much feebler, slightly more refrangible pair, but these have not the diffuse ghost-like character of those just alluded to.

These lines are phenomena of the high potential discharge in which particles are torn off the electrodes with great violence and may well be thrown into a state of vibration which they will not assume by mere elevation of temperature.

There are two lines in the spark besides the well-known line at wave-length 4481 which have not been observed in the arc, but they are feeble and would be insignificant if it were not the fact that they, as well as wave-length 4481, all short lines seen generally only about the poles, appear to be present in the solar spectrum. In the sun we seem to have all the lines common to the flame, arc, and spark, and possibly the strong triplet of the flame at M. We have noticed that when the spark is taken in hydrogen, the line at wave-length 4570 appears stronger than that at wave-length 4703, while the reverse is the case when the atmosphere is nitrogen. It is possible then that the atmosphere may, besides the resistance it offers to the discharge, in some degree affect the vibrations of the metallic particles.

The substantive result of the investigation is to prove that the chemical atoms of magnesium are capable of taking up a great variety of vibrations, and by mutual action on each other, or on particles of matter of other kinds, give rise to a great variety of vibrations of the luminiferous ether; and to trace satisfactorily the precise connection between the occurrence of the various vibrations and the circumstances under which they occur, will require an extended series of observations.

On the Spectrum of Water.

In our observations "On the Spectrum of the Compounds of Carbon," we noticed that a remarkable series of lines, extending over the region between the lines S and R of the solar spectrum, were developed in the flame of coal-gas burning in oxygen.* The arrangement of lines and bands, of which this spectrum consists, is shown in the Pl. II., Fig. 3. It begins at the more refrangible end with two strong bands, with wave-lengths about 3062, 3068, and extends up to about the wave-length 3210. It is well developed in the flame of hydrogen as well as of hydrocarbons, burning in oxygen, and less strongly in the flames of non-hydrogenous gases, such as carbonic oxide and cyanogen, if burnt in moist oxygen. The same spectrum is given by the electric spark taken, without condenser, in moist hydrogen, oxygen, nitrogen, and carbonic acid gas, but it disappears if the gas and apparatus be thoroughly dried. We are led to the conclusion that the spectrum is that of water. The plate, Fig. 3, is a general view of this spectrum. It was necessary to pass a current of dry gas for fully an hour through the warmed sparking apparatus before the moisture was sufficiently absorbed by the dehydrating agents. When this was done, photographs of the spark showed either no trace, or only the faintest traces, of the spectrum above described. On introducing a drop of water, and letting it spread over a plug of asbestos placed in the current of gas, the spectrum above described at once imprinted itself on the photographic plate. The effect was the same, whether the gas used was hydrogen, oxygen, nitrogen, or carbonic acid. In the case of nitrogen, some of the channelled bands due to that gas overlap the water spectrum, and partly obscure it, but not so much but that it can be still very distinctly recognised. When a condenser is used, the water spectrum disappears. The same spectrum appears in the De Meritens arc, but is less fully developed. The spectrum we have figured does not by any means exhaust the ultra-violet spectra of the flames we have observed. In writing of this and other spectra which we have traced to compounds, we abstain from speculating upon the particular molecular condition or stage of combination of decomposition, which may give rise to such spectra. The fact of an ultra-violet spectrum of water occurring in spectra of flames opens up

---

* This we recorded in a Note of date June 8, 1880, see 'Proc. Roy. Soc.' No. 205, p. 5. Dr. Huggins discovered the same spectrum independently, and communicated the same on June 16, 1880. Our paper on this special spectrum bears date 17th June. Both papers were read at the same meeting of the Society.

questions as to the emissive power for radiation of short wave-lengths of gaseous substances at comparatively low temperatures.

Such facts completely modify the inferences which have been drawn as to the continuity of flame spectra and the character of the specific absorption of the vapour of water.

Identity of Spectral Lines.

In Kirchhoff's 'Researches on the Spectra of the Chemical Elements,' p. 10, the following reference is made to the apparent identity of wave-length of some spectral lines.

"If we compare the spectra of the different metals with each other several of the bright lines appear to coincide. This is especially noticeable in the case of an iron and magnesium line at $1655\cdot6$ (b_4), and with an iron line and calcium line at $1522\cdot7$ (E). It seems to me to be a question of great interest to determine, whether these and other similar coincidences are real or only apparent; whether the lines in question actually fall one upon the other, or whether they lie very close together. I believe that my method of observation does not possess the requisite accuracy for the purpose of answering this question with any degree of probability, and I think that a large number of prisms and an increased intensity of light will prove necessary."*

The subsequent investigations of Ångström and Thalén increased the number of apparent coincidences amongst the spectral lines of different elements.

The question of the identity of spectral lines exhibited by different elements is one of great interest, because it is very improbable that any single molecule should be capable of taking up all the immense variety of vibrations indicated by the complex spectrum of iron or that of titanium, and it might therefore be expected that such substances consist of heterogeneous molecules, and that some molecules of the same kind as occur in these metals should occur in more than one of the supposed elements. Further, the supposed identity of certain lines in the spectra of more than one element has been made by Mr. Lockyer the ground of an argument in support of a theory as to the dissociation of chemical elements into still simpler constituents, and in reference to this he wrote: † "The 'basic' lines recorded by Thalén will require special study, with a view to determine whether their existence in different spectra can be explained or not on the supposition that they represent the vibrations of forms, which, at an early stage of the planet's history, entered into combination with other forms, differing in proximate origin, to produce different 'elements.'"

Young, on examining with a spectroscope of high dispersion the

---

* 'Researches on the Spectra of the Chemical Elements,' by G. Kirchhoff, p. 10.   1862.
† 'Proc. Roy. Soc.' vol. xxx. p. 31.

70 lines given in Angström's map as common to two or more substances, has found that 56 are double or treble, 7 more doubtful, and only 7 appear definitely single, and he remarks:* "The complete investigation of the matter requires that the bright line spectra of the metals in question should be confronted with each other and with the solar spectrum under enormous dispersive power, in order that we may determine which of the components of each double line belongs to one and which to the other element." It is this confronting of the bright line spectra of some of the terrestrial elements which we have attempted, and of which we now give an account. For the dispersion we have used a reflecting grating similar to that used by Young, with 17,296 lines to the inch, and a ruled surface of about $3\frac{1}{4}$ square inches; telescope and collimator, each with an aperture of $1\frac{1}{2}$ inch and focal length 18 inches, the lenses being of quartz, cut perpendicularly to the axis and unachromatised, giving a very good definition with monochromatic light. The chromatic aberration is in this case an advantage, for when the telescope is in focus for lines in the spectrum of any given order, the overlapping parts of spectra of different orders are out of focus, and their brightness consequently more or less enfeebled. We have sometimes used green or blue glasses to enhance this result. The telescope and the collimator were generally fixed at about 45°, the collimator being more nearly normal to the grating than the telescope, and the grating moved to bring in successive parts of the spectra. For the parts of the spectra less refrangible than the Fraunhöfer line E the spectrum of the third order was employed, for the more refrangible rays that of the fourth order. The source of light was the electric arc taken in a crucible of magnesia or lime, the image of the arc being focussed on the slit; and, for the examination of any supposed coincidence, first one metal was introduced into the crucible, and the line to be observed placed on the pointer of the eye-piece; the second metal was then introduced, and then in most cases, as detailed below, *two* lines were seen where only one was visible before, and the pointer indicated which of the two belonged to the metal first introduced. In some cases where both metals were already in the crucible, we had to reinforce the spectrum of one of the metals by the introduction of more of that metal, which generally brought out the spectrum of that metal more markedly than the other, and enabled us to distinguish the lines with a high degree of probability. Thus the crucibles of magnesia, or the carbons, always contain sufficient lithium to show the orange line and the calcium line heretofore supposed coincident with it (wave-length 6101·9), but we observed these lines quite distinct and separated by a distance, estimated by the eye in comparison with the distance of neighbouring titanium lines, at about one division of Angström's scale. On dropping a minute piece of lithium carbonate into the crucible, the less refrangible line was seen to expand and

---

* American Journal of Science,' vol. xx. p. 353.

for a short time to be reversed, the other line remaining narrow and quite unaltered. When the lithium had evaporated, and both lines were again narrow, a small piece of Iceland spar was dropped into the crucible, which immediately caused the expansion, and on one occasion the reversal, of the more refrangible line, while now the less refrangible line was unaffected.

In this way we satisfied ourselves that the calcium line is the more refrangible of the two, and is probably represented by the line at wave-length 6101·9 in Angström's normal solar spectrum, while the lithium line appears to be unrepresented.

In the case of iron, which gives such a multitude of lines, it was à priori probable that some lines would be coincident, or nearly so, with lines of other elements; and in fact we find that in five-sixths of the supposed coincidences lines of iron are involved. We have, therefore, chiefly directed our attention to iron lines. A complete account of the separate resolutions will be found in the 'Proceedings of the Royal Society,' May, 1881.

Pl. II., Fig. 5, shows the appearance of the magnesium group of the solar spectrum as observed in spectroscopes used by different observers. The lines marked b^3 and b^4, which appear to be single lines in the maps of Angström and Kirchhoff, are resolved into double lines by the greater dispersion employed by Thollon. The following table shows the relative dispersion and number of lines seen by different observers when powerful instruments are directed to the same solar group:—

GROUP E OF SOLAR SYSTEM.

	Number of Lines.	Dispersion.
Angström	11	800
Kirchhoff	12	1400
Pickering	29	2000
Young	36	2720

The indium line 4101·2 we found very difficult to separate from the hydrogen line (h), as the latter had to be observed from a tube with a spark, and it is both faint and diffuse; but several observations all led to the conclusion that the indium line is very slightly less refrangible than that of hydrogen.

We have also directly compared the iron line at 5316·07 with the solar spectrum, and found that the iron line corresponds with the less refrangible of the two solar lines at this place, so that the chromospheric line is in all probability the other line of the pair.

There are still a few cases of supposed coincidences which we have not examined. The results which we have recorded strongly confirm Young's observations, and leave, we think, little doubt that the few as yet unresolved coincidences either will yield to a higher dispersion, or are merely accidental. It would indeed be strange if, amongst all the variety of chemical elements and the still greater variety of vibrations which some of them are capable of taking up, there were no two which could take up vibrations of the same period. We certainly should have supposed that substances like iron and titanium, with such a large number of lines, must each consist of more than one kind of molecule, and that not single lines, but several lines of each, would be found repeated in the spectra of some other chemical elements. The fact that hardly single coincidences can be established is a strong argument that the materials of iron and titanium, even if they be not homogeneous, are still different from those of other chemical elements. The supposition that the different elements may be resolved into simple constituents, or into a single one, has long been a favourite speculation with chemists; but however probable this hypothesis may appear *à priori*, it must be acknowledged that the facts derived from the most powerful method of analytical investigation yet devised give it scant support.

[J. D.]

PLATE I.

Fig. 1.

Fig. 2.

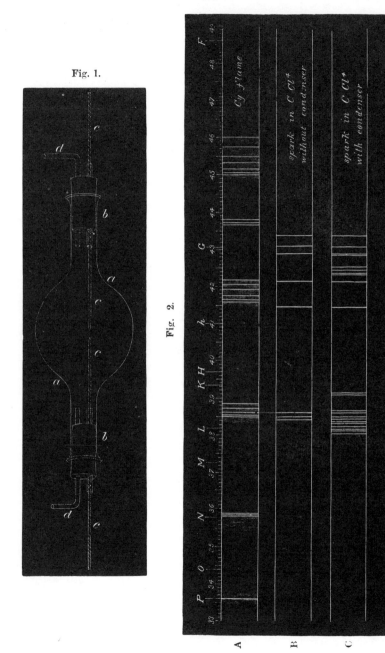

PLATE II.

Fig. 3.

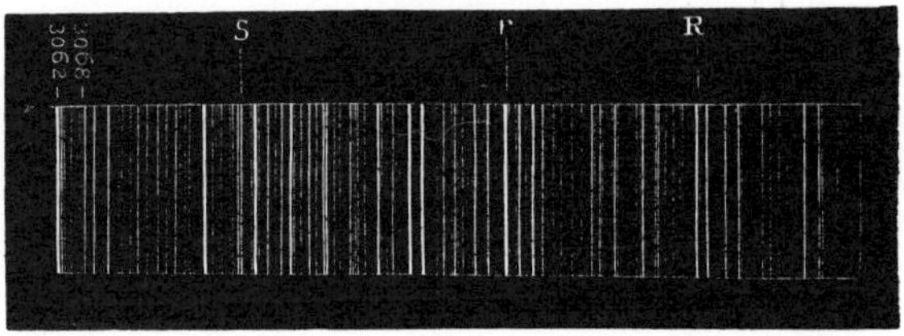

Fig. 5.

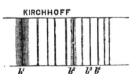

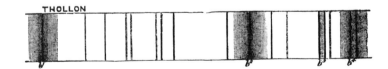

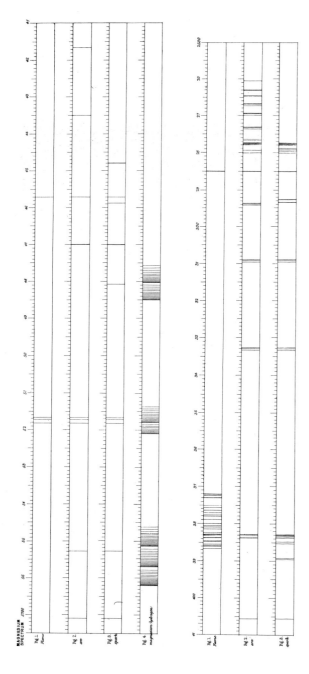

Plate III.

Friday, March 17, 1882.

GEORGE BUSK, Esq. F.R.S. Treasurer and Vice-President,
in the Chair.

CAPTAIN W. DE W. ABNEY, R.E. F.R.S.

Spectrum Analysis in the infra red of the Spectrum.

AT the Royal Institution it would be almost an impertinence on my part were I to attempt to prove the existence of dark rays which lie below the red of the spectrum: Professor Tyndall has made you all so thoroughly acquainted with these dark rays that I may assume not only your acquaintance with them, but also your interest in them. The most accessible means hitherto of exploring this region of the spectrum has been by a study of its heating effect, as shown by the thermopile; and more particularly when an integration of the effect produced by its component rays is required. Beautifully delicate is the thermopile, but it must not be forgotten that it depends for its delicacy largely on the area of its face. A little consideration will show that if you wish to examine the spectrum by its means, you must be prepared for a drawback.

In the thermopile I have here, the elements composing it are arranged in a line, and I can allow any fineness of beam to penetrate to its surface by means of this adjustable slit. Unfortunately, however, the narrower the slit the smaller is the heating effect on the thermopile; so that, if I wish to measure the heating effect of a very narrow beam of radiation which may find its way between the jaws of the slit, the heating effect will be so small that it may escape detection. In the visible solar spectrum, as you are all aware, there are breaks of continuity presenting to the eye the appearance of fine dark lines, and showing a diminished radiation. Now, suppose for an instant that the part of the spectrum which is visible to us was dark, not exciting vision, it is manifest, even were the energy of this part of the spectrum very much greater than it is, that to ascertain the existence of these fine lines by means of the thermopile would be next to impossible; since the slit would have to be closed to an excessive degree of fineness, and when so closed the thermopile would be insensible to radiations when such existed. Now this is precisely the case with which we have to deal in the spectrum below the red. We know of its existence; but with any source of radiation, such as the sun for instance, the thermopile would stand but a small chance of finding any narrow breaks in its continuity. I need only refer to Lamansky's thermogram (Fig. 1) of the solar spectrum, taken by a linear thermopile, in which, after taking extraordinary precautions, he found the existence of three breaks. The thermopile, with its galvanometer, takes no cognizance of time; given a constant and steady flux of radiation striking it during a certain time, the electrical current generated, which is shown by the deflection of the galvanometer needle, remains constant, and no increase of time gives a greater deflection. If we could imagine a thermopile the current generated by which, by the efflux of time, would proportionally increase the deflection of the needle, then the most limited radiation

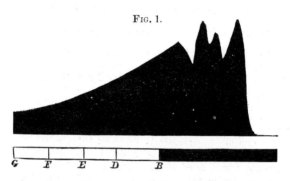

FIG. 1.

striking the pile could be measured and calculated. At present there seems to be no method capable of taking account of time for this purpose except photography, and until recently it seemed chimerical to apply it. I should like to show you how photography takes cognizance of time, and also why it seemed unsuitable for investigating the dark rays below the red end of the spectrum. I propose to photograph the spectrum on an ordinary photographic compound.

A piece of paper has been coated with silver bromide, and this compound, when placed in the spectrum, is acted upon by the ultra violet, the violet and the blue rays. If I cover up $\frac{2}{3}$ of the slit of the lantern and allow an exposure of the paper to the spectrum of two seconds, then with the next $\frac{1}{3}$ of the slit an exposure of ten seconds, and with the remaining $\frac{1}{3}$ give an exposure of thirty seconds, it will be seen when I develop the image that the length of the spectrum varies, and also the darkening of the different portions. Thus the longest exposure will show the greatest intensity of action and the greatest length of spectrum. [Shown.] This experiment serves a double purpose, for whilst it makes it clear that photography takes cognizance of time, yet it seemingly shows that it is unfitted for exploring the infra-red region, since ordinary photographic compounds are unacted upon by them. Could a compound be found which was sensitive to the dark rays, it is manifest that the battle would be won, and that investigations full of interest might be the outcome.

Some eight years ago I tried my hand at the matter, and after several years of experimenting it was my good fortune to find a compound which was chemically acted upon by the dark radiations.

I will not weary you with the various experiments undertaken; suffice it to say that silver bromide was selected as the salt to work upon. My aim was to prepare an emulsion of bromide of silver in collodion (an emulsion being silver bromide in a fine state of division suspended in collodion) which should transmit green-blue light. Let me show you why.

The spectrum on the screen, when unabsorbed by any medium, shows every colour. If, however, a piece of green glass is inserted before the slit, it is seen at once that the violet is absorbed and also the lowest part of the red. Inferentially it may be supposed that the infra-red rays are also absorbed. Orange is the usual colour of the silver bromide, and a piece of orange glass placed in front of the slit cuts off from the spectrum all the most refrangible part of the spectrum and none of the least refrangible.

On the principle of conservation of energy, where radiation is absorbed, there work must be done by the absorbing body and show itself as heat or chemical action. Heat with the thermopile, for instance, and chemical action with the salt of silver. Thus, with the orange bromide we should expect, as we have already seen is the case, that the violet and blue rays would do work on it whilst the other rays would be passive.

The green state was attained after much labour. The colour of this new preparation of silver bromide, and that of the old, are now shown by means of the lantern on the screen.

Now you will see that if the work done in the green bromide was chemical decomposition, the problem was solved, and that the unknown might be made to write down, in hieroglyphics perhaps, but still in a manner capable of being deciphered, its character and peculiarities.

I will endeavour to experimentally illustrate that the green compound *is* acted upon by the dark rays. It will be in your recollection that Professor Graham Bell's recently introduced photophone is in reality an instrument consisting of a perforated disc rotating in front of a source of radiation, and by means of a lens the radiant energy is focussed on the surface of a selenium cell, to which a telephone is attached, and that by this means a musical sound is produced in the telephone.

Professor Bell showed that the same effect was produced when a piece of ebonite was introduced between the source of light and the selenium cell. Dr. Huggins proposed to me that I should try the permeability of the ebonite by the dark rays; and this was done, with the result that the spectrum was taken through it, showing an impression on the green bromide of the dark rays. An image of the incandescent carbon points of the electric light are now formed on a piece of ebonite, and behind it is a glass plate covered with the bromide; an exposure of twenty seconds will suffice to impress the image of the points by their dark rays. [The image was developed and subsequently shown.] It will be seen that the bromide in this state is somewhat sluggish to respond to the vibrations of the dark rays. I will now make an experiment to show how different is the behaviour of the orange bromide. Behind this rotating disc,

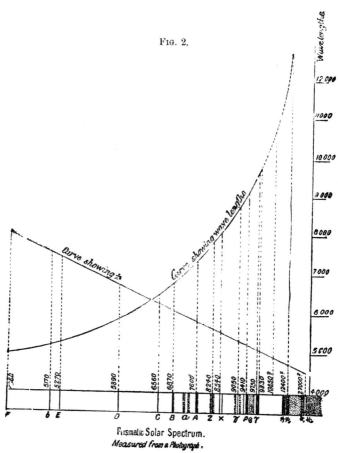

FIG. 2.

Prismatic Solar Spectrum.
Measured from a Photograph.

which is made up of alternate transparent and opaque sectors, is a plate prepared with the orange bromide. A spark $\frac{1}{5}$ of an inch in length from a battery of Leyden jars is sufficient to impress a sharp image of the sectors on the plate though they are rapidly rotating. [The exposure was made to the spark whilst the disc was rotating; and developed before the audience and subsequently the photograph was shown.] The exposure is estimated by Cazin as $\frac{47}{1000000}$ of a second. It would require twenty such sparks to impress the red end of the spectrum on a pure bromide plate. I should wish to show you one more remarkable example of the action of the dark rays on the blue-green silver bromide. On the screen we have a slide showing certain opaque discs and triangles. They were produced in the following manner: A card was perforated with such discs and triangles, and placed $\frac{1}{8}$ of an inch above a green bromide plate; above this was suspended a kettle of boiling water, and the radiation from the kettle acted on the plate through these holes, with the result that, after considerable exposure, an image of the holes was developed on the bromide below. This shows that this particular form of silver salt responds to waves of very low refrangibility.

The first application of the new compound was to the solar spectrum, and on the screen we have the first impression of the infra-red region ever taken. On the diagram (Fig. 2) there are bands ϕ and ψ drawn which do not appear in the photograph; only on two occasions have they been impressed, for reasons which will be explained. To show you how far our knowledge of this region is extended, a photograph is shown on the screen of the spectrum obtained photographically by Draper, which he obtained by indirect means.

FIG. 3.

When a grating of large dispersion replaces the prism, the bands are broken up into lines; and very beautiful lines they are in some cases. From such photographs a wave-length map was made. [Shown]. The line of greatest wave-length impressed in the spectrum is 22,000. Now the visible part of the spectrum extends from $\lambda\, 3800$ to $\lambda\, 7600$; thus the invisible spectrum, as photographed, is *five times* longer than the visible spectrum.

In the visible portion of the solar spectrum, most of the lines have been traced to the absorption of different metallic or other vapours existent in the solar or our own atmosphere. The cause of the absorptions in the invisible part of the spectrum are as yet untraced, except in one or two instances to which I shall have to allude presently. With the exception of sodium and calcium, no metallic vapours seem to have what would be bright lines, were they visible, in the infra-red portion of the spectrum; hence we are almost bound to suppose that absorption lines in this region are really due to compound bodies of some description. Knowing the results that Professor Tyndall had got with the hydrocarbon and other vapours and liquids in the infra-red region by thermopile integration, Colonel Festing and myself determined to see if we could disintegrate Professor Tyndall's integrations, and locate in the invisible spectrum the absorptions which he had noted.

We commenced with water, and were delighted to find that water gave a very definite spectrum; and I propose to show the method adopted for this research. In front of the slit of the spectroscope, which has three prisms, was placed a tube of water or other liquid, in some cases of the length of two feet, but more generally of six inches. The crater or bright luminous patch from the positive pole of the electric light was projected on the slit, the rays having to traverse the liquid. The image of the spectrum was then received on a sensitive plate. In this manner I propose to take the spectrum of a two-foot length of water. [The photograph was taken and subsequently shown on the screen.] Before proceeding further, I will again show you the superior sensitiveness of the orange form of bromide for blue rays over the green bromide for the dark rays. The same length of spark as before shall be used, and the light from it projected by means of a lens through a couple of prisms, and the image be focussed on an orange bromide film. The spark passes, and the most refrangible end of the spectrum will be found to be impressed. [The photograph was developed before the audience, and at the close of the lecture thrown upon the screen.] Our next attempts were with alcohol and ether, and in these we got many definite absorptions, some lined and some banded, the bands being more or less shaded. On trying ethyl iodide, however, we came upon a spectrum which was composed of fine lines and bands with comparatively sharp edges, differing in this respect from the two former spectra. The difference in composition between alcohol and ethyl iodide is shown in the following diagram.

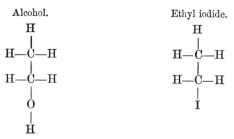

The prime difference is the presence of oxygen in the one and its absence in the other. We next tried methyl iodide, and got a simpler form of spectrum than the ethyl iodide. It now struck us that if we could diminish the hydrogen we might have something again different.

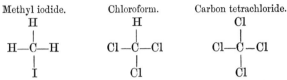

|Methyl iodide.|Chloroform.|Carbon tetrachloride.|

We therefore tried chloroform, and much to our surprise all the broad bands had vanished, and we had a linear spectrum. What would happen if we removed the last hydrogen? In carbon tetrachloride the last hydrogen is removed, and we got no absorption spectrum at all; indicating that hydrogen had a most important bearing on the absorptions. Carbon disulphide and cyanogen gave us the same results. On the other hand, when we spectroscoped hydrochloric acid we had a linear spectrum, as we had with ammonia, sulphuric acid, and nitric acid. Even water was found not to be free from lines, as the boundary of each band was a line. I think, then, that this satisfactorily settles the point that hydrogen gives the initiative to all the special absorptions we noticed. The introduction of oxygen gives shaded bands; but on measurement it was found that shades were made up of step by step absorptions between two or more positions of hydrogen lines. Again, what is called the radical of each group of compounds was found to have a definite absorption in a definite locality; hence this spectroscopic method became a means of qualitatively determining the composition of an unknown compound and its molecular structure. I would point out also how the absorptions found by the photographic method go hand in hand with those found by Professor Tyndall. In the annexed table we have the value of the absorptions found by him, and following the absorption spectra of the same bodies through six inches of liquid. The coincidence is remarkable and worthy of attention.

ABSORPTION OF HEAT BY LIQUIDS.

(*Source of Heat a Platinum Spiral raised to Bright Redness by a Voltaic Current.*)

Liquid.	Thickness in Parts of an Inch.	
	0·02.	0·27.
Carbon disulphide	5·5	17·3
Chloroform	16·6	44·8
Methyl iodide	36·1	68·6
Ethyl iodide	38·2	71·5
Benzine	43·4	73·6
Amylene	58·3	82·3
Ether	63·3	85·2
Alcohol	67·3	89·1
Water	80·7	91·0

Your attention should be drawn more especially to the water spectrum, in which it will be seen what a large proportion of the infra red is cut off by even a small thickness. Aqueous vapour absorbs in the same locality as the water; hence you will see

FIG. 4.

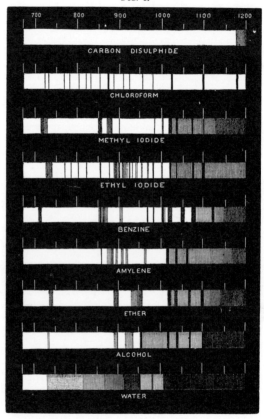

how it is that I have photographed the bands ϕ and ψ in the solar spectrum but seldom (see Fig. 2). When they were impressed a very dry and biting north-east wind was blowing, which enabled this part of the spectrum to find its way through our atmosphere.

On comparing the solar spectrum with the absorption spectra of the above organic bodies, we found that the principal lines of the benzine and ethyl series found a place in the solar absorptions; and for reasons which I have not time to enter into now, we were induced to locate these bodies outside our atmosphere. In what part of space they exist is a moot point, but there is no doubt that they are somewhere present in it. That alcohol is to be found in the sun would be perhaps to stretch a point too far; and it would be unwise to wish to find it there, as it might rouse the animosity of a small section of the community against this branch of spectrum analysis. The ingredients to make it are there, however, without any doubt.

In regard to these same class of spectra it is interesting to find that there is a marked difference in bodies containing the same relative proportions of carbon, hydrogen, and oxygen, but molecularly different. Take for example aldehyde and paraldehyde (Fig. 5). A

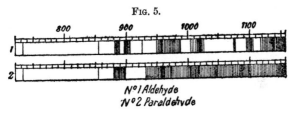

Fig. 5.

Nº 1 Aldehyde
Nº 2 Paraldehyde

molecule of the latter contains three molecules of the former, and we see that their spectra differ materially. If such be the case in organic compounds, we may surely expect to find the same difference in the spectra of the elements, if there is a different molecular grouping of them at different temperatures. Whether the changes in metallic spectra are due to this cause has still to be proved, though it seems probable that it may be so.

The latest point in our research in this subject which will be of interest to chemists appears to be the possibility of distinguishing between para- and ortho-organic compounds. Our experiments so far demonstrate that this can be done. If on further research it prove to be the case, this branch of spectrum analysis would be worthy of study by chemists for this reason alone.

[W. de W. A.]

Friday, March 31, 1882.

GEORGE BUSK, Esq. F.R.S. Treasurer and Vice-President, in the Chair.

W. SPOTTISWOODE, Esq. LL.D. Pres. R.S. *M.R.I.*

Matter and Magneto-Electric Action.

THE late Professor Clerk Maxwell, in his work on 'Electricity and Magnetism' (vol. ii. p. 146), lays down as a principle that "the mechanical force which urges a conductor carrying a current across the lines of magnetic force, acts, not on the electric current, but on the conductor which carries it. If the conductor be a rotating disk or a fluid it will move in obedience to this force, and this motion may or may not be accompanied with a change of position of the electric current which it carries. But if the current itself be free to choose any path through a fixed solid conductor or a network of wires, then, when a constant magnetic force is made to act on the system, the path of the current through the conductors is not permanently altered, but after certain transient phenomena, called induction currents, have subsided, the distribution of the current will be found to be the same as if no magnetic force were in action. The only force which acts on electric currents is electromotive force, which must be distinguished from the mechanical force which is the subject of this chapter."

In the investigation on electric discharges, on which Mr. Moulton and myself have been long engaged, we have met with some phenomena of which the principle above enunciated affords the best, if not the only, explanation. But whether they be regarded as facts arising out of that investigation, or as experimental illustrations of a principle laid down by so great a master of the subject as Professor Clerk Maxwell, I have ventured to hope that they may possess sufficient interest to form the subject of my present discourse.

The experiments to which I refer, and of which I now propose to offer a summary, depend largely upon a special method of exciting an induction coil. This method was described in two papers, published in the 'Philosophical Magazine' (November 1879) and in the 'Proceedings of the Royal Society' (vol. xxx. p. 173), respectively; but as its use appears to be still mainly confined to my own laboratory, and to that of the Royal Institution, I will, with your permission, devote a short time to a description of it, and to an exhibition of its general effects.

The method consists in connecting the primary circuit directly with a dynamo- or magneto-machine giving alternate currents. In the present case, I use one of M. de Meritens' excellent machines driven by an Otto gas engine. The speed of the de Meritens' machine, so driven, is about 1100 revolutions per minute.

In this arrangement the currents in the secondary are of course alternately in one direction and in the other, and equal in strength; so that the discharge appears to the eye, during the working of the machine, to be the same at both terminals.

The currents in the primary are also alternately in one direction and in the other, and consequently, at each alternation, their value passes through zero. But they differ from those delivered in the primary coil with a direct current and contact breaker in an important particular, namely, that while the latter, at breaking, fall suddenly from their full strength to zero, and then recommence with equal suddenness, the former undergo a gradual although very rapid change from a maximum in one direction through zero to a maximum in the opposite direction. The ordinary currents with a contact breaker would be represented by a figure of this kind, while those from the alternate machine approximately by a curve of the following form. The rise and fall of the latter are, however, sufficiently rapid to induce currents of high tension and of great quantity in the secondary.

From these considerations it follows: first, that as the machine effects its own variations in the primary current, no contact breaker is necessary; secondly, that as there is no sudden rupture of current, there is no tendency in the extra current to produce a spark or any of the inconveniences due to an abrupt opening of the circuit, and consequently that the condenser may be dispensed with; thirdly, that the variations in the primary, and consequently the strength and period of delivery of the secondary currents are perfectly regular; fourthly, that the strength of the currents in the secondary is very great. With a 26-inch coil by Apps I have obtained a spark about 7 inches in length, of the full thickness of an ordinary cedar pencil. But for a spark of thickness comparable at least with this, and of 2 inches in length, an ordinary 4-inch coil is sufficient.

Owing to the double currents, the appearance of the discharge is that of a bright point at each terminal, and a tongue of the yellow flame, such as is usually seen with thick sparks from a large coil, issuing from each. This torrent of flame (which, owing to the rapidity with which the currents are delivered by the machine, is apparently continuous) may be maintained for any length of time. The sparks resemble those given by my great coil (exhibited in this theatre on Friday, April 13th, 1877, and described in the 'Philosophical Magazine,' 1877, vol. iii. p. 30) with large battery-power and with a mercury break; but with that instrument it is doubtful whether such thick sparks could be produced at short intervals, or in a rapid shower, as in this case.

In order to contrast the effects of the two methods, I will excite the coil, first with a battery, and secondly with the alternating machine. You will notice that with the battery we can obtain either long, bright, and thin sparks, or short and comparatively thick discharges; but, unless the latter are made very short, they occur only at comparatively long and even perceptible intervals of time. On the other hand, with the alternate machine, although the method does not lend itself so readily to the production of long and bright sparks, we can produce a perfect torrent of discharges more rapid and more voluminous than by any other means yet devised. Long bright sparks can, however, be obtained by interrupting the flow of the currents from the machine, and by allowing only single currents to pass at comparatively long intervals. It may be interesting to know that the number of currents given out by the machine, and consequently the number of discharges issuing from the coil, is no less than 35,200, that is, 17,600 in each direction, per minute. The number may be determined by the pitch of the note which always accompanies the action of an alternate machine.

A comparison of the two methods may also be made when a Leyden jar is used as a secondary condenser. This application of the jar is well known as a valuable aid in spectroscopic research; and the employment of the alternating machine so materially heightens the effects that, judging from some experiments made in the presence of Mr. Lockyer, and from others of a different character in the presence of Professor Dewar, I am led to hope from it a further extension of our knowledge in this direction. In order that you may form, at all events, some rough idea of the nature of such discharges, I venture, at the risk of causing some temporary inconvenience from the noise, to project the spectrum of this spark.

I will detain you with only one more instance of comparison. The ordinary effect of an induction coil in illuminating vacuum tubes is well known. The result is usually rather unsteady. Several instruments have been devised to obviate this inconvenience, e. g. the rapid breakers described in the 'Proceedings' of the Royal Society (vol. xxiii. p. 455, and vol. xxv. p. 547), or the break called the "Trembleur" of Marcel Deprez (see 'Comptes Rendus,' 1881, I. Semestre, p. 1283). The use of the alternating machine, however, not only gives all the regularity in period, and uniformity in current, aimed at in these instruments, but also at the same time supplies currents of great strength. The result is a discharge of great brilliancy and steadiness, and it is perhaps not too much to say that the effects are comparable to those obtained with Mr. De La Rue's great chloride of silver battery. The configuration of the discharge produced in this way can also be controlled by a suitable shunt applied to the secondary circuit; for example, one formed by a column of glycerine and water, or the one consisting of a film of plumbago spread upon a slab of slate, constructed by my assistant Mr. P. Ward, and here exhibited.

One test of the strength of current passing through a tube is the amount of surface of negative terminal which it will illuminate with a bright glow. I here have a tube with terminals in the form of rings, each of which would be regarded of ample size for currents obtained in the ordinary way. These are now all connected together so as to form one grand negative terminal; and it will be found that with the currents from the alternate machine the whole system is readily illuminated at once.

It should perhaps be here remarked that, while the strength of the secondary currents passing through the tube is partly due directly to the strength of the primary currents from the machine, it is probably also in part due to the rapidity with which the secondary currents follow one another. Owing to the latter circumstance the column of gas maintains a warmer and more conductive condition than would prevail if the interval between the discharge was longer; and in consequence of this a larger portion of the discharges can make its way through than would otherwise be the case.

Before leaving the instrumental part of my discourse, I desire to bring under your notice a modification of the machine which we have thus far used for producing, by the intervention of the induction coil, currents of high tension. This consists of a machine of the same general construction as the other, but having the armatures wound with a much greater number of convolutions of much finer wire. The result is a machine giving off currents of sufficient tension to effect, by direct action, discharges through vacuum tubes, and even in air. The currents are of course alternate; but by diminishing the size of one of the terminals to a mere point, as well as by other methods described elsewhere, it is possible to shut off the currents in one direction, leaving only those in the other direction to discharge themselves through the tube. I hope on some future occasion to give a fuller account of this remarkable machine, which has only quite recently been completed.

Returning to the discharge in air, it will be noticed that when the terminals are set horizontally the torrent of thick discharges assumes the appearance of a flame, which takes the form of an inverted V. This is the result of convection currents due to the heat given off by the discharges themselves. The discharges are by their nature as it were fixed at each end, but within the limits of discharging distance free to move about and to extend themselves in space, especially in their central part. Further, it may be observed that the length of the spark which can be maintained is greater than that over which it will leap in the first instance. The explanation of this is to be sought in the fact that when the sparks follow very rapidly in succession, the whole path of each discharge remains so far in a heated state as to assist the passage of the next; and, further, that in the middle part of the discharge or apex of the Λ, where the heat is greatest, the heat prevails to such an extent as to render a portion of the path highly conductive. This may be illustrated by holding a gas jet near the path of the discharge. The flames will then leap to the two ends of the jet, which will perform the part of a conductor; and the real length of the discharge will be that traversed from terminal to terminal, minus the length of the intervening flame. The permanently heated part of the flame will act in the same manner in extending the effective length of the discharge.

The discharge which we are now examining is not homogeneous throughout, but consists of more than one layer. The flame, which, from the fact of its forming the outer sheath of the discharge, is the most prominent feature, consists mainly of heated but solid particles emanating from the terminals. That this is the case may be inferred in a general way from the colours which the flame assumes when different substances are placed upon the terminals; for example, lithium or sodium. The spectrum of the flame appears to be always continuous. A convenient substance to affix to the terminals is boron glass, on account of the brilliancy to which it gives rise in the discharge; this will enable us to project the phenomenon. Within this sheath of flame, the discharge consists of the pink light characteristic of air, and in the centre of all the true bright spark. There is reason to think that, under certain circumstances, there are more layers to be seen; but the above division is sufficient for our present purpose. In this somewhat complicated structure, the pink light corresponds to the arc, and the flame to a similar accompaniment which is seen playing about the upper carbon in electric lamps when a current of great strength is used.

From this account of the methods here employed I now turn to the main question. In the investigation, to which allusion was made at the beginning of this lecture, it occurred to us that an examination of the effects of a magnetic field on discharges of this character through air or other gases at atmospheric pressure, and a comparison with those obtained at lower pressures, might throw some fresh light on the nature of electrical discharges in general. It is these phenomena to which I now propose to ask your attention.

When the discharge, originally in the form of a vertical spindle, is submitted to the action of a magnet whose poles are horizontal, it spreads out into two nearly semicircular disks, one due to the discharges in one direction, and the other to those in the opposite direction. As the magnetism is strengthened, the flame retreats towards the edge of the disks, and ultimately disappears. The disk then consists mainly of the pink discharge; but with a still stronger magnetic field, it is traversed at intervals by bright semicircular sparks at various distances from the centre. In every case, bright sparks pass directly between the terminals at the opening of each separate discharge.

In order further to disentangle the parts of this phenomenon, recourse was had in the original experiments to a revolving mirror. The light in the disks is insufficient to allow of a projection of the effects, but the accompanying diagrams represent the appearances seen in the mirror. Fig. 1 shows the arrangement of the terminals and the magnetic poles; Fig. 2 the appearance of the discharges in a plane at right angles to that of Fig. 1; Fig. 3 the appearance of three

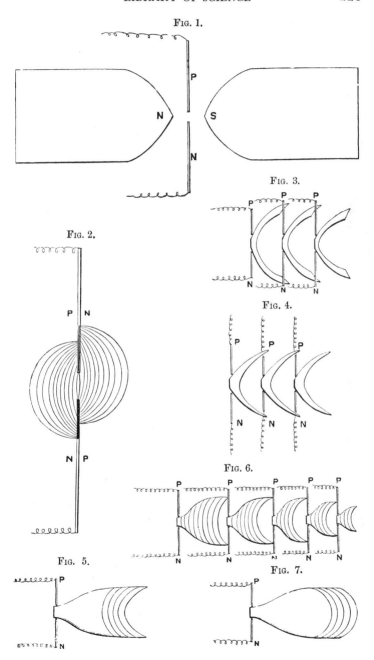

successive discharges (in the same direction) with a weak magnetic field and a slowly revolving mirror; Fig. 4 the same, with a slightly more rapid rate of revolution; Fig. 5 a single discharge, with a stronger field and greater speed of mirror; Fig. 6 a single discharge in a strong field, with a still greater speed of mirror. It should be mentioned that in all these figures the images to the left are to be regarded as anterior to those on the right, and that they represent various phases of the left-hand discharge in Fig. 2.

If, however, we observe the right-hand discharges with a mirror revolving in the same direction as before, it is clear that the actual curvature of the discharge will be turned in the opposite direction (with reference to the motion of the mirror) to that in the case of the left-hand discharges. The consequence will be that the appearance in the mirror, when the rate of revolution is not too great, will be something like Fig. 7, instead of Fig. 6. As the speed of the mirror is increased, the convexity will diminish, and ultimately be replaced by a concavity of the same kind, although not so marked, as that in the case of the left-hand discharges.

These diagrams show that each coil discharge commences with a bright spark passing directly between the terminals; that this spark is in general followed by the pink light or arc discharge, which passes first in the immediate neighbourhood of the initial spark, and gradually extends like an elastic string in semicircular loops outwards; and that the flame proper is a phenomenon attendant on the close of the entire discharge. It should be added that observations with a mirror revolving on a horizontal axis, and with a horizontal slit in front of the discharge, show that the disk is not simultaneously illuminated throughout, but that it is a locus of a curvilinear discharge which moves outwards and expands in its dimensions from the centre.

The mechanism of the discharge would therefore seem to be as follows: In the first place, as soon as the tension is sufficient, the electricity from the terminals breaks through the intervening air, but with such rapidity that the fracture is like that of glass, or other rigid substance. This opens a path, along which, if there remains sufficient electricity of sufficient tension, the discharge will continue to flow. During such continuance the gas becomes heated, and behaves like a conductor carrying a current; and upon this the magnet can act according to known laws. As long as the electricity continues to flow, the heat will at each moment determine the easiest, although not the shortest path for its subsequent passage. In this way the gas, which acts at one moment as the conductor of the discharge, and at the next as the path for it, will be carried further and further out until the supply of the electricity from the coil fails, and the whole discharge ceases. We are, in fact, led by these experiments to the conclusion that it is the gas in the act of carrying the current, and not the current moving freely in the gaseous space, upon which the magnet acts.

This explanation of the magnetic displacement of a discharge receives strong support from the phenomena represented in Figs. 5, 6, and 7. The successive bright lines there shown must be due to successive falls and revivals of tension within a single coil discharge. The existence of such alternations in coil discharges of large quantity is otherwise known. When the fall in temperature is such that the conductivity of the gas is insufficient to maintain the arc, the discharge can make its way through the air only by a fresh rent of the same kind as the first fracture. But how can this be reconciled with the fact that the tension can never reach its original degree, and must, on the whole, be gradually falling, and that, in addition, the paths represented by these various sparks are successively longer and longer? The answer to this question is to be found plainly written in the phenomena themselves. Any irregularity in one of these bright lines is always found to be accurately repeated in all of the same series. Now, it is scarcely to be conceived that, at successive instants of time and in different portions of space, irregularities in the discharge itself, and in the distribution of the gas, so precisely the same, would constantly and for certain recur; and we are therefore driven to the conclusion that it is the same portion of gas which at first occupied the centre of the field, with its same yet unhealed rent, which is moved outward under the action of the magnet. If this be so, we have in this repetition of minute details nothing more than what would necessarily follow from successive reopenings of the weak parts of the gas, which would be surely found out by the electricity in its struggle to pass.

The view here taken of the material character of the luminous discharge is further borne out by the fact that the spindle of light is capable of being diverted by a blast of air. When the blast is gentle, the discharge becomes curvilinear, approximately semicircular, and the yellow flame may be seen playing about the outer edge, in the same way as in a weak magnetic field. When the blast is stronger, the sheet of light becomes irregular in form, and it is traversed by a series of bright lines, all of which follow, even in their minute details, the configuration of the sheet. The analogy between this and the phenomena produced in a strong magnetic field needs no further remark. If the strength of the blast be still further increased, the flame and the sheet of light both disappear, and nothing remains but bright sparks passing directly, and undisturbed, between the terminals. In this case the air is both displaced and cooled so rapidly by the blast, that it no longer offers a practicable conductive path for the remainder of the electricity, coming from the coil, to follow. Of this a succession of disruptive sparks is a necessary consequence.

The effect thus produced by a very strong blast is in fact similar to that observed when a jar is used as a secondary condenser. In this case the electricity, instead of flowing gradually from the coil, passes in one or more instantaneous discharges with finite intervals of time between them. Each of these has to break its way through the air; and, that done, it ceases. Hence, neither a magnet, nor a blast of air will have any effect in diverting such a discharge.

As a last stage of the phenomena, it may be mentioned that, if the interval between the terminals be near the limit of striking distance, either a blast of air, or the setting up of a magnetic field will alike extinguish the discharge.

Our experiments have been thus far carried on in air at atmospheric pressure; but there is nothing in this pressure which is essential to them or to the conclusions to which we have been led. We may therefore repeat them in air, or any other gaseous medium, at any pressure we please. This consideration leads us into the region (so fertile in an experimental point of view) of discharges in vacuum tubes.

Commencing with a tube of moderate diameter and of very slight exhaustion, we can at once recognize our former phenomena slightly changed. Proceeding to another tube, of larger diameter and of moderate exhaustion, and placing it axially or equatorially in a magnetic field, we see not only that the discharge (or rather the conductor carrying it) is displaced, but also that the displaced part is spread out into a sheet or ribbon, showing that the discharge is affected gradually, exactly in the same way as was found in the open air.

When the exhaustion is carried further, the phenomena become rather more complicated. At an early stage there is a distinct separation between the "negative glow" and the rest of the luminous column; and at a more advanced stage the column itself is broken into separate luminosities or striæ. When this is the case, it is usually said that the negative glow follows the lines of magnetic force, while the luminous column distributes itself according to Ampère's law.

It will, however, be found that when completely analysed the action of the magnet upon the striæ, taken individually, is the same as that upon the negative glow, due allowance being made for the differences in local circumstances subsisting between the one and the other. We have elsewhere shown that the negative glow is in reality as truly a stria as any other individual member of the luminous column; but with this difference, that it is anchored to, and dependent for its form on, a rigid metallic terminal, whereas each of the others is dependent on the variable form and position of the stria immediately next in order, reckoning from the negative end of the tube. The action of a magnet in throwing the negative glow into a sheet of light, which is the locus of the lines of force passing through the terminal, and which consequently varies with the position of the tube in the field, is a phenomenon so well known that we need repeat only a single experiment by way of reminder.

Although it is not altogether so easy to show that the other striæ are directly affected by a magnetic field in the same way as is the anchored stria, we may still satisfy ourselves that it is the fact, from the consideration that when the striæ are well developed and the magnetic field is strong, it is quite possible to form a magnetic arch at any part of the column. In this experiment it will be noticed that for the formation of the arch in mid-column it is necessary that both poles of the magnet should act upon one and the same stria. This, in fact, means that the pole nearest the negative end anchors the stria, and thereby brings it into conditions similar to those of the negative glow. When this is effected the two exhibit similar modifications in the magnetic field.

In support of this view, we may adduce another and quite independent method of anchoring a stria, and of thereby producing a magnetic arch elsewhere than at the negative terminal. It was noticed by Goldstein and others that if the negative terminal of a tube be enveloped by an insulating surface of any form pierced with a number of holes, or if a diaphragm similarly pierced be placed anywhere in the tube, that the pierced surface will act as a negative terminal. He also found that the finer and closer the holes, the more complete the resemblance to the action of a negative terminal. But even when the substance is metallic, and when the holes are neither very small nor very numerous, a perforated diaphragm will so far act like a negative terminal as to serve as a point of departure of a stria. There is, however, this difference, that the blank space immediately adjoining the diaphragm, as it is usually called, is not generally so large as that at the true terminal; and the striæ thus artificially formed always lie close up to the holes. The diaphragm, in fact, anchors the stria, and renders it susceptible of the same magnetic effect as was shown in the cases studied before.

The action of a diaphragm in a magnetic field gives rise to many other interesting and remarkable results; some of which would further illustrate the views now submitted for your consideration. But these must be reserved for another occasion.

In the foregoing experiments, and in the remarks which have accompanied them, I have endeavoured to illustrate, by reference to gaseous media, the principle enunciated at the outset, that in the displacement of the discharge in a magnetic field, the subject of the magnetic action is the material substance or medium which conveys the discharge. I have shown also that, even when the discharge takes place in media so attenuated as to produce the phenomena of striæ, the same principle applies not only to the discharge as a whole, but also to each component stria or unit; and, lastly, that the apparent diversity of effect on the various striæ is due to local circumstances, and not to any fundamental difference between the "negative glow" and the members of the "positive column."

Seeing now that the magnetic displacement of the luminous discharge means displacement of the matter in a luminous condition, and that a crowding of such luminous matter involves an increase of luminosity, may we not infer with a high degree of probability that the striæ are themselves aggregations of matter, and that the dark spaces between them are comparatively vacuous.

It is true that such a view of the case would seem to imply that, in gaseous media, the better the vacuum the more easily can the electricity pass; and that this might at first sight appear to be at variance with the known fact that the resistance of a tube decreases with the pressure until a minimum, determinate for each kind of gas, and then increases. But it has been suggested by Edlund ('Annales de Chemie et de Physique,' 1881, tom. iii. p. 199) that the resistance of a tube may really consist of two parts, first that due to the passage of the electricity through the gas itself and, secondly, that due to its passage from the terminals to the gas; and also that the former decreases, while the latter increases, as the pressure is lowered. On this supposition the observed phenomena may be explained, without assigning any limit to the facility with which electricity may traverse the most vacuous space.

We may even carry the suggestion of a resistance of the second kind a little further, and suppose that there is a resistance due to the passage of electricity from a medium of one density to that of another, or from layer to layer of different degrees of pressure. And from this point of view we may regard the striæ as expressions of resistance due to the varying pressure in different parts of the tube. Into the question, whence this variation of pressure, I am not at present prepared to enter; it must suffice for this evening to have shown that the conclusions which we have drawn from our experiments are not in disaccordance with other known phenomena of the electrical discharge.

[W. S.]

Friday, February 2, 1883.

GEORGE BUSK, Esq. F.R.S. Treasurer and Vice-President,
in the Chair.

Sir WILLIAM THOMSON, LL.D. F.R.S.

The Size of Atoms.

FOUR lines of argument founded on observation have led to the conclusion that atoms or molecules are not inconceivably, not immeasurably small. I use the words "inconceivably" and "immeasurably" advisedly. That which is measurable is not inconceivable, and therefore the two words put together constitute a tautology. We leave inconceivableness in fact to metaphysicians. Nothing that we can measure is inconceivably large or inconceivably small in physical science. It may be difficult to understand the numbers expressing the magnitude, but whether it be very large or very small there is nothing inconceivable in the nature of the thing because of its greatness or smallness, or in our views and appreciation and numerical expression of the magnitude. The general result of the four lines of reasoning to which I have referred, founded respectively on the undulatory theory of light, on the phenomena of contact electricity, on capillary attraction, and on the kinetic theory of gases, agrees in showing that the atoms or molecules of ordinary matter must be something like the 1-10,000,000th, or from the 1-10,000,000th to the 1-100,000,000th of a centimetre in diameter. I speak somewhat vaguely, and I do so, not inadvertently, when I speak of atoms and molecules. I must ask the chemists to forgive me if I even abuse the words and apply a misnomer occasionally. The chemists do not know what is to be the atom; for instance, whether hydrogen gas is to consist of two pieces of matter in union constituting one molecule, and these molecules flying about; or whether single molecules each indivisible, or at all events undivided in chemical action, constitute the structure. I shall not go into any such questions at all, but merely take the broad view that matter, although we may conceive it to be infinitely divisible, is not infinitely divisible without decomposition. Just as a building of brick may be divided into parts, into a part containing 1000 bricks, and another part containing 2500 bricks, and those parts viewed largely may be said to be similar or homogeneous; but if you divide the matter of a brick building into spaces of nine inches thick, and then think of subdividing it farther, you find you have come to something which is atomic, that is, indivisible without destroying the elements of the structure. The question of the molecular structure of a building does not necessarily involve the question, Can a brick be divided into parts? and can those parts be divided into much smaller parts? and so on. It used to be a favourite subject for metaphysical argument amongst the schoolmen whether matter is infinitely divisible, or whether *space* is infinitely divisible, which some maintained, whilst others maintained only that *matter* is not infinitely divisible, and demonstrated that there is nothing inconceivable in the infinite subdivision of space. Why, even time was divided into moments (time-atoms!), and the idea of continuity of time was involved in a halo of argument, and metaphysical—I will not say absurdity—but metaphysical word-fencing, which was no doubt very amusing for want of a more instructive subject of study. There is in sober earnest this very important thing to be attended to, however, that in chronometry, as in geometry, we have absolute continuity, and it is simply an inconceivable absurdity to suppose a limit to smallness whether of time or of space. But on the other hand, whether we can divide a piece of glass into pieces smaller than the 1-100,000th of a centimetre in diameter, and so on without breaking it up, and making it cease to have the properties of glass, just as a brick has not the property of a brick wall, is a very practical question, and a question which we are quite disposed to enter upon.

I wish in the beginning to beg you not to run away from the subject by thinking of the exceeding smallness of atoms. Atoms are not so exceedingly small after all. The four lines of argument I have referred to make it perfectly certain that the molecules which constitute the air we breathe are not very much smaller, if smaller at all, than 1-10,000,000th of a centimetre in diameter. I was told by a friend just five minutes ago that if I give you results in centimetres you will not understand me. I do not admit this calumny on the

Fig. 1.

One centimetre.   One millimetre.

Royal Institution of Great Britain; no doubt many of you as Englishmen are more familiar with the unhappy British inch; but you all surely understand the centimetre, at all events it was taught till a few years ago in the primary national schools. Look at that diagram (Fig. 1), as I want you all to understand an inch, a centimetre, a millimetre, the 1-10th of a millimetre, and the 1-100th of a millimetre, the 1-1000th of a millimetre, and the 1-1,000,000th of a millimetre. The diagram on the wall represents the metre; below that the yard; next the decimetre, and a circle of a decimetre diameter, the centimetre and a circle of a centimetre, and the millimetre, which is 1-10th of a centimetre, or in round numbers 1-40th of an inch. We will adhere however to one simple system, for it is only because we are in England that the yard and inch are put before you at all, among the metres and centimetres. You see on the diagram then the metre, the centimetre, the millimetre, with circles of the same diameter. Somebody tells me the millimetre is not there. I cannot see it, but it certainly is there, and a circle whose diameter is a millimetre, both accurately painted in black. I say there is a millimetre, and you cannot see it. And now imagine *there* is 1-10th of a millimetre, and *there* 1-100th of a millimetre and 1-1000th of a millimetre, and *there* is a round atom of oxygen 1-1,000,000th of a millimetre in diameter. You see them all.

Now we must have a practical means of measuring, and optics supply us with it for thousandths of a millimetre. One of our temporary standards of measurement shall be the wave-length of light; but the wave-length is a very indefinite measurement, because there are wave-lengths for different colours of light, visible and invisible, in the ratio of 1 to 16. We have, as it were—borrowing an analogy from sound—four octaves of light that we know of. How far the range in reality extends above and below the range hitherto measured, we cannot even guess in the present state of science. The table before you (Table I.) gives you an idea of magnitudes of length,

TABLE I.—DATA FOR VISIBLE LIGHT.

Line of Spectrum.	Wave-length in Centimetres.	Wave Frequency, or Number of Periods per Second.
A	$7 \cdot 604 \times 10^{-5}$	$395 \cdot 0 \times 10^{12}$
B	$6 \cdot 867$,,	$437 \cdot 3$,,
C	$6 \cdot 562$,,	$457 \cdot 7$,,
D_1	$5 \cdot 895$,,	$509 \cdot 7$,,
D_2	$5 \cdot 889$,,	
E	$5 \cdot 269$,,	$570 \cdot 0$,,
b	$5 \cdot 183$,,	
F	$4 \cdot 861$,,	$617 \cdot 9$,,
G	$4 \cdot 307$,,	$697 \cdot 3$,,
H_1	$3 \cdot 968$,,	$756 \cdot 9$,,
H_2	$3 \cdot 933$,,	$763 \cdot 6$,,

and again of small intervals of time. In the column on the left you have the wave-length of light in fractions of a centimetre; the unit in which these numbers to the left is measured is the 1-100,000th (or 10^{-5}) of a centimetre. We have then, of visible light, wave-lengths from $7\frac{1}{2}$ to 4 nearly, or $3 \cdot 9$. You may say then roundly, that for the wave-lengths of visible light, which alone is what is represented on that table, we have wave-lengths of from 4 to 8 on our scale of 1-100,000th of a centimetre. The 8 is invisible radiation a little below the red end of the spectrum. The lowest, marked by Fraunhofer with the letter A, has for wave-length 7½-100,000th of a centimetre. On the model before you I will now show you what is meant by a "wave-length;" it is not length along the crest, such as we sometimes see well marked in a breaking wave of the sea, on a long straight beach; it is distance from crest to crest of the waves. [This was illustrated by a large number of horizontal rods of wood connected together and suspended bifilarly by two threads in the centre hanging from the ceiling;* on moving the lowermost rod, a wave was propagated up the series.] Imagine the ends of those rods to represent particles. The rods themselves let us suppose to be invisible, and merely their ends visible, to represent the particles acting upon one another mutually with elastic force, as if of indiarubber bands, or steel spiral springs, or jelly, or elastic material of some kind. They do act on one another in this model through the central mounting. Here again is another model illustrating waves (Fig. 2).† The white circles on the wooden rods represent pieces of matter—I will not say molecules at present, though we shall deal with them as molecules afterwards. Light consists of vibrations transverse to the line of propagation, just as in the models before you.

---

* The details of this bifilar suspension need not be minutely described, as the new form, with a single steel pianoforte wire to give the required mutual forces, described below and represented in Fig. 2, is better and more easily made.

† This apparatus, which is represented in the woodcut, Fig. 2, is of the following dimensions and description. The series of equal and similar bars (B) of which the ends represent molecules of the medium, and the pendulum bar (P), which performs the part of exciter of vibrations, or of kinetic store of vibrational energy, are pieces of wood each 50 centimetres long, 3 centimetres broad, and 1·5 centimetres thick. The suspending wire is steel pianoforte wire No. 22 B. W. G. (·07 of a cm. diameter), and the bars are secured to it in the following manner. Three brass pins of about ·4 of a centimetre diameter are fitted loosely in each bar in the position as indicated; i.e. forming the corners of an isosceles triangular figure, with its base parallel to the line of the suspending wire, and about 1 mm. to one side of it. The suspending wire, which is laid in grooves cut in the pins, is passed under the upper pin, outside the pin at the apex of the triangle, over the upper side of the lower pin, and thence down to the next bar. The upper end of this wire is secured by being taken through a hole in the supporting beam and several turns of it put round a pin placed on one side of the hole, as indicated in the diagram. To each end of the pendulum bar is made fast a steel spiral spring as shown; the upper ends of these springs being secured to short cords which pass up through holes in the supporting beam, and are fastened by two or three turns taken round the pins. These steel springs serve as potential stores of vibrational energy alternating in each vibration with the kinetic store constituted by the pendulum bar. The ends of the vibrating bars (B) are loaded with masses of lead attached to them. The much larger masses of lead seen on the pendulum bar, which are adjustable to different positions on the bar, are, in the diagram, shown at the smallest distance apart. The lowermost bar carries two vanes of tin projecting downwards, which dip into viscous liquid (treacle diluted with water) contained in the vessel (c). A heavy weight resting on the bottom of this vessel, and connected to the lower end of the suspending wire by a stretched indiarubber band, serves to keep the lower end of the apparatus in position. The period of vibration of the pendulum bar is adjustable to any desired magnitude by shifting in or out the attached weights, or by tightening or relaxing the cords which pull the upper ends of the spiral springs.

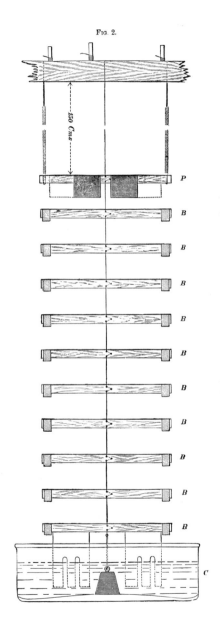

Fig. 2.

Now in that beautiful experiment well known as Newton's rings we have at once a measure of length in the distance between two pieces of glass to give any particular tint of colour. The wave-length you see, in the distance from crest to crest of the waves travelling up the long model when I commence giving a simple harmonic oscillation to the lowest bar. I have here a convex lens of very long focus, and a piece of plate glass with its back blackened. When I press the piece of glass against the glass blackened behind, I see coloured rings; the phenomenon will be shown to you on the screen by means of the electric light reflected from the space of air between the two pieces of glass. This phenomenon was first observed by Sir Isaac Newton, and was first explained by the undulatory theory of light. [Newton's rings are now shown on the screen before you by reflected electric light.] If I press the glasses together, you see a dark spot in the centre; the rings appear round it, and there is a dark centre with irregularities. Pressure is required to produce that spot. Why? The answer generally given is, because glass repels glass at a distance of two or three wave-lengths of light; say at a distance of 1-5000th of a centimetre. I do not believe that for a moment. The seeming repulsion comes from shreds or particles of dust between them. The black spot in the centre is a place where the distance between them is less than a quarter of a wave-length. Now the wave-length for yellow light is about 1-17,000th of a centimetre. The quarter of 1-17,000th is about 1-70,000th. The place where you see the middle of that black circle corresponds to air at a distance of less than 1-70,000th of a centimetre. Passing from this black spot to the first ring of maximum light, add half a wave-length to the distance, and we can tell what the distance between the two pieces of glass is at this place; add another half wave-length, and we come to the next maximum of light again; but the colour prevents us speaking very definitely because we have a number of different wave-lengths concerned. I will simplify that by reducing it all to one colour, red, by interposing a red glass. You have now one colour, but much less light altogether, because this glass only lets through homogeneous red light, or not much besides. Now look at what you see on the screen, and you have unmistakable evidence of fulcrums of dust between the glass surfaces. When I put on the screw, I whiten the central black spot by causing the elastic glass to pivot, as it were, round the innumerable little fulcrums constituted by the molecules of dust; and the pieces of glass are pressed not against one another, but against these fulcrums. There are innumerable—say thousands—of little particles of dust jammed between the glass, some of them of perhaps 1-3000th of a centimetre in diameter, say 5 or 6 wave-lengths. If you lay one piece of glass on another, you think you are pressing glass on glass, but it is nothing of the kind; it is glass on dust. This is a very beautiful phenomenon, and my first object in showing this experiment was simply because it gives us a linear measure bringing us down at once to 1-100,000th of a centimetre.

Now I am just going to enter a very little into detail regarding the reasons that those four lines of argument give us for assigning a limit to the smallness of the molecules of matter. I shall take contact electricity first, and very briefly. If I take these two pieces of zinc and copper and touch them together at the two corners, they become electrified, and attract one another with a perfectly definite force, of which the magnitude is ascertained from absolute measurements in connection with the well-established doctrine of contact electricity. I do not feel it, because the force is very small. You may do the thing in a measured way; you may place a little metallic knob or projection on one of them of 1-100,000th of a centimeter, and lean the other against it. Let there be three such little metal feet put on the copper; let me touch the zinc plate with one of them, and turn it gradually down till it comes to touch the other two. In this position, with an air-space of 1-100,000th of a centimetre between them, there will be positive and negative electricity on the zinc and copper surfaces respectively, of such quantities as to cause a mutual attraction amounting to 2 grammes weight per square centimetre. The amount of work done by the electric attraction upon the plates while they are being allowed to approach one another with metallic connection between them at the corner first touched, till they come to the distance of 1,100,000th of a centimetre, is 2-100,000ths of a centimetre-gramme, supposing the area of each plate to be one square centimetre.

I will now read you a statement from an article which was published thirteen years ago in 'Nature.' *

" Now let a second plate of zinc be brought by a similar process to the other side of the plate of copper; a second plate of copper to the remote side of this second plate of zinc, and so on till a pile is formed consisting of 50,001 plates of zinc and 50,000 plates of copper, separated by 100,000 spaces, each plate and each space 1-100,000th of a centimetre thick. The whole work done by electric attraction in the formation of this pile is two centimetre-grammes.

" The whole mass of metal is eight grammes. Hence the amount of work is a quarter of a centimetre-gramme per gramme of metal. Now 4030 centimetre-grammes of work, according to Joule's dynamical equivalent of heat, is the amount required to warm a gramme of zinc or copper by one degree Centigrade. Hence the work done by the electric attraction could warm the substance by only 1-16,120th of a degree. But now let the thickness of each piece of metal and of each intervening space be 1-100,000,000th of a centimetre, instead of 1-100,000th. The work would be increased a millionfold unless 1-100,000,000th of a centimetre approaches the smallness of a molecule. The heat equivalent would therefore be enough to

---

* See article "On the Size of Atoms," published in 'Nature,' vol. i. p. 551; printed in Thomson and Tait's 'Natural Philosophy,' second edition, 1883, vol. i. part 2, Appendix F.

raise the temperature of the material by 62°. This is barely, if at all, admissible, according to our present knowledge, or, rather, want of knowledge, regarding the heat of combination of zinc and copper. But suppose the metal plates and intervening spaces to be made yet four times thinner, that is to say, the thickness of each to be 1-400,000,000th of a centimetre. The work and its heat equivalent will be increased sixteenfold. It would therefore be 990 times as much as that required to warm the mass by one degree Centigrade, which is very much more than can possibly be produced by zinc and copper in entering into molecular combination. Were there in reality anything like so much heat of combination as this, a mixture of zinc and copper powders would, if melted in any one spot, run together, generating more than heat enough to melt each throughout; just as a large quantity of gunpowder if ignited in any one spot burns throughout without fresh application of heat. Hence plates of zinc and copper of 1-300,000,000th of a centimetre thick, placed close together alternately, form a near approximation to a chemical combination, if indeed such thin plates could be made without splitting atoms."

In making brass, if we mix zinc and copper together we find no very manifest signs of chemical affinity at all; there is not a great deal of heat developed; the mixture does not become warm, *it does not explode*. Hence we can infer certainly that contact-electricity action ceases, or does not go on increasing according to the same law, when the metals are subdivided to something like 1-100,000,000th of a centimetre. Now this is an exceedingly important argument. I have more decided data as to the actual magnitude of atoms or molecules to bring before you presently, but I have nothing more decided in *giving for certain a limit to supposable smallness*. We cannot reduce zinc and copper beyond a certain thickness without putting them into a condition in which they lose their properties as wholes, and in which, if put together, we should *not* find the same attraction as we should calculate upon from the thicker plates. I think it is impossible, consistently with the knowledge we have of chemical affinities and of the effect of melting zinc and copper together, to admit that a piece of copper or zinc could be divided to a thinness of much less, if at all less, than 1-100,000,000th of a centimetre without separating the atoms or dividing the molecules, or doing away with the composition which constitutes as a whole the solid metal. In short, the structure as it were of bricks, or molecules, or atoms, of which copper and zinc are built up, cannot be much, if at all, less than 1-100,000,000th of a centimetre in diameter, and may be considerably greater.

Similar conclusions result from that curious and most interesting phenomenon, the soap-bubble. Philosophers old and young, who occupy themselves with soap-bubbles, have one of the most interesting subjects of physical science to admire. Blow a soap-bubble and look at it,—you may study all your life perhaps, and still learn lessons in physical science from it. You will now see on the screen the image of a soap-film in a ring of metal. The light is reflected from the film filling that ring, and focused on the screen. It will show, as you see, colours analogous to those of Newton's rings. As you see the image it is upside down. The liquid streams down (up in the image), and thins away from the highest point of the film. First we see that brilliant green colour. It will become thinner and thinner there, and will pass through beautiful gradations of colour till you see, as now, a deep red, then much lighter, till it becomes a dusky, yellowish-white, then green, and blue, and deep violet, and lastly black, but after you see the black spot it very soon bursts. The film itself seems to begin to lose its tension, when it gets considerably less than a quarter of the wave-length of yellow light, which is the thickness for the dusky white, preceding the final black. When you are washing your hands, you may make and deliberately observe a film like this, in a ring formed by the forefingers and thumbs of two hands, and watch the colours. Whenever you begin to see a black spot or several black spots, the film soon after breaks. The film retains its strength until we come to the black spot, where the thickness is clearly much less than 1-60,000th of a centimetre, which is the thickness of the dusky white.*

Newton, in the following passage in his 'Optics' (pp. 187 and 191 of edition 1721, Second Book, Part I.), tells more of this important phenomenon of the black spot than is known to many of the best of modern observers.

"Obs. 17.—If a bubble be blown with water, first made tenacious by dissolving a little soap in it, it is a common observation that after a while it will appear tinged with a variety of colours. To defend these bubbles from being agitated by the external air (whereby their colours are irregularly moved one among another so that no accurate observation can be made of them), as soon as I had blown any of them I covered it with a clear glass, and by that means its colours emerged in a very regular order, like so many concentric rings encompassing the top of the bubble. And as the bubble grew thinner by the continual subsiding of the water, these rings dilated slowly and overspread the whole bubble, descending in order to the bottom of it, where they vanished successively. In the meanwhile, after all

---

* Since this lecture was delivered a paper "On the Limiting Thickness of Liquid Films," by Professors Reinold and Rücker, has been communicated to the Royal Society, and an abstract has been published in the 'Proceedings,' No. 225, 1883. The authors give the following results for the thickness of a black film of the liquids specified:—

Liquid.	Method.	Mean Thickness.
Plateau's "Liquide Glycérique."	Electrical.	$\cdot 119 \times 10^{-5}$ cm.
	Optical.	$\cdot 107$,,
Soap Solution.	Electrical.	$\cdot 117$,,
	Optical.	$\cdot 121$,,

The thickness, therefore, of a film of the liquide glycérique and that of a film of a soap solution containing no glycerine are nearly the same, and about 1-50th of the wave-length of sodium light.

the colours were emerged at the top, there grew in the centre of the rings a small round black spot like that in the first observation, which continually dilated itself till it became sometimes more than one-half or three-quarters of an inch in breadth before the bubble broke. At first I thought there had been no light reflected from the water in that place, but observing it more curiously I saw within it several smaller round spots, which appeared much blacker and darker than the rest, whereby I knew that there was some reflection at the other places which were not so dark as those spots. And by farther trial I found that I could see the images of some things (as of a candle or the sun) very faintly reflected, not only from the great black spot, but also from the little darker spots which were within it.

"Obs. 18.—If the water was not very tenacious, the black spots would break forth in the white without any sensible intervention of the blue. And sometimes they would break forth within the precedent yellow, or red, or perhaps within the blue of the second order, before the intermediate colours had time to display themselves."

Now I have a reason, an irrefragable reason, for saying that the film cannot keep up its tensile strength to 1-100,000,000th of a centimetre, and that is, that the work which would be required to stretch the film a little more than that would be enough to drive it into vapour.

The theory of capillary attraction shows that when a bubble—a soap-bubble, for instance—is blown larger and larger, work is done by the stretching of a film which resists extension as if it were an elastic membrane with a constant contractile force. This contractile force is to be reckoned as a certain number of units of force per unit of breadth. Observation of the ascent of water in capillary tubes shows that the contractile force of a thin film of water is about 16 milligrammes weight per millimetre of breadth. Hence the work done in stretching a water film to any degree of thinness, reckoned in millimetre-milligrammes, is equal to sixteen times the number of square millimetres by which the area is augmented, provided the film is not made so thin that there is any sensible diminution of its contractile force. In an article "On the Thermal Effect of Drawing out a Film of Liquid," published in the 'Proceedings' of the Royal Society for April 1858, I have proved from the second law of thermodynamics that about half as much more energy, in the shape of heat, must be given to the film, to prevent it from sinking in temperature while it is being drawn out. Hence the intrinsic energy of a mass of water in the shape of a film kept at constant temperature increases by 24 milligramme-millimetres for every square millimetre added to its area.

Suppose, then, a film to be given with the thickness of a millimetre, and suppose its area to be augmented ten thousand and one fold: the work done per square millimetre of the original film, that is to say per milligramme of the mass, would be 240,000 millimetre-milligrammes. The heat equivalent to this is more than half a degree

Centigrade (0·57°) of elevation of temperature of the substance. The thickness to which the film is reduced on this supposition is very approximately 1-10,000th of a millimetre. The commonest observation on the soap-bubble shows that there is no sensible diminution of contractile force by reduction of the thickness to 1-10,000th of a millimetre; inasmuch as the thickness which gives the first maximum brightness, round the black spot seen where the bubble is thinnest, is only about 1-8000th of a millimetre.

The very moderate amount of work shown in the preceding estimates is quite consistent with this deduction. But suppose now the film to be farther stretched until its thickness is reduced to 1-10,000,000th of a millimetre (1-100,000,000th of a centimetre). The work spent in doing this is two thousand times more than that which we have just calculated. The heat equivalent is 280 times the quantity required to raise the temperature of the liquid by 1° Centigrade. This is far more than we can admit as a possible amount of work done in the extension of a liquid film. It is more than half the amount of work which, if spent on the liquid, would convert it into vapour at ordinary atmospheric pressure. The conclusion is unavoidable, that a water-film falls off greatly in its contractile force before it is reduced to a thickness of 1-10,000,000th of a millimetre. It is scarcely possible, upon any conceivable molecular theory, that there can be any considerable falling off in the contractile force as long as there are several molecules in the thickness. It is therefore probable that there are not several molecules in a thickness of 1-10,000,000th of a millimetre of water.

Now when we are considering the subdivision of matter, look at those beautiful colours which you see in this little casket, left, I believe, by Professor Brande to the Royal Institution. It contains polished steel bars, coloured by having been raised to different degrees of heat, as in the process of annealing hard-tempered steel. These colours, produced by heat on other polished metals besides steel, are due to thin films of transparent oxide, and their tints, as those of the soap-bubble and of the thin space of air in "Newton's rings," depend on the thickness of the film, which, in the case of oxidisable metals, forms, by combination with the oxygen of the air under the influence of heat, a true surface-burning.

You are all familiar with the brilliant and beautifully distributed fringes of heat-colours on polished steel grates and fire-irons escaping that unhappy rule of domestic æsthetics which too often keeps those articles glittering and cold and useless, instead of letting them show the exquisite play of warm colouring naturally and inevitably brought out when they are used in the work which is their reason for existence. The thickness of the film of oxide which gives the first perceptible colour, a very pale orange or buff tint, due to the enfeeblement or extinction of violet light and enfeeblement of blue, and less enfeeblement of the other colours in order, by interference of the reflections from the two surfaces of the film, is about 1-100,000th of a centimetre, being something less than a quarter wave-length of violet light in the oxide.

The exceedingly searching and detective efficacy of electricity comes to our aid here, and by the force, as it were, spread through such a film, proves to us the existence of the film when it is considerably thinner than that 1-100,000th of a centimetre, when in fact it is so very thin as to produce absolutely no perceptible effect on the reflected light, that is to say, so thin as to be absolutely invisible. If in the apparatus for measuring contact electricity, of which the drawing is before you ('Nature,' vol. xxiii. p. 567), two plates of freshly polished copper be placed in the Volta condenser, a very perfect zero of effect is obtained. If, then, one of the plates be taken out, heated slightly by laying it on a piece of hot iron, and then allowed to cool again and replaced in the Volta condenser, it is found that negative electricity becomes condensed on the surface thus treated, and positive electricity on the bright copper surface facing it, when the two are in metallic connection. If the same process be repeated with somewhat higher temperatures, or somewhat longer times of exposure to it, the electrical difference is augmented. These effects are very sensible before any perceptible tint appears on the copper surface as modified by heat. The effect goes on increasing with higher and higher temperatures of the heating influence, until oxide tints begin to appear, commencing with buff, and going on through a ruddier colour to a dark-blue slate colour, when no farther heating seems to augment the effect. The greatest contact-electricity effect which I thus obtained between a bright freshly polished copper surface and an opposing face of copper, rendered almost black by oxidation, was such as to require for the neutralising potential in my mode of experimenting * about one-half of the potential of a Daniell's cell.

Some not hitherto published experiments with polished silver plates, which I made fifteen years ago, showed me very startlingly an electric influence from a quite infinitesimal whiff of iodine vapour. The effect on the contact-electricity quality of the surface seems to go on continuously from the first lodgment, to all other tests quite imperceptible, of a few atoms or molecules of the attacking substance (oxygen, or iodine, or sulphur, or chlorine, for example), and to go on increasing until some such thickness as 1-30,000th or 1-40,000th of a centimetre is reached by the film of oxide or iodide, or whatever it may be that is formed.

The subject is one that deserves much more of careful experimental work and measurement than has hitherto been devoted to it. I allude to it at present to point out to you how it is that by this

---

* First described in a letter to Joule, published in the 'Proceedings of the Literary and Philosophical Society of Manchester' of Jan. 21, 1862, where also I first pointed out the demonstration of a limit to the size of molecules from measurements of contact-electricity. The mode of measurement is more fully described in the article of 'Nature' (vol. xxiii. p. 567), referred to above.

electric action we are enabled as it were to sound the depth of the ocean of molecules attracted to the metallic surface by the vapour or gas entering into combination with it.

When we come to thicknesses of considerably less than a wave-length we find solid metals becoming transparent. Through the kindness of Prof. Dewar I am able to show you some exceedingly thin films of measured thicknesses of platinum, gold, and silver, placed on glass plates. The platinum is of $1\cdot9 \times 10^{-5}$ cm. thickness, and is quite opaque; but here is a gold film of about the same thickness, which is transparent to the electric light, as you see, and transmits the beautiful green colour which you see on the screen. The thickness of this gold ($1\cdot9$, or nearly 2) is just half the wave-length of violet light in air. This transparent gold, transmitting green light to the screen as you see, at the same time reflects yellow light to the ceiling. Now I will show you the silver. It is thinner, being only $1\cdot5 \times 10^{-5}$ of a centimetre thick, or $\frac{3}{8}$ths of the air-wave-length of violet light. It is quite opaque to the electric light so far as our eyes allow us to judge, and reflects all the light up to the ceiling. It is not wonderful that it should be opaque; we might wonder if it were otherwise; but there is an invisible ultra-violet light of a small range of wave-lengths, including a zinc line of air wave-length $3\cdot4 \times 10^{-5}$, which this silver film transmits. For that particular light the silver film of $1\cdot5 \times 10^{-5}$ thickness is transparent. The image which you now see on the screen is a magic lantern representation of the self-photographed spectrum of light that actually came through that silver. You see the zinc line very clear across it near its middle. Here then we have gold and silver transparent. The silver is opaque for all except that very definite light of wave-lengths from about $3\cdot07$ to $3\cdot32$.

The different refrangibility of different colours is a result of observation of vital importance in the question of the size of atoms. You now see on the screen before you a prismatic spectrum, a well-known phenomenon produced by the differences of the refractions of the different colours in traversing the prism. The explanation of it in the undulatory theory of light has taxed the powers of mathematicians to the utmost. Look first, however, to what is easy and made clear by that diagram (Fig. 3) before you, and you will easily understand that refraction depends on difference of velocity of propagation of light in the two transparent mediums concerned. The angles in the diagram are approximately correct, for refraction at an interface between air or vacuum and flint glass; and you see that in this case the velocity of propagation is less in the denser medium. The more refractive medium (not always the denser) of the two has the less velocity for light transmitted through it. The "refractive index" of any transparent medium is the ratio of the velocity of propagation in the ether to the velocity of propagation in the transparent substance.

Now that the velocity of the propagation of light should be different in different mediums, and should in most cases be smaller in the denser than in the less dense medium, is quite what we should, according to dynamical principles, expect from any conceivable constitution of the luminiferous ether and of palpable transparent substance. But that the velocity of propagation in any one transparent substance should be different for light of different colours, that is to say, of different periods of vibration, is not what we should expect, and could not possibly be the fact if the medium is homogeneous, without any limit as to the smallness of the parts of which the qualities are compared. The fact that the velocity of propagation *does* depend on the period, gives what I believe to be irrefragable proof that the substance

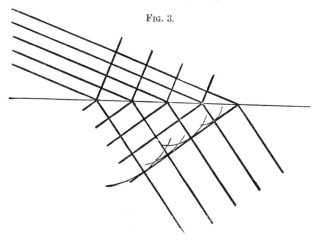

Fig. 3.

Diagram of Huyghen's construction for wave front of refracted light.
Drawn for light passing from air to flint glass.

of palpable transparent matter, such as water, or glass, or the bisulphuret of carbon of this prism, whose spectrum is before you, is not infinitely homogeneous; but that, on the contrary, if contiguous portions of any such medium, any medium in fact which can give the prismatic colours, be examined at intervals not incomparably small in comparison with the wave-lengths, utterly heterogeneous quality will be discovered; such heterogeneousness as that which we understand, in palpable matter, as the difference between solid and fluid; or between substances differing enormously in density; or such heterogeneousness as differences of velocity and direction of motion, in different positions of a vortex ring in an homogeneous liquid; or such differences of material occupying the space examined, as we find in a great mass of brick building when we pass from brick to brick through mortar (or through *void*, as we too often find in Scotch-built domestic brick chimneys).

Cauchy was, I believe, the first of mathematicians or naturalists to allow himself to be driven to the conclusion that the refractive dispersion of light can only be accounted for by a finite degree of molecular coarse-grainedness in the structure of the transparent refracting matter; and as, however we view the question, and however much we may feel compelled to differ from the details of molecular structure and molecular inter-action assumed by Cauchy, we remain more and more surely fortified in his conclusion, that finite grainedness of transparent palpable matter is the cause of the difference of the velocity of different colours of light propagated through it, we must regard Cauchy as the discoverer of the dynamical theory of the prismatic colours.

But now we come to the grand difficulty of Cauchy's theory.* Look at this little Table (Table II.), and you will see in the heading the formula which gives the velocity, in terms of the number of particles to the wave-length, supposing the medium to consist of equal particles arranged in cubic order, and each particle to attract its six

TABLE II.—VELOCITY (V) ACCORDING TO NUMBER (N) OF PARTICLES IN WAVE-LENGTH.

N.	$V \left(= 100 \frac{\sin(\pi/N)}{\pi/N} \right).$
2	63·64
4	90·03
8	97·45
12	98·86
16	99·36
20	99·59
∞	100·00

nearest neighbours, with a force varying directly as the excess of the distance between them, above a certain constant line (the length of which is to be chosen, according to the degree of compressibility possessed by the elastic solid, which we desire to represent by a crowd of mutually interacting molecules). If you suppose particles of real matter arranged in the cubic order, and six steel wire spiral springs, or elastic indiarubber bands, to be hooked on to each particle and stretched between it and its six nearest neighbours, the postulated force may be produced in a model with all needful accuracy; and if we could but successfully wish the theatre of the Royal Institution conveyed to the centre of the earth and kept there for five minutes, I should have great pleasure in showing you a model of an elastic solid thus constituted, and showing you waves propa-

---

* For an account of the dynamical theory of the "Dispersion of Light," see 'View of the Undulatory Theory as applied to the Dispersion of Light,' by the Rev. Baden Powell, M.A., &c. (London, 1841.)

gated through it, as are waves of light in the luminiferous ether. Gravity is the inconvenient accident of our actual position which prevents my showing it to you here just now. But instead, you have

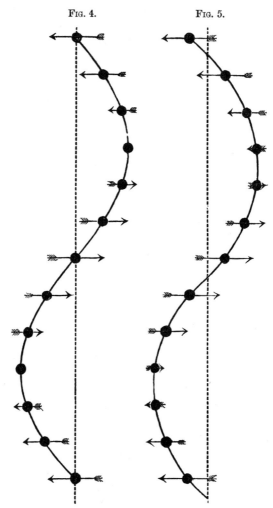

Twelve particles in Wave-Length.

these two wave-models (see Fig. 2), each of which shows you the displacements and motions of a line of particles in the propagation of a wave through our imaginary three-dimensional solid, the line of molecules chosen being those which in equilibrium are in one direct

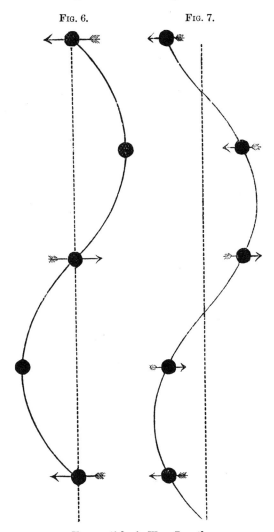

Four particles in Wave-Length.

straight line of the cubic arrangement, and the supposed wave having its wave front perpendicular to this line, and the direction of its

244   LIBRARY OF SCIENCE

vibration the direction of one of the other two direct lines of the cubic arrangement.

You have also before you this series of diagrams (Figs. 4 to 9) of

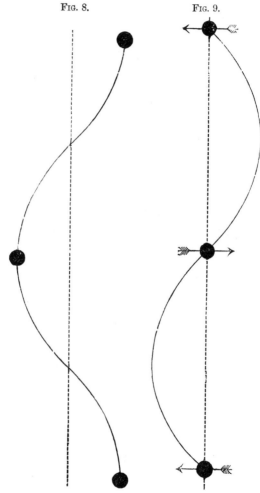

FIG. 8.   FIG. 9.

Two particles in Wave-Length.

waves in a molecularly-constituted elastic solid. These two diagrams (Figs. 4 and 5) illustrate a wave in which there are twelve molecules in the wave-length; this one (Fig. 4) showing (by the length and position of the arrows) the magnitude and direction of velocity of each molecule at the instant when one of the molecules is on the crest of the wave, or has reached its maximum displacement; that one (Fig. 5) showing the magnitude and direction of the velocities after the wave has advanced such a distance as (in this case equal to 1-24th of the wave-length) to bring the crest of the wave to midway between two molecules. This pair of diagrams (Figs. 6 and 7) shows the same for waves having four molecules in the wave-length, and this pair (Figs. 8 and 9) for a wave having two molecules in the wave-length.

The more nearly this critical case is approached, that is to say, the shorter the wave-length down to the limit of twice the distance from molecule to molecule, the less becomes the difference between the two configurations of motion constituted by waves travelling in opposite directions. In the extreme or critical case the difference is annulled, and the motion is not a wave motion, but a case of what is often called "standing vibration." Before I conclude this evening I hope to explain in detail the kind of motion which we find instead of wave-motion (become mathematically imaginary), when the vibrational period of the exciter is anything less than the critical value, because this case is of extreme importance and interest in physical optics, according to Stokes's hitherto unpublished explanation of phosphorescence.

This supposition of each molecule acting with direct force only on its nearest neighbour is not exactly the postulate on which Cauchy works. He supposes each molecule to act on all around it, according to some law of rapid decrease as the distance increases; but this must make the influence of coarse-grainedness on the velocity of propagation smaller than it is on the simple assumption realised in the models and diagrams before you, which therefore represents the extreme limit of the efficacy of Cauchy's unmodified theory to explain dispersion.

Now, by looking at the little table (Table II.) of calculated results, you will see that, with as few as 20 molecules in the wave-length, the velocity of propagation is $99\frac{1}{2}$ per cent. of what it would be with an infinite number of molecules; hence the extreme difference of propagational velocity, accountable for by Cauchy's unmodified theory in its idealised extreme of mutual action limited to nearest neighbours, amounts to 1-200th. Now look at this table (Table III.) of refractive indices, and you see that the difference of velocity of red light A, and of violet light H, amounts in carbon disulphide to 1-17th; in dense flint glass to nearly 1-30th; in hard crown glass to 1-73rd; and in water and alcohol to rather more than 1-100th. Hence, none of these substances can have so many as 20 molecules in the wave-length, if dispersion is to be accounted for by Cauchy's unmodified theory, and by looking back to the little table of calculated results (Table II.), you will see that there could not be more than 12 molecules in the wave-length of violet light in water or alcohol;

TABLE III.—TABLE OF REFRACTIVE INDICES.

Line of Spectrum.	Material.				
	Hard Crown Glass.	Extra dense Flint Glass.	Water at 15° C.	Carbon Disulphide at 11° C.	Alcohol at 15° C.
A	1·5118	1·6391	1·3284	1·6142	1·3600
B	1·5136	1·6429	1·3300	1·6207	1·3612
C	1·5146	1·6449	1·3307	1·6240	1·3621
D	1·5171	1·6504	1·3324	1·6333	1·3638
E	1·5203	1·6576	1·3347	1·6465	1·3661
b	1·5210	1·6591	..	..	..
F	1·5231	1·6442	1·3366	1·6584	1·3683
G	1·5283	1·6770	1·3402	1·6836	1·3720
h	1·5310	1·6836	..	..	..
H	1·5328	1·6886	1·3431	1·7090	1·3751

The numbers in the first two columns were determined by Dr. Hopkinson, those in the last three by Messrs. Gladstone and Dale. The index of refraction of air for light near the line E is 1·000294.

say 10 in hard crown glass; 8 in flint glass; and in carbon disulphide actually not more than 4 molecules in the wave-length, if we are to depend upon Cauchy's unmodified theory for the explanation of dispersion. So large coarse-grainedness of ordinary transparent bodies, solid or fluid, is quite untenable. Before I conclude, I intend to show you, from the kinetic theory of gases, a *superior limit* to the size of molecules, according to which, in glass or in water, there is probably something like 600 molecules to the wave-length, and almost certainly *not fewer* than 2, or 3, or 400. But even without any such definite estimate of a superior limit to the size of molecules, there are many reasons against the admission that it is probable or possible there can be only four, or five, or six, to the wave-length. The very drawing, by Nobert, of 4000 lines on a breadth of a millimetre, or at the rate of 40,000 to the centimetre, or about two to the ether wave-length of blue (F) light,[*] seems quite to negative the idea of any such possibility of only five or six molecules to the wave-length, even if we were not to declare against it from theory and observation of the reflection of light from polished surfaces.

We must then find another explanation of dispersion. I believe there is another explanation. I believe that, while giving up Cauchy's unmodified theory of dispersion, we shall find that the same general principle is applicable, and that by imagining each molecule to be loaded in a certain definite way by elastic connection with heavier matter—each molecule of the ether to have, in palpable transparent matter, a small fringe so to speak of particles, larger and larger in

---

[*] Loschmidt, "quoting from the Zollvereins department of the London International Exhibition of 1862, p. 83, and from Harting 'On the Microscope,' p. 881," 'Sitzungsberichte der Wiener Akademie Math. Phys.,' 1865. Vol. iii.

their successive order, elastically connected with it—we shall have a rude mechanical explanation, realisable by the notably easy addition of the proper appliances to the dynamical models before you, to account for refractive dispersion in an infinitely fine-grained structure. It is not seventeen hours since I saw the possibility of this explanation. I think I now see it perfectly, but you will excuse my not going into the theory more fully under the circumstances.* The difficulty of Cauchy's theory has weighed heavily upon me when thinking of bringing this subject before you. I could not bring it before you and say there are only four particles in the wave-length, and I could not bring it before you without saying there is some other explanation. I believe another explanation is distinctly to be had in the manner I have slightly indicated.

Now look at those beautiful distributions of colour on the screen before you. They are diffraction spectrums from a piece of glass ruled with 2000 lines to the inch. And again look, and you see one diffraction spectrum by reflection from one of Rutherford's gratings, in which there are 17,000 lines to the inch on polished speculum-metal. The explanation by "interference" is substantially the same as that which the undulatory theory gives for Newton's rings of light reflected from the two surfaces, which you have already seen. Where light-waves from the apertures between the successive bars of the grating reach the screen in the same phase, they produce light; there, again, where they are in opposite phases, they produce darkness.

The beautiful colours which are produced depend on the places of conspiring and opposing vibrations on the screen, being different for light-waves of different wave-lengths; and it is by the measurements of the dimensions of a diffraction spectrum such as the first set you saw (or of finer spectrums from coarser gratings) that Fraunhofer first determined the wave-lengths of the different colours.

I have now, closely bearing on the question of the size of atoms, thanks to Dr. Tyndall, a most beautiful and interesting experiment to show you—the artificial "blue sky," produced by a very wonderful effect of light upon matter, which he discovered. We have now an empty glass tube—it is "optically void." A beam of electric light passes through it now, and you see nothing. Now the light is stopped, and we admit vapour of carbon disulphide into the tube. There is now introduced some of this vapour to about 3 inches pressure, and there is also introduced, to the amount of 15 inches pressure, air impregnated with a little nitric acid, making in all rather less than the atmospheric pressure. What is to be illustrated here is the presence of molecules of substances produced by the decomposition of carbon disulphide by the light. At present you see nothing in the tube; it still continues to be, as before the admission of the vapours,

---

* Farther examination has seemed to me to confirm this first impression; and in a paper on the Dynamical Theory of Dispersion, read before the Royal Society of Edinburgh, on the 5th of March, I have given a mathematical investigation of the subject.—W. T., March 16, 1883.

optically transparent; but gradually you will see an exquisite blue cloud. That is Tyndall's "blue sky." You see it now. I take a Nicol's prism, and by looking through it I find the azure light coming from the vapours in any direction perpendicular to the exciting beam of light to be very completely polarised in the plane through my eye and the exciting beam. It consists of light-vibrations in one definite direction, and that, as finally demonstrated by Professor Stokes, it seems to me beyond all doubt, through reasoning on this phenomenon of polarisation,* which he had observed in various experimental arrangements giving minute solid or liquid particles scattered through a transparent medium, must be the direction perpendicular to the plane of polarisation.

What you are now about to see, and what I tell you I have seen through the Nicol's prism, is due to what I may call secondary or derived waves of light diverging from very minute liquid spherules, condensed in consequence of the chemical decomposing influence exerted by the beam of light on the matter in the tube, which was all gaseous when the light was first admitted.

To understand these derived waves, first you must regard them as due to motion of the ether round each spherule; the spherule being almost absolutely fixed, because its density is enormously greater than

---

* Extract from Professor Stokes's paper "On the Change of Refrangibility of Light," read before the Royal Society, May 27th, 1852, and published in the 'Transactions' for that date:—

"§ 183. Now this result appears to me to have no remote bearing on the question of the directions of the vibration in polarised light. So long as the suspended particles are large compared with the waves of light, reflection takes place as it would from a portion of the surface of a large solid immersed in the fluid, and no conclusion can be drawn either way. But if the diameters of the particles be small compared with the length of a wave of light, it seems plain that the vibrations in a reflected ray cannot be perpendicular to the vibrations in the incident ray. Let us suppose for the present, that in the case of the beams actually observed, the suspended particles were small compared with the length of a wave of light. Observation showed that the reflected ray was polarised. Now all the appearances presented by a plane polarised ray are symmetrical with respect to the plane of polarisation. Hence we have two directions to choose between for the direction of the vibrations in the reflected ray, namely, that of the incident ray, and a direction perpendicular to both the incident and the reflected rays. The former would be necessarily perpendicular to the directions of vibration in the incident ray, and therefore we are obliged to choose the latter, and consequently to suppose that the vibrations of plane polarised light are perpendicular to the plane of polarisation, since experiment shows that the plane of polarisation of the reflected ray is the plane of reflection. According to this theory, if we resolve the vibrations in the [horizontal] incident ray horizontally and vertically, the resolved parts will correspond to the two rays, polarised respectively in and perpendicularly to the plane of reflection, into which the incident ray may be conceived to be divided, and of these the former alone is capable of furnishing a . . . . . ray reflected vertically upwards [to be seen by an eye above the line of the incident ray, and looking vertically downwards]. And, in fact, observation shows that, in order to quench the dispersed beam, it is sufficient, instead of analysing the reflected light, to polarise the incident light in a plane perpendicular to the plane of reflection."

that of the ether surrounding it. The motion that the ether had in virtue of the exciting beam of light alone, before the spherules came into existence, may be regarded as being compounded with the motion of the ether relatively to each spherule, to produce the whole resultant motion experienced by the ether when the beam of light passes along the tube, and azure light is seen proceeding from it laterally. Now this second component motion is clearly the same as the whole motion of the ether would be, if the exciting light were annulled and each spherule kept vibrating in the opposite direction, to and fro through the same range as that which the ether in its place had, in virtue of the exciting light, when the spherule was not there.

Supposing now, for a moment, that without any exciting beam at all, a large number of minute spherules are all kept vibrating through very small ranges * parallel to one line. If you place your eye in the plane through the length of the tube and perpendicular to that line, you will see light from all parts of the tube, and this light which you see will consist of vibrations parallel to that line. But if you place you eye *in* the line of the vibration of a spherule, situated about the middle of the tube, you will see no light in that direction; but keeping your eye in the same position, if you look obliquely towards either end of the tube, you will see light fading into darkness, as you

---

* In the following question of the recent Smith's Prize Examination at Cambridge (paper of Tuesday, Jan. 30, 1883), the dynamics of the subject, and particularly the motion of the ether produced by keeping a single spherule embedded in it vibrating to and fro in a straight line, are illustrated in parts (*a*) and (*d*):—

"8. (*a*) From the known phenomenon that the light of a cloudless blue sky, viewed in any direction perpendicular to the sun's direction, is almost wholly polarised in the plane through the sun, assuming that this light is due to particles of matter of diameters small in comparison with the wave-length of light, prove that the direction of the vibrations of plane polarised light is perpendicular to the plane of polarisation.

"(*b*) Show that the equations of motion of a homogeneous isotropic elastic solid of unit density, are

$$\frac{d^2 a}{d t^2} = (k + \tfrac{1}{3} n) \frac{d \delta}{d x} + n \nabla^2 a,$$
$$\frac{d^2 \beta}{d t^2} = (k + \tfrac{1}{3} n) \frac{d \delta}{d y} + n \nabla^2 \beta,$$
$$\frac{d^2 \gamma}{d t^2} = (k + \tfrac{1}{3} n) \frac{d \delta}{d z} + n \nabla^2 \gamma,$$

where k denotes the modulus of resistance to compression; n the rigidity-modulus; a, β, γ, the components of displacement at (x, y, z, t); and

$$\delta = \frac{d a}{d x} + \frac{d \beta}{d y} + \frac{d \gamma}{d z},$$
$$\nabla^2 = \frac{a^2}{d x^2} + \frac{d^2}{d y^2} + \frac{d^2}{d z^2}.$$

"(*c*) Show that every possible solution is included in the following:—

$$a = \frac{d \phi}{d x} + u, \quad \beta = \frac{d \phi}{d y} + v, \quad \gamma = \frac{d \phi}{d z} + w,$$

turn your eye from either end towards the middle. Hence, if the exciting beam be of plane polarised light—that is to say, light of which all the vibrations are parallel to one line—and if you look at the tube in the direction perpendicular to this line and to the length of the tube, you will see light of which the vibrations will be parallel to that same line. But if you look at the tube in any direction parallel to this line, you will see no light; and the line along which you see no light is the direction of the vibrations in the exciting beam; and this direction, as we now see, is the direction perpendicular to what is technically called the plane of polarisation of the light. Here, then, you have Stokes's *experimentum crucis* by which he has answered, as seems to me beyond all doubt, the old vexed question—Whether is the vibration *perpendicular to*, or *in* the plane of polarisation? To show you this experiment, instead of using unpolarised light for the exciting beam, as in the previous experiment, and holding a small Nicol's prism in my hand and telling you what I saw when I looked through it, I place, as is now done, this great Nicol's prism in the course of the beam of light before it enters the tube. I now turn the Nicol's prism into different directions and turn the apparatus round, so that, sitting in all parts of the theatre, you may all see the tube in the proper direction for the successive phenomena of "light," and "no light." You see them now exactly fulfilling the description which I gave you in anticipation. If each of you had a Nicol's prism in your hand, you would learn that when you see light at all, its plane of polarisation is in the plane through your eye and the axis of the tube; and I hope you all now perfectly understand the proof that the direction of vibration is perpendicular to this plane.

Now I want to bring before you something which was taught me

---

where u, v, w are such that

$$\frac{du}{dx} + \frac{dv}{dy} + \frac{dw}{dz} = 0.$$

"Find differential equations for the determination of ϕ, u, v, w. Find the respective wave-velocities for the ϕ-solution, and for the (u, v, w)-solution.

"(d) Prove the following to be solutions, and interpret each for values of $r\,[\sqrt{(x^2 + y^2 + z^2)}]$ very great in comparison with λ (the wave-length).

(1) $\begin{cases} \alpha = \dfrac{d\phi}{dx}, \quad \beta = \dfrac{d\phi}{dy}, \quad \gamma = \dfrac{d\phi}{dz} \\ \text{where } \phi = \dfrac{1}{r}\sin\dfrac{2\pi}{\lambda}[r - t\sqrt{(k + \tfrac{4}{3}n)}]. \end{cases}$

(2) $\begin{cases} \alpha = 0, \quad \beta = -\dfrac{d\psi}{dz}, \quad \gamma = \dfrac{d\psi}{dy} \\ \text{where } \psi = \dfrac{1}{r}\sin\dfrac{2\pi}{\lambda}[r - t\sqrt{n}]. \end{cases}$

(3) $\alpha = \left(\dfrac{2\pi}{\lambda}\right)\psi + \dfrac{d^2\psi}{dx^2}, \quad \beta = \dfrac{d^2\psi}{dx\,dy}, \quad \gamma = \dfrac{d^2\psi}{dx\,dz}.$

a long time ago by Professor Stokes; and year after year I have begged him to publish it, but he has not done so, and so I have asked

Fig. 10.

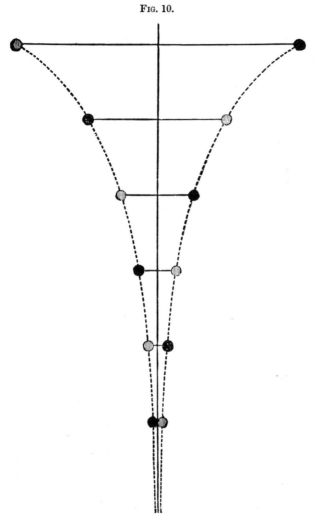

Diagram showing the different amplitudes of vibration of a row of particles oscillating in a period less than their least wave-period.

him to allow me to speak of it to-night. It is a dynamical explanation of that wonderful phenomenon called fluorescence or phosphorescence.

The principle is mechanically represented by this model (described above with reference to Fig. 2). A simple harmonic motion is, as you now see, sustained by my hand in the uppermost bar, in a period of about four seconds. You see that a regular wave-motion travels down the line of molecules represented by these circular disks on the ends of the bars, and the energy continually given to the top bar, by my hand, is continually consumed in heating the basin of treacle and water at the foot. I now remove my hand and leave the whole system to itself. The very considerable sum of kinetic and potential energies of the large masses and spiral springs, attached to the top bar, is gradually spent in sending the diminishing series of waves down the line, and is ultimately converted into heat in the treacle and water. You see that about half of the amplitude of vibration, and therefore three-fourths of the energy, is lost in half a minute.

You will see on quickening the oscillation how very different the result will be. The quick oscillations which I now give to the top bar (the period having been reduced to about one and a half seconds), is incapable of sending waves along the line of molecules; and it is that rapid oscillation of the particles which, according to Stokes, constitutes latent or stored-up light. Remark now that when I remove my hand from the top bar, as no waves travel down the line, no energy is spent in the treacle; and the vibration goes on for ever (or, to be more exact, say for one minute) as you see, with *no loss* (or, to be quite in accordance with what we see, let me say scarcely any sensible loss). This is a mechanical model correctly illustrating the dynamical principle of Stokes's explanation of phosphorescence or stored-up light, stored as in the now well-known luminous paint, of which you see the action in this specimen, and in the phosphorescent sulphides of lime in these glass tubes kindly lent by Mr. De La Rue. (Experiment shown.)

Now I will show you Stokes's phenomenon of *fluorescence* in a piece of uranium glass. I hold it in the beam from the electric lamp dispersed by the prism as you see. You see the uranium glass now visible by being illuminated by invisible rays. The rays by which it is illuminated even before it comes into the visible rays are manifestly invisible so far as the screen receiving the spectrum is a test of visibility; because the uranium glass, and my hands holding it, throw no shadow on the screen. Also you see the uranium glass which I hold in my hand in the ultra-violet light, while you do not see my hand. I now bring it nearer the place where you see the air (or rather the dust in it) illuminated by the violet light: still no shadow on the screen, but the uranium glass in my hand glowing more brilliantly with its green light of very mixed constitution, consisting of waves of longer periods than that of the ultra-violet, which the incident light, of shorter period than that of violet light, causes the particles of the uranium glass to emit. This light is altogether unpolarised. It was the absolute want of polarisation, and the fact of its periods being all less than those of the exciting light, that led Stokes to distinguish this illumination, which you see in the uranium glass,* from the mere molecular illumination (always polarised partially if not completely, and always of the same period as that of the exciting light) which we were looking at previously in Dr. Tyndall's experiment.

Stokes gave the name of fluorescence to the glowing with light of larger period than the exciting light, because it is observed in fluor spar, and he wished to avoid all hypothesis in his choice of a name. He pointed out a strong resemblance between it and the old known phenomenon of phosphorescence; but he found some seeming contrasts between the two, which prevented him from concluding fluorescence to be in reality a case of phosphorescence.

In the course of a comparison between the two phenomena (sections 221 to 225 of his 1852 paper), the following statement is given:—"But by far the most striking point of contrast between the two phenomena consists in the apparently instantaneous commencement and cessation of the illumination, in the case of internal dispersion when the active light is admitted and cut off. There is nothing to create the least suspicion of any appreciable duration in the effect. When internal dispersion is exhibited by means of an electric spark, it appears no less momentary than the illumination of a landscape by a flash of lightning. I have not attempted to determine whether any appreciable duration could be made out by means of a revolving mirror." The investigation here suggested has been actually made by Edmund Becquerel, and the question—Is there any appreciable duration in the glow of fluorescence?—has been answered affirmatively by this beautiful and simple little machine before you, which he invented for the purpose. The experiment giving the answer is most interesting, and I am sure you will see it with pleasure. It consists of a flat circular box, with two holes facing one another in the flat sides near the circumference; inside are two disks, carried by a rapidly revolving shaft, by which the holes are alter-

---

* The same phenomenon is to be seen splendidly in sulphate of quinine. An interesting experiment may be made by writing on a white paper screen, with a finger or a brush dipped in a solution of sulphate of quinine. The marking is quite imperceptible in ordinary light; but if a prismatic spectrum be thrown on the screen, with the ultra-violet invisible light on the part which had been written on with the sulphate of quinine, the writing is seen glowing brilliantly with a bluish light, and darkness all round. The phenomenon presented by sulphate of quinine and many other vegetable solutions, and some minerals, as, for instance, fluor spar, and various ornamental glasses, as a yellow Bohemian glass, called in commerce "canary glass" (giving a dispersed greenish light), had been discovered by Sir David Brewster ('Transactions,' Royal Society of Edinburgh, 1833, and British Association, Newcastle, 1838), and had been investigated also by Sir John Herschel, and by him called "epipolic dispersion" ('Phil. Trans.,' 1845). A complete experimental analysis of the phenomenon, showing precisely what it was that the previous observers had seen, and explaining many singularly mysterious things which they had noticed, was made by Stokes, and described in his paper, "On the Change of Refrangibility of Light" ('Phil. Trans.,' May 27, 1852).

nately shut and opened; one open when the other is closed, and *vice versâ*. A little piece of uranium glass is fixed inside the box between the two holes, and a beam of light from the electric lamp falls upon one of the holes. You look at the other.

Now when I turn the shaft slowly you see nothing. At this instant the light falls on the uranium glass through the open hole far from you, but you see nothing, because the hole next you is shut. Now the hole next you is open, but you see nothing, because the hole next the light is shut, and the uranium glass shows no perceptible after-glow as arising from its previous illumination. This agrees exactly with what you saw when I held the large slab of uranium glass in the ultra-violet light of the prismatic spectrum. As long as I held the uranium glass there you saw it glowing; the moment I took it out of the invisible light it ceased to glow. The "moment" of which we were then cognisant may have been the tenth of a second. If the uranium glass had continued to glow sensibly for the twentieth or the fiftieth of a second, it would have seemed to our slow-going sense of vision to cease the moment it was taken out. Now I turn the wheel at such a rate that the hole next you is open about a fiftieth of a second after the uranium glass was bathed in light; still you see nothing. I turn it faster and faster, and it now begins to glow, when the hole next you is open about the two-hundredth of a second after the immediately preceding admission of light by the other hole. I turn it faster and faster, and it glows more and more brightly, till now it is glowing like a red coal; further augmentation of the speed shows, as you see, but little difference in the glow.

Thus it seems that fluorescence is essentially the same as phosphorescence; and we may expect that substances will be found continuously bridging over the difference of quality between this uranium glass, which glows only for a few thousandths of a second, and the luminous sulphides which glow for hours or days or weeks after the cessation of the exciting light.

The most decisive and discriminating method of estimating the size of atoms I have left until my allotted hour is gone—that founded on the kinetic theory of gases. Here is a diagram (Fig. 11) of a crowd of atoms or molecules showing, on a scale of 1,000,000 to 1, all the molecules of air, of which the centres may at any instant be in a space of a square of 1-10,000th of a centimetre side and 1-100,000,000th of a centimetre thick. The side of the square you see in the diagram is a metre, and represents 1-10,000th of a centimetre. The diagram shows just 100 molecules, being 1-10,000th of the whole number of particles (10^6) in the cube of 1-10,000th centimetre, or all the molecules in a slice of 1-10,000th of the thickness of that cube. Think of a cube filled with particles, like these glass balls,* scattered at random

---

* The piece of apparatus now exhibited, illustrated the collisions taking place between the molecules of gaseous matter and the diffusion of one gas into another. It consisted of a board of about one metre square, perforated with

through a space equal to 1000 times the sum of their volumes. Such a crowd may be condensed (just as air may be condensed) to 1-1000th of its volume, but this condensation brings the molecules into contact. Something comparable with this may be imagined to be the condition of common air of ordinary density, as in our atmosphere. The diagram with size of each molecule, which, if shown in it to scale, would be 1 millimetre (or too small to be seen by you), to represent

FIG. 11.

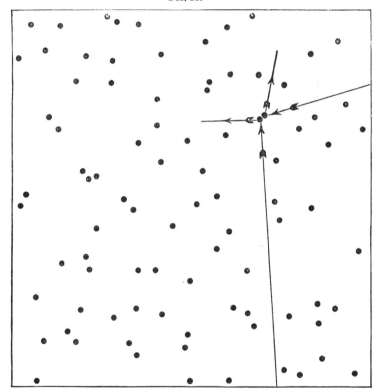

Diagram illustrating the number of molecules in a space of 1-10,000th of a centimetre square and 1-100,000,000th of a centimetre thick.

---

100 holes in ten rows of ten holes each. From each hole was suspended a cord five metres long. To the lower end of each cord, in five contiguous rows, there was secured a blue coloured glass ball of four centimetres diameter; and similarly to each cord of the other five rows, a red coloured ball of the same size. A ball from one of the outer rows was pulled aside, and, being set free, it plunged in amongst the others, causing collisions throughout the whole plane in which the suspended balls were situated.

an actual diameter 1-10,000,000th of a centimetre, represents a gas in which a condensation of 1 to 10 linear, or 1 to 1000 in bulk, would bring the molecules close together.

Now you are to imagine the particles moving in all directions, each in a straight line until it collides with another. The average length of free path is 10 centimetres in our diagram, representing 1-100,000th of a centimetre in reality. And to suit the case of atmospheric air of ordinary density and at ordinary pressure you must suppose the actual velocity of each particle to be 50,000 centimetres per second, which will make the average time from collision to collision 1-5,000,000,000th of a second.

The time is so far advanced that I cannot speak of the details of this exquisite kinetic theory, but I will just say that three points investigated by Maxwell and Clausius, viz. the viscosity or want of perfect fluidity of gases, the diffusion of gases into one another, and the diffusion of heat through gases—all these put together give an estimate for the average length of the free path of a molecule. Then a beautiful theory of Clausius enables us, from the average length of the free path, to calculate the magnitude of the atom. That is what Loschmidt has done,* and I, unconsciously following in his wake, have come to the same conclusion; that is, we have arrived at the absolute certainty that the dimensions of a molecule of air are something like that which I have stated.

The four lines of argument which I have now indicated lead all to substantially the same estimate of the dimensions of molecular structure. Jointly they establish, with what we cannot but regard as a very high degree of probability, the conclusion that, in any ordinary liquid, transparent solid, or seemingly opaque solid, the mean distance between the centres of contiguous molecules is less than the 1-5,000,000th, and greater than the 1-1,000,000,000th of a centimetre.

To form some conception of the degree of coarse-grainedness indicated by this conclusion, imagine a globe of water or glass, as large as a football,† to be magnified up to the size of the earth, each constituent molecule being magnified in the same proportion. The magnified structure would be more coarse-grained than a heap of small shot, but probably less coarse-grained than a heap of footballs.

[W. T.]

---

* *Sitzungsberichte* of the Vienna Academy, Oct. 12, 1865, p. 395.
† Or say a globe of 16 centimetres diameter.

Friday, March 9, 1883.

SIR FREDERICK POLLOCK, Bart. M.A. Vice-President, in the Chair.

Professor GEORGE D. LIVEING, M.A. F.R.S.

The Ultra-Violet Spectra of the Elements.

IT seems probable that the range of our vision as regards colour is closely connected with the intensity of that part of the solar radiation which reaches us on the earth, for Langley's observations on the intensity of the sun's rays in different parts of the spectrum bring out the fact that the region of greatest intensity falls nearly in the middle of the visible spectrum, and includes those colours to which our eyes are most sensitive. The ultra-violet rays, those which lie beyond the violet on the more refrangible side, are not, however, absolutely invisible, for, by carefully excluding light of lower refrangibility, Herschel found that he could see some distance beyond the Fraunhofer line H, into what he called the lavender-grey; and Helmholtz has succeeded in seeing nearly all the strong lines in the solar spectrum almost or quite up to its limit. Still these rays may fairly be said to be beyond ordinary vision; and from their power of chemical action they used to be distinguished as "actinic" rays. We know now that they have no monopoly of chemical activity, and we recognise no difference between luminous and actinic rays, the visible and the ultra-violet, except in their oscillation frequencies; that is, in the rate at which the successive pulsations of the ray succeed one another, and in the colour and refrangibility which are directly dependent on that rate. That the ultra-violet part of the solar spectrum extended at least as far above the line H as F is below it, has been known since the time of Wollaston, who observed its effect in blackening silver salts; but it is only about twenty years since Stokes made known to us the great length and intensity of the ultra-violet spectrum of the electric spark. Stokes used his own invention, a fluorescent screen, for observing the rays; and at the very time when Stokes published his discovery, W. A. Miller published photographs of the spectra of sparks taken between various metallic electrodes. Both these methods, that of fluorescent screens and that of photography, have been used by Professor Dewar and me in our researches. For the method of fluorescence we have used a modification of Soret's eye-piece, substituting for the uranium glass-plate a wedge-shaped vessel full of a solution of æsculine, placed with its edge horizontal so that we look down on the fluorescent liquid. The wedge form of the vessel has the advantage of refracting out of the line of vision all the rays except those which produce fluorescence, a matter of no small importance when faint light is to be observed (see Fig. 1).

Now, although the intensity of the sun's rays falls away rapidly beyond the Fraunhofer line H, and comes to nothing about as far above H as F is below it, it is far otherwise with the radiation of our terrestrial elements when heated up in the electric spark or arc, or even in some cases in flames; some of those elements which we know to be abundant in the sun, such as iron and magnesium, exhibit their most intense radiation, their strongest and most persistent rays, in the ultra-violet region, in waves which succeed one another at the shortest intervals. Indeed those metals so readily take up certain ultra-violet vibrations, that when there is much metal in the arc, and it is confined in a crucible of lime or magnesia, they often give their characteristic lines strongly reversed, dark absorption-bands being produced by the slightly cooled vapour which is outside the arc. This is seen in the photographic plate Nos. 1 to 3. No. 2 shows the strongest magnesium line, in a region beyond the limit of the solar spectrum, at wave length 2852, expanded and reversed. No. 1 shows it enormously expanded, its bright wings reversing iron lines up to S. No. 3 shows a strong group of iron lines, still more refrangible, also expanded and reversed by putting iron wire into the arc. The dark bands in the photograph are due to absorption by the metallic vapour, and in their places strong bright lines appear when less metal is present. The spectrum of iron is of all metals the most complicated, and those of the other elements which are most closely related to iron in chemical characters come next to it in the number and complication of their ultra-violet lines. Manganese and chromium are especially remarkable for showing many groups of closely-set lines. No. 2 shows a group of chromium lines between the solar lines S and U. It is probably not without significance that this group of elements

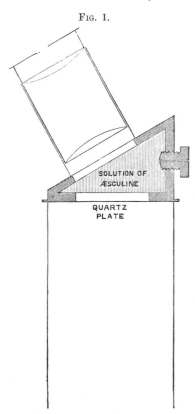

FIG. 1.

which exhibit the greatest variety in their chemical relations, and produce combinations of the greatest number of types, and the most complicated spectra, are also those which produce the most highly-coloured compounds. In marked contrast to the thick-set ranks of iron, manganese, and chromium lines, are the few scattered rays exhibited by those metals which form their combinations each chiefly on a single type, such as aluminium, and the alkali and alkaline earth metals. These spectra are probably even simpler than at first sight they seem to be. That of lithium is the simplest (Plate II. fig. 3): a series of single lines succeeding one another at decreasing intervals, and with diminishing intensity, closely resembling in these respects the spectrum of hydrogen. In the case of hydrogen, we know that the oscillation frequencies of some of its rays are related in a simple harmonic ratio. We are not able to say that the relation is so simple in the case of lithium; but still the whole series are probably overtones of a fundamental vibration, not so simply related as the harmonics of a uniform stretched string, but, like the overtones of a string which is not of uniform thickness, or is loaded at different points, similarly related in origin, though not exact harmonics. That the different rays are in many cases so related as overtones of a fundamental vibration appears more plainly, perhaps, when not single lines but groups of two, three, or four lines recur. Potassium shows a series of pairs to which the well-known violet pair, and perhaps that in the red also, belong. Calcium, magnesium, and zinc, each show a series of triplets, which are alternately sharply defined and diffuse (see photographs 4 and 5 and Plate II. figs. 5 and 6). In other cases the same characters may be traced, though less readily, because there is sometimes more than one such series of lines or groups. The alkali metals have each one such series in the visible spectrum, and another in the ultra-violet. It may happen in other cases that two or more such series overlap, and it may then be very difficult to distinguish and separate them.

In some cases elements show at a lower temperature a far more complicated spectrum than they do at higher temperatures further removed from their points of liquefaction. This has been observed by Roscoe and Schuster in the case of the alkali metals potassium and sodium, which give at temperatures only a little above their boiling-points absorption spectra which consist of closely-set fine lines, producing an appearance of shaded bands quite unlike their emission spectra at higher temperatures. In some few cases we have observed similar "fluted" or "venetian blind" spectra, as they have been called, in the ultra-violet, as, for example, one produced by tin; but in general the temperature of the arc, which we have chiefly used in our observations on metals, is high enough to carry the metals beyond the stage in which their vibrations are constrained by the state approaching to liquefaction.

But though metals do not often show spectra of this class at the high temperature of the arc, it is otherwise with metalloids and with compounds. Nitrogen gives in the arc as well as in the spark a channelled spectrum of singular beauty, extending with but short breaks almost to the extremity of the ultra-violet region which we have examined. (See photograph 7.) These multitudinous lines of nitrogen constantly present in the arc taken in air, help to make the problem of unravelling the spectrum of the arc, and assigning each line to its proper source, far more difficult than it might at first sight be supposed. Carbon, which in the arc frequently gives a channelled spectrum in the visible region, gives only a limited number of lines in the ultra-violet; but cyanogen gives one set of flutings near the line L, and another near N, which are so brilliant in the arc as to obscure the metallic lines in their neighbourhood (photograph 4). To the same class we may refer the spectrum of water, of which the most brilliant portion is given in photograph 6.

The series of lines produced by the same element, which I have spoken of as overtones of a fundamental vibration, have been likened to these channellings, but in reality they are very different. In the series, which I have supposed to have a sort of harmonic relation, the successive lines or groups of lines invariably become nearer to one another as the wave-lengths become shorter, and at the same time they diminish in strength and sharpness; whereas in the channelled spectra the strongest lines are at the end where they are most closely set, and they generally diminish in strength as they get further apart. Also increase of distance between the lines of channelled spectra is sometimes towards the less, sometimes towards the more, refrangible end of the spectrum.

I have before observed that a great part of the ultra-violet spectra of the elements which we have observed lies entirely beyond the limit of the solar spectrum; that limit is the line U at wave-length 2947. But though this is the limit of the solar radiation which reaches us on the earth, we can hardly suppose that the sun itself, or the photosphere, emits no radiation of shorter wave-length. We know that there is plenty of iron and magnesium in the sun, and the strongest radiations at high temperatures of these elements are of shorter wave-length than U. Moreover, the continuous spectra of incandescent solids in many cases extend far beyond U. The continuous spectrum of burning magnesium reaches quite up to the wave-length 2380, that of the flame of carbon disulphide mixed with hydrogen and fed with oxygen reaches even further, that of lime heated with an oxyhydrogen blowpipe, though feeble beyond the limit of the solar spectrum, extends up to wave-length 2680. The temperature of the sun cannot be less than that of any of these sources of heat, so that we are forced to suppose that the radiation, more refrangible than U, which leaves the body of the sun, is stopped somewhere either in our atmosphere, or in planetary space, or in the atmosphere of the sun himself. Now Cornu has found that when the thickness of our atmosphere traversed by the sun's rays is diminished as much as possible by taking the sun at its greatest altitude, and making the observation from an elevated station (the Riffelberg), the solar spectrum only reaches to wave-length 2932, that is, only a very trifle beyond U. We must therefore suppose that the absorbent substance, whatever it be, is not in our atmosphere. The same reason will lead us to reject the notion that the absorption can be due to matter in planetary space, for it is not easy to suppose that the gases which pervade that space in extreme tenuity can differ much from those in our atmosphere, because the earth in its annual

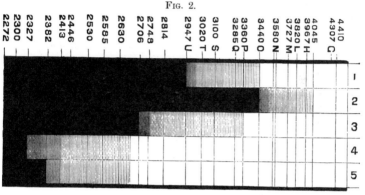

No. 1. Solar spectrum.
No. 2. Candle flame.
No. 3. Lime light.
No. 4. Carbon disulphide and hydrogen flame.
No. 5. Magnesium flame.

course must pick them up whatever they are, and they must then diffuse into our atmosphere, and we must in time have them in a more condensed state in our atmosphere than in planetary space. The absorbent is therefore probably neither in our atmosphere nor in planetary space, and we must look for it in the solar atmosphere. When we notice how much of the radiation of our terrestrial elements is of shorter wave-length than the solar line U, we might almost fancy that the blotting out of the sun's light beyond that point is simply due to an increase in the number and breadth of the Fraunhofer lines. Indeed we have frequently observed the strong magnesium line, wave-length 2852, expanded so that the dark absorption band in its middle reached quite up to U on one side (see photograph No. 1) and equally far on the other side, and this, together with such expansion of the strong iron lines beyond as we have occasionally observed, would go a long way towards completely hiding all light above U. But such expansions of iron and magnesium lines, high in the scale of refrangibility, do not occur without a considerable expansion of the lines of the same elements lower in the scale, expansions far exceeding what we actually observe in the Fraunhofer lines. Moreover the Fraunhofer lines, though dark by comparison with the brightness of the photosphere, are themselves luminous, even bright, when there is no other still brighter light wherewith to contrast them, so that if there were no other absorbent action the solar spectrum would be continued by the emitted rays of the metallic vapours which produce these lines. Probably then the absorbent is something at a lower temperature, higher in the solar atmosphere. A change of temperature may, and in some cases certainly does imply such a change of state that there may be a corresponding change in the particular vibrations which can be most easily taken up.

The metals in the liquid and solid states are so very opaque that we should hardly be able to discern their absorption spectra; nevertheless in very thin films they are translucent in different degrees. Gold leaf, as is well-known, transmits a green light, and we have found that a thin film of gold chemically deposited on a plate of quartz is fairly transparent for all the ultra-violet rays, so that its selective absorption is almost wholly of the less refrangible rays. Silver deposited in a similar way produces a very different effect. It is almost wholly opaque, except for one rather narrow band which begins a little below the solar line P and extends with diminishing transparency to about S. Cornu has before noticed this property of silver, but placed the transparent band at wave-length 270 instead of 330. Dr. W. A. Miller had observed that the light reflected by gold is equally distributed all through the ultra-violet, but feebler than that reflected by other metals; while that reflected by silver is characterised by giving a sudden cessation of the photographic image for a certain distance. These characters of the reflected rays he attributed to absorption by the metal.

When we examine the absorption produced by the haloid elements, we find that chlorine absorbs a wide band in the ultra-violet with its centre near the solar line P, extending, when the chlorine is in small quantity, from N to T, increasing in width on both sides when the quantity of chlorine is increased, but still leaving the rays above wave-length 2550 unabsorbed.

Bromine vapour shows an absorption band which begins in the visible spectrum, and extends, when the bromine is in small quantity, up to L, and when the bromine is in greater quantity up to P. From that point, up to about wave-length 2500, the vapour is transparent, but beyond it is again absorbent, the absorption increasing gradually with the refrangibility of the rays.

Iodine vapour, when thin, is transparent for ultra-violet rays, but produces strong absorption in the violet region. With thicker vapour this absorption extends nearly to H, but the vapour is still transparent for rays more refrangible than H.

Lecoq de Boisbaudran has observed that in the spectra of similar elements we may trace a shifting of similar lines, or groups of lines, towards the less refrangible side as the atomic weight is increased. Thus the violet pair of lines given by potassium is represented by an indigo pair in the case of rubidium, and a blue pair in the case of cæsium ; and the indigo line of calcium is represented by a blue line in the spectrum of strontium, and by a green line in the spectrum of barium.

We may observe something of the same kind in regard to the haloid elements: the absorption band which in the case of the element of lowest atomic weight, namely chlorine, is altogether ultra-violet, is shifted towards the less refrangible side in the case of bromine, and lies altogether in the visible region in the case of iodine, the element of highest atomic weight.

FIG. 3.

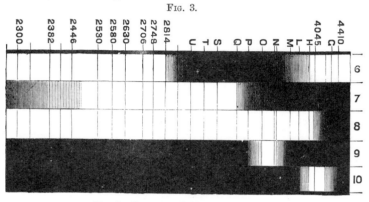

No. 6. Absorption of chlorine.
No. 7.   ,,   ,,   bromine vapour.
No. 8.   ,,   ,,   iodine vapour.
No. 9.   ,,   ,,   bromine liquid.
No. 10. ,,   ,,   iodine solution.

It is remarkable that bromine in the liquid state and iodine in solution show absorptions quite different from those of their vapours. A thin film of liquid bromine between two quartz plates is transparent for a band which ends just where the transparency of the vapour begins, while the film is opaque for rays both above and below this band. Iodine dissolved in carbon disulphide is also transparent for a certain distance, but the band is shifted to a less refrangible region lying between G and H.

Compound gases and vapours show, as might be expected, various absorptions of ultra-violet rays. The absorbent action of coal-gas begins at about the wave-length 2680, and above 2580 it is nearly complete. Sulphurous acid has an absorption band extending from about R (3179) to rays of wave-length 2630, with a weaker absorption extending some way beyond these limits on both sides. Sulphuretted hydrogen produces a pretty complete obliteration of all rays above wave-length 2580. Vapour of carbon disulphide in very small quantity produces an absorption extending from P to T, shading away at each end. With more vapour this band widens, and a second absorption band begins at about the wave-length 2580. Chlorine peroxide gives a succession of nine shaded bands at nearly equal intervals between M and S, while in the highest regions of the spectrum it seems to be quite transparent.

I mentioned at the outset the probable connection between the intensity of the solar radiation and the sensitiveness of our eyes to rays of different colours. The consideration of ultra-violet absorption spectra leads to the mention of another fact connected with vision, or rather with the construction of the eyes of the higher animals. Soret has investigated, and recently Chardonnet has more fully examined, the limits of transparency of the crystalline, cornea, and vitreous humour of the eyes of various animals and man, and found them all more or less transparent for ultra-violet rays. The limit of transparency in many cases approaches, but never exceeds, the limit of the solar spectrum. Chardonnet places the limit of transparency of the crystalline of the human eye as low as M, which is not consistent with the observations of Herschel and Helmholtz before mentioned, but this inconsistency is probably due to alterations which had taken place after death in the eyes experimented on by Chardonnet. That the transparency of the materials of the eye does not extend beyond the solar line U, Chardonnet regards as a provision of nature to protect the retina from the extreme radiations of artificial lights; but I venture to offer a different explanation, which is, that the selection of the materials of the eye has been determined not by what they will absorb but by what they will transmit. If the materials in question were in any great degree opaque to the ultra-violet solar rays, these rays must be absorbed and must either be used in heating the absorbent or do work upon it in some form, perhaps alter it chemically, and so impair its efficiency as part of an optical instrument. I see, then, in the selection of these materials for our eyes an instance, one amongst many, of the marvellous adaptation of our organisation to the natural, rather than to the artificial surroundings in which we are placed. [G. D. L.]

DESCRIPTION OF THE PLATES.

In the photographic plate:—

No. 1 shows an expansion of the magnesium line at wave-length 2852, while it is also so strong y reversed as to produce a complete obliteration of all the lines above U within the range of the photograph, while its bright wing reverses the iron lines near T.

No. 2 shows the same line at b much less expanded but still self-reversing. The lines at a are also magnesium lines, wave-lengths 2795 and 2801. Most of the lines between b and S are chromium lines.

No. 3 shows iron lines reversed by putting iron wire into the arc.

No. 4 shows calcium lines in recurring triplets; also the cyanogen bands between K and M and at N.

No. 5 shows three of the zinc triplets.

No. 6 is the brightest part of the ultra violet spectrum of water.

No. 7 shows part of the channelled spectrum of nitrogen in the ultra violet.

The lithographic plate gives the position of the ultra violet lines of several metals to a scale of wave lengths.

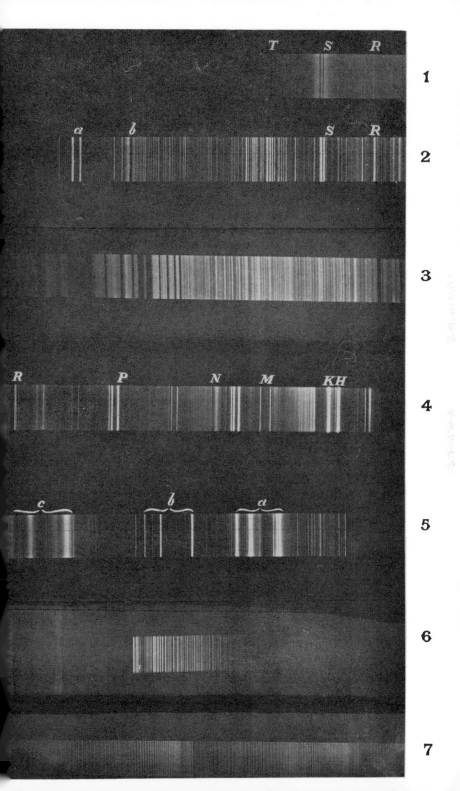

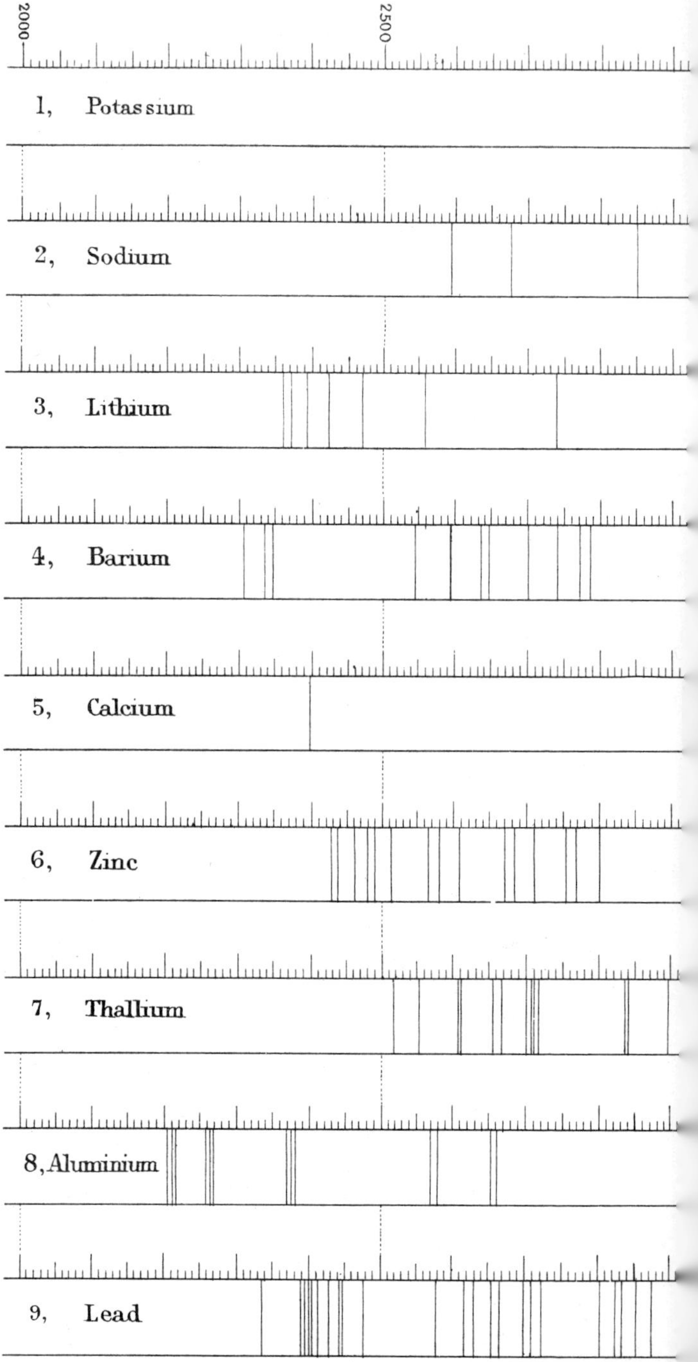

1, Potassium

2, Sodium

3, Lithium

4, Barium

5, Calcium

6, Zinc

7, Thallium

8, Aluminium

9, Lead

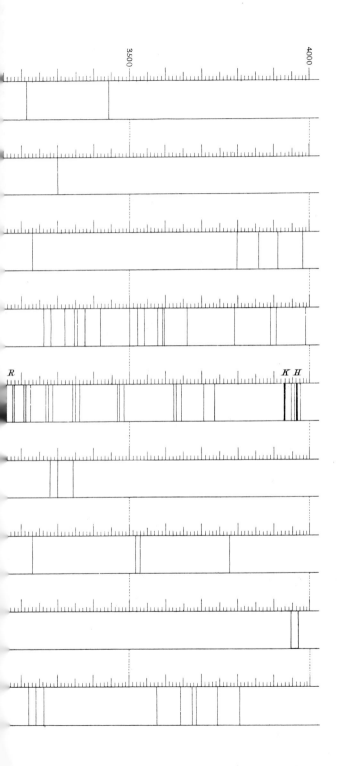

Friday, March 16, 1883.

WILLIAM BOWMAN, Esq. LL.D. F.R.S. Honorary Secretary and Vice-President, in the Chair.

PROFESSOR TYNDALL, D.C.L. F.R.S. *M.R.I.*

Thoughts on Radiation, Theoretical and Practical.

SCIENTIFIC discoveries are not distributed uniformly in time. They appear rather in periodic groups. Thus, in the two first years of this century, among other gifts presented by men of science to the world, we have the Voltaic pile; the principle of Interference, which is the basis of the undulatory theory of light; and the discovery by William Herschel of the dark rays of the sun.

Directly or indirectly, this latter discovery heralded a period of active research on the subject of radiation. Leslie's celebrated work on the Nature of Heat was published in 1804, but he informs us, in the preface, that the leading facts which gave rise to the publication presented themselves in the spring of 1801. An interesting but not uncommon psychological experience is glanced at in this preface. The inconvenience of what we call ecstacy, or exaltation, is that it is usually attended by undesirable compensations. Its action resembles that of a tidal river, sometimes advancing and filling the shores of life, but afterwards retreating and leaving unlovely banks behind. Leslie, when he began his work, describes himself as "transported at the prospect of a new world emerging to view." But further on the note changes, and before the preface ends he warns the reader that he may expect variety of tone, and perhaps defect of unity in his disquisition. The execution of the work, he says, proceeded with extreme tardiness; and as the charm of novelty wore off, he began to look upon his production with a coolness not usual in authors.

The ebb of the tide, however, was but transient; and to Leslie's ardour, industry, and experimental skill, we are indebted for a large body of knowledge in regard to the phenomena of radiation. In the prosecution of his researches he had to rely upon himself. He devised his own apparatus, and applied it in his own way. To produce radiating surfaces, he employed metallic cubes, which to the present hour are known as Leslie's cubes. The different faces of these cubes he coated with different substances, and filling the cubes with boiling water, he determined the emissive powers of the substances thus heated. These he found to differ greatly from each other. Thus, the radiation from a coating of lampblack being called 100, that from the uncoated metallic surface of his cube was only 12. He pointed out the reciprocity existing between radiation and absorption, proving that those substances which emit heat copiously absorb it greedily. His thermoscopic instrument was the well-known differential-thermometer invented by himself. In experiment Leslie was very strong, but in theory he was not so strong. His notions as to the nature of the agent whose phenomena he investigated with so much ability are confused and incorrect. Indeed, he could hardly have formed any clear notion of the physical meaning of radiation before the undulatory theory of light, which was then on its trial, had been established.

A figure still more remarkable than Leslie occupied the scientific stage at the same time, namely, the vigorous, original, and practical Benjamin Thompson, better known as Count Rumford, the originator of the Royal Institution. Rumford traversed a great portion of the ground occupied by Leslie, and obtained many of his results. As regards priority of publication, he was obviously discontented with the course which things had taken, and he endeavoured to place both himself and Leslie in what he supposed to be their right relation to the subject of radiant heat. The two investigators were unknown to each other personally, and their differences never rose to scientific strife. There can hardly, I think, be a doubt that each of them worked independently of the other, and that where their labours overlap, the honour of discovery belongs equally to both.

The results of Leslie and Rumford were obtained in the laboratory; but the walls of a laboratory do not constitute the boundary of its results. Nature's hand specimens are always fair samples, and if the experiments of the laboratory be only true, they will be ratified throughout the universe. The results of Leslie and Rumford were in due time carried from the cabinet of the experimenter to the open sky by Dr. Wells, a practising London physician. And here let it be gratefully acknowledged that vast services to physics have been rendered by physicians. The penetration of Wells is signalised among other things by the fact recorded by the late Mr. Darwin, that forty-five years before the publication of the 'Origin of Species,' the London doctor had distinctly recognised the principle of Natural Selection, and that he was the first to recognise it. But Wells is principally known to us through his 'Theory of Dew,' which, prompted by the experiments of Leslie and Rumford, and worked out by the most refined and conclusive observations on the part of Wells himself, first revealed the cause of this beautiful phenomenon. Wells knew that through the body of our atmosphere invisible aqueous vapour is everywhere diffused. He proved that grasses and other bodies on which dew was deposited were powerful emitters of radiant heat; that when nothing existed in the air to stop their radiation, they became self-chilled; and that while thus chilled they condensed into dew the aqueous vapour of the air around them. I do not suppose that any theory of importance ever escaped the ordeal of assault on its first enunciation. The theory of Wells was thus assailed; but it has proved immovable, and will continue so to the end of time.

The interaction of scientific workers causes the growth of science to resemble that of an organism. From Faraday's tiny magneto-electric spark, shown in this theatre half a century ago, has sprung the enormous practical development of electricity at the present time. Thomas Seebeck in 1822 discovered thermo-electricity, and eight years subsequently bars of bismuth and antimony were first soldered together by Nobili so as to form a thermo-electric pile. In the self-same year Melloni perfected the instrument and proved its applicability to the investigation of radiant heat. The instrumental appliances of science have been well described as extensions of the senses of man. Thus the invention of the thermopile vastly augmented our powers over the phenomena of radiation. Melloni added immensely to our knowledge of the transmission of radiant heat through liquids and solids. His results appeared at first so novel and unexpected that they excited scepticism. He waited long in vain for a favourable Report from the Academicians of Paris; and finally, in despair of obtaining it, he published his results in the Annales de Chimie. Here they came to the knowledge of Faraday, who, struck by their originality, brought them under the notice of the Royal Society, and obtained for Melloni the Rumford medal. The medal was accompanied by a sum of money from the Rumford fund; and this, at the time, was of the utmost importance to the young political exile, reduced as he was to penury in Paris. From that time until his death, Melloni was ranked as the foremost investigator in the domain of radiant heat.

As regards the philosophy of the thermopile, and its relation to the great doctrine of the conservation of energy, now everywhere accepted, a step of singular significance was taken by Peltier in 1834. Up to that time it had been taken for granted that the action of an electric current upon a conductor through which it passed, was always to generate heat. Peltier, however, proved that, under certain circumstances, the electric current generated cold. He soldered together a bar of antimony and a bar of bismuth, end to end, thus forming of the two metals one continuous bar. Sending a current through this bar, he found that when it passed from antimony to bismuth across the junction, heat was always there developed, whereas when the direction of the current was from bismuth to antimony, there was a development of cold. By placing a drop of chilled water upon the junction of the two metals, Lenz subsequently congealed the water to ice by the passage of the current.

The source of power in the thermopile is here revealed, and a relation of the utmost importance is established between heat and electricity. Heat is shown to be the nutriment of the electric current. When one face of a thermopile is warmed, the current produced, which is always from bismuth to antimony, is simply heat consumed and transmuted into electricity.

Long before the death of Melloni, what the Germans call "Die Identitäts Frage," that is to say, the question of the identity of light and radiant heat, agitated men's minds and spurred their inquiries. In the world of science people differ from each other in wisdom and penetration, and a new theoretic truth has always at first the minority on its side. But time, holding incessantly up to the gaze of inquirers the unalterable pattern of nature, gradually stamps that pattern on the human mind. For twenty years Henry Brougham was able to quench the light of Thomas Young, and to retard, in like proportion, the diffusion of correct notions regarding the nature and propagation of radiant heat. But such opposing forces are, in the end, driven in, and the undulatory theory of light being once established, soon made room for the undulatory theory of radiant heat. It was shown by degrees that every purely physical effect manifested by light was equally manifested by the invisible form of radiation. Reflection, refraction, double refraction, polarization, magnetization, were all proved true of radiant heat, just as certainly as they had been proved true of light. It was at length clearly realised that radiant heat, like light, was propagated in waves through that wondrous luminiferous medium which fills all space, the only real difference between them being a difference in the length and frequency of the ethereal waves. Light, as a sensation, was seen to be produced by a particular kind of radiant heat, which possessed the power of exciting the retina.

And now we approach a deeper and more subtle portion of our subject. What, we have to ask, is the origin of the ether waves, some of which constitute light, and all of which constitute radiant heat? The answer to this question is that the waves have their origin in the vibrations of the ultimate particles of bodies. But we must be more strict in our definition of ultimate particles. The ultimate particle of water, for example, is a *molecule*. If you go beyond this molecule and decompose it, the result is no longer water, but the discrete *atoms* of oxygen and hydrogen. The molecule of water consists of three such atoms tightly held together, but still capable of individual vibration. The question now arises: Is it the molecules vibrating as wholes, or the shivering atoms of the molecules that are to be considered as the real sources of the ether waves? As long as we were confined to the experiments of Leslie, Rumford, and Melloni, it was difficult to answer this question. But when it was discovered that gases and vapours possessed—in some cases to an astonishing extent—the power both of absorbing and radiating heat, a new light was thrown upon the question.

You know that the theory of gases and vapours, now generally accepted, is that they consist of molecular or atomic projectiles darting to and fro, clashing and recoiling, endowed, in short, with a motion not of vibration but of translation. When two molecules clash, or when a single molecule strikes against its boundary, the first effect is to deform the molecule, by moving its atoms out of their places. But gifted as they are with enormous resiliency, the atoms immediately recover their positions, and continue to quiver in consequence of the shock. Held tightly by the force of affinity, they resemble a string stretched to almost infinite tension, and therefore capable of generating tremors of almost infinite rapidity. What we call the heat of a gas is made up of these two motions—the flight of the molecules through space, and the quivering of their constituent atoms. Thus does the eye of science pierce to what Newton called "the more secret and noble works of Nature," and make us at home amid the mysteries of a world lying in all probability vastly further beyond the range of the microscope than the range of the microscope, at its maximum, lies beyond that of the unaided eye.

The great principle of radiation, which affirms that all bodies absorb the same rays that they emit, is now a familiar one. When, for example, a beam of white light is sent through a yellow sodium flame, produced by a copious supply of sodium vapour, the yellow constituent of the white beam is stopped by the yellow flame, and if the beam be subsequently analysed by a prism, a black band is found in the place of the intercepted yellow band of the spectrum. We have been led, as you know, to our present theoretic knowledge of light by a close study of the phenomena of sound, which in the present instance will help us to a conception of the action of the sodium flame. The atoms of sodium vapour synchronize in their vibrations with the particular waves of ether which produce the sensation of yellow light. The vapour, therefore, can take up or absorb the motion of those waves, as a stretched piano-string takes up or absorbs the pulses of a voice pitched to the note of the string. I will now show you the action of sodium vapour, in a way and with a result which startled and perplexed me on first making the experiment, more than twenty years ago. You know that the spectra of incandescent metallic vapours are not continuous, but formed of brilliant bands. I wished, in 1861, to obtain the brilliant yellow band produced by incandescent sodium vapour. To this end, I placed a bit of sodium in a carbon crucible, and volatilized it by a powerful voltaic current. A feeble spectrum overspread the screen, from which I thought the sodium band would stand out with dominant brilliancy. To my surprise, at the very point where I expected this brilliant band to appear, a band of darkness took its place. By humouring the voltaic arc a little, the darkness vanished, and in the end I obtained the bright band which I had sought at the beginning. On reflection the cause was manifest

The first ignition of the sodium was accompanied by the development of a large amount of sodium vapour, which spread outwards and surrounded, as a cool envelope, the core of intensely heated vapour inside. By the cool vapour the rays from the hot were intercepted, but on lengthening the arc the outer vapour in great part was dispersed, and the rays passed to the screen. This relation as to temperature was necessary to the production of the black band; for were the outside vapour as hot as the inside, it would, by its own radiation, make good the light absorbed.

An extremely beautiful experiment of this kind was made here last week by Professor Liveing, with rays which, under ordinary circumstances, are entirely invisible. Professor Dewar and Professor Liveing have been long working with conspicuous success at the ultra-violet spectrum, and with Professor Dewar's aid I will now show you this spectrum, as it was shown last week by Professor Liveing. Using prisms and lenses of a certain kind, and a powerful dynamo machine to volatilize our metals, we cast a spectrum upon the screen. You notice the terminal violet of this spectrum. Far beyond that violet, waves are now impinging upon the screen, which have no sensible effect upon the organ of vision; they constitute what we call the ultra-violet spectrum. Professor Stokes has taught us how to render this invisible spectrum visible, and it is by a skilful application of Stokes' discovery that Liveing and Dewar bring the hidden spectrum out with wondrous strength and beauty.

You notice here a small second screen, which can be moved into the ultra-violet region. Felt by the hand, the surface of this screen resembles sandpaper, being covered with powdered uranium glass, a highly fluorescent body. Pushing the moveable screen towards the visible spectrum, at a distance of three or four feet beyond the violet, light begins to appear. On pushing in the screen, the whole ultra-violet spectrum falls upon it, and is rendered visible from beginning to end. The spectrum is not continuous, but composed for the most part of luminous bands derived from the white-hot crucible in which the metals are to be converted into vapour. I beg of you to direct your attention on one of these bands in particular. Here it is, of fair luminous intensity. My object now is to show you the reversal, as it is called, of that band which belongs to the vapour of magnesium, exactly as I showed you a moment ago the reversal of the sodium band. An assistant will throw a bit of magnesium into the crucible, and you are to observe what first takes place. The action is rapid, so that you will have to fix your eyes upon this particular strip of light. On throwing in the magnesium, the luminous band belonging to its vapour is cut away, and you have, for a second or so, a dark band in its place. I repeat the experiment three or four times in succession, with the same unfailing result. Here, as in the case of the sodium, the magnesium surrounded itself for a moment by a cool envelope of its own vapour, which cut off the radiation from within, and thus produced the darkness.

And now let us pass on to an apparently different, but to a really similar result. Here is a feebly luminous flame, which you know to be that of hydrogen, the product of combustion being water vapour. Here is another flame of a rich blue colour, which the chemists present know to be the flame of carbonic oxide, the product of combustion being carbonic acid. Let the hydrogen flame radiate through a column of ordinary carbonic acid—the gas proves highly transparent to the radiation. Send the rays from the carbonic oxide flame through the same column of carbonic acid—the gas proves powerfully opaque. Why is this? Simply because the radiant, in the case of the carbonic oxide flame, is hot carbonic acid, the rays from which are quenched by the cold acid exactly as the rays from the intensely heated sodium vapour were quenched a moment ago by the cooler envelope which surrounded it. Bear in mind the case is always one of synchronism. It is because the atoms of the cold acid vibrate with the same frequency as the atoms of the hot, that the pulses sent forth from the latter are absorbed.

Newton, though probably not with our present precision, had formed a conception similar to that of molecules and their constituent atoms. The former he called corpuscles, which, as Sir John Herschel says, he regarded "as divisible groups of atoms of yet more delicate kind." The molecules he thought might be seen if microscopes could be caused to magnify three or four thousand times. But with regard to the atoms, he made the remark already alluded to:—"It seems impossible to see the more secret and nobler works of nature within the corpuscles, by reason of their transparency."

I have now to ask your attention to an illustration intended to show how radiant heat may be made to play to the mind's eye the part of the microscope, in revealing to us something of the more secret and noble works of atomic nature. Chemists are ever on the alert to notice analogies and resemblances in the atomic structures of different bodies. They long ago pointed out that a resemblance exists between that evil-smelling liquid, bisulphide of carbon, and carbonic acid. In the latter substance, we have one atom of carbon united to two of oxygen, while in the former we have one atom of carbon united to two of sulphur. Attempts have been made to push the analogy still further by the discovery of a compound of carbon and sulphur which should be analogous to carbonic oxide, where the proportions, instead of one to two, are one to one, but hitherto, I believe, without success. Let us now see whether a little physical light cannot reveal an analogy between carbonic acid and bisulphide of carbon more occult than any hitherto pointed out. For all ordinary sources of radiant heat the bisulphide, both in the liquid and vaporous form, is the most transparent, or diathermanous, of bodies. It transmits, for example, 90 per cent. of the radiation from our hydrogen flame, 10 per cent.

only being absorbed. But when we make the carbonic oxide flame our source of rays, the bisulphide shows itself to be a body of extreme opacity. The transmissive power falls from 90 to about 25 per cent., 75 per cent. of the radiation being absorbed. To the radiation from the carbonic oxide flame the bisulphide behaves like the carbonic acid. In other words, the group of atoms constituting the molecule of the bisulphide vibrate in the same periods as those of the atoms which constitute the molecule of the carbonic acid. And thus we have established a new, subtle, but most certain resemblance between these two substances. The time may come when chemists will make more use than they have hitherto done of radiant heat as an explorer of molecular condition.

The term " theoretical radiation " introduced into the title of this discourse is, I hope, thus justified. The conception of these quivering atoms is a theoretic conception, but it is one which gives us a powerful grasp of the facts, and enables us to realise mentally the mechanism on which radiation and absorption depend. We will turn in a moment to what I have called practical "radiation." It is pretty well known that for a long series of years I conducted an amicable controversy with one of the most eminent experimenters of our time, as regards the action of the earth's atmosphere on solar and terrestrial radiation. My contention was that the great body of our atmosphere—its oxygen and nitrogen—had but little effect upon either the rays of the sun coming to us, or the rays of the earth darting away from us into space, but that mixed with the body of our air there was an attenuated and apparently trivial constituent which exercised a most momentous influence. That body, as many of you know, is aqueous vapour, the amount of which does not exceed 1 per cent. of the whole atmosphere. Minute, however, as its quantity is, the life of our planet depends upon that vapour. Without it, in the first place, the clouds could drop no fatness. In this sense the necessity for its presence is obvious to all. But it acts in another sense as a preserver. Without it as a covering, the earth would soon be reduced to the frigidity of death. Observers were, and are, slow to take in this fact, which nevertheless is a fact, however improbable it may at first sight appear. The action of aqueous vapour upon radiant heat has been established by irrefragable experiments in the laboratory; and these experiments, though not unopposed, have been substantiated by some of the most accomplished meteorologists of our day.

I wished much to instruct myself a little by actual observation on this subject, under the open sky, and my first object was, to catch, if possible, states of the weather which would enable me to bring 'my views to a practical test. Thanks to an individual who devotes her life to taking care of mine, a little iron hut, embracing a single room, has been placed for my benefit, upon the wild moorland of Hind Head. From the plateau on which the hut stands, there is a free outlook in all directions. Here, amid the heather, I had two stout poles fixed firmly in the ground eight feet asunder, and a stout cord stretched from one to the other. From the centre of this cord a thermometer is suspended with its bulb four feet above the ground. On the ground is placed a pad of cotton wool, and on this cotton wool a second thermometer, the object of the arrangement being to determine the difference of temperature between the two thermometers, which are only four feet vertically apart.

Permit me at the outset to deal with the subject in a perfectly elementary way. In comparison with the cold of space, the earth must be regarded as a hot body, sending its rays, should nothing intercept them, across the atmosphere into space. The cotton wool is chosen because it is a powerful, though not the most powerful, radiator. It pours its heat freely into the atmosphere, and by reason of its flocculence, which renders it a non-conductor, it is unable to derive from the earth heat which might atone for its loss. Imagine the cotton wool thus self-chilled. The air in immediate contact with it shares its chill, and the thermometer lying upon it partakes of the refrigeration. In calm weather the chilled air, because of its greater density, remains close to the earth's surface, and in this way we sometimes obtain upon that surface a temperature considerably lower than that of the air a few feet above it. The experiments of Wilson, Six, and Wells have made us familiar with this result. On the other hand, the earth's surface during the day receives from the sun more heat than it loses by its own radiation, so that when the sun is active, the temperature of the surface exceeds that of the air.

These points will be best illustrated by describing the course of temperature for a day, beginning at sunrise and ending at 10.20 P.M. on March 4. The observations are recorded in the annexed table, at the head of which is named the place of observation, its elevation above the sea, and the state of the weather. The first column in the Table contains the times at which the two thermometers were read. The column under "Air" gives the temperatures of the air, the column under "Wool" gives the temperatures of the wool, while the fourth column gives the differences between the two temperatures. It is seen at a glance that from sunrise to 9.20 A.M. the cotton wool is colder than the air; at 9.30 the temperatures are alike. This is the hour of "intersection," which is immediately followed by "inversion." Throughout the day and up to 4 P.M. the wool is warmer than the air. At 4.5 P.M. the temperatures are again alike; while from that point downwards the loss by terrestrial radiation is in excess of the gain derived from all other sources, the refrigeration reaching a maximum at 7.30 P.M., when the difference between the two thermometers amounted to 10° Fahr. When the observations are continued throughout the night, the greater cold of the surface is found to be maintained until sunrise, and for some hours beyond it. Had the air been perfectly still during the observations, the nocturnal chilling of the surface would have been in this case greater; for you can readily understand that even a light wind sweeping over the surface, and mixing the chilled with the warmer air, must seriously interfere with the refrigeration.

HIND HEAD, Elevation, 850 feet.

Course of Temperature, March 4th, 1883.

Sky cloudless. Hoar frost. Wind light from north-east.

Time.	Air.	Wool.	Difference.
6.50 A.M. (sunrise)	31°	25°	6°
7.20	32½	24½	8
7.40	34	25	9
8.5	35	27	8
8.20	35	30	5
9.15	40	38	2
9.20	41	40	1
9.30 (intersection)	41	41	0
9.40 (inversion)	41	42	1
10.15	42½	45	2½
11.	45	52	7
11.30	47	55	8
12. noon	50	58	8
12.30 P.M.	50	59½	9½
1.	50	57½	7½
2.	49	60	11
2.30	48	58	10
3.	49	56	7
3.30	48	52	4
4.	47	48	1
4.5 (intersection)	47	47	0
4.10 (inversion)	47	45	2
4.15	47	43	4
4.30	46	41	5
7.	35	26	9
7.30	35	25	10
8·30	34	24½	9½
9.40	33	24½	8½
10·20	32	24	8

Glacial wind from north-east. Stars very bright.

Various circumstances may contribute to lessen, or even abolish, the difference between the two thermometers. Haze, fog, cloud, rain, snow, are all known to be influential. These are visible impediments to the outflow of heat from the earth; but my position for some time has been that a very powerful obstacle to that outflow exists which is entirely invisible. The pure vapour of water, for example, is a gas as invisible as the air itself. It is everywhere diffused through the air; but, unlike the oxygen and nitrogen of the atmosphere, it is not constant in quantity. We have now to examine whether meteorological observations do not clearly indicate its influence on terrestrial radiation.

With a view to this examination, I will choose a series of observations made during the afternoon and evening of a day of extraordinary calmness and serenity. The visible condition of the atmosphere at the time was that which has hitherto been considered most favourable to the outflow of terrestrial heat, and therefore best calculated to establish a large difference between the air and wool thermometers. The 16th of last January was a day of this kind, when the observations recorded in the annexed table were made.

January 16th.—Extremely serene. Air almost a dead calm. Sky without a cloud. Light south-westerly air.

Time.	Air.	Wool.	Difference.
P.M.	°	°	°
3.40	43	37	6
3.50	42	35	7
4.	41	35	6
4.15	40	34	6
4.30	38	32	6
5.	37	28	9
5.30	37	30	7
6.	36	32	4
6.30	36	31	5
7.	36	28	8
7.30	$35\frac{1}{2}$	28	$7\frac{1}{2}$
8.	35	26	9
8.30	34	25	9
9.	35	27	8
10.	35	28	7
10.30	35	29	6

During these observations there was no visible impediment to terrestrial radiation. The sky was extremely pure, the moon was shining; Orion, the Pleiades, Charles's Wain, including the small companion star at the bend of the shaft, the North Star, and numbers of others, were clearly visible. After the last observation, my note-book contains the remark, "Atmosphere exquisitely clear; from zenith to horizon cloudless all round."

A moment's attention bestowed on the column of differences in the foregoing table will repay us. Why should the difference at 6 P.M. be fully 5° less than at 5 P.M.; and again 5° less than at 8 and at 8.30 respectively? There was absolutely nothing in the aspect of the atmosphere to account for the approach of the two thermometers at 6 o'clock—nothing to account for their preceding and subsequent divergence from each other. Anomalies of this kind have been observed by the hundred, but they have never been accounted for, and they did not admit of explanation until it had been proved that the intrusion of a perfectly invisible vapour was competent to check the radiation, while its passing away re-opened a doorway into space.

It is well to bear in mind that the difference between the two thermometers on the evening here referred to varied from 4° to 9°, the latter being the maximum.

Such observations might be multiplied, but, with a view to saving space, I will limit the record. On the evening of January 30th, the atmosphere was very serene; there was no moon, but the firmament was powdered with stars. At 7.15 p.m. the difference between the two thermometers was 6°; while at 9.30 p.m. it was 4°, the wool thermometer being in both cases the colder of the two. On February 3rd observations were made under similar conditions of weather, and with a similar result. At 7.15 p.m. the difference between the thermometers was 6°; while at 8.25 p.m. it was 4°. On both these evenings the sky was cloudless, the stars were bright, while the movement of the air was light, from the south-west.

In all these cases the air passing over the plateau of Hind Head had previously grazed the comparatively warm surface of the Atlantic Ocean, where it had charged itself with aqueous vapour to a degree corresponding to its temperature. Let us contrast its action with that of air coming to Hind Head from a quarter less competent to charge it with aqueous vapour. We were visited by such air on the 10th of last December, when the movement of the wind was light from the north-east, the temperature at the time, moreover, was very low, and hence calculated to lessen the quantity of atmospheric vapour. Snow a foot deep covered the heather. At 8.5 a.m. the two thermometers were taken from the hut, having a common temperature of 35°. The one was rapidly suspended in the air, and the other laid upon the wool. I was not prepared for the result. A single minute's exposure sufficed to establish a difference of 5° between the thermometers; an exposure of five minutes produced a difference of 13°; while after ten minutes' exposure the difference was found to be no less than 17°. Here follow some of the observations.

December 10th.—Deep snow; low temperature; sky clear; light north-easterly air.

Time.	Air.	Wool.	Difference.
A.M.	°	°	°
8.10	29	16	13
8.15	29	12	17
8.20	27	12	15
8.30	26	11	15
8.40	26	10	16
8.45	27	11	16
8.50	29	11	18

During these observations, a dense bank of cloud on the opposite ridge of Blackdown, virtually retarded the rising of the sun. It had, however, cleared the bank during the last two observations, and, touching the air thermometer with its warmth, raised its temperature from 26° to 27° and 29°. The very large difference of 18° is in part to be ascribed to this raising of the temperature of the air thermometer. I will limit myself to citing one other case of a similar kind. On the evening of the 31st of March, though the surface temperature was far below the dew point, very little dew was deposited. The air was obviously a dry air. The sky was perfectly cloudless, while the barely perceptible movement of the air was from the north-east. At 10 p.m. the temperature of the air thermometer was 37°, that of the wool thermometer was 20°, a refrigeration of 17° being therefore observed on this occasion.

From the behaviour of a smooth ball when urged in succession over short grass, over a gravel walk, over a boarded floor, and over ice, it has been inferred that, were friction entirely withdrawn, we should have no retardation. In a similar way, under atmospheric conditions visibly the same, we observe that the refrigeration of the earth's surface at night markedly increases with the dryness of the atmosphere: we may infer what would occur if the invisible atmospheric vapour were entirely withdrawn. I am far from saying that the body of the atmosphere exerts no action whatever upon the waves of terrestrial heat; but only that its action is so small that, when due precautions are taken to have the air pure and dry, laboratory experiments fail to reveal any action. Without its vaporous screen, our solid earth would practically be in the presence of stellar space; and with that space, so long as a difference existed between them, the earth would continue to exchange temperatures. The final result of such a process may be surmised. If carried far enough, it would infallibly extinguish the life of our planet.

[J. T.]

Friday, February 29, 1884.

Sir William Bowman, Bart. LL.D. F.R.S. Honorary Secretary and Vice-President, in the Chair.

Professor D. E. Hughes, F.R.S. *M.R.I.*

Theory of Magnetism.

The theory of magnetism, which I propose demonstrating this evening, may be termed the mechanical theory of magnetism, and, like the now well-established mechanical theory of heat, replaces the assumed magnetic fluids and elementary electric currents by a simple, symmetrical, mechanical motion of the molecules of matter and ether.

That magnetism is of a molecular nature has long been accepted, for it is evident that, no matter how much we divide a magnet, we still have its two poles in each separate portion, consequently we can easily imagine this division carried so far that we should at last arrive at the molecule itself possessing its two distinctive poles, consequently all theories of magnetism attempt some explanation of the cause of this molecular polarity, and the reason for apparent neutrality in a mass of iron.

Coulomb and Poisson assume that each molecule is a sphere containing two distinct magnetic fluids, which in the state of neutrality are mixed together, but when polarised are separated from each other at opposite sides; and, in order to explain why these fluids are kept apart as in a permanent magnet, they had to assume, again, that each molecule contained a peculiar coercive force, whose functions were to prevent any change or mixing of these fluids when separated.

There is not one experimental evidence to prove the truth of this assumption; and as regards coercive force, we have direct experimental proof opposing this view, as we know that molecular rigidity or hardness, as in tempered steel, and molecular freedom or softness, as in soft iron, fulfil all the conditions of this assumed coercive force.

Ampère's theory, based upon the analogy of electric currents, supposes elementary currents flowing around each molecule, and that in the neutral state these molecules are arranged hap-hazard in all directions, but that magnetisation consists in arranging them symmetrically.

The objections to Ampère's theory are numerous. 1st. We have no knowledge or experimental proof of any elementary electric currents continually flowing without any expenditure of energy. 2nd. If we admit the assumption of electric currents around each molecule, the molecule itself would then be electro-magnetic, and the question still remains, What is polarity? Have the supposed electric currents separated the two assumed magnetic fluids contained in the molecule, as in Poisson's theory? or are the electric currents themselves magnetic, independent of the iron molecule?

In order to produce the supposed heterogeneous arrangement of neutrality, Ampère's currents would have either to change their position upon the molecule, and have no fixed axis of rotation, or else the molecule, with its currents and polarities, would rotate, and thus be acting in accordance with the theory of De la Rive. 3rd. This theory does not explain why (as in the case of soft iron) polarity should disappear whenever the exciting cause is removed, as in the case of transient magnetisation. It would thus require a coercive force in iron to cause exactly one-half of the molecules to instantly reverse their direction, in order to pass from apparent external polarity to that of neutrality.

The influence of mechanical vibrations and stress upon iron in facilitating or discharging its magnetism, as proved by Matteucci, 1847, in addition to the discovery by Page, 1837, of a molecular movement taking place in iron during its magnetisation, producing audible sounds, and the discovery by Dr. Joule, 1842, of the elongation of iron when magnetised, followed by the discoveries of Guillemin, that an iron bar bent by a weight at its extremity would become straight when magnetised; also that magnetism would tend to take off twists or mechanical strains of all kinds—together with the researches of Matteucci, Marianini, De la Rive, Sir W. Grove, Faraday, Weber, Wiedemann, Du Moncel, and a host of experimenters, including numerous published researches by myself—all tend to show that a mechanical action takes place whenever a bar of iron is magnetised, and that the combined researches demonstrate that the movement is that of molecular rotation.

De la Rive was the first to perceive this, and his theory, like those of Weber, Wiedemann, Maxwell, and others, is based upon molecular rotation. Their theories, however, were made upon insufficient data, and have proved to be wrong as to the assumed state of neutrality, and right only where the experimental data clearly demonstrated rotation.

I believe that a true theory of magnetism should admit of complete demonstration, that it should present no anomalies, and that all the known effects should at once be explained by it.

From numerous researches I have gradually formed a theory of magnetism entirely based upon experimental results, and these have led me to the following conclusions:—

1. That each molecule of a piece of iron, as well as the atoms of all matter, solid, liquid, gaseous, and the ether itself, is a separate and independent magnet, having its two poles and distribution of magnetic polarity exactly the same as its total evident magnetism when noticed upon a steel bar-magnet.

2. That each molecule can be rotated in either direction upon its axis by torsion, stress, or by physical forces such as magnetism and electricity.

3. That the inherent polarity or magnetism of each molecule is a constant quantity like gravity; that it can neither be augmented nor destroyed.

4. That when we have external neutrality, or no apparent magnetism, the molecules arrange themselves so as to satisfy their mutual attraction by the shortest path, and thus form a complete closed circuit of attraction.

5. That when magnetism becomes evident, the molecules and their polarities have all rotated symmetrically, producing a north pole if rotated in a given direction, or a south pole if rotated in the opposite direction. Also, that in evident magnetism we have still a symmetrical arrangement, but one whose circles of attraction are not completed except through an external armature joining both poles.

6. That we have permanent magnetism when the molecular rigidity, as in tempered steel, retains them in a given direction, and transient magnetism whenever the molecules rotate in comparative freedom, as in soft iron.

Experimental Evidences.

In the above theory the coercive force of Poisson is replaced by molecular rigidity and freedom; and as the effect of mechanical vibrations, torsion, and stress upon the apparent destruction and facilitation of magnetism is well known, I will, before demonstrating the more serious parts of the theory, make a few experiments to prove that molecular rigidity fulfils all the requirements of an assumed coercive force.

I will now show you that if I magnetise a soft iron rod, the slightest mechanical vibration reduces it to zero; whilst in tempered steel or hard iron, the molecules are comparatively rigid, and are but slightly affected. The numerous experimental evidences which I shall show prove that whilst the molecules are not completely rigid in steel, they are comparatively rigid when compared with the extraordinary molecular freedom shown in soft iron. (*Experiments shown.*)

If I now take a bottle of iron filings, I am enabled to show how completely rigid they appear if not shaken; but the slightest motion allows these filings to rotate and short circuit themselves, thus producing apparent neutrality. Now I will restore the lost magnetism by letting the filings slowly fall on each other under the influence of the earth's magnetic force; and here we have an evident proof of rotation producing the result, as we can ourselves perceive the arrangement of the filings. (*Experiment shown.*)

If I take this extremely soft bar of iron, you notice that the slightest mechanical tremor allows molecular rotation, and consequent loss or change of polarity; but if I put a slight strain on this bar, so as to fasten each molecule, they cannot turn with the same freedom as before, and they now retain their symmetrical polarity like tempered steel, even when violently hammered. (*Experiment shown.*)

We can only arrive at one conclusion from this experiment, viz. that the retention of apparent magnetism is simply due to a frictional resistance to rotation; and whenever this frictional resistance is reduced, as when we take off a mechanical strain, or by making the bar red hot, the molecules then rotate with an almost inconceivable freedom from frictional resistance.

Conduction.

You notice that if I place this small magnet at several inches' distance from the needle, it turns in accordance with the pole presented. How is the influence transmitted from the magnet to the needle? It is through the atmosphere and the ether, which is the intervening medium. I have made a long series of researches on the subject, involving new experimental methods, the results of which are not yet published. One result, however, I may mention. We know that iron cannot be magnetised beyond a certain maximum, which we call its saturation point. It has a well-defined curve of rise to saturation, agreeing completely with a curve of force produced by the rotation of a bar magnet, the force of which was observed from a fixed point. I have completely demonstrated by means of my magnetic balance (*shown in the Library*) that our atmosphere, as well as Crooke's vacuum, has its saturating point exactly similar in every respect to that of iron: it has the same form through every degree. We cannot reduce nor augment the saturating point of ether; it is invariable, and equals the finest iron. We may, however, easily reduce that of iron by introducing frictional resistance to the free motion of its molecules.

From consideration of the ether having its saturating point, I am forced to the conclusion that it could only be explained by a similar rotation of its atoms as demonstrable in iron.

Reflection would teach us that there cannot be two laws of magnetism, such as one of vibrations in the ether and rotations in iron. We cannot have two correct theories of heat, light, or magnetism; the mode of motion in the case of magnetism being rotation, and not vibration.

Let us observe this saturation point of the atmosphere compared with iron. I pass a strong current of electricity in this coil. The coil is quite hot, so we are very near its saturation. I now place this coil at a certain distance from the needle (8 inches); we have now a deflection of 45° on the needle. I now introduce this iron core, exactly fitting the interior previously filled by the ether and atmosphere. Its force is much greater, so I gradually remove this coil to a distance, where I find the same deflection as before (45°). This happens to be at twice the distance, or 16 inches, so we know, according to the law of inverse squares, that the iron has four times the magnetic power of the atmosphere. But this is only true for this piece of iron: with extremely fine specimens of iron I have been enabled to increase the force of the coil forty times, whilst with manganese steel containing 10 per cent. of manganese it was only 30 per cent. superior. We see here that the atmosphere is extremely magnetic. Let us replace the solid bar by iron filings. We now only have twice the force. Replace this by a bottle of sulphate of iron in a liquid state: it is now a mere fraction superior to the atmosphere; and if we were still further to separate the iron molecules, as in a gaseous state, it is reasonable to suppose that if we could isolate the iron gas from that of ether, that iron gas would be strongly diamagnetic, or have far less magnetic capacity than ether, owing to the great separation of its molecules. These are assumptions, but they are based upon experimental evidences, which give it value.

Let us quit the domain of assumption to enter that of demonstration. Here I have a long bar of neutral iron. If I place this small magnet at one end, we notice that its pole has moved forward three inches, having a consequent point at that place. Let us now vibrate this rod, and you notice the slow but gradual creeping of the conduction until at the end of two seconds it has reached 14 inches. The molecules have been freed from frictional resistance by the mechanical vibrations, and have at once rotated all along the bar. (*Experiment shown.*) Let us repeat this experiment by heating the rod to red heat. You notice the gradual creeping or increased conduction as the heat allows greater molecular freedom. (*Experiment shown.*) Let us now again repeat this experiment by sending a current of electricity through the bar. You notice the instant that I touch the bar with this wire, conveying the current through it, that we have identically the same creeping forwards, no matter what direction of the current. (*Experiment shown.*) If you simply looked at the effects produced, you could not tell which method I had employed; either mechanical vibrations, heat vibrations, or electrical currents. Consequently, knowing the two first to be modes of motion, it is fair to assume that an electrical current is a mode of motion, the manner of which is at present unknown; but that there is a molecular disturbance in each case is evident from the experiments shown.

Neutrality.

If I take this bar of soft iron, introduce it in the coil, and pass a strong electric current though the coil, you notice that it is intensely magnetic, holding up this large armature of iron and strongly deflecting the observing needle. I now interrupt the current, the armature falls, and the needle only shows traces of the previous intense magnetisation. What has become of this polarity? or what has caused this sudden neutrality? Coulomb supposes that the magnetic fluids have become mixed in each molecule, thus neutralising each other. Ampère supposes that the elementary currents surrounding each molecule have become heterogeneous. De la Rive, Wiedemann, Weber, Maxwell, and all up to the present time have accounted for this disappearance as a case of mixture of polarities or heterogeneous arrangement.

My researches proved to me that neutrality was a symmetrical arrangement; I stated this in my paper upon the theory of magnetism to the Royal Society last year. I have since made a long series of researches upon this question, and my paper upon this subject will shortly be read at the Royal Society. This paper will demonstrate beyond question—1. That a bar of iron under the influence of a current or other magnetising force is more strongly polarised on the outside than in the interior; that its degree of penetration follows the well-defined law of inverse squares, up to the saturation point of each successive layer. 2. The instant that the current ceases, a reaction takes place, the stronger outside reacting upon the weaker inside, completely reversing it, until its reversed polarity exactly balances the external layers.

We might here suppose that there existed two distinct polarities at the same end of a neutral bar, but this is only partially true, as the rotation of the molecules from the inside to the exterior is a gradual, well-defined curve, perfectly marked, as shown in the diagrams. (*Diagrams explained.*) We see from these that in a large solid bar the reversed polarity would be in the interior, but in a thin bar under an intense field, the reversed polarity would be on the outside. Thus a bar which had previously strong north polarity under an external influence would, the instant it formed its neutrality, have a north polarity in the interior covered or rendered neutral by an equal south exterior, the sum of both giving the apparent neutrality that we notice. I must refer all interested upon this question to my paper shortly to be read, but I will make a few experiments to demonstrate this important fact.

If I take this piece of soft steel and magnetise it strongly, it has a strong remaining magnetism, or only partial neutrality. If I now heat this steel to redness, or put it into a state of mechanical vibration, the remaining magnetism almost entirely disappears, and we have apparent neutrality. This piece of steel being thin ($\frac{1}{2}$ millimetre), I know that the outside is reversed to its previous state. I place this piece of steel in a glass vase near the observing needle, and at present there seems no polarity. I now pour dilute nitric acid upon it, filling up the vase. The exterior is now being dissolved, and in a few minutes you will see a strong polarity in the steel, as the exterior reversed polarity is dissolved in the acid. (*Experiment shown.*)

Let us observe this by a different method. I take two strips of hard iron, and magnetise them both in the same direction.

If I place them together and then separate them, there seems no change, although in reality the mere contact produced a commencement of reversal. Let us vibrate them whilst together, allowing the molecules greater freedom to act as they feel inclined ; and now on separating we see that one strip has exactly the opposite polarity to the other, both extremely strong, but the sum of which, when placed together, is zero, or neutrality. (*Experiment shown.*)

Let us take two extremely soft strips placed together, and magnetised whilst together. On withdrawal of the inducing force, the rods are quite neutral. (*Experiment shown.*)

We now separate these strips, and find that one is violently polarised in one direction, whilst the other is equally strong in the reversed; the sum of both being again zero.

We might suppose that the reaction is due to having separate bars. I will now demonstrate that this is not the case by magnetising this large $\frac{3}{4}$-inch bar with a magnetising force just sufficient to render the rod completely neutral when held vertically or under the earth's magnetic influence. (*Experiment shown.*)

You notice that it is absolutely neutral, all parts as well as the ends showing not the slightest trace of polarisation. I reverse this bar, and you perceive that it is now intensely polarised. This is due to the fact that the earth's influence uncovers or reverses the outside molecules, and consequently they are now of the same polarity as its interior. Upon reversing this rod, the magnetism again disappears, and re-appears if turned as previously. We have thus a rod which appears intensely magnetic when one of its ends is lowermost, whilst if that same end is turned upwards all traces of magnetism disappear. These and several other demonstrations which I shall now show you (proving the enormous influence which thickness of a bar has in the production of neutrality or its retention of magnetism) are simple lecture demonstrations. For the complete proof of my discovery of neutral curves I must refer you to my forthcoming paper upon this subject. (*Experiments shown proving the great influence of a thickness of a bar upon its retentive and neutral powers.*)

Inertia.

I have remarked in my researches that the molecules have true inertia, that they resist being put in motion, and if put in motion will vanquish an opposing resistance by their simple momentum. To illustrate this, I take this large $\frac{3}{4}$-inch bar, magnetise it so that its south pole is at the lowest end. We know that the earth's influence is to make the lower end north. I now gently strike it with a wooden mallet, and the rod immediately falls to zero. I continue these blows, but the rod obstinately refuses to pass the neutral line to become north, the reason being in so doing it would have to change the whole internal reversed curve that I have discovered. It requires now extremely violent and repeated blows from the mallet to make it obey the earth's influence.

Let us repeat this experiment by starting the molecule rapidly in the first instance. The rod is now magnetised south as before. I give one single sharp tap; the molecules run rapidly round, pass through neutrality, breaking up its curve, and arrive at once to strong north polarity. (*Experiment shown.*)

A very extraordinary effect is shown if we produce this effect by electricity; it then almost appears as if electricity itself had inertia. I take this bar of hard iron and magnetise it to a fixed degree. On the passage of the current, you notice that the magnetism seems to be increased as the needle increases its arc, but this is caused by the deflection of the electric current in the bar. The current is now obliged to travel in spirals, as my researches have proved to me that electricity can only travel at right angles to the magnetic polar direction of a molecule, consequently in all permanent magnets the current must pass at right angles to the molecule, and its path will be that of a spiral. Let us replace this bar by one from a similar kind of iron well annealed. The molecules here are in a great state of freedom. We now magnetise this rod to the same degree as in the previous case; the electric current now, instead of being deflected, completely rotates the molecules, and the needle returns to zero, all traces of external magnetism having ceased. The electricity on entering this bar should have been forced to follow a tortuous circular route; its momentum was, however, too great for the molecules, and they elected to turn, allowing the electricity to pass in a straight line through the bar. Thus, in the first instant, magnetism was the master directing the course of the current; in the last, it became its servant, obeying by turning itself to allow a straight path to its electric master. (*Experiment shown.*)

Superposed Magnetism.

It is well known that we can superpose a weak contrary polarity upon an internal one of an opposite name. I have been enabled thus to superpose twenty successive stratas of opposite polarities upon a single rod, by simply diminishing the force at each reversal. I was anxious to prepare a steel wire so that in its ordinary state it would be neutral, but that in giving it a torsion to the right one polarity would appear, whilst a torsion to the left would produce the opposite polarity. This I have accomplished by taking ordinary soft steel drill wire and magnetising it strongly whilst under a torsion to the right, and more feebly with an opposite polarity when magnetised under torsion to the left.

The power of these wires, if properly prepared, is most remarkable, being able to reverse their polarity under torsion, as if they were completely saturated; and they preserve this power indefinitely if not touched by a magnet. It would be extremely difficult to explain the action of the rotative effects obtained in these wires under any other theory than that which I have advanced; and the absolute external neutrality that we obtain in them when the polarities are changing we know, from their structure, to be perfectly symmetrical.

I was anxious to show some mechanical movement produced by molecular rotation, consequently I have arranged two bells that are struck alternately by a polarised armature put in motion by the double polarised rod I have already described, but whose position, at three centimetres distant from the axis of the armature, remains invariably the same. The magnetic armature consists of a horizontal light steel bar suspended by its central axle; the bells are thin wineglasses, giving a clear musical tone loud enough, by the force with which they are struck, to be clearly heard at some distance. The armature does not strike these alternately by a pendulous movement, as we may easily strike only one continuously, the friction and inertia of the armature causing its movements to be perfectly dead-beat when not driven by some external force, and it is kept in its zero position by a strong directive magnet placed beneath its axle.

The mechanical power obtained is extremely evident, and is sufficient to put the sluggish armature in rapid motion, striking the bells six times per second, and with a power sufficient to produce tones loud enough to be clearly heard in all parts of the hall of the Institution.

There is nothing remarkable in the bells themselves, as they evidently could be rung if the armature was surrounded by a coil, and worked by an electric current from a few cells. The marvel, however, is in the small steel superposed magnetic wire producing by slight elastic torsions from a single wire, 1 millimetre in diameter, sufficient force from mere molecular rotation to entirely replace the coil and electric current. (*Experiment shown by ringing the bells by the torsion of a small $\frac{1}{16}$-inch wire placed 4 inches distant from bell-hammer.*)

Correlation of Forces.

There is at present a tendency to trace all physical forces to one, or rather a variation of modes of motion. In my last experiment the energy of my arm was transformed in the wire to molecular motion, producing evident polarity; this, again, acted upon the ether, putting the needle-hammer into mechanical motion. This by its impact upon the glass bells transformed its motions into sonorous vibrations; but this does not mean that we can convert directly sonorous vibrations into magnetism, or *vice versâ*.

Let us take this soft iron rod; it seems quite neutral, although we know that the earth's magnetism is trying to rotate its molecules to north polarity at its lowest extremity. We now put it in mechanical vibration by striking it gently with a wooden mallet; the molecules at once rotate, and we have the expected strong north polarity. Let us repeat this experiment by employing heat, and here, again, at red heat an equally strong north polarity appears.

Again we repeat, and simply pass an electric current of no matter what direction; again the same north pole appears. Thus these forces must be very similar in nature, and may be fairly presumed to be vibrations, or modes of motion, having no directive tendency except a slight one, as in the case of electricity. For the same three forces render the rod perfectly neutral, even when previously magnetised, when placed in a longitudinally neutral field, as east and west.

Motion of the molecules gives rise to external magnetism to a rod previously neutral, or renders it neutral when previously magnetised; in other words, it simply allows the molecules to obey an external directing influence; the only motion, therefore, is during a change of state or polarity. If there is constant polarity, there is no consequent motion of the molecules: in fact, the less motion of any kind that it can receive, the more perfect its retention of its previous position; consequently, constant magnetism cannot be looked upon as a mode of motion, neither vibratory nor rotatory; it is an inherent quality of each molecule, similar in its action to its chemical affinity, cohesion, or its polar power of crystallisation. A molecule of all kinds of matter has numerous endowed qualities; they are inherent, and special in degree to the molecule itself. I regard the magnetic endowed qualities of all matter or ether to be inherent, and that they are rendered evident by rotation to a symmetrical arrangement in which their complete polar attractions are not satisfied.

Time will not allow me to show how completely this view explains all the phenomena of electro-magnetism, diamagnetism, earth currents—in fact, all the known effects of magnetism—up to the original cause of the direction of the molecules of the earth. To explain the first cause of the direction of the molecules of the earth would rest altogether upon assumption as the first cause of the earth's rotation, and of all things down to the inherent qualities of the molecule itself.

The mechanical theory of magnetism which I have advocated seems to me as fairly demonstrable as the mechanical theory of heat, and it gives me great pleasure to have been allowed to present you with my views on the theory of magnetism.

[D. E. H.]

Friday, March 28, 1884.

SIR FREDERICK POLLOCK, Bart. M.A. Vice-President, in the Chair.

PROFESSOR OSBORNE REYNOLDS, M.A. F.R.S.

The Two Manners of Motion of Water.

IN commencing this discourse the author said :—

It has long been a matter of very general regret with those who are interested in natural philosophy, that in spite of the most strenuous efforts of the ablest mathematicians the theory of fluid motion fits very ill with the actual behaviour of fluids; and this for unexplained reasons. The theory itself appears to be very tolerably complete and affords the means of calculating the results to be expected in almost every case of fluid motion, but while in many cases the theoretical results agree with those actually obtained, in other cases they are altogether different.

If we take a small body such as a raindrop moving through the air, the theory gives us the true law of resistance; but if we take a large body such as a ship moving through the water, the theoretical law of resistance is altogether out. And what is the most unsatisfactory part of the matter is that the theory affords no clue to the reason why it should apply to the one class more than the other.

When, seven years ago, I had the honour of lecturing in this room on the then novel subject of vortex motion, I ventured to insist that the reason why such ill success had attended our theoretical efforts was because, owing to the uniform clearness or opacity of water and air, we can see nothing of the internal motion; and while exhibiting the phenomena of vortex rings in water rendered strikingly apparent by partially colouring the water, but otherwise as strikingly invisible, I ventured to predict that the more general application of this method, which I may call the method of colour-bands, would reveal clues to those mysteries of fluid motion which had baffled philosophy.

To-night I venture to claim what is at all events a partial verification of that prediction. The fact that we can see as far into fluids as into solids naturally raises the question why the same success should not have been obtained in the case of the theory of fluids as in that of solids? The answer is plain enough. As a rule, there is no internal motion in solid bodies; and hence our theory based on the assumption of relative internal rest applies to all cases. It is not, however, impossible that an, at all events seemingly, solid body should have internal motion, and a simple experiment will show that if a class of such bodies existed they would apparently have disobeyed the laws of motion.

These two wooden cubes are apparently just alike, each has a string tied to it. Now, if a ball is suspended by a string you all know that it hangs vertically below the point of suspension or swings like a pendulum. You see this one does so. The other you see behaves quite differently, turning up sideways. The effect is very striking so long as you do not know the cause. There is a heavy revolving wheel inside which makes it behave like a top.

Now what I wish you to see is, that had such bodies been a work of nature so that we could not see what was going on—if, for instance, apples were of this nature while pears were what they are—the laws of motion would not have been discovered; if discovered for pears they would not have applied to apples, and so would hardly have been thought satisfactory.

Such is the case with fluids: here are two vessels of water which appear exactly similar—even more so than the solids, because you can see right through them—and there is nothing unreasonable in supposing that the same laws of motion would apply to both vessels. The application of the method of colour-bands, however, reveals a secret: the water of the one is at rest, while that in the other is in a high state of agitation.

I am speaking of the two manners of motion of water—not because there are only two motions possible; looked at by their general appearance the motions of water are infinite in number; but what it is my object to make clear to-night is that all the various phenomena of moving water may be divided into two broadly distinct classes, not according to what with uniform fluids are their apparent motions, but according to what are the internal motions of the fluids which are invisible with clear fluids, but which become visible with colour-bands.

The phenomena to be shown will, I hope, have some interest in themselves, but their intrinsic interest is as nothing compared to their philosophical interest. On this, however, I can but slightly touch.

I have already pointed out that the problems of fluid-motion may be divided into two classes: those in which the theoretical results agree with the experimental, and those in which they are altogether different. Now what makes the recognition of the two manners of internal motion of fluids so important, is that all those problems to which the theory fits belong to the one class of internal motions.

The point before us to-night is simple enough, and may be well expressed by analogy. Most of us have more or less familiarity with the motion of troops, and we can well understand that there exists a science of military tactics which treats of the best manœuvres and evolutions to meet particular circumstances.

Suppose this science proceeds on the assumption that the discipline of the troops is perfect, and hence takes no account of such moral effects as may be produced by the presence of an enemy.

Such a theory would stand in the same relation to the movements of troops as that of hydrodynamics does to the movements of water. For although only the disciplined motion is recognised in military tactics, troops have another manner of motion when anything disturbs their order. And this is precisely how it is with water: it will move in a perfectly direct disciplined manner under some circumstances, while under others it becomes a mass of eddies and cross streams which may be well likened to the motion of a whirling, struggling mob where each individual particle is obstructing the others.

Nor does the analogy end here: the circumstances which determine whether the motion of troops shall be a march or a scramble, are closely analogous to those which determine whether the motion of water shall be direct or sinuous.

In both cases there is a certain influence necessary for order: with troops it is discipline; with water it is viscosity or treacliness.

The better the discipline of the troops, or the more treacly the fluid, the less likely is steady motion to be disturbed under any circumstances. On the other hand, speed and size are in both cases influences conducive to unsteadiness. The larger the army, and the more rapid the evolutions, the greater the chance of disorder; so with fluid the larger the channel, and the greater the velocity, the more chance of eddies.

With troops some evolutions are much more difficult to effect with steadiness than others, and some evolutions which would be perfectly safe on parade, would be sheer madness in the presence of an enemy. So it is with water.

One of my chief objects in introducing this analogy of the troops is to emphasise the fact, that even while executing manœuvres in a steady manner there may be a fundamental difference in the condition of the fluid. This is easily realised in the case of troops. Difficult and easy manœuvres may be executed in equally steady manners if all goes well, but the conditions of the moving troops are essentially different. For while in the one case any slight disarrangement would be easily rectified, in the other it would inevitably lead to a scramble. The source of such a change in the manner of motion under such circumstances, may be ascribed either to the delicacy of the manœuvre, or to the upsetting disturbance, but as a matter of fact, both of these causes are necessary. In the case of extreme delicacy an indefinitely small disturbance, such as is always to be counted on, will effect the change.

Under these circumstances we may well describe the condition of the troops in the simple manœuvre as stable, while that in the delicate manœuvre is unstable, i.e. will break down on the smallest disarrangement. The small disarrangement is the immediate source of the break-down in the same sense as the sound of a voice is sometimes the cause of an avalanche; but if we regard such disarrangement as certain to occur, then the source of the disturbance is a condition of instability.

All this is exactly true for the motion of water. Supposing no disarrangement, the water would move in the manner indicated in theory just as, if there is no disturbance, an egg will stand on its end; but as there is always slight disturbance, it is only when the condition of steady motion is more or less stable that it can exist. In addition then to the theories either of military tactics or of hydrodynamics, it is necessary to know under what circumstances the manœuvres of which they treat are stable or unstable. And it is in definitely separating these conditions that the method of colour-bands has done good service which will remove the discredit in which the theory of hydrodynamics has been held.

In the first place, it has shown that the property of viscosity or treacliness, possessed more or less by all fluids, is the general influence conclusive to steadiness, while, on the other hand, space and velocity are the counter influence; and the effect of these influences is subject to one perfectly definite law, which is that a particular evolution becomes unstable for a definite value of the viscosity divided by the product of the velocity and space. This law explains a vast number of phenomena which have hitherto appeared paradoxical. One general conclusion is, that with sufficiently slow motion all manners of motion are stable.

The effect of viscosity is well shown by introducing a band of coloured water across a beaker filled with clear water at rest. Now the water is quite still, I turn the beaker round about its axis. The glass turns but not the water, except that which is close to the glass. The coloured water which is close to the glass is drawn out into what looks like a long smear, but it is not a smear, it is simply a colour-band extending from the point in which the colour touched the glass in a spiral manner inwards, showing that the viscosity was slowly communicating the motion of the glass to the water within. To prove this I have only to turn the beaker back, and the colour band assumes its radial position. Throughout this evolution the motion has been quite steady—quite according to the theory.

When water flows steadily it flows in streams. Water flowing along a pipe is such a stream bounded by the solid surface of the pipe, but if the water be flowing steadily we can imagine the water to be divided by ideal tubes into a fagot of indefinitely small streams, any of which may be coloured without altering its motion, just as one column of infantry may be distinguished from another by colour.

If there is internal motion, it is clear that we cannot consider the whole stream bounded by the pipe as a fagot of elementary streams, as the water is continually crossing the pipe from one side to the other, any more than we can distinguish the streaks of colour in a human stream in the corridor of a theatre.

Solid walls are not necessary to form a stream: the jet from a fire hose, the falls of Niagara, are streams bounded by a free surface.

A river is a stream half bounded by a solid surface.

Streams may be parallel, as in a pipe; converging, as in a conical mouth-piece; or when the motion is reversed, diverging. Moreover, the streams may be straight or curved.

All these circumstances have their influence on stability in a manner which is indicated in the accompanying diagram:—

Circumstances conducive to

Direct or Steady Motion.	*Sinuous or Unsteady Motion.*
1. Viscosity or fluid friction which continually destroys disturbances. (Treacle is steadier than water.)	5. Particular variation of velocity across the stream, as when a stream flows through still water.
2. A free surface.	6. Solid bounding walls.
3. Converging solid boundaries.	7. Diverging solid boundaries.
4. Curvature with the velocity greatest on the outside.	8. Curvature with the velocity greatest on the inside.

It has for a long time been noticed that a stream of fluid through fluid otherwise at rest is in an unstable condition. It is this instability which gives rise to the talking-flame and sensitive-jet with which you have been long familiar in this room. I have here a glass vessel of clear water in front of the lantern, so that any colour-bands will be projected on the screen.

You see the ends of two vertical tubes one above the other. Nothing is flowing through these tubes, and the water in the vessel is at rest. I now open two taps, so as to allow a steady stream of coloured water to enter at the lower pipe, water flowing out at the upper. The water enters quite steadily, forms a sort of vortex ring at the end which proceeds across the vessel, and passes out at the lower tube. Now the coloured stream extends straight across the vessel, and fills both pipes. You see no motion; it looks like a glass rod. The water is, however, flowing slowly along it. The motion is so slow, that the viscosity is paramount, and hence the stream is steady.

I increase the speed, you see a certain wriggling sinuous action in the column; faster, the column breaks up into beautiful and well-defined eddies, and spreads out into the surrounding water, which, becoming opaque with colour, gradually draws a veil over the experiment.

The same is true of all streams bounded by standing water. If the motion is sufficiently slow, according to the size of the stream and the viscosity of the fluid, it is steady and stable. At a certain critical velocity, the which is determined by the ratio of the viscosity to the diameter of the stream, the stream becomes unstable. Under any conditions, then, which involve a stream flowing through surrounding water, the motion will be unstable if the velocity is sufficient.

Now, *one* of the most marked facts relating to experimental hydrodynamics is the difference in the way in which water flows along contracting and expanding channels; these include an enormously large class of the motions of water, but the typical phenomenon is shown by the simple conical tubes. Such a tube is now projected on the screen; it is surrounded with clear still water. The mouth of the tube at which the water enters is the largest part, and it contracts uniformly for some way down the channel, then the tube expands again gradually until it is nearly as large as at the mouth, and then again contracts to the tube necessary to discharge the water. I draw water through the tube, but you see nothing as to what is going on. I now colour one of the elementary streams outside the mouth; this colour-band is drawn in with the surrounding water, and will show us what is going on. It enters quite steadily, preserving its clear streak-like character until it has reached the neck where convergence ceases; now the moment it enters the expanding tube it is altogether broken up into eddies. Thus the motion is direct in the contracting tube, sinuous in the expanding.

The hydrodynamical theory affords no clue to the cause why; and even by the method of colour-bands the reason for the sinuosity is not at once obvious. If we start the current suddenly, the motion is at first the same in both tubes, its change in the expanding pipe seemed to imply that here the motion was unstable. If so, this ought to appear from the equations of motion. With this view this case was studied, I am ashamed to say how long, without any light. I then had recourse to the colour-bands again, to try and see how the phenomena came on. It all then became clear: there is an intermediate stage. When the tap is opened, the immediately ensuing motion is nearly the same in both parts; but while that in the contracting portion maintains its character, that in the expanding portion changes its character. A vortex ring is formed which, moving forward, leaves the motion behind that of a parallel stream through the surrounding water.

If the motion be sufficiently slow, as it is now, this stream is stable, as already explained. We thus have steady or direct motion in both the contracting and expanding parts of the tube, but the two motions are not similar: the first being one of a fagot of similar elementary contracting streams, the latter being that of one parallel stream through the surrounding fluid. The first of these is a stable form; the second an unstable form, and, on increasing the velocity, the first remains, while the second breaks down; and we have, as before, the expanding part filled with eddies.

This experiment is typical of a large class of motions. Wherever fluid flows through a narrow, as it approaches the neck it is steady, after passing, it is sinuous. The same effect is produced by an obstacle in the middle of a stream; and very nearly the same thing by the motion of a solid object through the water.

You see projected on the screen an object not unlike a ship. Here the ship is fixed, and the water flowing past it; but the effect would be the same if we had the ship moving through the water. In the front of the ship the stream is steady, and so till it has passed the middle, then you see the eddies formed behind the ship. It is these eddies which account for the discrepancy between the actual and theoretical resistance of ships. We see, then, that the motion in the expanding channel is sinuous because the only steady motion is that of a stream through water. Numerous cases in which the motion is sinuous may be explained in the same way, but not all.

If we have a perfectly parallel channel, neither contracting nor expanding, the steady moving stream will be a fagot of perfectly steady parallel elementary streams all in motion, but moving fastest at the centre. Here we have no stream through steady water. Now when this investigation began it was not known, or imperfectly known, whether such a stream was stable or not, but there was a well-known anomaly in the resistance to motion in parallel channels. In rivers, and all pipes of sensible size, experience had shown that the resistance increased as the square of the velocity, whereas in very small pipes, such as represent the smaller veins in animals, Poiseuille had proved the resistance increased as the velocity.

Now since the resistance would be as the square of the velocity with sinuous motion, and as the velocity, if direct, it seemed that the discrepancy could be accounted for if the motion could be shown to become unstable for a sufficiently large velocity. This suggested the experiment I am now about to produce before you.

You see on the screen a pipe with its end open. It is surrounded by clear water and by opening a tap I can draw water through it. This makes no difference to the appearance until I colour one of the elementary streams, when you see a beautiful streak of colour extend all along the pipe. The stream has so far been running steadily, and appears quite stable. I now merely increase the speed; it is still steady, but the colour-band is drawn down fine. I increase the colour and then again increase the speed. Now you see the colour-band at first vibrates and then mixes so as to fill the tube. This is at a definite velocity; if the velocity be diminished ever so little the band becomes straight and clear; increase it again, it breaks up. This critical speed depends on the size of the tube in the exact inverse ratio; the smaller the tube, the greater the velocity; also, the more viscous the water the greater the velocity.

We have then not only a complete explanation of the difference in the laws of resistance generally experienced and that found by Poiseuille, but also we have complete evidence of the instability of parallel streams flowing between or over solid surfaces. The cause of the instability is as yet not explained, but this much can be shown, that whereas lateral stiffness in the walls is unimportant, inextensibility or tangential rigidity is essential to the creation of eddies. I cannot show you this because the only way in which we can produce the necessary conditions without a solid channel is by a wind blowing over water. When the wind blows over water it imparts motion to the surface of the water just as a moving solid surface; moving in this way, however, the water is not susceptible of eddies. It is unstable, but the result of disturbance is waves. This is proved by an experiment long known, but which has recently attracted considerable notice.

If oil be put on the surface it spreads out into an indefinitely thin sheet which possesses only one of the characteristics of a solid surface, it offers resistance, very slight, but still resistance to extension and contraction. This, however, is sufficient to entirely alter the character of the motion. It renders the water unstable internally, and instead of waves, what the wind does is to produce eddies beneath the surface. This has been proved, although I cannot show you the experiments.

To those who have observed the phenomena of oil preventing waves, there is probably nothing more striking throughout the region of mechanics. A film of oil so thin that we have no means of illustrating its thickness, and which cannot be perceived except by its effect—which possesses no mechanical properties that can be made apparent to our senses—is yet able to entirely prevent an action which involves forces the strongest we can conceive, which upset our ships and destroy our coasts. This, however, becomes intelligible when we perceive that the action of the oil is not to calm the sea by sheer force, but merely, as by its moral force, to alter the manner of motion produced by the action of the wind from that of the terrible waves upon the surface into the harmless eddies below. The wind throws the water into a highly unstable condition, into what morally we should call a condition of great excitement. The oil by an influence we cannot perceive directs this excitement.

This influence, though insensibly small, is however now proved of a mechanical kind, and to me it seems that the phenomenon of one of the most powerful mechanical actions of which the forces of nature are capable, being entirely controlled by a mechanical force so slight as to be otherwise quite imperceptible, does away with every argument against the strictly mechanical sources of what we may call mental and moral forces.

But to return to the instability in parallel channels. This has been the most complete, as well as the most definite result of the colour-bands.

The circumstances are such as to render definite experiments possible. These have been made, and reveal a definite law of the instability, which law has been tested by reference to all the numerous and important experiments on the resistance in channels by previous observers; whereupon it is found that waters behave in exactly the same manner whether the channel, as in Poiseuille's experiment, is of the dimensions of a hair or whether it be the size of a water main or of the Mississippi; the only difference being that in order that the motions may be compared, the velocity must be inversely as the diameter of the pipe. But this is not the only point explained if we consider other fluids than water. Some fluids, like oil or treacle, apparently flow more slowly and steadily than water. This, however, is only in smaller channels; the critical velocity increases with the viscosity of the fluid. Thus, while water in comparatively large streams is always above its critical velocity, and the motion always sinuous, the motion of treacle in streams of such size as we see is below its critical velocity, and the motion direct. But if nature had produced rivers of treacle the size of the Thames, for instance, the treacle would have flowed just like water. Thus, in the lava streams from a volcano, although looked at close the lava has the consistence of a pudding, in the large and rapid streams down the mountain sides the lava flows as freely as water.

I have now only one circumstance left to which to ask your attention. This is the effect of curvature of the stream on the stability of the fluid.

Here again we see the whole effect altered by very slight causes.

If water be flowing in a bent channel in steady streams, the question as to whether it will be stable or not turns on the variation in the velocity from the inside to the outside of the stream.

In front of the lantern is a cylinder with glass ends, so that the light passes through in the direction of the axis. The disk of light on the screen being the light which passes through this water, and is bounded by the circular walls of the cylinder.

By means of two tubes temporarily attached, a stream of coloured water is introduced right across the cylinder extending from wall to wall; the motion is very slow, and the taps being closed, and the tubes removed, the colour-band is practically stationary. The vessel is now caused to revolve about its axis. At first, only the walls of the cylinder move, but the colour-band shows that the water gradually takes up the motion, the streak being wound off at the ends into a spiral thread, but otherwise remaining still and vertical. When the spirals meet in the middle, the whole water is in motion, but the motion is greatest at the outside, and is therefore stable. The vessel stops, and gradually stops the water, beginning at the outside. If the motion remained steady, the spirals would unwind, and the streak be restored. But the motion being slowest at the outside against the surface, you will see eddies form, breaking up the spirals for a certain distance towards the middle, but leaving the middle revolving steadily.

Besides indicating the effect of curvature, this experiment really illustrates the action of the surface of the earth on the air moving over it; the varying temperature having much the same influence as the curvature of the vessel on stability. The air is unstable for a few thousand feet above the surface, and the motion is sinuous, resulting in the mixing of the strata, and producing the heavy cumulus clouds; but above this the influence of temperature predominates, and clouds, if there are any, are of the stratus-form, like the inner spirals of colour. But it was not the intention of this lecture to trace the two manners of motion of fluids in the phenomena of Nature and Art, so I thank you for your attention. [O. R.]

Friday, June 6, 1884.

WARREN DE LA RUE, Esq. M.A. D.C.L. F.R.S. Manager and Vice-President, in the Chair.

WILLOUGHBY SMITH, Esq. *M.R.I.*

Volta-Electric and Magneto-Electric Induction.

THE subject which I shall bring before your notice this evening is "Volta-Electric and Magneto-Electric Induction"; and I propose to describe some of my experiments in connection with this phenomenon in electricity and magnetism, which was discovered, named, and for forty years fostered within this very building by our universally esteemed and beloved Michael Faraday.

It is now thirty years ago that Faraday gave me my first lesson in Volta-Electric Induction, at the same time impressing upon me the fact that physical science must ever be progressive and corrective, and that he was always pleased to learn that his experiments had been repeated by others, with a view to their verification or correction. In the remembrance of this, I have felt encouraged in coming before you this evening.

Doubtless all present are familiar with the fact, that about 1819 Professor Oersted made a discovery which has done much to advance our knowledge with regard to electricity and magnetism. This discovery arose from a very simple experiment, but it is none the less valuable on that account. I will now repeat this experiment, as the result obtained will lead up to, and enable you to better understand what I may show later on. Here is a length of copper wire, from the top surface and centre of which projects a metal pin, having, balanced on its point, a magnet. The position of the steel magnet is, as you perceive, parallel with the wire; and that position being north and south, it will thus remain until, by pressing down this spring, the length of copper wire is placed in metallic circuit with this battery, and you observe that the magnet immediately has a tendency to place itself at right angles to the copper wire, remaining thus displaced until the current ceases to flow, when the needle again obeys the influence of terrestrial magnetism, and returns to its former position. This was Oersted's experiment which led to the discovery that all bodies possess the qualities of a magnet whilst a current is passing through them. From this fact Faraday conceived the idea that, if a similar length of copper wire were placed parallel with the former one, and close to it, each time the circuit was made or unmade there would be an induced current flowing in this wire, provided it was part of a closed metallic or other conducting circuit. The first experiment was not successful, for the simple reason that the galvanometer was not sufficiently sensitive to be visibly affected by the very small current induced in so short a length of wire as that with which he made the experiment; nor was he prepared to find that induced currents were of such momentary duration. But failures with Faraday were merely stepping-stones to success; he repeated the experiments with spirals of insulated wire, instead of straight wires parallel to each other, thus getting comparatively long lengths in close proximity. By these means he gained the object of his search, the result of which he called "Volta-Electric Induction." Through the kindness of Dr. Tyndall, I have here the identical spirals which were made and used by Faraday on that memorable occasion. One of these Faraday connected in circuit with a galvanometer, and placed it on top of the other spiral, through which intermittent currents from a battery were sent at fixed intervals. On "making" the battery circuit, he noticed that the needle of the galvanometer was deflected in one direction, and on breaking the circuit, the needle was again deflected, but in the opposite direction. Faraday saw, in his mind's eye, each particle of the circuit through which the current was passing, acting as a centre of force, emitting its lines far from it, yet each of these lines returning to its own source; he in consequence made a series of experiments, by placing various substances in the path of the lines of force, to ascertain whether they would in any way be affected or intercepted by the substances so placed. For instance, he found that they were sensibly affected by iron, but with copper no satisfactory effects were perceived, although he felt sure the copper did in some way influence them, but so imperceptibly that he was unable to detect it. It was this conviction, and the doubt by Faraday of the result of his experiment with regard to copper, which led me to experiment in this direction. My apparatus and its arrangements being somewhat different from Faraday's, I will more fully describe them. Here are two flat spirals of fine silk-covered copper wire, about twelve inches in diameter, suspended spider-web fashion in separate frames, the two ends of each spiral being attached to terminals at the base of its own frame. These two spirals, which are marked respectively A and B, will now be placed a definite distance apart, and comparatively slow reversals from a battery of ten cells sent through spiral A. You will see the amount of the current induced in B by observing the deflection on the scale of the mirror reflecting galvanometer, which is in circuit with that spiral. These inductive effects vary inversely as the square of the distance between the two spirals when parallel to each other; the induced current in B being also proportional to the number of reversals of the battery current passing through spiral A, and also to the strength of the inducing current. Spiral A is so connected that reversed currents, at any desired speed per minute, can be passed through it from a battery; B is so connected to the galvanometer and a reverser as to show the deflections caused by the induced currents, which are momentary in duration, and, in the galvanometer circuit, all on the same side of zero; for, as the battery current, on making contact, produces an induced current in the reverse direction to itself, but in the same direction when broken, of course the one would neutralise the other, and the galvanometer remain unaffected. To obviate this, the galvanometer connections are reversed with each reversal of the battery current, and thus a steady deflection is produced.

Perhaps, for the information of those not acquainted with the construction of a galvanometer, I ought to explain it more fully. The one I am about to use consists of a coil of very fine silk-covered wire, in the centre of which is suspended a very small magnet; the ends of the coil of wire and the ends of spiral B are connected respectively together, thus forming a metallic circuit, one part of which is wound into a coil and the other into this spiral. Call to mind Oersted's experiment, and it will be readily understood that when a current of electricity is flowing in this metallic circuit, the magnet will be influenced in magnitude according to the amount of current thus flowing. The movements of the magnet, however, would be too small to be seen unless watched very closely; therefore fixed to it is a very small concave mirror, on to which a beam of light from the lamp is thrown, and the mirror reflecting this on to the scale, will, I hope, enable all to see that somewhat broad beam move in accordance with the movements of the magnet. Reversed currents at the speed of 100 per minute will now be passed through spiral A, and you will observe that the induced currents in B give about 28 divisions on the scale of the galvanometer: we will note this on the black-board. Now, we place this plate of iron midway between the two spirals, and you observe the deflection on the scale is reduced to about one-half, or in round numbers to 15, showing clearly that the presence of the iron plate has in some way influenced the previous effects. We now remove the iron, when you see the deflection returns to its original amount of 28 divisions; and if I now interpose a similar sheet of copper, the interposition does not alter the deflection. The results of this experiment are therefore as follows:—

Speed = 100 reversals per minute.
Induced current = 28° deflection.
Iron interposed = 15° ,,
Copper ,, = 28° ,,

I may here state that, up to this point, the results of my experiments confirm those of Faraday, viz. that all dielectrics and diamagnetic metals appear in no way to interrupt or interfere with the lines of force.

Now, let us repeat this experiment with the speed of the reversals increased ten times, or to 1000 per minute; the spirals are in the same position as before, and the deflection is now about 86. I have already said that the induced current is in direct proportion to the speed of the reversals, the battery and spirals remaining the same. It might therefore cause confusion were I not to explain that the deflection would be ten times as great as it was with the lower speed of reversals, but that the scale of the galvanometer not being sufficiently long to record this high deflection, we have what is termed "shunted" a part of the current; that is to say, between the two ends of the galvanometer coil we have inserted a length of copper wire, so that the current, on arriving at one terminal of the coil, divides, part going through the coil of the galvanometer, and the rest through the shunt; the two currents reunite at the other terminal of the galvanometer. By varying the resistance of the shunt, therefore, the desired amount of the current can be sent through the galvanometer. In this case sufficient current passes to keep the beam of light just on the scale at about 86 divisions. We now interpose the sheet of iron as before, and you see the deflection falls as before to about one-half. We withdraw the iron and the deflection returns to its former amount of 86. We now interpose the copper, when the deflection, instead of remaining stationary, as in the former experiment, actually falls to 17. We now obtain the following results, viz. :—

    Speed = 1000 reversals per minute.
 Induced current = 86° deflection.
 Iron interposed = 40° ,,
 Copper ,, = 17° ,,

Now, the question arose, why does copper, at the low speed of the reversals, apparently have no effect, while at the higher speed it plays so important a part in intercepting the lines of force? The only solution of the phenomenon which suggested itself to me was that the lines of force have first to polarise the molecules of substances placed in their path before they can pass through them, and in this process time is a very important element to be considered; for instance, at the slow speed there is sufficient time between the reversals for the copper to polarise before the next reversal takes place; whereas at the high speed the copper plate is unable to fully polarise before the next reversal arrives, and then the two induced effects partly blend, and being opposite in direction tend to cancel each other. Now, if this really be the case, the higher the speed the less should be the proportional deflection when experimenting with copper. The time at our disposal this evening will not allow of accurate measurements, or of other substances being experimented upon, but careful and reliable measurements have been made, the results of which are shown on the sheet before you marked 1. It will be seen by reference to these results that the percentage of inductive energy intercepted does not increase for different speeds of the reverser in the same rate with different metals, the increase with iron being very slight, whilst with copper the induced current set up is so long in duration that when the speed of the reverser is at all rapid, the current not having time to exhaust itself before the galvanometer is reversed, tends to produce a lower deflection. If the speed of the

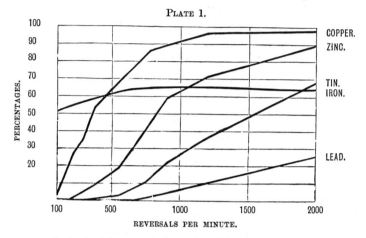

PLATE 1.

reverser is further increased, the induced current is received on the opposite terminal of the galvanometer, and thus a negative result is obtained.

My next object was to verify, if possible, by a different system of experiment, the correctness of this theory, and I could think of no better arrangements than those used by Faraday in some of his experiments on Magneto-Electric Induction. I was not, however, encouraged to proceed in that direction; for if my theory were correct, the results published by Faraday could not be so; and knowing what a careful experimentalist he was, I could not doubt that he was right. After long and careful thought on the subject, I ventured, however, to repeat some of his experiments, and I will again repeat them before you presently.

About sixty years ago Arago made the discovery in Electrical Science, that if a plate of copper be revolved close to a magnetic needle, or magnet suspended in such a way that the latter may rotate in a plane parallel to the former, the magnet tends to follow the motion of the plate; or if the magnet be revolved the plate tends to follow its motion. This simple apparatus will better illustrate the experiment. Here is a copper plate one-tenth of an inch thick, and seven and a half inches in diameter, fixed to a vertical spindle and enclosed in a wooden case having a glass cover; beneath the copper plate is a small grooved pulley around which passes an endless band; the band also passing round this horizontal wheel to which a handle is fixed, so that it may be conveniently revolved. A small brass disc is here provided, in the centre of which is fixed a pointed steel pin, and on this pin is balanced a steel magnet. I will place this on the glass cover over the centre of the copper plate. Now, if the copper plate be made to revolve, you will see that the magnet will revolve also in the direction of the copper disc. There it goes! No doubt you observed how sluggish its movements were at first, and that it was some time before it followed the movements of the disc; this was owing to the attraction of the earth's magnetism on the magnet, which held it in bondage until the speed of the disc was sufficient to overcome its attraction, then, once released, how merrily it appeared to obey the influence of a superior power. The disc now being at rest, the needle has returned into bondage; but if I judiciously use the influence of this magnet to partially neutralise the influence of the earth's attraction, you will observe how much more quickly it obeys the influence of the mysterious power of the revolving disc. There it goes! Apparently more readily than before. Were I lecturing on moral philosophy, I certainly should make use of the similes which might be drawn with advantage from experiments with this simple instrument; but my subject being of a different nature, I will resume without further digression. If the order were reversed and the magnet revolved, the copper disc would act in the same way as the magnet has just done. This is the phenomenon discovered by Arago, who also asserted that the effect takes place, not only with all metals, but with all substances. On this latter point there has always been a difference of opinion between experimentalists who have endeavoured to verify Arago's statement. As far as my experiments have gone, I have only obtained reliable results from good conductors of electricity; but I believe that, theoretically, Arago is right; for as all substances are, in a certain degree, conductors of electricity, it, I think, necessarily follows that we only want sufficiently sensitive instruments to develop the phenomenon, as Arago asserts, in every substance. It has been stated that all substances when subjected to a sufficiently strong magnetic force are found to give indications of polarity, and also, when magnetic force acts on any medium, whether magnetic, diamagnetic, or neutral, it produces within it a phenomenon called magneto-induction. If this be the case it materially strengthens the correctness of Arago's assertion. Arago's discovery that copper, a non-magnetic metal, was influenced by a rotating magnet, or that a magnet was, in the same way, affected by a rotating disc of copper, was looked upon at the time as a remarkable and new phenomenon in induced magnetism by philosophers both in England and other countries; but it was Faraday who, by the assistance of his knowledge of the evolution of electricity from magnetism, gave the true solution, by proving it to be the effect of electrical currents induced in the disc on account of its motion in a magnetic field. He not only proved by simple experiments the correct interpretation of the phenomenon, but he saw the way to making the discovery of Arago a new source of electricity, not despairing, by the aid of his knowledge of terrestrial-magneto-induction, of being able to construct a new magneto-electric machine. For this purpose he arranged an apparatus similar to the one I have here, which is simply a permanent magnet, so fixed that discs of metal or other substances can be rotated between its poles. Two wires leading one from each terminal of a galvanometer were applied to any desired part of the revolving disc, and the deflections on the galvanometer noted. The first experiments were made with a very large compound permanent magnet, and with what would now be termed a quantity astatic galvanometer. His discs of metal were twelve inches in diameter, and about one-fifth of an inch in thickness, fixed upon a brass axis. He experienced difficulty in making contact between the terminals of the galvanometer and the edge and other parts of the revolving disc, as also in maintaining uniform velocity of rotation. With a much smaller magnet and a more sensitive galvanometer, the results were more striking. Thus, with his accustomed simplicity, he demonstrated the production of a permanent current of electricity from an ordinary magnet, at the same time asserting that with powerful magnets and rapid rotation of a copper disc, very strong currents would be produced. What a wonderful lesson this apparently simple machine teaches, for it is able to exert the power, which has its origin within itself, on external matter, without in any way exhausting or diminishing that power!

The apparatus that I employed for my experiments is the same that I shall use this evening. On this stand is fixed an electromagnet, the poles of which are so placed that the rim of the disc under experiment can be freely revolved between them. The cores of the electro-magnet are $1 \cdot 75$ inch in diameter, and are wound with twelve layers of silk-covered copper wire of high conductivity, $\cdot 028$ of an inch in diameter, each layer having sixty-one turns; each core has 732 turns, making a total of 1464 turns on the two poles. The total resistance of the wire is $13 \cdot 75$ Ohms, and through this flows the current from twelve Leclanché cells. The same galvanometer is used as in the other experiments, being brought in circuit with the disc by means of the axis on which the disc revolves, and a metal brush which forms a rubbing contact on the rim of the disc; this, working on a fixed centre, can be readily shifted to any part of the circumference of the revolving disc. The disc is revolved by this cone-shaped pulley, worked lathe fashion, and connected by an endless band to a small grooved pulley fixed on the same axis as the metal disc. We will now make a few experiments with the copper disc, which is now revolving at the speed of 1000 revolutions per minute, and the connections are so made that the current will be taken from the rim of the disc just as it passes between the poles of the magnet. On pressing this spring the circuit is completed, and you observe the effect of the current on the scale of the galvanometer; it is about 100 divisions. We will now remove the contact to about one-fourth of the diameter of the disc or top position, so that the current will be taken at that distance from the poles of the magnet as the disc is approaching the poles; now, on completing the circuit, you observe the deflection is reduced to about fifteen divisions. The contact will now be removed to the opposite side, or bottom position of the disc, so that the current will be taken at the same distance from the poles, but after that part of the disc has just passed between them; on completing the circuit you see the deflection has increased to 33 divisions. Let us note these results as follows:—

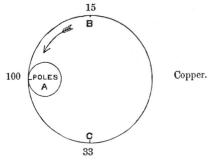

Copper.

We will now replace the copper disc with one of iron, and repeat precisely the same experiment. We find that the current from position A is so great that it is necessary to insert a resistance of 500 Ohms in the circuit to bring the beam of light on to the galvanometer scale: as before shown, by means of inserted resistance we can obtain any desired deflection. It is now about 98, and that we will take as the right measure at A. We take the top position, or B, and you observe that the deflection is only 7 divisions; the connection is now placed at position C, or bottom part, and we get but very little more current as the deflection is only increased by one division. The results with iron are therefore as follows:—

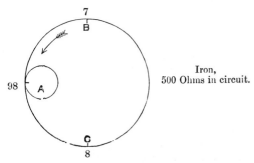

Iron,
500 Ohms in circuit.

If we compare the results of these experiments with those obtained with the same metals in volta-electric induction, we see that iron remains fairly uniform in its results in both cases, but that copper behaves differently. You must not, please, take the figures given as being absolutely correct, the experiments having been somewhat hurriedly made, and merely given with the view of enabling you to understand the results of carefully-made experiments with various metals. These results are graphically shown on the large sheet suspended before you, marked 2; and, as the measurements are to scale, they can be compared directly with each other. The large circles represent the revolving discs, the small ones show the position of the poles of the electro-magnet, and the arrows the direction of rotation. The coloured portions show the electro-motive force at every point round the revolving disc; thus in every case the strongest point is, as might be expected, directly in front of the poles of the exciting magnet, the strength gradually fading from thence on either side. The blue portions show the excess of inductive effect there produced, relatively to a point equidistant from the magnetic poles, but on the opposite side. If the current set up in the metal simply depended upon the intensity of that part of the magnetic field through which the metal was passing, then equidistant points on either side of the magnetic poles should produce equal deflections, because the magnetic intensity is equal; on inspecting the diagrams,

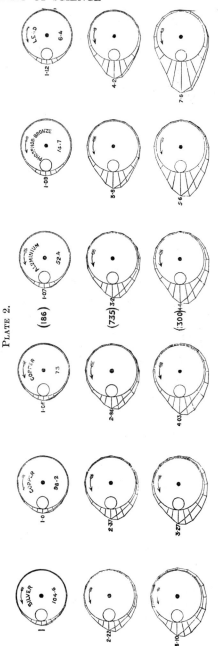

PLATE 2.

however, it appears as if a very sluggish inductive effect were produced in the diamagnetic metals of high conductivity, the sluggishness increasing with the conductivity of the metal, as though the atoms took a comparatively long time to accommodate themselves to the changes in the magnetic field through which they were passing. This, I think, confirms the theory I have ventured to advance with respect to copper while under the influence of induction.

If, as in the case of volta-electric induction, the lines of force are generated too quickly; or, as in the case of magneto-electric induction, the diamagnetic body passes too quickly through the magnetic field; the atoms of the substance, in each case, have not time to polarise with sufficient rapidity to radiate in the same time or place as when influenced more slowly by this force. This would account for the apparent interruption of the lines of force by the copper plate in the volta-electric experiments, and also for the "drag" or retardation on the disc of the same metal, as shown by the blue colour in the diagram whilst part of it is passing quickly through a magnetic field. That being the case, it is easy to perceive that a speed of the copper disc might be attained at which the current would be *nil*, through the atoms being unable to polarise in the allotted time. Not so with magnetic metals, as iron or nickel, in each of which the atoms appear to be very susceptible to magnetic influences, and to polarise very quickly, as shown by the way in which part of the disc is affected just before entering the poles, as indicated by the blue colour in the diagram marked 3.

PLATE 3.

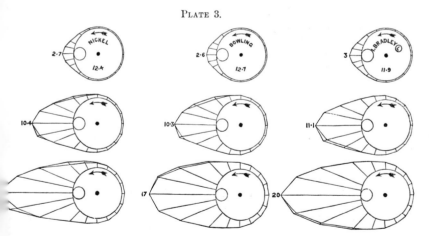

The atoms must as quickly depolarise, as you perceive there is but very little of what I have termed "drag" in those metals. The consequence of this is, that with iron the amount of current is nearly in direct proportion to the speed of the disc; whereas with copper the increase is very small, the current produced by the two metals being as 1 to 7, at a speed of about 1300 revolutions per minute. The higher the specific conductivity of the diamagnetic metal, the greater the "drag," and consequently the less the current at the poles, this being more manifest at the high speeds, which does not, however, agree with the results obtained by Faraday, for he found that the currents were proportionate in strength to the conducting power of the bodies experimented with, or, in other words, that the higher the conductivity the greater the current, whereas I find the reverse to be the case, as shown on the diagram before you.

Faraday also obtained much better results from copper than from iron, and thus he recommended copper for his new magneto-electric machine. Here, again, my results do not agree with his, for I find that iron gives much better results than copper, as shown on the diagram; and also that iron has the advantage that the current increases almost in direct proportion to the speed, whereas copper does not; in fact, as already stated, a speed might be obtained at which copper would give no current.

Faraday, in summing up the results of his experiments on Arago's phenomenon, says: "Nothing can be more clear, therefore, than that with iron and bodies admitting of ordinary magnetic induction, opposite poles on opposite sides of the edge of the plate neutralise each other's effects, while similar poles exalt the action. But with copper, and substances not sensible to ordinary magnetic impressions, *similar* poles on opposite sides of the plate neutralise each other, and opposite poles exalt the action." Perhaps you will more readily grasp the subject by reference to the diagram marked 4, in which P represents the plates of metal, and the other letters the respective poles of the magnet and their position.

PLATE 4.

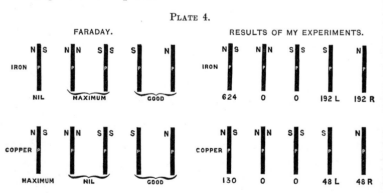

Here again it will be seen that the results obtained by me are opposed to those given by Faraday. I have given the actual figures obtained in my experiments, and it will be seen that the only difference between the metals copper and iron is that iron gives the higher current of the two, as it has done in all my experiments.

I have no doubt that we should be able to get approximately near for all practical purposes to the relative conductivity of metals by revolving discs of the metal under test in a magnetic field, and measuring the amount of current between the poles or the amount of "drag." The results given on diagram 5 show how suitable the method also is for obtaining the saturation point of the cores of electro-magnets, whether of different qualities of iron or different metals.

PLATE 5.

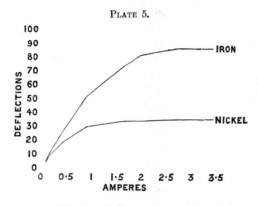

Very recently an Electrical Congress, held in Paris, agreed to an universal system of electrical units. I believe the value of some of the units to be adopted was determined by passing masses of metal at varying velocity through a magnetic field of uniform intensity, and noting the amount of current so produced. I understand that the results obtained by different experimentalists did not agree, and from what I have shown you this evening, you will readily understand the reason of the discrepancy. To obtain accurate results the influence of what I have called "drag" must certainly be taken into the calculations as an important factor. There are other matters to which I attach importance connected with the experiments I have endeavoured to make clear to you, especially with those on magneto-electric induction, which I publish for the first time this evening; but of the many good rules of this Institution, there is one which does not allow me to tax your patience for more than one hour, and, as I have already exhausted that time, I must not detain you longer, except to thank you, which I do most sincerely, for the kind attention you have given to my humble endeavours to advance our knowledge of volta-electric and magneto-electric induction.

[W. S.]

Friday, June 13, 1884.

WARREN DE LA RUE, Esq. M.A. D.C.L. F.R.S. Manager and Vice-President, in the Chair.

PROFESSOR JAMES DEWAR, M.A. F.R.S. *M.R.I.*

*Researches on Liquefied Gases.**

THE two Russian chemists, MM. Wroblewski and Olzewski, who have recently made such a splendid success in the production and maintenance of low temperature, have used in their researches an enlarged form of the well-known Cailletet apparatus; but for the purposes of lecture demonstration, which necessarily involves the projection on a screen of the actions taking place, the apparatus represented in the annexed woodcut is more readily and quickly handled, and enables comparatively large quantities of liquid oxygen to be produced. The arrangements will be at once understood on looking at the figure, which is taken from a photograph. The oxygen- or air-reservoir, C, is made of iron; it contains gas compressed for convenience to 150 atmospheres. A is the stopcock for regulating the pressure of the gas in the glass tube F, and D is the pressure-manometer, the fine copper tube which connects the gas-reservoir and the glass tube, F, being shown at I. The air-pump gauge is marked J, the tube leading to the double oscillating Bianchi being attached at H. The glass test-tube G, which contains the liquid ethylene, solid carbonic acid, or liquid nitrous oxide, which is to be boiled *in vacuo*, is placed in the middle of a larger tube. It has holes, shown at E, in the upper part, so that the cool vapours in their course to the air-pump are forced to pass round the outside of the vessel and help to guard it from external radiation. The lower part of the outer cylinder is covered with pieces of chloride of calcium, shown at K. If a thermometer is used and a continuous supply of ethylene maintained, the indiarubber cork through which the tube F passes has two additional apertures for the purpose of inserting the respective tubes. When the pump has reduced the pressure to 25 mm., the ethylene has a temperature of about $-140°$ C.; a pressure of between 20 and 30 atmospheres is then sufficient to produce liquid oxygen in the tube F. The tube F is 5 mm. in diameter and about 3 mm. thick in the walls, and when

---

* See Professor Dewar's discourse on the Liquefaction of Gases.—'Proceedings' of the Royal Institution, vol. viii. p. 657.

filled with fluid oxygen (for projection) holds at least 1·5 cubic centim. With such a quantity of fluid oxygen it is easy to show its ebullition at ordinary pressures, and by means of a thermo-junction to demonstrate the great reduction of temperature which is attendant on its change of state at atmospheric pressure.

Provided a supply of liquid ethylene can be had, there is no difficulty in repeating all the experiments of the Russian observers; but as this gas is troublesome to make in quantity, and cannot be bought like carbonic acid or nitrous oxide, such experiments necessitate a considerable sacrifice of time. It was therefore with considerable satisfaction that I observed the production of liquid oxygen

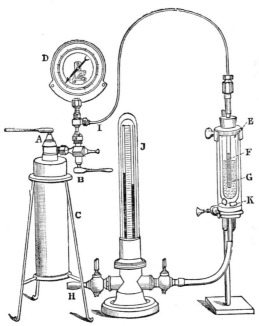

by the use of solid carbonic acid, or preferably liquid nitrous oxide. When these substances are employed and the pressure is reduced to about 25 mm., the temperature of $-115°$ C. may be taken as that of the carbonic acid, and $-125°$ C. as that of the nitrous oxide. As the critical point of oxygen, according to the Russian observers, is about $-113°$ C., both these cooling agents may be said to lower the temperature sufficiently to produce liquid oxygen, provided a pressure of the gas above the critical pressure, which is 50 atmospheres, is at command. In any case, however, the temperature is near that of the critical point; and as it is difficult to maintain the pressure below about an inch of mercury, the temperature is apt to be rather above the respective temperatures of $-115°$ C. and $-125°$ C. In order to get liquefaction conveniently with either of these agents, it is necessary to work at a pressure of oxygen gas from 80 to 100 atmospheres, and to have the means of producing a sudden expansion when the compressed gas is cooled to the above-mentioned temperatures. This is brought about by the use of an additional stopcock, represented in the figure at B. During the expansion the stopcock at A is closed and the pressure-manometer carefully observed. No doubt liquid nitrous oxide is the most convenient substance to use as a cooling agent; but as it is apt to get superheated during the reduction of pressure and boil over with explosive bursts of vapour, it is well to collect the fluid in a small flask of about 250 cub. centim. capacity, and to change it into the solid state by connecting the flask with the air-pump, and then to use the substance in this form. The addition of alcohol or ether to the solid nitrous oxide makes the body more transparent, and thereby favours the observations.

It is evident that this apparatus enables the observer to determine the density of the fluid gases condensed in the tube F; since he has only to measure the volume of fluid in F, and to collect, by means of the stopcock B, the whole volume of gas given by the fluid and condensed vapour, which gives an accurate determination of the total weight of substance distributed between fluid and vapour in the whole apparatus. The amount of substance which is required to produce the vapour is easily found by observing the vapour-pressure of the liquid gas before expanding it into gas for the volume measurement; and while keeping shut the stopcock B, by opening A suddenly until this pressure is just reached and then instantly shutting off the receiver. If this volume of gas is now measured by opening B as before, the difference between the two volumes thus collected will correspond to the real weight of substance in the liquid state. A rough experiment with oxygen near the critical point gave the density $0·65$.

As to the most convenient substance for use as a cooling agent, I am still of opinion that marsh-gas would be the best; and I may take the opportunity of pointing out that the employment of this body was suggested by me in a communication made to the Chemical Section of the British Association in 1883. The following extract from 'Nature,' of October 4, 1883, will prove that my experiments with liquid marsh-gas were made a year in advance of those made recently by M. Cailletet [*] and M. Wroblewski [†]:—

"Professor Dewar pointed out an important relation between the critical temperatures and pressures of volatile liquids and their molecular volumes. The ratio of the critical temperature to the

---

[*] "Sur l'emploi du Formène pour la production des très basses températures," *Comptes Rendus*, June 30, 1884.
[†] "Sur les propriétés du gaz des marais liquide, et sur son emploi comme réfrigérant," *Comptes Rendus*, July 21, 1884.

critical pressure is proportional to the molecular volume, so that the determination of the critical temperature and pressure of a substance gives us a perfectly independent measure of the molecular volumes. Prof. Dewar pointed out the great advantage of employing a liquid of low critical temperature and pressure such as liquid marsh-gas for producing exceedingly low temperature. He hoped to be able to approach the absolute zero by the evaporation of liquefied marsh-gas whose critical temperature was less than $-100°$ C., and whose critical pressure was only 39 atmospheres."

I ought to mention that the marsh-gas used in my experiments was made by the action of water on zinc methyl, and was therefore very pure, and that the observed critical pressure was not 39 atmospheres, but $47\cdot6$. The following table gives the values of the ratio of the absolute critical temperature to the critical pressure in the case of a number of substances. The values for ammonia, sulphuretted hydrogen, cyanogen, marsh-gas, and hydride of ethyl are new.

		T. Critical temperature.	P. Critical pressure.	$\dfrac{T}{P}$
Chlorine	Cl_2	141·0	83·9	5·0
Hydrochloric acid	HCl	52·3	86·0	3·7
Oxygen	O_2	−113·0	50·0	3·2
Water	H_2O	370·0	195·5	3·3
Nitrogen	N_2	−146·0	35·0	3·6
Hydrogen sulphide	H_2S	100·2	92·0	4·0
Ammonia	H_3N	130·0	115·0	3·5
Diethylamine	$(C_2H_5)_2HN$	220·0	38·7	15·4
Nitrous oxide	N_2O	35·4	75·0	4·1
Sulphurous acid	SO_2	155·4	78·9	5·4
Marsh-gas	CH_4	−99·5	50·0	3·5
Acetylene	C_2H_2	37·0	68·0	4·5
Ethylene	C_2H_4	10·1	51·0	5·5
Ethyl hydride	C_2H_6	35·0	45·2	6·8
Amylene	C_5H_{10}	191·6	33·9	13·7
Benzol	C_6H_6	291·7	60·4	9·3
Chloroform	$CHCl_3$	268·0	54·9	9·9
Carbon chloride	CCl_4	282·0	57·6	9·6
Carbonic acid	CO_2	31·9	77·0	4·0
Bisulphide of carbon	CS_2	277·7	78·1	7·0
Cyanogen	C_2N_2	124·0	61·7	6·4

A glance at the last column of the table shows that a large number of substances have at their respective critical temperatures simple volume relations. Thus hydrochloric acid, water, ammonia, and marsh-gas, the four chemical substances from which the great majority of chemical compounds may be derived by processes of substitution, have nearly the same volume; while the more complex derivatives show an increased volume which bears a simple ratio to that of the typical body. As the critical pressures are not known with any great accuracy at present, it would be useless to discuss the results with any severity. All that can be inferred is that the subject is worthy of further investigation and promises important generalisation. Sarrau (*Compt. Rend.* 1882) deduced the critical temperatures and pressures of hydrogen, oxygen, and nitrogen by the application of Clausius' formula to the experiments of Amagat; and it is interesting to compare his results with the experimental values.

Sarrau's Calculated Values.

	T. Critical temperature.	P. Critical pressure.	$\dfrac{T}{P}$.
Hydrogen	$-174°$	98·9	1·0
Oxygen	$-105·4$	48·7	3·4
Nitrogen	-124	42·1	3·5

It will be observed that the calculated critical temperatures of oxygen and nitrogen are remarkably near the truth, being respectively 8° and 22° too high. On the other hand, the values of the ratios of the calculated critical temperatures and pressures are almost identical with those obtained by direct experiment. The only peculiarity to be noted is in the case of hydrogen, which has such a high critical pressure, and therefore leads to a remarkably small molecular volume at the critical point. If the values of the $T \div P$ ratio be taken as proportional to the molecular volumes, then it is easy to infer the densities of the fluids at their respective critical temperatures, provided the density of one standard substance is known by experiment. The simple formula thus stated is

$$\frac{S'}{S} = \Psi \frac{W'}{W} \quad \Psi = \frac{V}{W},$$

where S and S' are the specific gravities of two bodies, W and W' their molecular weights, and V and V' their molecular volumes. It will be convenient to take the density of carbonic acid at the critical point as the standard density to which the others can be referred. The density of carbonic acid under such conditions may be taken as 0·65. Calculating with the above formula, the density of acetylene would be 0·32, whereas the experimental number of Ansdell is 0·36. In the same way the density of hydrochloric acid is found to be 0·6, the true value being 0·61. The density of oxygen would be 0·63, and that of nitrogen 0·45. The calculated density of hydrogen at its critical point would be 0·12, if we assume the correctness of Sarrau's values for the critical temperature and pressure. We may compare these values with the numbers obtained by Cailletet and Hautefeuille for the densities of oxygen, nitrogen, and hydrogen from their experiments on the density of liquid carbonic acid obtained from mixtures of this body with these gases. At the temperature of 0° C. the experiments found for oxygen, nitrogen, and hydrogen the respective values of 0·65, 0·37, and 0·025. It seems that the calculated values for oxygen and nitrogen are not very far wrong; but hydrogen is clearly incorrect. The explanation of this anomaly is probably to be found in the fact that the calculated molecular volume of hydrogen is wrong, and that instead of being unity on our scale it ought to be 3·5 like oxygen and nitrogen. In fact, the chemist would infer that, as the difference in the complexity of the molecular structure of hydrochloric acid, water, ammonia, and marsh-gas does not affect the molecular volume under the conditions we are discussing, in all probability the value for hydrogen would be identical with that of the above-mentioned bodies. If we adopt this view and change the value of the $T_c \div P_c$ to 3·5, then the density of the fluid would become 0·034, which is in accordance with the experimental number of Cailletet and Hautefeuille. An accurate determination of the critical temperature and pressure of hydrogen, for which, judging from the success of the experiments of M. Olzewski, chemists will not have to wait long, will thus be of great interest.

[J. D.]

Friday, March 20, 1885.

Sir Frederick Bramwell, F.R.S. Manager and Vice-President, in the Chair.

Professor A. W. Rücker, M.A. F.R.S.

Liquid Films.

The molecules in the interior of a liquid are surrounded on all sides by others which they attract, and by which they are themselves attracted, while those on the surface have neighbours on one side only. In consequence of this difference in their surroundings there is in all probability a difference in the grouping of the interior and exterior molecules which is attended by corresponding variations in the physical properties of the liquid of which they are constituent parts. Thus it was shown by M. Plateau that the viscosity of the surface of a liquid is in general different from that of its interior. The most striking example of this phenomenon is afforded by a solution of saponine. Two per cent. of this substance dissolved in water does not effect any marked change in the properties of the great mass of the liquid, but produces a most remarkable increase in the surface viscosity, so that forces which suffice to create rapid motion in bodies which are completely immersed, fail to produce any appreciable movement if they lie in the exterior surface. The first attempt to obtain a numerical estimate of the difference of the resistances experienced by a body oscillating in turn in the interior and in the surface of the liquid was made about two years ago by Messrs. Stables and Wilson, students in the Yorkshire College. In the case of a horizontal disc suspended in water, the logarithmic decrement diminishes to about one-half as the surface is approached. In a saponine solution, on the other hand, it is 125 times greater in the surface than in the interior, and about 38 times greater in the surface than at a depth of $0 \cdot 1$ mm. below it. Even in the latter case the greater part of the resistance is due, not to the friction between the disc and the liquid, but to that experienced by the supporting rod in the surface, so that in all probability the surface viscosity is more than 600 times greater than that of the mass of the liquid.

The immense change in the resistance which takes place when the disc is immersed to a depth of $0 \cdot 1$ mm. only confirms the general opinion that any peculiarity of grouping or arrangement due to proximity to the surface extends to a very small depth. A liquid must thus be conceived as surrounded by a very thin layer or skin, the properties of which are different from that of the liquid in the interior, and to which rather than to any ideal geometrical boundary the term surface might be applied. It may, however, prevent confusion if it is called the *surface-layer*.

Many attempts have been made to measure the thickness of the surface-layer. In particular, M. Plateau studied a thinning soap film with the view of determining whether or no the pressure exerted on the enclosed air by the film when very thin is the same as when it is comparatively thick. Had any such difference been observed it might but have been taken as *primâ facie* evidence that the tenuity was so great that all the interior portions of the film had drained away, and that the thickness did not exceed that of the two surface-layers.

This experiment has been criticised by Prof. Reinold and myself, but it is not intended in this lecture to enter upon the general question of the thickness of the surface-layer, or the interesting theoretical problems which are closely connected with it, as we are at present engaged in an investigation which we hope may throw further light upon the subject. There are, however, two preliminary questions on which we have arrived at definite conclusions.

In any experiments which have for their object the detection of small changes in the properties of a soap film as it becomes thinner, it is essential that we should be able to assert with certainty that no causes other than the increasing tenuity have been in play, by which the effect looked for might either be produced or masked. Changes in the temperature or composition of the film, must especially be prevented.

The liquid ordinarily employed for such investigations is the "liquide glycérique" of M. Plateau. In dry air some of the water of which it is in part composed would evaporate, while in moist air, in consequence of the hygroscopic properties of the glycerine, additional water would be absorbed. Though these facts were well known, and though they are evidently possible sources of error, no attempt (as far as I am aware) had been made before our own to determine what precautions it was necessary to take to prevent the results of experiments such as M. Plateau's being affected by them. The first question then that we set ourselves to answer, was—to what extent is the composition of a soap film altered by changes in the temperature or hygroscopic state of the air which surrounds it?

The method adopted in answering this inquiry was to measure the electrical resistance of soap films formed in an inclosed space containing a thermometer and hair hygrometer. If the observations led to the conclusion that the resistance of film varied inversely as its thickness, they would prove that no change in composition had taken place, and that the film at the thinnest had afforded no evidence of an approach to a thickness equal to that of the surface-layers. If the specific resistance was found to vary according to some regular law as the thickness altered, there would be a strong presumption, that the thickness was not much greater than, and was possibly even less than that of the two surface-layers. If, lastly, the changes were irregular, they might safely be ascribed to alterations in temperature or constitution.

To obtain the desired facts it was necessary (1) to devise a method of forming the films in a closed chamber, (2) to measure their thickness, and (3) to determine their electrical resistance.

The films were formed in a glass box at the lower extremity of a platinum ring which communicated by means of a tube with the outside. In the earlier experiments a cup of the liquid was raised by rackwork to the ring and then withdrawn, leaving a film behind it. The latter was blown out by air which had been dried and passed through tubes containing " liquide glycérique." When large enough it adhered to a second platinum ring placed vertically below the first, and on some of the air being withdrawn it assumed the cylindrical form.

The thickness was measured by means of the colours displayed, two independent determinations being obtained by two beams of light incident at different angles. Newton's Table of Colours was revised, and it was found that the differences between the thicknesses given by him and those determined by new experiment were far greater than the error of experiment of a single observer. Hence, if accurate measurements are required by means of Newton's scale, every experimenter must reconstruct that scale for himself.

At first the electrical resistance was determined by means of Wheatstone's Bridge. The edges of the film where it is close to its solid supports are often, however, the seat of phenomena which might affect the results. Thin rings of white or black appear which alter the resistance considerably, and which introduce errors for which it is almost impossible to make any accurate allowance. This fact, combined with the advantage of avoiding errors due to polarisation, and of being able to select any particular part of the film for examination instead of the whole, led us to adopt a different method. Gold wires attached to a movable support were thrust into the film, and the difference of potential between these when a current was passing through the film was compared with that between the extremities of a known resistance included in the same circuit.

The result of these observations was to prove that the specific resistance of the films altered in an irregular manner, varying between 200 and 137 ohms per cubic c.m. A closer inspection showed that abnormal results were always accompanied by abnormal variations in the thermometer or hygrometer. When those films were selected which had been observed when such variations were especially small, it was found that the range of variation of the specific resistances was only between 137 and 146, and that the mean value was 143, that of the liquid in mass being $140 \cdot 5$ (at the same temperature). It was also proved that between thicknesses varying from 1370 to 374 millionths of a millimetre, no regular change in specific resistance could be detected, the actual variations lying within $2 \cdot 5$ per cent.

The conclusion was thus arrived at that the specific resistance of the liquid of which a soap film is formed does not differ from that of the same liquid in mass, at all events when the thickness is greater than 374×10^{-6} mm, and that comparatively small changes in the temperature or hygroscopic state of the air in contact with the film are attended with great alterations in the specific resistance, which indicate a considerable change in composition.

The method of experiment made it possible to determine the amount of this change. Solutions were made up representing "liquide glycérique" which had lost or gained given percentages of water, their specific resistances were determined at various temperatures, and approximate formulæ obtained by which the percentage of water present could be calculated if the specific resistance and temperature were known.

The results of the application of this method of analysis to a film are shown in the accompanying figure. The abscissæ represent time, the ordinates of curve I. represent the average thickness of the film.

Fig. 1.

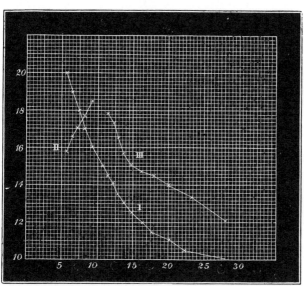

It will be observed that the film continued to get thinner during the whole time that it was under observation. The electrical observations however, proved that at first the product of the resistance and thickness steadily increased, indicating a continuous loss of water. Curve II. shows the number of parts of water in 100 of the solution lost at the times indicated by the abscissæ. After a while a piece of blotting paper which had been hung up inside the case was moistened with water. While this was being done the observations were interrupted. On their renewal it was found that although the film thinned as steadily as before, the product of the resistance and thickness diminished instead of increasing. Curve III. shows the steady absorption of water which followed the moistening of the air. These experiments proved that it is possible for a film to undergo great changes in composition without any indication of the fact being afforded by the colours it displays. They show that if the composition of the "liquide glycérique" is to be kept constant, all change in the temperature and hygrometric state of the air must be as far as possible prevented. In later experiments this condition has been secured by placing the film box in the centre of a water tank, and by keeping an endless band of linen hung up within the case, and which dips into the liquid, continually moistened. Observations made with this apparatus show that these precautions which are certainly necessary are also sufficient.

The second point to which special attention has hitherto been given by Prof. Reinold and myself is the measurement of the thickness of very thin films. If the thickness is less than a certain magnitude, the films appear black, and thus their colour gives only a limit to and not a measure of their thickness. Black films display many remarkable properties. In general there is a sudden change in thickness at the edge of the black indicated by the omission of several colours, or sometimes of one or two orders of colours. It is only under rare conditions that a gradual change in thickness can be observed from the white to the black of the first order.

To determine the thickness of the black its resistance was measured, and the thickness calculated *on the assumption that the specific resistance was the same as that of the liquid in mass.*

The observations were made in several different ways and proved that the thickness of the black portion remains constant in any given film, however much its area may alter. Thus, in the case of a group of films measured by Wheatstone's bridge, the average resistance of a black ring 1 mm. in breadth was $1·761$ megohms when the total breadth was 2 mm., and $1·760$ megohms when the total breadth lay between 10 and 12 mm.

Again, the resistance of the part of the film between the needles used in the electrometer method was practically the same when the black had extended over the whole film (40 mm. long) as it had been when only the upper 11 mm. were black. The final measurement differed from the mean by only $0·1$ per cent. Again, in another film the resistance of the black per millimetre remained the same to within $2·5$ per cent. for an hour and a half.

On the other hand the experiments also proved that the thickness of the black was different in different films. The values found varied between $7·2 \times 10^{-6}$ and $14·2 \times 10^{-6}$ mm. These differences are quite outside the possible error of experiment. If they were due to changes in the constitution of the liquid of which the films were formed, it is very improbable that the specific resistance of individual films would not have shown progressive changes. As has been stated, none such were observed. The mean thickness of the five films made of "liquide glycérique" which were observed was $11 \cdot 9 \times 10^{-6}$ mm., while that of 13 films made of soap solution without any glycerine was $11 \cdot 74 \times 10^{-6}$ mm.

The assumption made in these calculations that the specific resistance of a film, the thickness of which is ten or twelve millionths of a millimetre, is the same as that of the liquid in mass, is not justified by the previous experiments, which had proved it to hold good only to the much greater thickness of 370×10^{-6} mm. It was therefore desirable to check the results by an independent method. For this purpose 50 or 60 plane films were formed side by side in a glass tube which was placed in the path of one of the interfering beams in a Jamin's Interferential Refractometer. The compensator was adjusted so that it had to be moved through a large angle to cause one interference band to occupy the position previously held by its neighbour, i. e. to alter the difference of the paths of the interfering rays by one wave length. This angle was determined for the red light of known wave length transmitted by glass coloured with copper oxide. When the films had thinned to the black they were broken by means of a needle which had been included in the tube along with them, and which was moved, without touching the tube, by a magnet. The rupture of the films produced a movement of the interference fringes which was measured by the compensator, and from which, in accordance with well-known principles, the thickness of the films could be deduced.

The mean thickness given by seven experiments on films made of "liquide glycérique" was $10 \cdot 7 \times 10^{-6}$ mm., that obtained from nine experiments on films made of soap solution was $12 \cdot 1 \times 10^{-6}$ mm. The mean of these, or $11 \cdot 4 \times 10^{-6}$ mm., differed only by $0 \cdot 4 \times 10^{-6}$ mm. from the mean thickness deduced from the electrical experiments.

The last point to which reference is necessary is one which lies outside the main line of the enquiries above described, but which is nevertheless not without interest. In the course of the observations it was noticed that the rate of thinning of a film seemed to be affected by the passage of the electric current through it. Some experiments made on this point last year proved the fact beyond the possibility of doubt. The current appears to carry the matter of the film with it, so that it thins more rapidly if the current runs down, and less rapidly if the current runs up than if no current is passing. This may be shown as a lecture experiment.

A vertical rod which can be moved up and down by rackwork is passed through the centre of the cover of a glass film-box. To the lower extremity is attached a horizontal platinum wire, from which another similar horizontal wire is suspended by two silk fibres. A film is formed by lowering the whole into the liquid with which the lower part of the vessel is flooded. The light reflected from the film is passed through a lens, and an image formed upon a screen. When the bands of colour are seen descending from the upper part of the film a current from 50 Grove's cells is passed through it. If the current flows downwards the bands of colour move more quickly than before; if it flows upwards their motion is checked and they begin to ascend. The cause of this curious fact is still unknown. It may either be analogous to the phenomenon known as the "migration of the ions," or it may be a secondary effect due to a change in the surface tension.

The general relation of the results attained in these investigations as to the question of the size of molecules is interesting. Sir William Thomson has expressed the opinion that 2×10^{-6} mm. and 0.01×10^{-6} mm. are superior and inferior limits respectively to the diameter of a molecule. Van der Waals has been led, from considerations founded on the theory of gases, to give 0.28×10^{-6} mm. as an approximate value of the diameters of the molecules of the gases of which the atmosphere is composed. The number of molecules which could be placed side by side within the thickness of the thinnest soap film would, according to these various estimates, be 4, 26, and 720 respectively. The smallness of the first of these numbers, especially when it is remembered that the liquid used on some occasions was of a highly complex character, containing water, glycerine and soap, points to the conclusion that the diameter of a molecule is considerably less than 2×10^{-6} mm.

[A. W. R.]

Friday, January 29, 1886.

WILLIAM HUGGINS, Esq. D.C.L. LL.D. F.R.S. Vice-President, in the Chair.

SIR WILLIAM THOMSON, D.C.L. LL.D. F.R.S. *M.R.I.*

Capillary Attraction.

THE heaviness of matter had been known for as many thousand years as men and philosophers had lived on the earth, but none had suspected or imagined, before Newton's discovery of universal gravitation, that heaviness is due to action at a distance between two portions of matter. Electrical attractions and repulsions, and magnetic attractions and repulsions, had been familiar to naturalists and philosophers for two or three thousand years. Gilbert, by showing that the earth, acting as a great magnet, is the efficient cause of the compass needle's pointing to the north, had enlarged people's ideas regarding the distances at which magnets can exert sensible action. But neither he nor any one else had suggested that heaviness is the resultant of mutual attractions between all parts of the heavy body and all parts of the earth, and it had not entered the imagination of man to conceive that different portions of matter at the earth's surface, or even the more dignified masses called the heavenly bodies, mutually attract one another. Newton did not himself give any observational or experimental proof of the mutual attraction between any two bodies, of which both are smaller than the moon. The smallest case of gravitational action which was included in the observational foundation of his theory, was that of the moon on the waters of the ocean, by which the tides are produced; but his inductive conclusion that the heaviness of a piece of matter at the earth's surface, is the resultant of attractions from all parts of the earth acting in inverse proportion to squares of distances, made it highly probable that pieces of matter within a few feet or a few inches apart attract one another according to the same law of distance, and Cavendish's splendid experiment verified this conclusion. But now for our question of this evening. Does this attraction between any particle of matter in one body and any particle of matter in another continue to vary inversely as the square of the distance, when the distance between the nearest points of the two bodies is diminished to an inch (Cavendish's experiment does not demonstrate this, but makes it very probable), or to a centimetre, or to the hundred-thousandth of a centimetre, or to the hundred-millionth of a centimetre? Now I dip my finger into this basin of water; you see proved a force of attraction between the finger and the drop hanging from it, and between the matter on the two sides of any horizontal plane you like to imagine through the hanging water. These forces are millions of times greater than what you would calculate from the Newtonian law, on the supposition that water is perfectly homogeneous. Hence either these forces of attraction must, at very small distances, increase enormously more rapidly than according to the Newtonian law, or the substance of water is not homogeneous. We now all know that it is not homogeneous. The Newtonian theory of gravitation is not surer to us now than is the atomic or molecular theory in chemistry and physics; so far, at all events, as its assertion of heterogeneousness in the minute structure of matter apparently homogeneous to our senses and to our most delicate direct instrumental tests. Hence, unless we find heterogeneousness and the Newtonian law of attraction incapable of explaining cohesion and capillary attraction, we are not forced to seek the explanation in a deviation from Newton's law of gravitational force. In a little communication to the Royal Society of Edinburgh twenty-four years ago,* I showed that heterogeneousness does suffice to account for any force of cohesion, however great, provided only we give sufficiently great density to the molecules in the heterogeneous structure.

Nothing satisfactory, however, or very interesting mechanically, seems attainable by any attempt to work out this theory without taking into account the molecular motions which we know to be inherent in matter, and to constitute its heat. But so far as the main phenomena of capillary attraction are concerned, it is satisfactory to know that the complete molecular theory could not but lead to the same resultant action in the aggregate as if water and the solids touching it were each utterly homogeneous to infinite minuteness, and were acted on by mutual forces of attraction sufficiently strong between portions of matter which are exceedingly near one another, but utterly insensible between portions of matter at sensible distances. This idea of attraction insensible at sensible distances (whatever molecular view we may learn, or people not now born may learn after us, to account for the innate nature of the action), is indeed the key to the theory of capillary attraction, and it is to Hawksbee † that we owe it. Laplace took it up and thoroughly worked it out mathematically in a very admirable manner. One part of the theory which he left defective—the action of a solid upon a liquid, and the mutual action between two liquids—was made dynamically perfect by Gauss, and the finishing touch to the mathematical theory was given by Neumann in stating for liquids the rule corresponding to Gauss's rule for angles of contact between liquids and solids.

Gauss, expressing enthusiastic appreciation of Laplace's work, adopts the same fundamental assumption of attraction sensible only

---

* Proceedings of the Royal Society of Edinburgh, April 21, 1862 (vol. iv.).
† Royal Society Transactions, 1709-13.

at insensible distances, and, while proposing as chief object to complete the part of the theory not worked out by his predecessor, treats the dynamical problem afresh in a remarkably improved manner, by founding it wholly upon the principle of what we now call potential energy. Thus, though the formulas in which he expresses mathematically his ideas are scarcely less alarming in appearance than those of Laplace, it is very easy to translate them into words by which the whole theory will be made perfectly intelligible to persons who imagine themselves incapable of understanding sextuple integrals. Let us place ourselves conveniently at the centre of the earth so as not to be disturbed by gravity. Take now two portions of water, and let them be shaped over a certain area of each, call it A for the one, and B for the other, so that when put together they will fit perfectly throughout these areas. To save all trouble in manipulating the supposed pieces of water, let them become for a time perfectly rigid, without, however, any change in their mutual attraction. Bring them now together till the two surfaces A and B come to be within the one-hundred-thousandth of an inch apart, that is, the forty-thousandth of a centimetre, or 250 micro-millimetres (about half the wave-length of green light). At so great a distance the attraction is quite insensible: we may feel very confident that it differs, by but a small percentage, from the exceedingly small force of attraction which we should calculate for it according to the Newtonian law, on the supposition of perfect uniformity of density in each of the attracting bodies. Well-known phenomena of bubbles, and of watery films wetting solids, make it quite certain that the molecular attraction does not become sensible until the distance is much less than 250 micro-millimetres. From the consideration of such phenomena Quincke (Pogg. Ann., 1869) came to the conclusion that the molecular attraction does become sensible at distances of about 50 micro-millimetres. His conclusion is strikingly confirmed by the very important discovery of Reinold and Rücker[*] that the black film, always formed before an undisturbed soap bubble breaks, has a uniform or nearly uniform thickness of about 11 or 12 micro-millimetres. The abrupt commencement, and the permanent stability, of the black film demonstrate a proposition of fundamental importance in the molecular theory:—The tension of the film, which is sensibly constant when the thickness exceeds 50 micro-millimetres, diminishes to a minimum, and begins to increase again when the thickness is diminished to 10 micro-millimetres. It seems not possible to explain this fact by any imaginable law of force between the different portions of the film supposed homogeneous, and we are forced to the conclusion that it depends upon molecular heterogeneousness. When the homogeneous molar theory is thus disproved by observation, and its assumption of a law of attraction augmenting more rapidly than according to the Newtonian law when

---

[*] Proc. Roy. Soc., June 21, 1877; Trans. Roy. Soc., April 19, 1883.

the distance becomes less than 50 micro-millimetres is proved to be insufficient, may we not go farther and say that it is unnecessary to assume any deviation from the Newtonian law of force varying inversely as the square of the distance continuously from the millionth of a micro-millimetre to the distance of the remotest star or remotest piece of matter in the universe; and, until we see how gravity itself is to be explained, as Newton and Faraday thought it must be explained, by some continuous action of intervening or surrounding matter, may we not be temporarily satisfied to explain capillary attraction merely as Newtonian attraction intensified in virtue of intensely dense molecules movable among one another, of which the aggregate constitutes a mass of liquid or solid.

But now for the present, and for the rest of this evening, let us dismiss all idea of molecular theory, and think of the molar theory pure and simple, of Laplace and Gauss. Returning to our two pieces of rigidified water left at a distance of 250 micro-millimetres from one another. Holding them in my two hands, I let them come nearer and nearer until they touch all along the surfaces A and B. They begin to attract one another with a force which may be scarcely sensible to my hands when their distance apart is 50 micro-millimetres, or even as little as 10 micro-millimetres; but which certainly becomes sensible when the distance becomes one micro-millimetre, or the fraction of a micro-millimetre; and enormous, hundreds or thousands of kilogrammes' weight, before they come into absolute contact. I am supposing the area of each of the opposed surfaces to be a few square centimetres. To fix the ideas, I shall suppose it to be exactly thirty square centimetres. If my sense of force were sufficiently metrical I should find that the work done by the attraction of the rigidified pieces of water in pulling my two hands together was just about $4\frac{1}{2}$ centimetre-grammes. The force to do this work, if it had been uniform throughout the space of 50 micro-millimetres (five-millionths of a centimetre) must have been 900,000 grammes weight, that is to say, nine-tenths of a ton. But in reality it is done by a force increasing from something very small at the distance of 50 micro-millimetres to some unknown greatest amount. It may reach a maximum before absolute contact, and then begin to diminish, or it may increase and increase up to contact, we cannot tell which. Whatever may be the law of variation of the force, it is certain that throughout a small part of the distance it is considerably more than one ton. It is possible that it is enormously more than one ton, to make up the ascertained amount of work of $4\frac{1}{2}$ centimetre-grammes performed in a space of 50 micro-millimetres.

But now let us vary the circumstances a little. I take the two pieces of rigidified water, and bring them to touch at a pair of corresponding points in the borders of the two surfaces A and B, keeping the rest of the surfaces wide asunder (see Fig. 1). The work done on my hands in this proceeding is infinitesimal. Now, without at all altering the law of attractive force, let a minute film of the rigidified water become fluid all over each of the surfaces A and B; you see exactly what takes place. The pieces of matter I hold in my hands are not the supposed pieces of rigidified water. They are glass, with the surfaces A and B thoroughly cleaned and wetted all over each with a thin film of water. What you now see taking place is the same as what would take place if things were exactly according to our ideal supposition. Imagine, therefore, that these are really two pieces of water, all rigid, except the thin film on each of the surfaces A and B, which are to be put together. Remember also that the Royal Institution, in which we are met, has

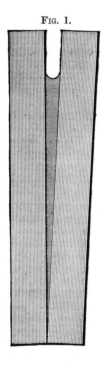

FIG. 1.

been, for the occasion, transported to the centre of the earth so that we are not troubled in any way by gravity. You see we are not troubled by any trickling down of these liquid films—but I must not say *down*, we have no up and down here. You see the liquid film does not trickle along these surfaces towards the table, at least you must imagine that it does not do so. I now turn one or both of these pieces of matter till they are so nearly in contact all over the surfaces A and B, that the whole interstice becomes filled with water. My metrical sense of touch tells me that exactly $4\frac{1}{2}$ centimetre-grammes of work has again been done; this time, however, not by a very great force, through a space of less than 50 micro-millimetres, but by a very gentle force acting throughout the large space of the turning or folding-together motion which you have seen, and now see again. We know, in fact, by the elementary principle of work done in a conservative system, that the work done in the first case of letting the two bodies come together directly, and in the second case of letting them come together by first bringing two points into contact and then folding them together, must be the same, and my metrical sense of touch has merely told me in this particular sense what we all know theoretically must be true in every case of proceeding by different ways to the same end from the same beginning.

Now in this second way we have, in performing the folding motion, allowed the water surface to become less by 60 square centimetres. It is easily seen that, provided the radius of curvature in every part of the surface exceeds one or two hundred times the extent of distance to which the molecular attraction is sensible, or, as we may say practically, provided the radius of curvature is everywhere greater than 5000 micro-millimetres (that is, the two-hundredth of a millimetre), we should have obtained this amount of work with the same diminution of water-surface, however performed. Hence our result is that we have found $4 \cdot 5/60$ (or $3/40$) of a centimetre-gramme of work per square centimetre of diminution of surface. This is precisely the result we should have had if the water had been absolutely deprived of the attractive force between water and water, and its whole surface had been coated over with an infinitely thin contractile film possessing a uniform contractile force of $3/40$ of a gramme weight, or 75 milligrammes, per lineal centimetre.

It is now convenient to keep to our ideal film, and give up thinking of what, according to our present capacity for imagining molecular action, is the more real thing—namely, the mutual attraction between the different portions of the liquid. But do not, I entreat you, fall into the paradoxical habit of thinking of the surface film as other than an ideal way of stating the resultant effect of mutual attraction between the different portions of the fluid. Look, now, at one of the pieces of water ideally rigidified, or, if you please, at the two pieces put together to make one. Remember we are at the centre of the earth. What will take place if this piece of matter resting in the air before you suddenly ceases to be rigid? Imagine it, as I have said, to be enclosed in a film everywhere tending to contract with a force equal to $3/40$ of a gramme or 75 milligrammes weight per lineal centimetre. This contractile film will clearly press most where the convexity is greatest. A very elementary piece of mathematics tells us that on the rigid convex surface which you see, the amount of its pressure per square centimetre will be found by multiplying the sum * of the curvatures in two mutually-perpendicular normal sections, by the amount of the force per lineal centimetre. In any place where the surface is concave the effect of the surface tension is to suck outwards—that is to say, in mathematical language, to exert negative pressure inwards. Now, suppose in an instant the rigidity to be annulled, and the piece of glass which you see, still undisturbed by gravity, to become water. The instantaneous effect of these unequal pressures over its surface will be to set it in motion. If it were a perfect fluid it would go on vibrating for ever with wildly-irregular vibrations, starting from so rude an initial shape as this which I hold in my hand. Water, as any other liquid, is in reality viscous, and therefore the vibrations will gradually subside, and the piece of matter will come to rest in a spherical figure, slightly warmed as the result of the work done by the forces of mutual attraction by which it was set in motion from the initial shape. The work done by these forces during the change of the body from any one shape to any other is in simple proportion to the diminution of the whole surface area; and the configuration of equilibrium, when there is no disturbance from gravity, or from any other solid or liquid body, is the figure in which the surface area is the smallest possible that can enclose the given bulk of matter.

I have calculated the period of vibration of a sphere of water † (a dew-drop!) and find it to be $\frac{1}{4} a^{\frac{3}{2}}$, where a is the radius measured in centimetres; thus—

For a radius of $\frac{1}{4}$ cm. the period is $\frac{1}{32}$ second.
,, 1 ,, ,, $\frac{1}{4}$,,
,, 2·54 ,, ,, 1 ,,
,, 4 ,, ,, 2 ,,
,, 16 ,, ,, 16 ,,
,, 36 ,, ,, 36 ,,
,, 1407 ,, ,, 13,200 ,,

The dynamics of the subject, so far as a single liquid is concerned, is absolutely comprised in the mathematics without symbols which I have put before you. Twenty pages covered with sextuple integrals could tell us no more.

Hitherto we have only considered mutual attraction between the parts of two portions of one and the same liquid—water for instance. Consider, now, two different kinds of liquid: for instance, water and carbon disulphide (which, for brevity, I shall call sulphide). Deal with them exactly as we dealt with the two pieces of water. I need

---

* This sum for brevity I henceforth call simply " the curvature of the surface " at any point.
† See paper by Lord Rayleigh in Proceedings of the Royal Society, No. 196, May 5, 1879.

not go through the whole process again; the result is obvious. Thirty times the excess of the sum of the surface-tensions of the two liquids separately, above the tension of the interface between them, is equal to the work done in letting the two bodies come together directly over the supposed area of thirty square centimetres. *Hence the interfacial tension per unit area of the interface is equal to the excess of the sum of the surface-tensions of the two liquids separately, above the work done in letting the two bodies come together directly so as to meet in a unit area of each.* In the particular case of two similar bodies coming together into perfect contact, the interfacial tension must be zero, and therefore the work done in letting them come together over a unit area must be exactly equal to twice the surface-tension; which is the case we first considered.

If the work done between two different liquids in letting them come together over a small area, exceeds the sum of the surface-tensions, the interfacial tension is negative. The result is an instantaneous puckering of the interface, as the commencement of diffusion and the well-known process of continued inter-diffusion follows.

Consider next the mutual attraction between a solid and a liquid. Choose any particular area of the solid, and let a portion of the surface of the liquid be preliminarily shaped to fit it. Let now the liquid, kept for the moment rigid, be allowed to come into contact over this area with the solid. The amount by which the work done per unit area of contact falls short of the surface-tension of the liquid is equal to the interfacial tension of the liquid. If the work done per unit area is exactly equal to the free-surface tension of the liquid, the interfacial tension is zero. In this case the surface of the liquid when in equilibrium at the place of meeting of liquid and solid is at right angles to the surface of the solid. The angle between the free surfaces of liquid and solid is acute or obtuse according as the interfacial tension is positive or negative; its cosine being equal to the interfacial tension divided by the free-surface tension. The greatest possible value the interfacial tension can have is clearly the free-surface tension, and it reaches this limiting value only in the, not purely static, case of a liquid resting on a solid of high thermal conductivity, kept at a temperature greatly above the boiling-point of the liquid; as in the well-known phenomena to which attention has been called by Leidenfrost and Boutigny. There is no such limit to the absolute value of the interfacial tension when negative, but its absolute value must be less than that of the free-surface tension to admit of equilibrium at a line of separation between liquid and solid. If minus the interfacial tension is exactly equal to the free-surface tension, the angle between the free surfaces at the line of separation is exactly 180°. If minus the interfacial tension exceeds the free-surface tension, the liquid runs all over the solid, as, for instance, water over a glass plate which has been very perfectly cleansed. If for a moment we leave the centre of the earth, and suppose ourselves anywhere else in or on the earth, we find the liquid running up, against gravity, in a thin film over the upper part of the containing vessel, and leaving the interface at an angle of 180° between the free surface of the liquid, and the surface of the film adhering to the solid above the bounding line of the free liquid surface. This is the case of water contained in a glass vessel, or in contact with a piece of glass of any shape, provided the surface of the glass be very perfectly cleansed.

When two liquids which do not mingle, that is to say, two liquids of which the interfacial tension is positive, are placed in contact and left to themselves undisturbed by gravity (in our favourite Laboratory in the centre of the earth suppose), after performing vibrations subsiding in virtue of viscosity, the compound mass will come to rest, in a configuration consisting of two intersecting segments of spherical surfaces constituting the outer boundary of the two portions of liquid, and a third segment of spherical surface through their intersection constituting the interface between the two liquids. These three spherical surfaces meet at the same angles as three balancing forces in a plane whose magnitudes are respectively the surface tensions of the outer surfaces of the two liquids and the tension of their interface. Figs. 2 to 5 (pp. 492, 493) illustrate these configurations in the case of bisulphide of carbon and water for several different proportions of the volumes of the two liquids. (In the figures the dark shading represents water in each case.) When the volume of each liquid is given, and the angles of meeting of the three surfaces are known, the problem of describing the three spherical surfaces is clearly determinate. It is an interesting enough geometrical problem.

If we now for a moment leave our gravitationless laboratory, and, returning to the Theatre of the Royal Institution, bring our two masses of liquid into contact, as I now do in this glass bottle, we have the one liquid floating upon the other, and the form assumed by the floating liquid may be learned, for several different cases, from the phenomena exhibited in these bottles and glass beakers, and shown on an enlarged scale in these two diagrams (Figs. 6 to 8, p. 494); which represent bisulphide of carbon floating on the surface of sulphate of zinc, and in this case (Fig. 8) the bisulphide of carbon drop is of nearly the maximum size capable of floating. Here is the bottle whose contents are represented in Fig. 8, and we shall find that a very slight vertical disturbance serves to submerge the mass of bisulphide of carbon. There now it has sunk, and we shall find when its vibrations have ceased that the bisulphide of carbon has taken the form of a large sphere supported within the sulphate of zinc. Now, remembering that we are again at the centre of the earth, and that gravity does not hinder us, suppose the glass matter of the bottle suddenly to become liquid sulphate of zinc, this mass would become a compound sphere like the one shown on that diagram (Fig. 3), and

FIG. 2.

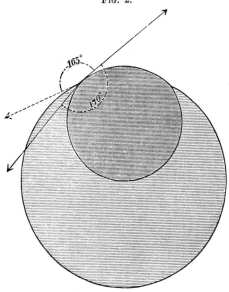

FIG. 3.

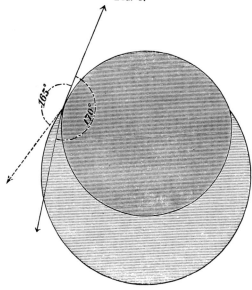

FIG. 4.

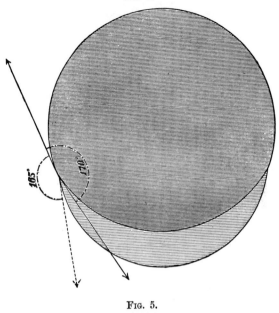

FIG. 5.

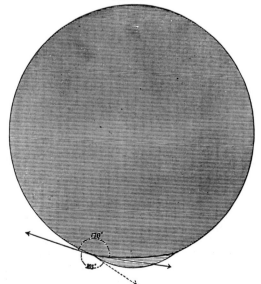

would have a radius of about 8 centimetres. If it were sulphate of zinc alone, and of this magnitude, its period of vibration would be about $5\frac{1}{2}$ seconds.

Fig. 6.

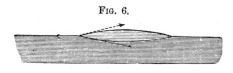

Fig. 7.

Fig. 8.

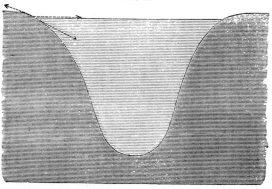

Fig. 9 shows a drop of sulphate of zinc floating on a wine-glassful of bisulphide of carbon.

In observing the phenomena of two liquids in contact, I have found it very convenient to use sulphate of zinc (which I find, by experiment, has the same free-surface tension as water) and bisulphide of carbon; as these liquids do not mix when brought together, and, for a short time at least, there is no chemical interaction between them. Also, sulphate of zinc may be made to have a density less than, or equal to, or greater than, that of the bisulphide, and the bisulphide may be coloured to a more or less deep purple tint by iodine, and this enables us easily to observe drops of any one of these liquids on the other. In the three bottles now before you the clear liquid is sulphate of zinc—in one bottle it has a density less than, in another equal to, and in the third greater than, the density of the sulphide—and you see how, by means of the coloured sulphide, all the phenomena of drops resting upon or floating within a liquid into which they do not diffuse may be observed, and, under suitable arrangements, quantitatively estimated.

When a liquid under the influence of gravity is supported by a solid, it takes a configuration in which the difference of curvature of the free surface at different levels is equal to the difference

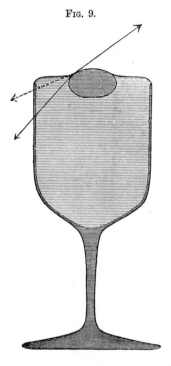

FIG. 9.

of levels divided by the surface tension reckoned in terms of weight of unit bulk of the liquid as unity; and the free surface of the liquid leaves the free surface of the solid at the angle whose cosine is, as stated above, equal to the interfacial tension divided by the free-surface tension, or at an angle of 180° in any case in which minus the interfacial tension exceeds the free-surface tension. The surface equation of equilibrium and the boundary conditions thus stated in words, suffice fully to determine the configuration when the volume of the liquid and the shape and dimensions of the solid are given. When I say determine, I do not mean unambiguously. There may of course be a multiplicity of solutions of the problem; as, for instance, when the solid presents several hollows in which, or projections hanging from which, portions of the liquid, or in or hanging from any one of which the whole liquid, may rest.

When the solid is symmetrical round a vertical axis, the figure assumed by the liquid is that of a figure of revolution, and its form is determined by the equation given above in words. A general solution of this problem by the methods of the differential and integral calculus transcends the powers of mathematical analysis, but the following simple graphical method of working out what constitutes mathematically a complete solution, occurred to me a great many years ago.

Draw a line to represent the axis of the surface of revolution. This line is vertical in the realisation now to be given, and it or any line parallel to it will be called vertical in the drawing, and any line perpendicular to it will be called horizontal. The distance between any two horizontal lines in the drawing will be called *difference of levels*.

Through any point, N, of the axis draw a line, N P, cutting it at any angle. With any point, O, as centre on the line N P, describe a very small circular arc through P P', and let N' be the point in which the line of O P' cuts the axis. Measure N P, N' P', and the difference of levels between P and P'. Denoting this last by δ, and taking a as a linear parameter, calculate the value of

$$\left(\frac{\delta}{a^2} + \frac{1}{OP} + \frac{1}{NP} - \frac{1}{N'P'}\right)^{-1}$$

Take this length on the compasses, and putting the pencil point at P', place the other point at O' on the line P' N', and with O' as centre, describe a small arc, P' P''. Continue the process according to the same rule, and the successive very small arcs so drawn will constitute a curved line, which is the generating line of the surface of revolution inclosing the liquid, according to the conditions of the special case treated.

This method of solving the capillary equation for surfaces of revolution remained unused for fifteen or twenty years, until in 1874 I placed it in the hands of Mr. John Perry (now Professor of Mechanics at the City and Guilds Institute), who was then attending the Natural Philosophy Laboratory of Glasgow University. He worked out the problem with great perseverance and ability, and the result of his labours was a series of skilfully executed drawings representing a large variety of cases of the capillary surfaces of revolution. These drawings, which are most instructive and valuable, I have not yet been able to prepare for publication, but the most characteristic of them have been reproduced on an enlarged scale, and are now on the screen before you.* Three of these diagrams, those to which I am now pointing (Figs. 10, 11, and 12), illustrate strictly theoretical solutions—that is to say, the curves there shown

FIG. 10.

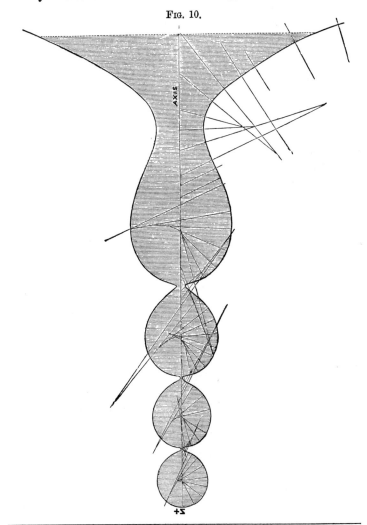

---

* The diagrams here referred to are now published in Figs. 10 to 24 of the present report of the lecture at the Royal Institution. These figures are accurate copies of Mr. Perry's original drawings, and I desire to acknowledge the great care and attention which Mr. Cooper, engraver to *Nature*, has given to the work.

Fig. 11.

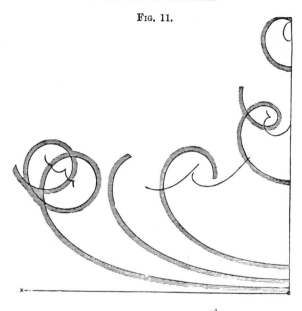

Fig. 12.

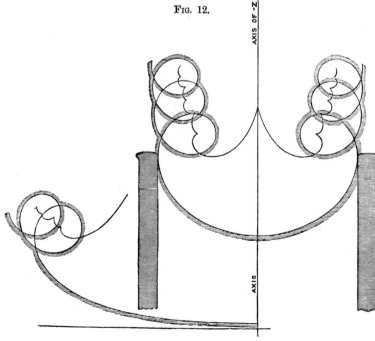

do not represent real capillary surfaces—but these mathematical extensions of the problem, while most interesting and instructive, are such as cannot be adequately treated in the time now at my disposal.

In these other diagrams, however (Figs. 13 to 28), we have certain portions of the curves taken to represent real capillary surfaces shown in section. In Fig. 13 a solid sphere is shown in four different positions in contact with a mercury surface; and again, in Fig. 14 we have a section of the form assumed by mercury resting

Fig. 13.

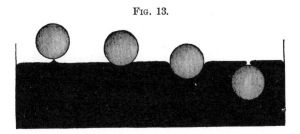

Mercury in contact with solid spheres (say of glass).

Fig. 14.

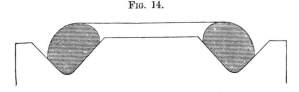

Sectional view of circular V-groove containing mercury.

in a circular V-groove. Figs. 15 to 28 (pp. 500–502) show water-surfaces under different conditions as to capillarity; the scale of the drawings for each set of figures is shown by a line the length of which represents 1 centimetre; the dotted horizontal lines indicate the positions of the free water-level. The drawings are sufficiently explicit to require no further reference here save the remark that *water* is represented by the lighter shading, and *solid* by the darker.

We have been thinking of our pieces of rigidified water as becoming suddenly liquefied, and conceiving them inclosed within ideal contractile films; I have here an arrangement by which I can exhibit on an enlarged scale a pendant drop, inclosed not in an *ideal* film, but in a *real* film of thin sheet indiarubber. The apparatus which you see here suspended from the roof is a stout metal ring of 60 centimetres diameter, with its aperture closed by a sheet of india-

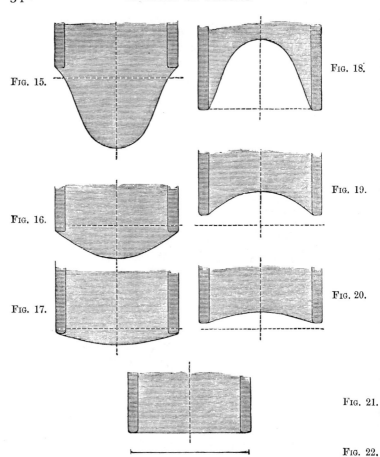

Water in glass tubes, the internal diameter of which may be found from Fig. 22, which represents a length of one centimetre.

Water resting in the space between a solid cylinder and a concentric hollow cylinder.

LIBRARY OF SCIENCE 343

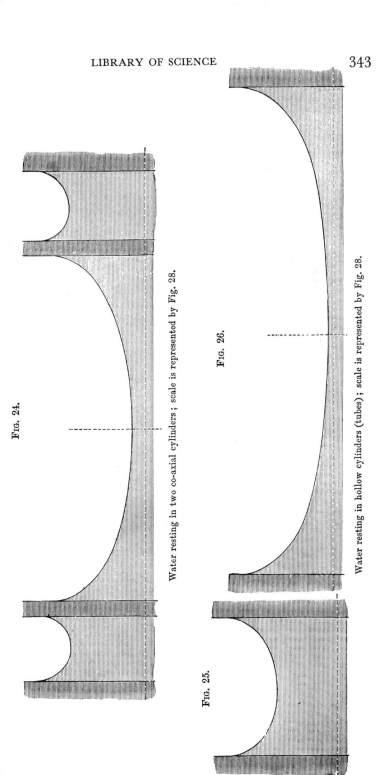

Fig. 24.

Fig. 25.

Water resting in two co-axial cylinders; scale is represented by Fig. 28.

Fig. 26.

Water resting in hollow cylinders (tubes); scale is represented by Fig. 28.

rubber tied to it all round, stretched uniformly in all directions, and as tightly as could be done without special apparatus for stretching it and binding it to the ring when stretched.

FIG. 27.

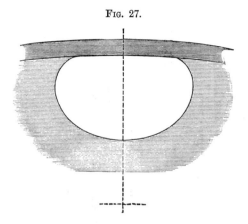

Section of the air-bubble in a level tube filled with water, and bent so that its axis is part of a circle of large radius; scale is represented in Fig. 28.

FIG. 28.

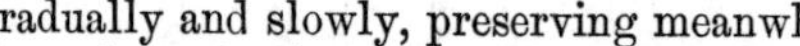

Represents a length of one centimetre for Figs. 24–27.

I now pour in water, and we find the flexible bottom assuming very much the same shape as the drop which you saw hanging from my finger after it had been dipped into and removed from the vessel of water (see Fig. 16). I continue to pour in more water, and the form changes gradually and slowly, preserving meanwhile the general form of a drop such as is shown in Fig. 15, until, when a certain quantity of water has been poured in, a sudden change takes place. The sudden change corresponds to the breaking away of a real drop of water from, for example, the mouth of a tea-urn, when the stopcock is so nearly closed that a very slow dropping takes place. The drop in the indiarubber bag, however, does not fall away, because the tension of the indiarubber increases enormously when the indiarubber is stretched. The tension of the real film at the surface of a drop of water remains constant, however much the surface is stretched, and therefore the drop breaks away instantly when enough of water has been supplied from above to feed the drop to the greatest volume that can hang from the particular size of tube which is used.

I now put this siphon into action, gradually drawing off some of the water, and we find the drop gradually diminishes until a sudden change again occurs and it assumes the form we observed (Fig. 16) when I first poured in the water. I instantly stop the action of the siphon, and we now find that the great drop has two possible forms of stable equilibrium, with an unstable form intermediate between them. Here is an experimental proof of this statement. With the drop in its higher stable form I cause it to vibrate so as alternately to decrease and increase the axial length, and you see that when the vibrations are such as to cause the increase of length to reach a certain limit there is a sudden change to the lower stable form, and we may now leave the mass performing small vibrations about that lower form. I now increase these small vibrations, and we see that, whenever, in one of the upward (increasing) vibrations, the contraction of axial length reaches the limit already referred to, there is again a sudden change, which I promote by gently lifting with my hands, and the mass assumes the higher stable form, and we have it again performing small vibrations about this form.

The two positions of stable equilibrium, and the one of unstable intermediate between them, is a curious peculiarity of the hydrostatic problem presented by the water supported by indiarubber in the manner of the experiment.

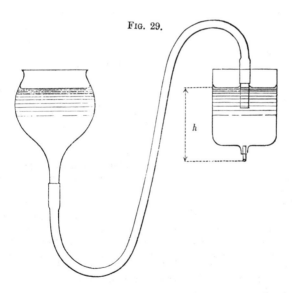

Fig. 29.

I have here a simple arrangement of apparatus (Figs. 29 and 30) by which, with proper optical aids, such as a cathetometer and a microscope, we can make the necessary measurements on real drops of water or other liquid, for the purpose of determining the values of the capillary constants. For stability the drop hanging from the open tube should be just less than a hemisphere, but for convenience it is shown, as in the enlarged drawing of the nozzle (Fig. 30),

FIG. 30.

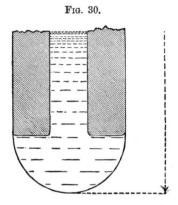

exactly hemispherical. By means of the siphon the difference of levels, h, between the free level surface of the water in the vessel to which the nozzle is attached, and the lowest point in the drop hanging from the nozzle, may be varied, and corresponding measurements taken of h and of r, the radius of curvature of the drop at its lowest point. This measurement of the curvature of the drop is easily made with somewhat close accuracy, by known microscopic methods. The surface-tension T of the liquid is calculated from the radius, r, and the observed difference of levels, h, as follows:—

$$\frac{2\,T}{r} = h\,;$$

for example, if the liquid taken be water, with a free-surface tension of 75 milligrammes per centimetre, and $r = \cdot 05$ cm., h is equal to 3 centimetres.

Many experiments may be devised to illustrate the effects of surface-tension when two liquids, of which the surface-tensions are widely different, are brought into contact with each other. Thus we may place on the surface of a thin layer of water, wetting uniformly the surface of a glass plate or tray, a drop of alcohol or ether, and so cause the surface-tension of the liquid layer to become smaller in the region covered by the alcohol or ether. On the other hand, from a surface-layer of alcohol largely diluted with water we may arrange to withdraw part of the alcohol at one particular place by promoting its rapid evaporation, and thereby increase the surface-tension of the liquid layer in that region by diminishing the percentage of alcohol which it contains.

In this shallow tray, the bottom of which is of ground glass resting on white paper, so as to make the phenomena to be exhibited more easily visible, there is a thin layer of water coloured deep blue with aniline; now, when I place on the water-surface a small quantity of alcohol from this fine pipette, observe the effect of bringing the alcohol-surface, with a surface-tension of only 25·5 dynes per lineal centimetre, into contact with the water-surface, which has a tension of 75 dynes per lineal centimetre. See how the water pulls back, as it were, all round the alcohol, forming a circular ridge surrounding a hollow, or small crater, which gradually widens and deepens until the glass plate is actually laid bare in the centre, and the liquid is heaped up in a circular ridge around it. Similarly, when I paint with a brush a streak of alcohol across the tray, we find the water drawing back on each side from the portion of the tray touched with the brush. Now, when I incline the glass tray, it is most interesting to observe how the coloured water with its slight admixture of alcohol flows down the incline—first in isolated drops, afterwards joining together into narrow continuous streams.

These and other well-known phenomena, including that interesting one, " tears of strong wine," were described and explained in a paper " On Certain Curious Motions Observable on the Surfaces of Wine and other Alcoholic Liquors," by my brother, Prof. James Thomson, read before Section A of the British Association at the Glasgow meeting of 1855.

I find that a solution containing about 25 per cent. of alcohol shows the "tears" readily and well, but that they cannot at all be produced if the percentage of alcohol is considerably smaller or considerably greater than 25. In two of those bottles the coloured solution contains respectively 1 per cent. and 90 per cent. of alcohol, and in them you see it is impossible to produce the " tears "; but when I take this third bottle, in which the coloured liquid contains 25 per cent. of alcohol, and operate upon it, you see—there—the " tears " begin to form at once. I first incline and rotate the bottle so as to wet its inner surface with the liquid, and then, leaving it quite still, I remove the stopper, and withdraw by means of this paper tube the mixture of air and alcoholic vapour from the bottle and allow fresh air to take its place. In this way I promote the evaporation of alcohol from all liquid surfaces within the bottle, and where the liquid is in the form of a thin film it very speedily loses a great part of its alcohol. Hence the surface-tension of the thin film of liquid on the interior wall of the bottle comes to have a greater and greater value than the surface-tension of the mass of liquid in the bottom, and where these two liquid surfaces, having different surface-tensions, come together we have the phenomena of " tears." There, as I hasten the evaporation, you see the horizontal ring rising up the side of the bottle, and afterwards collecting into drops which slip down the side and give a fringe-like appearance to the space through which the rising ring has passed.

These phenomena may also be observed by using, instead of alcohol, ether, which has a surface-tension equal to about three-fourths of that of alcohol. In using ether, however, this very curious effect may be seen.* I dip the brush into the ether, and hold it near to but not touching the water-surface. Now I see a hollow formed, which becomes more or less deep according as the brush is nearer to or farther from the normal water surface, and it follows the brush about as I move it so.

Here is an experiment showing the effect of heat on surface-tension. Over a portion of this tin plate there is a thin layer of resin. I lay the tin plate on this hot copper cylinder, and we at once see the fluid resin drawing back from the portion of the tin plate directly over the end of the heated copper cylinder, and leaving a circular space on the surface of the tin plate almost clear of resin, showing how very much the surface-tension of hot resin is less than that of cold resin.

Note of January 30, 1886.—The equations (8) and (9) on p. 59 of Clerk-Maxwell's article on "Capillary Attraction" in the ninth edition of the "Encyclopædia Britannica" do not contain terms depending on the mutual action between the two liquids, and the concluding expression (10), and the last small print paragraph of the page are wholly vitiated by this omission. The paragraph immediately following equation (10) is as follows:—

"If this quantity is positive, the surface of contact will tend to contract, and the liquids will remain distinct. If, however, it were negative, the displacement of the liquids which tends to enlarge the surface of contact would be aided by the molecular forces, so that the liquids, if not kept separate by gravity, would become thoroughly mixed. No instance, however, of a phenomenon of this kind has been discovered, for those liquids which mix of themselves do so by the process of diffusion, which is a molecular motion, and not by the spontaneous puckering and replication of the boundary surface as would be the case if T were negative."

It seems to me that this view is not correct; but that on the contrary there is this "puckering" as the *very beginning* of diffusion. What I have given in the lecture as reported in the text above seems to me the right view of the case as regards diffusion in relation to interfacial tension.

It may also be remarked that Clerk-Maxwell, in the large print paragraph of p. 59, preceding equation (1), and in his application of the term potential energy to E in the small print, designated by *energy* what is in reality exhaustion of energy or negative energy;

---

* See Clerk-Maxwell's article (p. 65) on "Capillary Attraction" ('Encyclopædia Britannica,' 9th edition).

and the same inadvertence renders the small print paragraph on p. 60 very obscure. The curious and interesting statement at the top of the second column of p. 63, regarding a drop of carbon disulphide in contact with a drop of water in a capillary tube would constitute a perpetual motion if it were true for a tube not first wetted with water through part of its bore—". . . if a drop of water and a drop of bisulphide of carbon be placed in contact in a horizontal capillary tube, the bisulphide of carbon will chase the water along the tube."

Additional Note of June 5, 1886.—I have carefully tried the experiment referred to in the preceding sentence, and have not found the alleged motion.

[W. T.]

Friday, February 12, 1886.

The Right Hon. LORD RAYLEIGH, M.A. D.C.L. LL.D. F.R.S. Manager and Vice-President, in the Chair.

PROFESSOR OSBORNE REYNOLDS, M.A. F.R.S.

Experiments showing Dilatancy, a property of Granular Material, possibly connected with Gravitation.

IN commencing this discourse, the author said, My principal object tonight is to show you certain experiments which I have ventured to think would interest you on account of their novelty, and of their paradoxical character. It is not, however, solely or chiefly on account of their being curious that I venture to call your attention to them. Let them have been never so striking, you would not have been troubled with them had it not been that they afford evidence of a fact of real importance in mechanical philosophy.

This newly recognised property of granular masses which I have called dilatancy will, it may be hoped, be rendered intelligible by the experiments, but it was not by these experiments that it was discovered.

This discovery, if I may so call it, was the result of an attempt to conceive the mechanical properties a medium must possess in order that it might fulfil the functions of an all-pervading æther—not only in transmitting waves of light, and refusing to transmit waves like those of sound, but in causing the force of gravitation between distant bodies, and actions of cohesion, elasticity, and friction between adjacent molecules, together with the electric and magnetic properties of matter, and at the same time allowing the free motion of bodies.

It will be well known to those who attend the lectures in this room, that although a vast increase has been achieved in knowledge of the actions called the physical properties of matter, we have as yet no satisfactory explanation as to the *prima causa* of these actions themselves; that to explain the transmission of light and heat it has been found necessary to assume space filled with material possessing the properties of an elastic jelly, the existence of which, though it accounts for the transmission of light, has hitherto seemed inconsistent with the free motion of matter, and failed to afford the slightest reason for the gravitation, cohesion, and other physical properties of matter. To explain these, other forms of æther have been invented, as in the corpuscular theory and the celebrated hypothesis of La Sage, the impossibilities of which hypotheses have been finally proved by the late Professor Maxwell, to whom we owe so much of our definite knowledge of the fundamental physics. Maxwell insisted on the fact, that even if each of the physical properties could be explained by a special æther, it would not advance philosophy, as each of these æthers would require another æther to explain its existence, *ad infinitum*. Maxwell clearly contemplated the existence of one medium, but it was a medium which would cause not one but all the physical properties of matter. His writings are full of definite investigations as to what the mechanical properties of this æther must be to account for the laws of gravitation, electricity, magnetism, and the transmission of light, and he has proved very clear and definite properties, although, as he distinctly states, he was unable to conceive a mechanism which should possess these properties.

As the result of a long-continued effort to conceive a mechanical system possessing the properties assigned by Maxwell, and further, which would account for the cohesion of the molecules of matter, it became apparent that the simplest conceivable medium—a mass of rigid granules in contact with each other—would answer not one but all the known requirements, provided the shape and mutual fit of the grains were such, that while the grains rigidly preserved their shape, the medium should possess the apparently paradoxical or anti-sponge property of swelling in bulk as its shape was altered.

I may here remark, that if æther is atomic or granular, that it should be a mass of grains holding each other in position by contact like the grains in the sack of corn is one of only two possible conceptions; the other being that of La Sage, or the corpuscular theory that the grains are free like bullets moving in space in all directions.

Nor, in spite of its paradoxical sound, is there any great difficulty of conceiving the swelling in bulk. When the grains are in contact, it appears at once that the mechanical properties of the medium must be to some extent affected by the shape and fit of the grains. And having arrived at the conclusion that in order to act the part of æther, this shape and fit must be such that the mass could not change its shape without changing its volume or space occupied, the next thing was to see what possible shape could be given to the grains, so that while these rigidly preserved their shape, the medium might possess this property of dilatancy.

It was obvious that the grains must so interlock, that when any change of shape of the mass occurred, the interstices between the grains should increase. This would be possessed by grains shaped to fit into each other's interstices in one particular arrangement.

In an ordinary mass of brickwork or masonry well bonded without mortar, the blocks fit so as to have no interstices; but if the pile be in any way distorted, interstices appear, which shows that the space occupied by the entire mass has increased (*shown by a model*).

At first it appeared that there must be something special and systematic, as in the brick wall, in the fit of the grain of æther, but subsequent consideration revealed the striking fact, *that a medium composed of grains of any possible shape possessed this property of dilatancy so long as one important condition was satisfied.*

This condition is that the medium should be continuous, infinite in extent, or that the grains at the boundary should be so held as to prevent a rearrangement commencing. All that is wanted is a mass of hard smooth grains, each grain being held by the adjacent grains, and the grains on the outside prevented from rearranging.

Smooth hard spheres arranged as an ordinary pile of shot are in their closest order, the interstices occupying a space about one-third that occupied by the spheres themselves. By forcing the outside shot so as to give the pile a different shape, the inside spheres are forced by those on the outside, and the interstices increase. Thus by shaping the outside of the pile, the interstices may be increased to any extent until they occupy about nine-tenths the volume of the spheres: this is the most open formation. A further change of shape in the same direction causes a contraction of the interstices until a minimum volume is reached, and then again an expansion, and so on. The point to be realised is that in any of these arrangements if the whole of the spheres on the outside of the group are fixed, those inside will be fixed also. (*Shown by a model.*)

An interior portion of a mass of smooth hard spheres therefore cannot have its shape changed by the surrounding spheres without altering the room it occupies, and the same is true for any granular mass, whatever be the shape of the grains.

Considering the generality of this conclusion, the non-discovery of this property as existing in tangible matter requires a word of explanation.

The physical properties of elasticity, adhesion, and friction so far render the molecules of ordinary matter incapable of behaving as a system of parts with the sole property of keeping their shape, and so prevents evidence of dilatancy in solids and fluids. This is quite consistent with dilatancy in the æther, for the properties of elasticity, cohesion, and friction in tangible matter are due to the presence of the æther, so that it would be illogical for the elementary atoms of the æther to possess these properties.

This although a sufficient reason why dilatancy has not been recognised as a property of solid and fluid matter, does not explain its non-existence in masses of solid, hard, free grains, as of corn, shot, and sand. To understand why it has not been observed in these, it must be remembered that to ordinary observation these present only an outside appearance, and that the condition essential for dilatancy, that the outside grains should not be free to rearrange, is seldom fulfilled. Also these granular forms of matter, though commonplace, have not been the subjects of physical research, and hence such evidence as they do afford has escaped detection.

Once, however, having recognised dilatancy as a universal property of granular masses, it was obvious that if evidence of it was to be sought from tangible matter, it must be sought in what have hitherto been the most commonplace and least interesting arrangements. That an important geometrical and mechanical property of a material system should have been hidden for thousands of years, even in sand and corn, is such a striking thought that it required no little faith in mechanical principles to undertake the search for it, and although finding nothing but what was strictly in accordance with the conclusions previously arrived at, the evidence obtained of this long-hidden property was as much a matter of visual surprise to the lecturer as it can be to any of the audience.

To render the dilatancy of a granular mass evident, it was necessary to accomplish two things: (1) the outside grains must be controlled so that they could not rearrange, and this without preventing change of shape and bulk of the mass; (2) the changes of bulk or volume of the mass, or of the interstices between the grains, must be rendered evident by some method of measurement which did not depend on the shape of the mass.

A very simple means—*a thin indiarubber* envelope or boundary—answered both these purposes to perfection. The thin indiarubber closed over the outside grains sufficiently to prevent their change of position, and the impervious character of the bag allowed of a continuous measure of the volume of the contents, by measuring the quantity of air or water necessary to fill the interstices.

Taking an indiarubber bag which will hold six pints of water, without stretching, and having only a small tubular aperture, getting it quite dry, and putting into it six pints of dry sea sand, such as will run in an hour-glass, sharp river sand, dry corn, shot, or glass marbles, it presents no very striking appearance, but all the same when filled with any of these materials, it cannot have its form changed as by squeezing between two boards without changing its volume. These changes of volume are not sufficient to be noticeable while the squeezing is going on, but they may be rendered apparent. It is sufficient to do this with the bag full of clean dry Calais sand, such as is used in an hour-glass.

The tube from the bag is connected with a mercurial pressure-gauge, so that the bag is closed by the mercury.

The actual volume occupied by the quartz grains is four and a half pints. The remaining space, one and a half pints, is occupied by the interstices between the grains in their closest order; these interstices are full of air, so that three-quarters of the bag are occupied by quartz, and one-quarter by air. Since the bag is closed and no more air can get in if interstices are increased from one pint and a half to two pints, the air must expand, and its pressure will fall from that of the atmosphere to three-quarters of an atmosphere. As soon as squeezing begins, the mercury rises on the side connected with the bag, and steadily rises as the bag flattens until it has risen seven inches, showing that the bag has increased in capacity by half a pint or one-twelfth of its initial capacity.

That by squeezing a porous mass like sand we should diminish the pressure of the air in the pores is paradoxical, and shows the anti-sponginess of the granular material; had there been a sponge in the bag, the pressure of the air would have increased with the squeezing.

This experiment has been mainly introduced to prevent a possible impression that the fluid filling the interstices has anything to do with the dilation besides measuring it.

Water affords a more definite measure of volume than air.

Taking a small indiarubber bottle with a glass neck full of shot and water, so that the water stands well into the neck. If instead of shot the bag were full of water or had anything of the nature of a sponge in it, when the bag was squeezed the water would be forced up the neck. With the shot the opposite result is obtained; as I squeeze the bag, the water decidedly shrinks in the neck.

This experiment, which you see is on a very small scale, was not designed to show to an audience; it was the original experiment which was made for my own satisfaction, when the idea of dilatancy first presented itself. The result but for the knowledge of dilatancy would appear paradoxical, not to say magical. When we squeeze a sponge between two planes, water is squeezed out; when we squeeze sand, shot, or granular material, water is drawn in.

Taking a larger apparatus, a bag which holds six pints of sand, the interstices of which are full of water without any air—the glass neck being graduated so as to measure the water drawn in. On squeezing the bag with a large pair of pincers, a pint of water is drawn from the neck into the bag. This is the maximum dilation; the grains of sand are now in the most open order into which they can be brought by this squeezing; further squeezing causes them to take closer order, the interstices diminish, and the water runs out into the vessel, and for still further squeezing is drawn back again, showing that as the change of form continues, the medium passes through maximum and minimum dilations.

This experiment may be repeated with granules of any size or shape, provided only they are hard, and shows the universality of dilatancy.

Although not more definite, perhaps more striking evidence of dilatancy is afforded by the means which the non-expansibility of water affords of limiting the volume of the bag. An impervious bag full of sand and water without air cannot have its contents enlarged without creating a vacuum inside it—the interstices of the sand are therefore strictly limited to the volume of the water inside it, unless forces are brought to bear sufficient to overcome the pressure of the atmosphere and create a vacuum. Since then, owing to this property of dilatancy, the shape of a granular mass at its greatest density cannot change without enlarging the interstices, preventing this enlargement by closing the bag we prevent change of shape.

Taking the same bag, the sand being at its closest order—closing the neck so that it cannot draw more water. A severe pinch is put on the bag, but it does not change its shape at all; the shape cannot alter without enlarging the interstices, these cannot enlarge without drawing more water, and this is prevented. To show that there is an effort to enlarge going on, it is only necessary to open a communication with a pressure-gauge, as in the experiment with air. The mercury rises on the side of the bag, showing when the pinch is hardest (about 200 lbs. on the planes) that the pressure in the bag is less by 27 inches of mercury than the pressure of the atmosphere; a little more squeezing and there is a vacuum in the bag. Without a knowledge of the property of dilatancy such a method of producing a vacuum would sound somewhat paradoxical. Opening the neck to allow the entrance of water, the bag at once yields to a slight pressure, changing shape, but this change at once stops when the supply is cut off, preventing further dilation.

In these experiments neither the thickness of the bag nor the character of the fluid has anything to do with the dilation of the contents considered as forming an interior group of a continuous medium, the bag merely controlling the outside members as they would be controlled by surrounding grains, and the fluid merely measuring or limiting the volume of the interstices.

It has, however, been absence of such control of the outside grains and such means of measuring the volume of the interstices that has prevented the dilatancy revealing itself as a general mechanical property of granular material, as a *mechanical* property, because dilatancy has long been known to those who buy and sell corn. It is seldom left for the philosopher to discover anything which has a direct influence on pecuniary interests; and when corn was bought and sold by *measure* it was in the interest of the vendor to make the interstices as large as possible, and of the vendee to make them as small; of the vendor to make the corn lie as lightly as possible, and of the vendee to get it as dense as possible. These interests are obvious; but the methods of getting corn dense and light are *paradoxical* when compared with the methods for other material. If we want to get any elastic material light we shake it up, as a pillow or a feather bed, or a basket of dried fruit; to get these dense we squeeze them into the measure. With corn it is the reverse; it is no good squeezing it to get it dense; if we try to press it into the measure we make it light—to get it dense we must shake it—which owing to the surface of the measure being free, causes a rearrangement in which the grains take the closest order.

At the present day the measure for corn has been replaced by the scales, but years ago corn was bought and sold by measure only, and measuring was then an art which is still preserved. It is understood that the corn is to be measured light, and the method employed is now seen to have made use of the property of dilatancy. The measure is filled over full and the top struck with a round pin called the strake or strickle. The universal art is to put the strake end on into the measure before commencing to fill it. Then when heaped full, to pull the strake gently out and strike the top; if now the measure be shaken it will be seen that it is only nine-tenths full.

Sand presents many striking phenomena well known but not hitherto explained, which are now seen to be simply evidence of dilatancy.

Every one who walks on the strand must have been painfully struck with the difference in the firmness and softness of the sand at different times; letting alone when it is quite dry and loose. At one time it will be so firm and hard that you may walk with high heels without leaving a footprint; while at others, although the sand is not dry, one sinks in so as to make walking painful. Had you noticed you would have found that the sand is firm as the tide falls, and becomes soft again after it has been left dry for some hours. The reason for this difference is exactly the same as that of the closed bags with water and air in the interstices of the sand. The tide leaves the sand, though apparently dry on the surface, with all its interstices perfectly full of water which is kept up to the surface of the sand by capillary attraction; at the same time the water is percolating through the sand from the sands above where the capillary action is not sufficient to hold the water. When the foot falls on this water-saturated sand, it tends to change its shape, but it cannot do this without enlarging the interstices—without drawing in more water. This is a work of time, so that the foot is gone again before the sand has yielded. If you stand still, you will find that your feet sink more or less, and that when you move, the sand becomes wet all round the space you stood on, which is the excess of water you have drawn in, set free by the sand regaining its densest form.

One phenomenon attending walking on firm sand is very striking; as the foot falls, the sand all round appears to shoot white or dry momentarily, soon becoming dark again. This is the suction into the enlarging interstices below the foot, which for the moment depresses the capillary surface of the water below that of the sand.

After the tide has left the sand for a sufficient time, the greater part of the water has run out of the interstices, leaving them full of air, which by expanding allows the interstices to enlarge, and the foot to sink in far enough to make walking unpleasant.

If we walk on sand under water, it is always more or less soft, for the interstices can enlarge, drawing in water from above.

The firmness of the sand is thus seen to be due to the interstices being full of water, and to the capillary action or surface tension of the water at the surface of the sand. This capillary action will hold the water up in the sand for some inches or feet, according to the fineness of the sand. This is shown by a somewhat striking experiment. If sand running in a stream from a small hole in the bottom of a vessel, as in an hour-glass, fall into a vessel containing a slight depth of water, the sand at first forms an island, which rises above the water. The sand which then falls on the top of this island is dry as it falls, but capillary action draws up the water which fills the interstices and gives the sand coherence. The island grows vertically, very fast, and assumes the form of a column, sometimes with branches like a tree or a fern, some inches or even a foot high. The strength of these consists in the surface tension of the water preventing air from being drawn in to enlarge the interstices, which therefore cannot change shape; it is therefore another evidence of dilatancy.

By substituting an impervious envelope for the surface of water, firmness of sand saturated with water may be rendered very striking.

Thin indiarubber balloons, which may be easily expanded with the mouth, afford an almost transparent envelope.

Taking one with about six pints of sand and water closed without air, there being more water than will fill the interstices at the densest, but not enough to allow them to enlarge to the full extent. When standing on the table, the elasticity of the envelope given is a rounded shape. The sand has settled down to the bottom, and the excess of water appears above the sand, the surface of which is free. The bag may be squeezed and its shape altered, apparently as though it had no firmness, but this is only so long as the surface is free. But taking it between two vertical plates and squeezing, at first it submits, apparently without resistance, when all at once it comes to a dead stop. Turning it on to its side, a 56-lb. weight produces no further alteration of shape; but on removing the weight, the bag at once returns to its almost rounded shape.

Putting the bag now between two vertical plates, and slightly shaking while squeezing, so as to keep the sand at its densest, while it still has a free surface, it can be pressed out until it is a broad flat plate. It is still soft as long as it is squeezed, but the moment the pressure is removed, the elasticity of the bag tends to draw it back to its rounded form, changing its shape, enlarging the interstices, and absorbing the excess of water; this is soon gone, and the bag remains a flat cake with peculiar properties. To pressures on its sides it at once yields, such pressures having nothing to overcome but the elasticity of the bag, for change of shape in that direction causes the sand to contract. To radial pressures on its rim, however, it is perfectly rigid, as such pressures tend further to dilate the sand; when placed on its edge, it bears one cwt. without flinching.

If, however, while supporting the weight it is pressed sufficiently on the sides, all strength vanishes, and it is again a rounded bag of loose sand and water.

By shaking the bag into a mould, it can be made to take any shape; then, by drawing off the excess of water and closing the bag, the sand becomes perfectly rigid, and will not change its shape without the envelope be torn; no amount of shaking will effect a change. In this way bricks can be made of sand or fine shot full of water and the thinnest indiarubber envelope, which will stand as much pressure as ordinary bricks without change of shape; also permanent casts of figures may be taken.

I have now shown as fully as time will allow, the experiments which afford evidence of the existence of the property of dilatancy, and how it explains natural phenomena hitherto but little noticed.

Beyond affording evidence of the existence of the property of dilatancy, these experiments have no direct connection with gravitation or the physical properties of matter.

These properties cannot be deduced by direct experiment on granular material, for the simple reason that the grains of the medium which constitutes the æther must be free from friction, while the grains with which we work are subject to friction. These properties can only be deduced by mathematical reasoning, into which I will not drag you to-night. I will merely show you one or two or three facts which may serve to convey an idea of how dilatancy should have such a bearing on the foundation of the universe.

If you look at this diagram, you see it represents a ball surrounded by a continuous mass of grain, the density of the grains being indicated by the depth of colour. If that ball were to grow in volume, it would have to push out the medium on all sides, and in that way it would distort the groups of grains or change their form, causing the interstices to increase; those nearer the ball would be distorted more than those further away. Then the interstices of these would grow the most rapidly, and those adjacent to the ball would first come to their openest order for further growth; these would contract somewhat, those a little further away would reach the openest order, and if the process of growth steadily continued, we should have a series of undulations of density commencing at the ball and moving outwards; the first of these waves of open order would not, however, get beyond half the diameter of the ball away. The diagram represents the interstices that would result if a single grain of the material had grown to the size of the ball, pushing the medium out before it. It is not necessary that the ball should have grown, to produce this result; however the ball were originally placed, if it were moved away from its original place it would assume this arrangement, and with this arrangement it would be free to move. Now, although I cannot attempt to enter upon the relation between the density of the medium and the force of attraction between two bodies in it, I may call your attention to this fact, that the dilation as calculated varies exactly as the force of gravitation, inversely as the square of the distance from an infinite distance till close to the ball, and then goes through several undulations, corresponding exactly to the variations in the attraction of bodies necessary to explain the elasticity and cohesion of molecules. As is shown in the other diagrams, these undulations in density, which may be experimentally produced, not only appear to afford a clear explanation of cohesion, but are the only suggestion of an explanation ever made. And further, similar undulations have been found necessary to explain one of the phenomena of light. My reason for calling your attention to them was partly an experiment, which, although not the most striking, is the most advanced experiment in the direction of dilatancy.

The apparatus is that represented in the diagram; the medium is contained in the large elastic bag; in the middle of this bag is a small hollow elastic ball, which can be expanded by water forced in through a tube passing through the medium and outside ball; the quantity of water which passes in is measured by a mercury gauge, the water being forced in by the pressure of the mercury. The medium between the two balls is sand and water, and is connected with a gauge, the water drawn from which measures the dilation.

The full pressure of 30 inches is on the interior ball, but produces no expansion, because the medium outside cannot dilate as the supply of water is now cut off; opening the tap to admit water to the outer ball, it at once draws water. It has now drawn 3 oz.; in the meantime the mercury has fallen, showing that an ounce and a half was admitted to the interior ball, the expansion of which drew the water into the outer envelope. This experiment is not striking, but it is definite, and enables us to measure the dilation consequent on a given distortion.

It is impossible for me to go further into this explanation, so I will merely state that the ability of the grains of a medium to slide over a smooth surface has been experimentally shown to produce phenomena closely resembling the conduction of electricity, to complete which it is only necessary to construct the medium of two different sorts of grains, different in size or different in shape, the separation of which would afford the two electricities and be a simple way out of the difficulty hitherto found in explaining the non-exhaustibility of the electricity in a body. Hitherto the two electric fluids have been supposed to reside together in the matter of the machine, which, however much has been withdrawn, has never shown signs of exhaustion. In the dilatant hypothesis these electricities are the two constituents of the æther which the machine separates, and it is worth noticing that the ordinary electrical machine resembles in all essential particulars the machines used by seedsmen for separating two kinds of seed, trefoil and ryegrass, which grow together: as long as there is a supply of the mixture, the machine is never exhausted.

This dilatant hypothesis of æther is very promising, although it cannot be put forward as proved until it has been worked out in detail, which will take long. In the meantime it is put forward mainly to excite interest in the property of dilatancy to the discovery of which it has led. This property, now that it has once been recognised, is quite independent of any hypothesis, and offers a new field for philosophical and mathematical research quite independent of the æther.

[O. R.]

Friday, April 16, 1886.

Sir WILLIAM BOWMAN, Bart. LL.D. F.R.S. Vice-President, in the Chair.

PROFESSOR SIR HENRY E. ROSCOE, M.P. LL.D. F.R.S.

On Recent Progress in the Coal-tar Industry.

THOSE who have read Goethe's episodes from his life, known as 'Wahrheit und Dichtung,' will remember his description of his visit in 1741 to the burning hill near Dutweiler, a village in the Palatinate. Here he met old Stauf, a coal philosopher, *philosophus per ignem*, whose peculiar appearance and more peculiar mode of life, Goethe remarks upon. He was engaged in an unsavoury process of collecting the oils, resin, and tar, obtained in the destructive distillation of coal carried on in a rude form of coke oven. Nor were his labours crowned with pecuniary success, for he complained that he wished to turn the oil and resin into account, and save the soot, on which Goethe adds that in attempting to do too much, the enterprise altogether failed. We can scarcely imagine, however, what Goethe's feelings would have been could he have foreseen the beautiful and useful products which the development of the science of a century and a half has been able to extract from Stauf's evil smelling oils. With what wonder would he have regarded the synthetic power of modern chemistry, if he could have learnt that not only the brightest, the most varied colours of every tone and shade can be obtained from this coal tar, but that some of the finest perfumes can, by the skill of the chemist, be extracted from it. Nay, that from these apparently useless oils, medicines which vie in potency with the rare vegeto-alkaloids can be obtained, and lastly, perhaps most remarkable of all, that the same raw material may be made to yield an innocuous principle, termed *saccharine*, possessed of far greater sweetness than sugar itself. The attainment of such results might well be regarded as savouring of the chimerical dreams of the alchemist, rather than expressions of sober truth, and the modern chemist may ask a riddle more paradoxical than that of Samson, " Out of the burning came forth coolness, and out of the strong came forth sweetness"; and by no one could the answer be given who had not ploughed with the heifer of science, " What smells stronger than tar and what tastes sweeter than saccharine ? " That these are matters of fact we may assure ourselves by the most convincing of all proofs—their money value, and we learn that the annual value of the products now extracted from an unsightly and apparently worthless material, amounts to several millions sterling, whilst the industries based upon these results give employment to thousands of men.

Sources of the Coal-tar products.—In order to obtain these products, whether colours, perfumes, antipyretic medicines, or sweet principle, a certain class of raw material is needed, for it is as impossible to get nutriment from a stone as to procure these products from wrong sources. All organic compounds can be traced back to certain hydrocarbons, which may be said to form the skeletons of the compounds, and these hydrocarbons are divisible into two great classes: (1) the paraffinoid, and (2) the benzenoid hydrocarbons. The chemical differences both in properties and constitution between these two series are well marked. One is the foundation of the fats, whilst the other class gives rise to the essences or aromatic bodies. Now all the colours, finer perfumes, and antipyretic medicines referred to, are members of the latter of these two classes. Hence if we wish to construct these complicated structures, we must employ building materials which are capable of being cemented into a coherent edifice, and therefore we must start with hydrocarbons belonging to the benzenoid series, as any attempt to build up the colours directly from paraffin compounds would prove impracticable. Of all the sources of hydrocarbons, by far the largest is the natural petroleum oils. But these consist almost entirely of paraffins, and hence this source is commercially inapplicable for the production of colours. We have, however, in coal itself, a raw material which by suitable treatment may be made to yield oils of a valuable character. Of these treatments, that followed out in the process of gas-making is the most important, for in addition to illuminating gas in abundant supply, tar is produced which contains principally that benzenoid class of substances already referred to, and which, to use the words of Hofmann, " is one of the most wonderful productions in the whole range of chemistry." The production of these latter as distinguished from the paraffinoid group appears to depend upon a high temperature being employed, to effect the necessary decomposition.

The quantity of coal made into coke for use in the blast furnace is larger than that distilled for gas-making, no less than between eleven and twelve million tons of coal being annually consumed in the blast furnaces of this country in the form of coke, and capable of yielding two million tons of volatile products. Up to recent times, however, the whole of these volatile products has been burnt and lost in the coke ovens. But lately, various processes have been devised for preventing this loss, and for obtaining the oils, which might be made available as colour-producing materials. It is, moreover, a somewhat remarkable fact that only in one or two cases have the conditions been complied with which render it possible to obtain the necessary benzenoid substances. In the ordinary coking ovens, as well as in the blast furnaces, although the temperature ultimately reached is far in excess of that needed to form the colour-giving hydrocarbons, yet the heating process is carried on so gradually that the volatile products from the coal are obtained in the form of paraffinoid bodies mainly, and hence are useless for colour-making purposes. Amongst the few coking processes in which the heat is suddenly applied, and consequently a yield of colour-giving hydrocarbons is obtained, may be mentioned the patented process of Simon-Carvès, the use of which is now spreading in England and abroad. The tar obtained in this process is almost identical in composition with the average gas-works tar, whilst the coke also appears to be equal for iron-smelting purposes to that derived from other coke-ovens. A third source of these oils yet remains to be mentioned, viz. those obtained as a by-product in blast furnaces fed with coal.

Another condition has, in addition, to be considered in this industry, and that is the nature of the coal employed for distillation. It is a well-known fact that if Lancashire cannel be exclusively employed in gas-making a highly luminous gas is obtained, but the tar is too rich in paraffins to be a source of profit to the tar-distiller, whilst, on the other hand, coal of a more anthracitic character, like that from Newcastle or Staffordshire, yields a tar too rich in one constituent, viz. naphthalene, and too poor in another, viz. benzene. It is also known to those engaged in carbonising coal principally for the sake of the tar that the coal from different measures, even in the same pit, yields tars of very different constitution. That under these varying conditions products of varying composition are obtained is a result that will surprise no one who considers the complicated chemical changes brought about in the process of the destructive distillation of coal.

History of Benzene and its derivatives.—Having thus sketched the principles upon which the formation of these valuable tar colours depends, we should do wrong to pass over the history of the discovery of benzene (C_6H_6), which contributed so much to the unlocking of the coal-tar treasury.

Faraday in 1825 discovered two new hydrocarbons in the oils obtained from portable gas. One of these was found to be butylene (C_4H_8); to the other Faraday gave the name of bicarburet of hydrogen, as he ascertained its empirical formula to be C_2H ($C = 6$). By exploding its vapour with oxygen, he observed that one volume contains 36 parts by weight of carbon to 3 parts by weight of hydrogen, and its specific gravity compared with hydrogen is therefore 39.*

Mitscherlich, in 1834, obtained the same hydrocarbon by distillation of benzoic acid, $C_7H_6O_2$, with slaked lime, and termed it benzin. He assumed that it is formed from benzoic acid simply by removal of carbon dioxide. Liebig denied this, adding the following editorial note to Mitscherlich's memoir:—" We have changed the name of the body obtained by Professor Mitscherlich by the dry distillation of benzoic acid and lime, and termed by him benzin, into benzol, because the termination 'in' appears to denote an analogy between strychnine, quinine, &c., bodies to which it does not bear the slightest

---

* 'Phil. Trans.,' 1825, p. 440.

resemblance, whilst the ending in 'ol' corresponds better to its properties and mode of production. It would have been perhaps better if the name which the discoverer, Faraday, had given to this body had been retained, as its relation to benzoic acid and benzoyl compounds is not any closer than it is to that of the tar or coal from which it is obtained."

Almost at the same time Péligot found that the same hydrocarbon occurs, together with benzone, $C_{13}H_{10}O$ (diphenylketone $CO(C_6H_5)_2$), in the products of the dry distillation of calcium benzoate.

The different results obtained by Mitscherlich and Péligot are represented by the following formulæ:—

$$C_7H_6O_2 + CaO = C_6H_6 + CaCO_3.$$
$$(C_7H_5O_2)_2Ca = C_{13}H_{10}O + CaCO_3.$$

Péligot obtained benzene only as a by-product, exactly as in the preparation of acetone (dimethylketone) from calcium acetate, a certain quantity of marsh gas is always formed.

It is not clear how Liebig became acquainted with the fact that benzene is formed by the dry distillation of coal, as his pupil Hofmann, who obtained it in 1845 from coal-tar, observes: "It is frequently stated in memoirs and text-books that coal-tar oil contains benzene. I am, however, unacquainted with any research in which this question has been investigated." It is, however, worthy of remark that about the year 1834, at the time when Mitscherlich had converted benzene into nitrobenzene, the distillation of coal-tar was carried out on a large scale in the neighbourhood of Manchester; the naphtha which was obtained was employed for the purpose of dissolving the residual pitch, and thus obtaining black varnish. Attempts were made to supplant the naphtha obtained from wood-tar, which at that time was much used in the hat factories at Gorton, near Manchester, for the preparation of "lacquer," by coal-tar naphtha. The substitute, however, did not answer, as the impure naphtha left, on evaporation, so unpleasant a smell, that the workmen refused to employ it. It was also known about the year 1838, that wood-naphtha contained oxygen, whilst that from coal-tar did not, and hence Mr. John Dale attempted to convert the latter into the former, or into some similar substance. By the action of sulphuric acid and potassium nitrate, he obtained a liquid possessing a smell resembling that of bitter almond oil, the properties of which he did not further investigate. This was, however, done in 1842 by Mr. John Leigh, who exhibited considerable quantities of benzene, nitrobenzene, and dinitrobenzene, to the Chemical Section of the British Association meeting that year in Manchester. His communication is, however, so printed in the Report, that it is not possible from the description to identify the bodies in question.

Large quantities of benzene were prepared in 1848, under Hofmann's direction, by Mansfield, who proved that the naphtha in coal-tar contains homologues of benzenes, which may be separated from it by fractional distillation. On the 17th of February, 1856, Mansfield was occupied with the distillation of this hydrocarbon, which he foresaw would find further applications, for the Paris Exhibition, in a still. The liquid in the retort boiled over and took fire, burning Mansfield so severely that he died in a few days.

The next step in the production of colours from benzene and toluene is the manufacture of nitrobenzene, $C_6H_5NO_2$ and nitrotoluene, $C_7H_7NO_2$. The former compound, discovered in 1834 by Mitscherlich, was first introduced as a technical product by Collas under the name of artificial oil of bitter almonds, and Mansfield in 1847 patented a process for its manufacture. It is now used for perfuming soap, but mainly for the manufacture of aniline ($C_6H_5NH_2$) for aniline blue and aniline black and for magenta. It is made on a very large scale by allowing a mixture of well-cooled fuming nitric acid and strong sulphuric acid to run into benzene contained in cast-iron vessels provided with stirrers.

To prepare aniline from nitrobenzene, this compound is acted upon with a mixture of iron turnings and hydrochloric acid in a cast-iron vessel. Commercial aniline is a mixture of this compound with toluidine obtained from toluene contained in commercial benzene. Some idea of the magnitude of this industry may be gained from the fact that in one aniline works near Manchester no less than 500 tons of this material are manufactured annually. From the year 1857, after Perkin's celebrated discovery [*] of the aniline colours, up to the present day, the history of the chemistry of the tar products has been that of a continued series of victories, each one more remarkable than the last.

Coal-tar Colours.—To even enumerate the different chemical compounds which have been prepared during the last thirty years from coal-tar would be a serious task, whilst to explain their constitution and to exhibit the endless variety of their coloured derivatives which are now manufactured would occupy far more time than is placed at my disposal. Of the industrial importance of these discoveries, the speaker reminded his audience of the wonderful potency of chemical research, as shown by the fact that the greasy material which in 1869 was burnt in the furnaces or sold as a cheap waggon grease at the rate of a few shillings a ton, received two years afterwards, when pressed into cakes, a price of no less than one shilling per pound, and this revolution was caused by Gräbe and Liebermann's synthesis of alizarin, the colouring matter of madder,[†] which is now manufactured

---

[*] See Lectures by Professor Hofmann, F.R.S., 'On Mauve and Magenta,' April 11, 1862, and W. H. Perkin, F.R.S., 'On the Newest Colouring Matters,' May 14, 1869, Proc. Roy. Inst.; also President's Address (Dr. Perkin, F.R.S.), 'Journal of Society of Chemical Industry,' Vol. IV., July 1884, on Coal Tar Colours.

[†] 'On the Artificial Production of Alizarine, the Colouring Matter of Madder,' by Prof. H. E. Roscoe, Proc. Roy. Inst., April 1, 1870; also Dr. Perkin, F.R.S., 'On the History of Alizarine,' Journal Society of Arts, May 30, 1879.

from anthracene at a rate of more than two millions sterling per annum; and it is stated that an offer was once made, in the earlier stages of its history, by a manufacturer of anthracene to the Paris authorities to take up the asphalt used in the streets for the purpose of distilling it, in order to recover the crude anthracene.

Again, we have in the azo-scarlets derived from naphthalene a second remarkable instance of the replacement of a natural colouring matter, that of the cochineal insect, by artificial tar-products, and the naphthol-yellows are gradually driving out the dyes obtained from wood extracts and berries. It is, however, true that some of the natural dye-stuffs appear to withstand the action of light better than their artificial substitutes, and our soldiers' red coats are still dyed with cochineal.

The introduction of these artificial scarlets has, it is interesting to note, greatly diminished the cultivation of cochineal in the Canaries, where, in its place, tobacco and sugar are now being largely grown.

Let us next turn to inquire as to the quantities of these various products obtainable by the distillation of one ton of coal in a gas-retort. The six most important materials found in gas-tar from which colours can be prepared, are :—

1. Benzene.
2. Toluene.
3. Phenol.
4. Metaxylene (from solvent naphtha).
5. Naphthalene.
6. Anthracene.

The average quantity of each of these six raw materials obtainable by the destructive distillation of one ton of Lancashire coal is seen in Table I. Moreover, this table shows the average amount of certain colours which each of these raw materials yields, viz. :—

1.
2. } Magenta 0·623 lb.
3. Aurin 1·2 lb.
4. (*Xylidine* 0·07 lb.)
5. Vermilline scarlet 7·11 lbs.
6. Alizarin 2·25 lbs. (20%).

Further it shows the dyeing power of the above quantities of each of these colours, all obtained from one ton of coal, viz.:—

1 and 2. Magenta, 500 yards of flannel.
    3. Aurin, 120 yards of flannel 27 in. wide.
    4.
    5. } Vermilline scarlet, 2560 yards of flannel.
    6. Alizarin, 255 yards Turkey red cloth.

Lastly, to point out still more clearly these relationships, the dyeing-power of one pound of coal is seen in the lowest horizontal column, and here we have a particoloured flag, which exhibits the exact amount of colour obtainable from one pound of Lancashire coal.

Let us moreover remember, in this context, that no less than ten million tons of coal are used for gas-making every year in this country, and then let us form a notion of the vast colouring power which this quantity of coal represents.

TABLE I.—ONE TON OF LANCASHIRE COAL YIELDS WHEN DISTILLED IN GAS RETORTS ON AN AVERAGE

Gas (cubic feet).	Ammoniacal Liquor, 5° Tw.	Equal to Ammonium Sulphate.	Coal (Gas) Tar, sp. gr. 1·16.	Coke.
10,000	20 to 25 gallons.	30 lbs.	12 gallons = 139·2 lbs.	13 hundredweights.

TWELVE GALLONS OF GAS TAR YIELD (Average of Manchester and Salford Tar).

Benzene.	Toluene.	Phenol proper.	Solvent Naphtha for Indiarubber, containing the three Xylenes.	Heavy Naphtha.	Naphthalene.	Creosote.	Heavy Oil.	Anthracene.	Pitch.
lb. 1·10 = Aniline 1·10 = Magenta 0·623 or 1·10 lb. Aniline yield 1·23 lb. Methyl Violet.	lb. 0·90 = Toluidine 0·77	lb. 1·5 Aurin 1·2	lb. 2·44 yielding 0·12 Xylene = 0·07 Xylidine	lb. 2·40	lb. 6·30 = α Naphthylamine 5·25 = α or β Naphthol 4·75 = Vermilline Scarlet, RRR 7·11 or = Naphthol Yellow * 9·50	lb. 17·0	lb. 14	lb. 0·46 Alizarin 20 %. 2·25.	lb. 69·6

DYEING POWER OF COLOURS FROM 1 TON OF LANCASHIRE COAL.

0·623 Magenta or 1·23 Methyl Violet dye 500 yards 27 in. wide Flannel a full shade.	1·23 Methyl Violet dye 1000 yards 27 in. wide Flannel a full Violet.	9·50 Naphthol Yellow or 7·11 Vermilline dye 3800 yards 27 in. wide Flannel a full Yellow.	7·11 Vermilline dye 2560 yards 27 in. wide Flannel a full Scarlet.	1·2 Aurin dye 120 yards 27 in. wide Flannel a full Orange.	2·25 Alizarin 20 % dye 255 yds Printer's cloth a full Turkey Red.

DYEING POWER OF COLOURS FROM 1 LB. OF LANCASHIRE COAL.

Magenta or Violet a piece of Flannel 8 in. by 27 in.	Violet a piece of Flannel 24 in. by 27 in.	Yellow a piece of Flannel 61 in. by 27 in.	or Scarlet a piece of Flannel 41 in. by 27 in.	Orange a piece of Flannel 1·93 in. by 27 in.	Turkey Red a piece of Flannel 4 in. by 27 in.

* The Naphthol Yellow is a representative colour from α Naphthol, while the Vermilline Scarlet is a representative colour from the combination of

The several colours here chosen as examples are only a few amongst a very numerous list of varied colour derivatives of each group. Thus we are at present acquainted with about sixteen distinct yellow colours; about twelve orange; more than thirty red colours; about fifteen blues, seven greens, and nine violets; also a number of browns and blacks, not to speak of mixtures of these several chemical compounds, giving rise to an almost infinite number of shades and tones of colour. These colours are capable of a rough arrangement according as they are originally derived from one or other of the hydrocarbons contained in the coal-tar. The fifty specimens of different colours exhibited may thus be classified, but in this Table, for the sake of brevity, only the commercial names and not the chemical formulæ of these compounds is given.

Azo-colours.—Amongst the most important of the artificial colouring matters may be classed the so-called azo-colours. These colours are chiefly bright scarlets, oranges, reds, and yellows, with a few blues and violets. They owe their existence to the discovery by Griess, in 1860, of the fact that the so-called azo-group $-N=N-$ can replace hydrogen in phenols and amido compounds. But it is to Dr. O. N. Witt that is due the honour of having given the first start in a practical direction to the chrysoidine class of azo-colours by the discovery of chrysoidine, and perhaps still more so by the suggestions contained in a paper read before the Chemical Society. Dr. Caro, of Mannheim, was also acquainted with several compounds which belong to this class at the time Witt published his results, but it does not appear that he made practical use of them until Witt introduced the chrysoidines and tropeolines. To Roussin, of the firm of Poirrier of Paris, is due the credit of having first brought into the market some of the beautiful azo-derivatives of naphthol. Griess, therefore, as the original discoverer of the typical compounds and reactions by which the azo-colours are obtained, may be considered as the grandfather, whilst Roussin and Witt are really the fathers, of the azo-colour industry. Nor must it be forgotten that it is to Perkin we owe the recognition of the value of the sulpho group in relation to azo-colours, a discovery patented in 1863. Moreover it is interesting to note that changes in colour from yellow to red and claret are effected by the increase in the molecular weights of the radicals introduced as well as by the relative positions occupied by these groups.

Indophenol.—Witt is also the discoverer of a new blue dye-stuff termed indophenol, which has been used as a substitute for indigo. Certain difficulties, however, have arisen in the adoption of this colour on the large scale. The most important use indophenol is at present put to is for producing dark blues on reds dyed with azo-colours, both on wool and cotton. The piece goods are dyed a uniform red first, and then printed with indophenol white; for like indigo itself indophenol yields a colourless body on reduction, and this being a very powerful reducing agent destroys the azo-colour, being itself transformed into indophenol blue. The process works with surprising

368 LIBRARY OF SCIENCE

TABLE II.

	1. Benzene.	2. Toluene.	3. Phenols.	4. Xylene.	5. Naphthalene.	6. Anthracene.
Browns Yellows	Orange Yellow, or Acid Yellow, Metanil Yellow, Auramine	..	Picric Acid	.. .. ..	Manchester Yellow (Dinitronaphthol) Naphthol Yellow ..	Alizarin (pure) Anthrapurpurin
	Brown, Y	Bismarck Brown, R	..	..	..	Flavopurpurin
Oranges	Diphenylamine Orange (Blackley Orange) Chrysoidine, Y Orange I. (mixture of 1 and 5) Orange II. (1 and 5) Orange III. (Helianthine) Orange IV.	Chrysoidine, R	Aurin	.. ..	.. ..	
Reds	Safranin	Magenta, R Magenta, B	Eosin Safrosin Cyanosine Rose Bengal Phloxin Erythrosin	.. .. (Mixture of Xylene and Naphthalene) Xylidine Scarlet (Mixture of Cumene and Naphthalene) Cumidine Scarlet	Bordeaux Vermilline Scarlet, R Vermilline Scarlet, R R R Vermilline Scarlet, B B B Roccellin (Mixture 1 and 5) New Red Biebrich Scarlet Crocein Scarlet ,	
Blues	Diphenylamine Blue Methylene Blue Indulin (Campbelline)	Blackley Blue, R Blackley Blue, I Alkali Blue, R Alkali Blue, 6 B	.. ..	.. ..	(Mixtures of 1 and 5) Victoria Blue, 1 Victoria Blue, 5	
Violets	Methyl Violet, 6 B Methyl Violet, R	.. ..	.. ..	.. ..	.. ..	
Greens	Malachite Green Brilliant Green Acid Green (Acid Green)	.. ..	.. ..	.. ..	.. ..	

nicety and is very cheap. The blue is formed and the red discharged with such precision that patterns can be produced in which the blue discharge covers a great deal more space than the original red. This new printing process was devised by Mr. H. Koechlin, of Lorrach. The reds used for the purpose are in the case of wool, the usual azo-scarlets, for cotton congo-red.

Artificial Indigo.—About five years ago the speaker had the honour of bringing before this audience * the remarkable discovery made by Baeyer of the artificial production from coal-tar products of indigo blue. Since that time but little progress has been made in this manufacture, as the cost of the process, unlike the case of alizarin, has as yet proved too serious to enable the artificial to compete successfully in the market with the natural indigo.

Through the kindness of a number of eminent colour manufacturers in this country and on the Continent, the speaker was enabled to illustrate his subject by a most complete series of specimens both of the colours themselves and of their application to the dyeing and printing of fabrics of all kinds. His thanks are especially due to his friend, Mr. Ivan Levinstein, of Manchester, for the interesting series of samples of cloth dyed with known quantities of fifty different coal-tar colours, each having a different chemical composition; also to the same gentleman, and to Messrs. Burt, Boulton, and Haywood, of London, for the interesting and unique series of specimens indicating the absolute quantities of products obtainable from *one ton of coal*, as well as for much assistance on the part of Mr. Levinstein in the preparation of the experimental illustrations for this discourse. To Dr. Martius of Berlin for a valuable series of colours, especially the well-known Congo red, made by his firm, including samples of wool dyed therewith, he is also much indebted. For the interesting details concerning indophenol and its applications the speaker owes his thanks to Dr. Witt and M. Koechlin.

Coal-tar Antipyretic Medicines.—Next in importance to the colour industry comes the still more novel discovery of the synthetical production of antipyretic medicines.

Up to this time quinine has held undisputed sway as a febrifuge and antiperiodic, but the artificial production of this substance has as yet eluded the grasp of the chemist. Three coal-tar products have, however, been recently prepared which have been found to possess strong febrifuge qualities, which if still in some respects inferior to the natural alkaloids, yet possess most valuable qualities, and are now manufactured in Germany at Höchst and at Ludwigshafen in large quantity. And here it is well to call to mind that the first tar colouring-matter discovered by Perkin (mauve) was obtained in 1856 during the prosecution of a research which had for its object the artificial production of quinine.

---

* 'On Indigo and its Artificial Production,' Proc. Roy. Inst., May 27th, 1881.

In considering the historical development of this portion of his subject, the speaker added that it is interesting to remember that the initiative in the production of artificial febrifuges was given by Professor Dewar's discovery in 1881 that quinoline, the basis of these antipyretic medicines, is an aromatic compound, as from it he obtained aniline. Moreover, that Dewar and McKendrick were the first to observe that certain pyridine salts act as febrifuges. So that these gentlemen may be said to be the fathers of the antipyretic medicines, as Witt and Roussin are of the azo-colour industry.

Kairine, the first of these, was discovered by Prof. O. Fischer, of Munich, in the year 1881, whilst engaged on his investigations of the oxyquinolines. The febrifuge properties of this substance were first noticed by Prof. Filehne, of Erlangen. Kairine is manufactured from quinoline, a basic product derived from aniline by heating it with glycerin and nitrobenzene by the following process. When treated with sulphuric acid, SO_4H_2, it forms quinoline sulphonic acid, and this when fused with caustic soda yields *oxyquinoline*, which is then reduced by tin and hydrochloric acid into tetrahydroxyquinoline, and this again on treatment with C_2H_5Br yields ethyl-tetraoxyquinoline or kairine. The lowering of the temperature of the body by this compound is most remarkable, though, unfortunately, the action is of much shorter duration than that effected by quinine itself; but on the other hand, with the exception of its burning taste, it exerts no evil effects such as are often observed after administration of large doses of quinine. The commercial article is the hydrochloride, the price is 85*s*. per lb., and the quantity manufactured has lately diminished owing to the discovery of the second artificial febrifuge, antipyrine.

The following graphical formula shows the constitution of kairine:—

$$C_6H_3(OH) < \begin{matrix} CH_2CH_2 \\ N(CH_3)CH_2 \end{matrix}$$
$$\wedge$$
$$H\ Cl$$

Antipyrine, the second of these febrifuges, was discovered in 1883 by Dr. L. Knorr in Erlangen, and its physiological properties were investigated by Prof. Filehne of Erlangen. The materials used in the manufacture of antipyrine are aniline and aceto-acetic ether. The aniline is first converted into phenylhydrazine, a body discovered by Emil Fischer in 1876. This body combines directly with aceto-acetic ether, with separation of water and alcohol, to form a body called pyrazol ($C_{10}H_{10}N_2O$). The methyl derivative of pyrazol derived by treating it with iodide of methyl, is *antipyrine*, its composition being $C_{11}H_{12}N_2O$. As a febrifuge, antipyrine is superior in many respects to kairine and even to quinine itself. It equals kairine in the certainty of its action whilst in its duration it resembles quinine. It is almost tasteless and odourless, is easily soluble in cold water, and takes the form of a white crystalline powder. Its use as a medicine is accompanied by no drawbacks. It occurs in commerce in the free state. The production of antipyrine, in spite of these valuable qualities, is as yet small, its chief employment being in Germany, where it has been successfully used in cases of typhoid epidemic. The price is 6s. per pound.

The following equations explain the formation and constitution of this interesting body. The foregoing febrifuges are manufactured at Höchst under the superintendence of Dr. Pauli, to whose kindness the speaker is indebted for an interesting series of specimens illustrative of the manufacture of antipyrine.

$$\underset{\text{Acetoacetic ether}}{CH_3.CO.CH_2.CO_2C_2H_5} + \underset{\text{Phenylhydrazine}}{C_6H_5.NH.NH_2}$$

$$= H_2O + C_2H_5.OH + \underset{\text{Pyrazol}}{C_{10}H_{10}N_2O}$$

$$C_{10}H_{10}N_2O + ICH_3 = \underset{\text{Antipyrine-hydriodide}}{IH.C_{10}H_9(CH_3)N_2O}$$

Dr. Knorr formulates pyrazol thus:

$$C_6H_4 \begin{array}{c} N\text{---}NH \\ | \quad \backslash \\ C\text{---}CH_3 \\ | \\ CO\text{---}CH_2 \end{array}$$

And antipyrine is

$$C_6H_4 \begin{array}{c} N\text{---}N\text{---}CH_3 \\ | \quad \backslash \\ C\text{---}CH_3 \\ | \\ CO\text{---}CH_2. \end{array}$$

The antipyretic effect of this compound is strikingly shown in the following temperature readings in a case of typhoid kindly communicated to the speaker by his friend Dr. Dreschfeld of Manchester. Each of the second set of readings was made two hours after a dose of 30 grains of antipyrine had been administered.

I.	II.	Diff.
105·0°	103·0°	2·0°
103·5	100·2	3·2
103·8	100·8	3·0
105·2	101·4	3·8
104·4	100·6	3·8

Thalline.—The third of the artificial febrifuges is *thalline*, which is offered as the tartrate and sulphate. It is manufactured by the Badische Company. Thalline is said to be used as an antidote for yellow fever. Its scientific name is tetrahydroparaquinanisol, and it was first prepared by Skraup by the action of methyl iodide and potash on paroxyquinoline.

We must, however, bear in mind that none of these synthetical febrifuges are antiperiodics, and therefore cannot be employed instead of the natural alkaloid quinine in cases of ague or intermittent fevers.

Coal-tar Aromatic Perfumes.—A third group of no less interest comprises the artificial aromatic essences, and of these may here be mentioned, in the first place, *cumarin*, $C_9H_6O_2$, the crystalline solid found in the sweet woodruff, in Tonka bean, and in certain sweet-scented grasses. This is now artificially prepared by acting upon sodium salicyl aldehyde with acetic anhydride by the reaction which is associated with the name of Dr. Perkin, and is used in the manufacture of the perfume known as "extract of new-mown hay."

A second interesting case of a production of a naturally occurring flavour, is the artificial production of *vanillin*, the crystalline principal of vanilla. Vanilla is the stalk of the *Vanilla planifolia*, which incloses in its tissues prisms of crystalline vanillin, to which substance it owes its fragrance. Tiemann and Harrmann showed that vanillin is the aldehyde of methyl protocatechuic acid

$$C_6H_3(OH)(OCH_3)CHO, [CHO : OCH_3 : OH = 1 : 3 : 4].$$

The chief seats of the vanilla productions are on the slopes of the Cordilleras north-west of Vera Cruz in Mexico, also the island of Réunion, and in the Mauritius. Since the discovery of the artificial production of vanillin, the growth of the vanilla has been very much restricted.

A variety of vanilla, termed vanillon, obtained in the East Indies, has long been used in perfumery for preparing "essence of heliotrope." This contains vanillin together with an oil, which is probably oil of bitter almonds. The essence of white heliotrope is now entirely prepared by synthetical operations. It is manufactured by adding a small quantity of artificial oil of bitter almonds to a solution of artificial vanillin; when these substances are allowed to remain for some time in contact, the mixture assumes an odour closely resembling that of natural heliotrope. Through the kindness of Mr. Rimmel the speaker was able to render the fragrance of this coal-tar perfume perceptible to his audience. Nor must we forget to mention the so-called essence of mirbane (nitrobenzene), of which about 150 tons per annum are used for perfuming soap; and artificial oil of bitter almonds, employed as a flavour in place of the natural oil.

Coal-tar Saccharine.—Of all the marvellous products of the coal-tar industry, the most remarkable is perhaps the production of a sweet principle surpassing sugar in its sweetness *two hundred and twenty* times. This substance is not a sugar, it contains carbon, hydrogen, sulphur, oxygen, and nitrogen. Its formula is

$$C_6H_4{<}^{CO}_{SO_2}{>}NH,$$

and its chemical name is benzoyl sulphonic imide, or for common use, saccharine. It does not act as a nutriment, but is non-poisonous, and passes out of the body unchanged. The following is a concise statement of its properties, and mode of production from the toluene of coal-tar. It should, however, be first mentioned that the compound benzoyl sulphonic imide (saccharine) was first discovered by Constantin Fahlberg and Remsen, in America. But no patent was taken out for a commercial process till recently, and it is now patented in this country.

STEP I.—Toluene is treated with fuming sulphuric acid in the cold, or it is heated with ordinary sulphuric acid of $168\frac{1}{2}°$ Twaddell on the water-bath, or not above $100°$ C. The latter method is the better. The acid is best caused to act upon the toluene in closed vessels rotating on horizontal axles.

$$C_6H_5CH_3 + SO_4H_2 = C_6H_4\begin{Bmatrix}CH_3\\SO_2.OH\end{Bmatrix} + H_2O.$$
Toluene.   Toluene sulphonic acids (ortho and para).

STEP II.—After all toluene (which as toluene is insoluble in the acid) has disappeared, the contents of the agitating vessel are run into wooden tanks in part filled with cold water, and the whole liquid is stirred up with chalk to neutralise the excess of sulphuric acid used and to obtain the two isomeric toluene sulphonic acids as calcium salts.

$$2\left(C_6H_4\begin{Bmatrix}CH_3\\SO_2.OH\end{Bmatrix}\right) + SO_4H_2 + 2(CaCO_3) =$$
Toluene, ortho- and para-sulphonic acids

$$\left(C_6H_4\begin{Bmatrix}CH_3\\SO_3\end{Bmatrix}\right)_2Ca + CaSO_4 + 2CO_2 + 2H_2O.$$
Calcium toluene ortho- and para-sulphonates

The neutralised mass is filtered through a filter-press to separate therefrom the precipitate of gypsum, which is washed with hot water, and the washings added to the filtrate.

STEP III.—The calcium salts are now treated with carbonate of sodium, to obtain the sodium salts, with precipitation of carbonate of calcium. The precipitate is removed by means of a filter-press from the solution containing the sodium ortho- and para-sulphonates.

$$\left(C_6H_4\begin{Bmatrix}CH_3\\SO_3\end{Bmatrix}\right)_2Ca + Na_2CO_3 = CaCO_3 + 2C_6H_4\begin{Bmatrix}CH_3\\SO_2.ONa\end{Bmatrix}.$$
The sodium toluene sulphonates

STEP IV.—The solution of the sodium salts from III. is evaporated either in an open- or in a vacuum-pan so far that a portion taken out will solidify on cooling. The contents of the pan are then run into moulds of wood or iron, and allowed to cool and solidify. The lumps are at length taken from the moulds, broken up small, and dried in a drying-room, and subsequently in a drying apparatus heated with steam, until quite desiccated.

STEP V.—The sodium sulphonate salts are now converted into their corresponding sulphonic chlorides. This is effected as follows:— The dried sulphonates are thoroughly mixed with phosphorus trichloride, itself as dry as possible. The mixture is then placed in lead-lined iron vessels, and a current of chlorine is passed over the mixture till the reaction is ended. The temperature generated by the reaction must be properly regulated by cooling the apparatus with water. The phosphorus oxychloride resulting from the decomposition is driven off, collected, and utilised for developing chlorine from bleaching powder for the chlorinating process, phosphate of lime being precipitated, which can be used in manures. For this purpose the oxychloride is treated with water, and the mixture, now containing hydrochloric and phosphoric acids, is brought into contact with the chloride of lime.

The reaction by which the ortho- and para-toluene sulphonic chlorides are produced is indicated by the following equation:—

$$C_6H_4\{{CH_3 \atop SO_3Na}} + (PCl_3 + 2Cl) = \underset{\text{Toluene sulphonic chlorides}}{C_6H_4\{{CH_3 \atop SO_2Cl}}} + POCl_3 + NaCl$$

The two sulphonic chlorides remaining in the apparatus are allowed to cool slowly, when the solid one (the para compound) is deposited in large crystals, so that the liquid one can be easily removed by the aid of a centrifugal machine. The crystalline residue is freed from all the liquid sulphonic chloride by washing with cold water. Only the liquid orthotoluene sulphonic chloride is capable of yielding saccharine, and the liquid product above separated is cooled with ice to crystallise out the last traces of the crystalline compound. The solid parasulphonic chloride obtained as by-product, is decomposed into toluene, hydrochloric, and sulphurous acids by mixing it with carbon, moistening the mixture, and subjecting it under pressure to the action of superheated steam. The total change proceeds in two stages:—

1. $C_6H_4\{{CH_3 \atop SO_2Cl}} + H_2O = C_6H_4\{{CH_3 \atop SO_2OH}} + HCl.$

2. $2\left(C_6H_4\{{CH_3 \atop SO_2OH}}\right) + C = 2(C_6H_5.CH_3) + CO_2 + SO_2.$

The toluene is then used again in Step I., and the hydrochloric and sulphurous acids in Step VII.

STEP VI.—The liquid orthotoluene sulphonic chloride is now converted into the orthotoluene sulphonic amide by treating the former with solid ammonium carbonate in the required proportions, and subjecting the resulting thick pulpy mixture to the action of steam.

Carbonic acid is set free, and a mixture of orthotoluene sulphonic amide and ammonium chloride remains.

$$C_6H_4\begin{Bmatrix}CH_3\\SO_2Cl\end{Bmatrix} + (NH_4)_2CO_3 = C_6H_4\begin{Bmatrix}CH_3\\SO_2.NH_2\end{Bmatrix} + NH_4Cl + H_2O + CO_2.$$
Toluene sulphonic chloride   Toluene sulphonic amide

As the mixture is very liable to solidify on cooling, cold water is at once added to prevent this, and to dissolve out the ammonium chloride, the amide remaining in the solid state. The liquid is separated by centrifugating.

STEP VII.—The orthotoluene sulphonic amide is now oxidised, preferably by means of potassium permanganate. The result of this will be, precipitated manganese dioxide, free alkali and alkaline carbonate, and an alkaline orthosulphamido-benzoate. The alkaline liquid requires careful neutralisation during the oxidising process, and especially before evaporating, with a mineral acid, or else the sulphamido-benzoate formed would be again split up into orthosulphonic benzoate and free ammonia, thus:—

$$C_6H_4\begin{Bmatrix}CO.ONa\\SO_2.NH_2\end{Bmatrix} + NaOH = C_6H_4\begin{Bmatrix}CO.ONa\\SO_2.ONa\end{Bmatrix} + NH_3.$$

The oxidation process itself is thus represented:—

$$C_6H_4\begin{Bmatrix}CH_3\\SO_2.NH_2\end{Bmatrix} + 3O + NaOH = C_6H_4\begin{Bmatrix}CO.ONa\\SO_2.NH_2\end{Bmatrix} + 2H_2O.$$
Sodium orthotoluene sulphamido-benzoate

By precipitation with dilute mineral acids, such as hydrochloric or sulphurous acids, the pure benzoyl sulphonic imide is at once precipitated:—

$$C_6H_4\begin{Bmatrix}CO.ONa\\SO_2.NH_2\end{Bmatrix} + HCl = NaCl + H_2O + C_6H_4\begin{Bmatrix}CO\\SO_2\end{Bmatrix}NH.$$
"Saccharine," or benzoyl sulphonic imide

Saccharine possesses a far sweeter taste than cane sugar, and has a faint and delicate flavour of bitter almonds. It is said to be 220 times sweeter than cane sugar, and to possess considerable antiseptic properties. On this account, and because of its great sweetness, it is possible that it may be useful in producing fruit preserves or jams, consisting of almost the pure fruit alone; the small percentage of saccharine necessary for sweetening these preserves being probably sufficient to prevent mouldiness. Saccharine has been proved by Stutzer, of Bonn, to be quite uninjurious when administered in considerable doses to dogs, the equivalent as regards sweetness in sugar administered, being comparable to over a pound of sugar each day. Stutzer found, moreover, that saccharine does not nourish as sugar does, but that it passes off in the urine unchanged. It is proposed thus to use it for many medical purposes, where cane sugar is excluded from the diet of certain patients, as in cases of "diabetes mellitus," and in this respect it may prove a great boon to suffering humanity, although we must remember that, as certain of the aromatic compounds if administered for a length of time are known to exert a physiological effect, especially on the liver, it will be desirable to use caution in the regular use of saccharine until its harmless action on the human body has been ascertained beyond doubt.

Saccharine is with difficulty soluble in cold water, from hot aqueous solutions it is easily crystallised. Alcohol and ether easily dissolve it. Hence from a mixture of sugar and saccharine, ether would easily separate the saccharine by solution, leaving the sugar. It melts at about 200° C. with partial decomposition.

The taste is a very pure sweet one, and in comparison with cane sugar it may be said that the sensation of sweetness is much more rapidly communicated to the palate, on contact with saccharine, than on contact with sugar. The speaker expressed his thanks to the discoverer of saccharine, Dr. Fahlberg, of Leipzic, for a complete and interesting series of preparations illustrating the domestic and medicinal uses of this remarkable compound, and also to his friend Mr. Watson Smith for the kind aid afforded him in the experimental illustration of his discourse.

[H. E. R.]

Friday, May 7, 1886.

WILLIAM HUGGINS, Esq. D.C.L. LL.D. F.R.S. Vice-President,
in the Chair.

FREDERICK SIEMENS, Esq.

On Dissociation Temperatures with special reference to Pyrotechnical questions.

IN bringing the subject of dissociation before the Royal Institution of Great Britain, I wish it to be understood that I propose to confine myself to its influence on combustion and heating, that is to say, to its effects on combustible gases and the products of combustion, and on furnace work generally. My researches have been made for the most part in connection with large gas furnaces constructed according to my new system of working with radiated heat, or what may be otherwise called free development of flame. In perfecting this system of furnace the principle of which is in many respects the reverse of that generally accepted, both as regards construction and working, I had to examine into the accuracy of certain scientific theories which could not be brought into harmony with the actual results I obtained.

In order that I may be clearly understood it is necessary to describe shortly my system of furnaces before entering upon the theory which alone appears to explain satisfactorily the practical results obtained by its means. These furnaces have of late been largely introduced and are now extensively applied. I first described them in a paper read at the Meeting of the Iron and Steel Institute held at Chester in September 1884; their main peculiarity consists in the arrangements by which the heat is abstracted in two different ways, and at two different periods. In the first, or active, stage of combustion the flame passes through a large combustion chamber (all contact with its surfaces being avoided), and parts with its heat by radiation only; while in its second stage the products of combustion are brought into direct contact with the surfaces and materials to be heated, by which means the remainder of its heat is abstracted. This, in a few words, is a description of the method of heating with free development of flame, and it now only remains to explain how to construct fireplaces and furnaces on this principle.

As regards its principal application hitherto, namely, to regenerative gas furnaces, the two successive stages of heating are, by radiation in the furnace chamber, and by contact in the regenerators. The flame during active combustion heats the furnace chamber and material placed therein by radiation only, and as soon as this stage is completed the fully burnt gases enter the regenerative chambers and deposit their remaining heat by coming into contact with the loose brickwork filling them. As it is essential that the flame during its first stage, while still in chemical action, should give up heat by radiation only, it is found absolutely necessary that it should not touch the sides or walls of the furnace chamber, or any material contained in the furnace. The sides and arches of the furnace and the flues leading into it must therefore be so arranged that *the flame does not touch anything*, and its length must be sufficient to allow time for complete combustion before the flame leaves it. When regenerative furnaces are arranged so as to fulfil these conditions the heat developed by the flame is much more intense than otherwise, while notwithstanding the higher temperature and increased working power attained, their durability is largely augmented.

Intensity of temperature and durability are two advantages of the greatest importance which were formerly seldom found combined in furnaces. The manner in which these advantages are insured in the radiation furnace may be thus explained:—

Adopting the generally accepted theory of combustion, according to which a flame consists of a chemically excited mixture of gases, whose particles are in violent motion, either oscillating to and from each other or rotating around one another, it follows that any solid substance brought into contact with gases, thus agitated, must necessarily have an impeding effect on their motion. Motion being the primary condition of combustion, the latter will be more or less interfered with, according to the greater or less extent of the surfaces which impede the action of the particles forming the flame; in the immediate neighbourhood of such surfaces the combustion of the gases will cease altogether, because the attractive influence of the surfaces will entirely prevent their motion; further off, their combustion will be partial, and only at a comparatively great distance the particles of gas will be free to continue unimpeded the motion required to maintain combustion. On the other hand, the surfaces themselves must suffer from the motion of the particles of gas producing the flame, for however small these particles may be, they produce, while in such violent motion, an amount of energy which acting constantly will in time destroy the surfaces opposed to them, just as "continual dropping wears away stone." This circumstance fully accounts for the fact that the inner sides of furnaces, and the materials they contain, are soon destroyed, not by heat, but by the mechanical, and perhaps also by the chemical action of the flame. It would seem strange that the heating power of a large volume of flame should be so much interfered with by the contact of its outer parts only with the inner sides of a large furnace chamber, if there was not another cause besides imperfect combustion to reduce the heating effect of a flame, which touches the surfaces to be heated. A flame when in the state of combustion radiates heat not only from its outer surface, but also from its interior by allowing the heat to radiate through its mass. In this manner every particle of flame sends its rays in all directions, but if the flame itself touches anywhere combustion ceases there, free carbon is liberated and produces smoke which envelopes that part and prevents the rays of heat of the other portions of the flame from reaching it.

Radiation plays a much greater part in all heating operations than has been hitherto acknowledged, consequently any cause which tends to lessen the radiating power of flame, or to screen its rays, reduces also the amount of heat which can be thus utilized.

If the flame is not allowed to come into contact with bodies to be heated, combustion is improved, while full advantage is also gained of its heat-radiating power, which would otherwise be diminished more or less, as already explained. The ordinary mode of applying flame, by allowing it to impinge directly upon the surfaces to be heated, causes imperfect combustion, prevents the rays of heat from reaching them, and also destroys, or tends to destroy them; this is particularly the case when hydrocarbons and carbonic oxide are used. These statements are fully borne out by the results which I have obtained in practice with the new and old form of regenerative furnace respectively, and they also fully agree with the theoretical explanation I have suggested; that theory, however, is still incomplete, as it does not deal with the subject of dissociation, a subject to which for various reasons I have avoided referring until recently, although it has been brought forward by several writers, and used as an argument against my new system of furnace; as according to these writers it would appear to be impossible to produce such exceedingly high temperatures as I claim to reach. I have long held the opinion that appearances of dissociation not being observable in furnaces heated by radiation, but occurring in furnaces in which the flame is allowed to come into contact with surfaces, must be due to the action on the flame of those surfaces at high temperature. I was led to this conclusion partly from my own observations, and partly from descriptions of dissociation observed by others, amongst whom was my brother, the late Sir William Siemens, who described a case of dissociation (see lecture delivered March 3rd, 1879, at the Royal United Service Institution, entitled " On the production of Steel and its application to military purposes ") which occurred in a regenerative gas furnace constructed according to our old views of combustion and heating. *The conclusion at which I have arrived is, that solid surfaces, besides obstructing active combustion, must also at high temperatures have a dissociating influence on combustible gases and on the products of combustion.*

In order to obtain information on this subject I examined the laws and theory of dissociation, and endeavoured to bring the various results obtained by scientific authorities into agreement with one another, and with my own experience, but failed entirely in doing so. The temperatures of dissociation of carbonic acid and steam, the two principal gases forming the products of combustion when ordinary fuel is used, vary very much according to these observers, and the results I have obtained in practice are different from most of them. I hope to prove that the temperature at which dissociation sets in, is, in most cases, much higher than generally admitted; and that the authorities I am about to refer to have omitted in almost all the experiments they have made to take into proper consideration one element which is liable to alter materially the results obtained by them. *This element is the surface, form, and material of the apparatus used for those experiments.*

In considering the question of dissociation, I propose to commence with Deville, who first discovered and called attention to the dissociation of gases at high temperatures. He made numerous experiments with various gases, dissociating steam, carbonic acid, and also carbonic oxide (in the latter case producing carbonic acid and carbon), and fixed certain temperatures at which he found that either complete or partial dissociation took place. Without going into details I may mention that Deville required to use vessels and tubes of definite dimensions, material, and structure, in order to obtain the results stated. One experiment had to be made with a porous tube, another required the use of a vessel with rough interior surfaces, or containing some rough or smooth material. In this way Deville arrived at a great variety of results, and although he does not state that the rough surfaces, or porous tubes, or the solid material placed inside the vessels which he employed, had any particular influence on the temperature at which dissociation took place, yet it would appear that he could not obtain his results without having recourse to those means. Deville's results depended very much upon the various kinds of surfaces he used in his experiments, if they were not entirely brought about by them; these experiments, moreover, were of a very complicated nature, so I propose to pass on to more modern authorities whose experiments are of simpler character, and less open to objection.

The most important experiments, which modify those of Deville, are due to Bunsen. Bunsen observed the dissociation of steam and carbonic acid by employing small tubes filled with an explosive mixture of these gases, to which suitable pressure gauges were attached. On igniting the gaseous mixture explosion took place, and a high momentary pressure was produced within the tube; from the pressure developed Bunsen calculated the temperature at which the explosion took place, and found that it varied with the mixtures employed. He records the circumstance that only about one-third of the combustible gases took part in the explosion, from which circumstance he concluded that the temperature attained was the limit at which combustion occurred. To prove this, Bunsen allowed the gases sufficient time to cool, after which a second explosion was produced, and even a third explosion when time was allowed for the gases to cool down again. Bunsen's theory seems very plausible, besides which he obtains much higher temperatures for his limits of dissociation than other physicists, so that I might have accepted the figures at which he arrives; these are for steam about 2400° C., and for carbonic acid about 3000° C. These temperatures are probably higher

LIBRARY OF SCIENCE

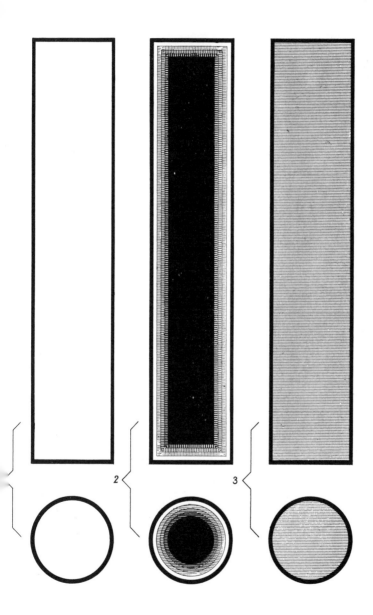

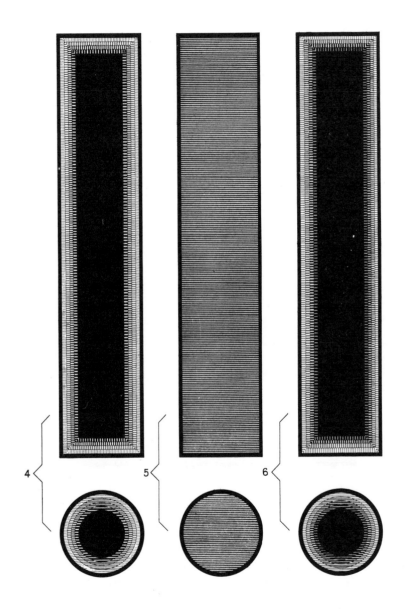

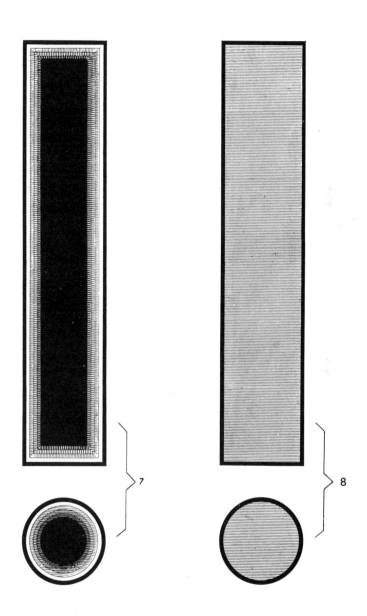

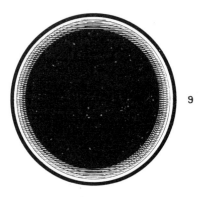

9

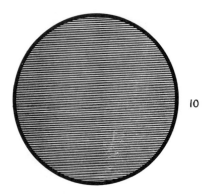

10

than are reached in the arts, as materials used in furnace-building would not withstand such temperatures for any length of time; but still I must call attention to the circumstance that if the influence of the inner surfaces of the tubes on the combustion of the gases therein could be removed, the dissociation temperatures arrived at would be found still higher. I cannot admit that Bunsen's explanation of the cause of the second and third explosions is quite satisfactory, as it is not the cooling of the gases alone which renders the subsequent explosions possible, but also the thorough re-mixture of the gases by diffusion after each explosion. This I will illustrate by means of the diagrams exhibited, Figs. 1 to 6, which represent:—

1. A tube filled with an explosive gas mixture which is shown white.

2. The same tube immediately after an explosion has taken place, the white margin indicating the unexploded mixture close to the sides, and the deep-red, towards the middle of the tube, the exploded gases. The white is shown as merging into deep-red by degrees, because close up to the sides the surfaces prevent explosion or combustion altogether; nearer the middle partial combustion takes place, whilst only in the middle of the tube the gases find sufficient space for complete combination.

3. The same tube after the burnt and unburnt gases have mixed by means of diffusion, which is coloured light-red.

4. The same tube immediately after the second explosion, coloured light-red at the sides, turning into deep-red by degrees towards the middle.

5. The same tube after diffusion has done its work a second time, coloured a deeper shade of red.

6. The same tube after the third explosion, coloured nearly deep-red throughout, but still a lighter shade on the sides.

In Bunsen's mode of determining dissociation at high temperatures we have only to deal with the obstruction which surfaces offer to combustion, leaving out their dissociating influence at high temperatures which affect most of Deville's results. For that reason Bunsen arrives at much higher dissociation temperatures than Deville, and his mode of experimenting possesses the advantage that it may lead to a proper settlement of the question of temperatures at which dissociation would set in when taking place in a space unencumbered by surfaces.

I should wish some one more experienced than I am with purely physical investigation to make the following experiment:—

Take a narrow tube of about the same size as Bunsen used for his experiments, and a hollow sphere of the same capacity, in both of which Bunsen's experiment should be repeated. The sphere offering less surface than the tube in proportion to the quantity of gas it contains, the dissociation temperature should be found higher in the former than in the latter, if my views are correct. The results that would, in my opinion, be obtained are shown approximately by the red and white coloured surfaces in the diagram (Figs. 7 to 10).

Figs. 1–10 (originally printed in red and black) are reproduced here in one colour only.

From the temperatures thus obtained, in each case, the real dissociation temperature, if no surfaces were present to influence the result, might be approximately calculated.

Bunsen's method of experimenting, according to my view of the matter, should form the foundation of further research to determine the dissociation temperatures of products of combustion. Even if means were found for eliminating the influence of surfaces, no known material at our disposal could withstand the very high temperature to which the vessels or tubes would be subjected if experiments were carried out according to Deville's method.

That the surfaces of highly heated vessels or tubes either produce, or tend to produce, dissociation, has been corroborated lately by two Russian experimentalists, Menschutkin and Kronowalow. These gentlemen found that dissociation of carbonic acid and other gases was much facilitated when the vessels used for the experiments were filled with material offering rough surfaces, such as asbestos or broken glass.

My view of the theory of dissociation caused or influenced by surfaces, may be given as follows:—Increase of temperature producing expansion of gases will reduce the attractive tendency of the atoms towards one another, or, in other words, diminish their chemical affinity. In the same ratio as the temperature is increased the repelling tendency of the atoms must increase also, until at last decomposition, or what is called dissociation, takes place. This being admitted, it will follow that the adhesive or condensing influence of surfaces on the atoms of the gas, which action will increase at high temperatures, will assist this decomposition by increasing the repelling tendency of the atoms.

Victor Meyer, who at first disputed the accuracy of the results obtained by the two physicists I have mentioned, ultimately accepted them. This circumstance I was very pleased to learn as their experiments confirmed the results I arrived at in practical work with furnaces. Thus the question may be considered nearly settled, the more so as Meyer is himself a great authority in questions of dissociation, having carried out many interesting experiments. Meyer, for instance, proved dissociation by dropping melted platinum into water, and finds that oxygen and hydrogen are evolved from the steam produced. There can be no doubt on this point, but the question arises whether heat is the sole agent that brings about the dissociation of steam in this case. In the first place the dissociating influence of the highly heated surfaces of platinum on steam has to be taken into consideration, and secondly the chemical affinity which platinum has for oxygen, and still more for hydrogen. The same remarks apply to Meyer's experiment of passing steam or carbonic acid through heated platinum tubes, in which case he obtains only traces of dissociation, the temperature being much lower. Other experiments might be mentioned, but none lead to a different conception of the question.

There is one other circumstance connected with dissociation, proved by experiment, which however, requires explanation. It is considered as a sure sign that dissociation is going on when a flame whose temperature is raised becomes longer; this it is said can only be accounted for by dissociation taking place. I agree with this conclusion, but the experiments by which it has been proved have been made, like others referred to, in narrow tubes or passages in which the dissociating action of the heated surfaces must come into play. It is not alone the heat to which the gases are raised that in these cases causes dissociation and increases the length of the flame, but also the influence of the heated surfaces in contact with the combustible gases, more especially if these gases contain hydrocarbons. The extension of the flame is also partly due to the obstruction which the surfaces offer to the recombustion of the dissociated gases through want of space. If the same flame be allowed free development in a space unencumbered by surfaces, as in my radiation furnace, no such extension of its length would be observed; but, on the contrary, it would get shorter with increase of temperature. This action can be best observed in a regenerative gas-burner whose flame is shorter the greater the intensity of the temperature, and therefore of the light produced. On the other hand, flame may be extended almost to any length if conducted through narrow passages; this may be seen in regenerative furnaces which will send the flame to the top of the chimney if the reversing valves are so arranged that the flame, instead of passing through the furnace chamber, is made to burn directly down into the regenerators. No proper combustion can then take place in the brick checkerwork of the regenerative chambers, and the flame will consequently continue to extend until cooled down below a red heat, being ultimately converted into dark smoke; thus in this case, the extensive surfaces offered by regenerators will act both ways, by preventing combustion, and by assisting dissociation.

It will now be understood that regenerative furnaces themselves offer special opportunities for making experiments, most questions being best settled by the results obtained in actual work. If dissociation sets in we see the consequences in want of heat, reduced output, and in destruction of furnace and material. If the causes of dissociation are removed we immediately become aware of the circumstance by a rise in temperature, increased output, longer furnace life, and saving of material.

Similar results may be obtained with other furnaces, but the beneficial action will not be so great as in the case of the regenerative furnace, because the intensity of heat obtainable in them is much lower.

In applying the principle of heating by radiation, or free development of flame, to boilers, it is necessary to prevent the flame in its active stage of combustion from touching either the sides of the boiler or its brickwork setting. The flame is allowed free space to burn in, and thus good combustion is obtained, after which the products of combustion are brought into intimate contact with the surfaces to be heated. While combustion is going on in the open space heat is transmitted by radiation only, but after active combustion is completed it is transmitted by contact, and it is in this manner that flame must be applied to boilers, and may be applied equally well to nearly all other heating operations.

In heating a boiler the intensity of heat produced is not very great, because the relatively cold surfaces abstract heat from the flame very eagerly, thus preventing its temperature from rising above a certain point which is below that of dissociation. But although no dissociation of the products of combustion can take place in boiler firing, the detrimental effect on combustion of the surfaces of the boiler is nevertheless very great, perhaps, indeed, greater than in any other application of firing and heating. The cold surface of the boiler has the power of extinguishing flame altogether, especially if brought into actual contact with it, because, besides the peculiar influence of surfaces on combustion, the cooling in this case is so great that the necessary temperature for combustion cannot be maintained. Thus it seems clear that heating by radiation must be most advantageous for firing boilers, but particular care should be taken that the products of combustion, as distinguished from the flame, are brought as much as possible into contact with their surfaces. Galloway tubes are preferable to bafflers for this purpose, but it will be necessary to be careful that combustion is complete before the products of combustion are allowed to come into contact with these tubes or bafflers, as otherwise they would interfere with combustion at that point.

In the paper I read before the Iron and Steel Institute, to which I have already referred, I described a boiler heated on the radiation principle, fired with the producer gas used in our regenerative gas furnaces, and with that boiler no smoke is produced. I will now describe a boiler worked on the same principle, fired with common coal, by the use of which great saving of fuel is effected, and very little or no smoke is produced. In this boiler one end of the internal flue is lined with brickwork, and contains an ordinary fire-grate, while the longer part is furnished with rings of cast iron or fire-clay, which prevent the flame from striking on the inner boiler surface. The products of combustion, after leaving the inner flue, where the flame has not come into contact with the boiler plates, are conducted underneath and at the sides of the boiler, and in these channels they may be directed by means of bafflers against the boiler sides and bottom. If the internal flue of the boiler is so long that the flame ceases before reaching its extremity, bafflers, in the form of cones, may also be placed at its far end for the purpose of causing the products of combustion to strike against its sides. Instead of cones, cross tubes of the Galloway type may be used with advantage, but it is absolutely necessary that active combustion should have ceased before the products of combustion come into contact with either bafflers or tubes. In a boiler so arranged and constructed that the flame heats mostly by radiation in its first, and by contact in its second stage, a great saving of fuel is effected, and almost no smoke produced. To avoid altogether the production of smoke in this boiler, the fuel should be charged on the grate in a uniform manner. It is quite impossible to avoid producing smoke and waste if fuel is charged unequally on the grate and at irregular intervals, however well the boiler may have been arranged and constructed. Various kinds of automatic and mechanical coal-feeding arrangements have been suggested, and some have been applied, but none have given full satisfaction. At the London Smoke Abatement Exhibition numerous apparatus of this kind were to be seen, and apparently worked successfully, but when tested at other places, and under different conditions, they have been found wanting, owing to the faulty manner in which the flame and products of combustion were dealt with. These clever appliances would, in my opinion, have worked more satisfactorily if firing and heating had been carried out in them in two successive stages.

There is a very simple way of firing, which I have employed, that may possibly not be quite new, but answers very well, and does not require complicated constructions and appliances, always more or less objectionable. It depends upon the following considerations. When fresh coal is charged upon incandescent fuel, as is the case in the usual mode of firing boilers, the volatile gases of the fresh fuel are rapidly evolved, filling the fire-box to such an extent as to prevent the ingress of air through the grate, and this occurs at the very time the air-supply should be considerably increased. The result is imperfect combustion and consequent waste of the very best combustible gases, viz. the hydro-carbons, which cannot burn for want of air to combine with; free carbon is thus liberated from these gases, and smoke is produced. In order to avoid smoke, and consequent loss of fuel, any sudden production of volatile gases, either during or after firing, must be prevented; and sufficient air should always be introduced, and so distributed, as to burn those gases as quickly as they are produced. This can be done in the following manner:—

Before putting on fresh coal the burning fuel should be pushed back from the front part of the grate and distributed on the incandescent fuel behind, care being taken that this portion of the grate is entirely free from hot fuel. When the front part of the grate has become comparatively cool owing to cold air passing through it, fresh coal is distributed thereon. The freshly charged fuel lying on the cool grate with cold air passing through it will be heated by radiation only, partly from the incandescent fuel behind, partly by the flame from its own gases, and partly by the surrounding hot brickwork. The volatile gases will consequently be liberated at a comparatively slow rate, and will combine with the air which entering through the interstices in the fuel on the cool part of the grate will be evenly distributed over its surface. Gas and air will thus be supplied in nearly the proper proportions for complete combustion of the fuel, and as the production of volatile gases diminishes, the air passing through the front part of the grate, will enter into combustion with the fuel thereon which has been deprived of nearly all its volatile constituents. By means of this simple arrangement the sudden production of a large volume of volatile gases is avoided, and air in a well divided state is always present to consume the gases liberated; thus smokeless combustion and saving of fuel are realised. Care must be taken that the fresh fuel is charged at regular intervals of time and in equal quantities. It still remains to be considered in what manner the clinkers and ashes may be most easily removed, but by the use of a movable pocket at the far end of the grate to collect them in, and a hooked bar to draw it forward at intervals, good practical results would be obtained.

Having dealt so fully with the subject of heating furnaces by radiation, I wish to be allowed to bring before your notice an apparatus for warming rooms by the same means, which I have found to be both satisfactory and economical, and England, I believe, is the country in which it is likely to be fully appreciated, as heating by radiation is almost exclusively used for domestic purposes here.

It must be borne in mind that the regenerative flame radiates much more heat than an ordinary fire or gas flame, because most of the heat which passes away from ordinary flames is in this case employed to increase the temperature, or to accumulate heat, the intensity of the flame, and consequently its radiating power, are thus much increased, or in other words, the heat ordinarily passing away from the flame with the products of combustion is converted into radiant heat.

The apparatus to which I refer is a stove provided with a *regenerative burner*, supplied with *ordinary illuminating or retort gas*, and is intended to warm apartments mainly by the radiated heat of the intensely hot flame produced; it was fully described by me in a paper read at the meeting of the Gas Institute held in Manchester last year. We find in nature that direct heating is effected by radiation exclusively; every organism, as well as all mankind, owe their existence and development to the radiated heat from the sun, and we should try to imitate nature in our methods of obtaining artificial heat. The wind which produces that change of air we require for our well being is another result of the action of the sun. Both heating and ventilation I have endeavoured to supply by means of this stove. Although there is only a small amount of heat passing away from the flame after having heated the air required for combustion it is entirely utilised for the sake of economy; but in cases where economy is not the primary object, but hygienic considerations are paramount, the regenerative gas flame is placed at the foot of the chimney in front of the grate in place of the ordinary coal fire. The background of the stove is of china or other white material, to act as a reflector, by which means the heat, otherwise lost by being radiated backwards or sideways, is recovered to assist in warming

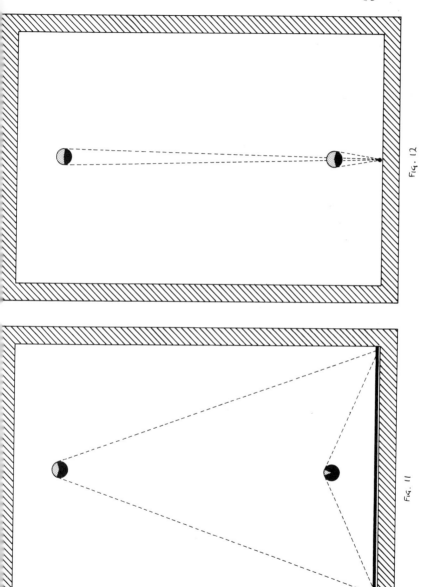

Fig. 11

Fig. 12

the room. Owing to the high temperature of the regenerative gas flame and the employment of a reflector, which would be quite impossible with an ordinary fire, the economy of fuel attained by the use of this stove, and in a less degree by the use of the chimney fire, is considerable. The maximum consumption of gas in the stove exhibited is about 12 cubic feet per hour. Considering that the gas need not burn constantly, or may be lowered as required, its daily consumption may be set down at 50 to 150 cubic feet of gas for an ordinary room of about 4000 to 5000 cubic feet capacity.

There is still one very important point which remains to be mentioned in connection with a regenerative gas flame of high intensity, provided with a reflecting background; that is, the better distribution of the radiated heat. I find that a room warmed by means of a stove or open fire, such as described, is of a more uniform temperature than when warmed by an ordinary fire or by a gas and coke fire, such as my brother was engaged in introducing into this country shortly before his death.

This, in my opinion, is mainly due to the fact that a source of radiant heat of low intensity but of large surface, sending out its rays at various angles, heats an object in its vicinity very much more than is the case with a smaller source of radiant heat of greater intensity, whose rays strike the object from one direction only, notwithstanding that both sources radiate the same quantity of heat. This action is illustrated by means of the two diagrams exhibited, Figs. 11 and 12, which represent two rooms, the one, Fig. 12, heated by a small flame of high intensity, and the other, Fig. 11, by a large flame of low intensity, both radiating the same quantity of heat. In each room two objects, globes or spheres, are represented, the one close to, and the other at a distance from the source of heat. The object in the one room near to the source having the large heating surface is almost enveloped in rays, while that in the second receives rays only in one direction, the former therefore being much more heated than the latter. This difference does not occur when the two globes at a distance from the two sources of heat are compared. The law that the rays of heat are diminished in the inverse ratio of the square of the distance is only correct as regards small but intense sources of heat, whilst the decrease of radiant heat takes place in a much higher proportion, in the case of large sources of heat of low intensity. This clearly proves that for the purpose of warming rooms by means of radiation, it is important that the heat should be concentrated in an intensely hot focus, as is the case in nature, our earth being warmed in this way by the radiant action of the sun.

From various considerations I am led to believe that the question of sanitary and economical warming is one which commands a great deal of attention in this country. Not many years ago I had to report to my own Government on the Smoke Abatement Exhibition, held in this city, and I understand that a Smoke Abatement Institution has since been inaugurated. There seems to be a general feeling that something will before long have to be used instead of the present fireplace with its smoky chimney, especially now that people are massed together in enormous cities in which cleanliness and pure air are of the greatest importance. Under these circumstances I would venture to draw the attention of authorities in sanitary science to the method of warming dwellings to which I have shortly referred, as resting on a scientific basis, being cleanly and easy of application, demanding little or no attention, and fulfilling all sanitary and economical requirements, and finally as being entirely free from all dissociating influences owing to the free development of the flame.

[F. S.]

Friday, May 14, 1886.

HENRY POLLOCK, Esq. Treasurer and Vice-President, in the Chair.

PROFESSOR JOHN MILLAR THOMSON, F.C.S.

Suspended Crystallisation.

THE phenomena attending the ordinary solution of metallic salts in water have been so often and ably brought before the consideration of this audience, that I have determined to confine myself this evening to certain considerations relating to the formation of and deposition from so-called supersaturated solutions of these salts. At the same time it is necessary for me to remind you of one or two points connected with the ordinary solution of a salt in water, which I may enumerate as follows.

The solubility of a salt in water depends:

(*a*) On the mass of salt presented to the water for solution, and the state of aggregation in which that mass may be at the time of solution.

(*b*) The temperature at which the solution of the salt is carried out; rise in temperature generally producing an increase in the solubility of the salt, although there are certain exceptions to this rule.

(*c*) Each salt has its own definite rate or amount of solubility; some being extremely soluble, as calcium chloride or sodium acetate; others having a very low amount of solubility, as calcium sulphate.

(*d*) On the cooling of a hot solution containing large quantities of salt, a deposition of the salt takes place, which deposition is known under the term "crystallisation."

Certain salts, however, which generally present abnormal phenomena in their solution show no tendency to be deposited from their solutions on cooling, provided such solutions are kept covered from the access of the outside air. Such solutions are said to have become supersaturated.

This branch of the subject is the one which will engage our attention.

It may be divided into two classes. (1) That which occurs in the presence of the undissolved salt; and (2) That which occurs in the absence of the undissolved salt; this latter class being the one we shall examine.

Glauber's salt (sulphate of soda) affords a very good example of this. If the crystallised sulphate be added to boiling water in a flask, as long as it is dissolved the water will take up nearly twice its weight of salt. If this solution be now allowed to cool in an open vessel an abundant deposition of crystals will take place, as the water when cold will only dissolve about one-third of its weight of crystallised sulphate. But if the flask be tightly corked or stoppered with cotton wool whilst the solution is boiling, it may be kept for several days without crystallising, although moved about from place to place. On withdrawing the plug, however, the air on entering the flask will produce a slight disturbance of the surface of the fluid, and from that point beautiful prismatic crystals shoot through the solution until the whole has become perfectly solid.

It is not necessary in preparing such solutions that the flask be closed completely with a stopper, as we find that plugging the neck with cotton wool exercises the same effect in preserving the solution, and that the air which enters the flask during the cooling becomes thoroughly deprived of its crystallising influence, evidently undergoing a filtering process. A very good instance of the inability of filtered air to start crystallisation is afforded by a solution of alum saturated at 194° F., and allowed to cool in a flask stoppered with cotton wool in the manner previously described. On withdrawing the plug of cotton wool the crystallisation, which in this case takes a much longer time to commence than with the sulphate of soda, will be seen beginning at various points on the surface of the liquid, and will spread slowly from these, octahedral crystals of alum half an inch or more in length being built up in a few seconds.

It is evident from these two experiments that the cause of this sudden crystallisation is to be sought for in the peculiar action of the air when it comes in contact with the solution; but that it is not due to the action of the air alone is shown by the fact that the air in the flask plugged with cotton wool must enter during the cooling, and that therefore the action of the cotton by filtering the air in some way deprives it of its active property.

But it will be found that there are other means of destroying this activity without filtering the air through cotton wool. Thus, if the solution of sodium sulphate containing two-thirds of its weight of crystallised salt be allowed to cool in a flask closed by a cork, furnished with two tubes plugged with cotton wool; on removal of the cotton plugs air may be blown from the lungs through the longer tube without causing the crystallisation, apparently from the fact that the air has been deprived of the nucleus which induces the crystallisation by passage through the lungs. On blowing air, however, from a bellows, after a few strokes the solution will be found to solidify almost instantaneously.

The earliest ideas concerning this sudden solidification were naturally those which supposed that the mere entrance of air into the flask upon opening it started the crystallisation. The late Dr. Graham, and Professor Thomas Thomson, held this view; and the former of these chemists carried out a considerable series of investigations on the action of different gases in determining the crystallisation. These ideas were followed by the statement of Ziz, that not only air but also solids were capable of acting as nuclei when dry, but that when wet, or boiled with the solution, or placed in it when hot and allowed to cool at the same time with it, they lost their effect. The activity of air, however, as in itself a nucleus, has been subsequently shown by Löwel to be incorrect; but at the same time he admits that solid bodies after exposure to air become active.

In 1851 two chemists, Selmi and Goskynski, introduced the explanation that dry air is active by virtue of its getting rid of the water at the surface, thus producing small crystals which continue the action. This seems to be a repetition of the theory propounded by Gay Lussac, who held that the air absorbed at the surface precipitated a portion of the salt in the same way that one salt precipitates another, and that this deposition continued the crystallising action. By an elaborate series of investigations, conducted in 1866, and also during later years, the French savants, MM. Gernez and Violette, came to the conclusion that there is only one nucleus for a supersaturated solution, and that is a crystal of the body itself; also indicating in certain experiments that substances which possess the same crystalline form and chemical structure may be found active to supersaturated solutions of each other. They have conclusively shown at the same time that heated or washed air, or bodies of different constitution but chemically clean, remain perfectly inactive as regards their supersaturated solutions.

That disruption of the solution may take place without causing crystallisation when the body added is perfectly clean may be easily shown by carefully removing the cotton plug from a solution of sodium sulphate containing only two-thirds its weight of crystals. On introducing a perfectly clean platinum wire no crystallisation is induced, but on removing the wire and touching it with a substance which has not previously been rendered chemically clean, and on again introducing it in the flask, the moment it touches the liquid crystallisation is at once induced.

This question of the nuclear action of substances upon these solutions has received considerable attention from English chemists, most notably from Mr. Tomlinson, Professor Liversidge, and Professor Grenfell, all of whom have conducted elaborate researches on the subject, leading, however, to different conclusions with each investigator. The arguments held by Tomlinson are strongly in support of some physical cause for the phenomenon, and that the result may be brought about by the action of substances possessing no chemical relations to the solutions experimented on. Professor Liversidge, on the other hand, has tried the action of several substances with the most careful precautions, that their addition to the solutions should be effected without contamination from, or access of, the external air to the flask during the experiments. The result of his inquiries, so far as the substances he has experimented with are concerned, confirms those of Gernez, and limits the number of bodies capable of acting as nuclei within very narrow limits. In fact, he concludes that the only body capable of causing the crystallisation of such solutions is a crystal of the substance itself, of exactly the same composition as it possesses when in a state of supersaturation. Thus Glauber's salt, which exists in a supersaturated solution, combined with 10 parts of water, will only crystallise by the addition of a crystal of the substance also containing 10 parts of water, and is perfectly inactive to crystals of the same salt which contain 7 proportions of water, or those which are anhydrous.

As it is almost impossible to conceive that our atmosphere is laden with minute particles of the many different metallic salts which we are acquainted with, some hesitation in accepting such a limited explanation may be excusable; but when we consider that it has been shown by Dr. Angus Smith and others that the air, especially in the vicinity of large manufacturing towns, is filled with small particles, more especially of Glauber's salt, we are not surprised that solutions of this body at least crystallise at once on the removal of the filtering medium. And the probability of the explanation may be further strengthened by the fact that these solutions when opened in the still air of country places may retain their liquid condition for considerable lengths of time.

At the same time the limit fixed by Professor Liversidge may appear a somewhat narrow one, and experiments made some time ago and published in the 'Journal of the Chemical Society of London' (May 1879, and September 1882), have confirmed a few experiments first indicated by Gernez, and go a considerable length in showing that bodies possessing not only the same crystalline form but an identical chemical structure are active nuclei in causing the crystallisation of supersaturated solutions.

In these experiments two methods for the addition of the nuclei intended to excite crystallisation were adopted. The first of these consisted of a flask and bulb tube, the supersaturated solution of the salt to be experimented on being placed in the flask, whilst the small bulb was filled with a solution of the body intended to act as nucleus. The solution in the bulb having been thoroughly boiled, the tube is stoppered with cotton wool and then introduced through the centre of a second cotton plug into the neck of the flask. The contents of the flask are now heated, the contents of the bulb receiving a second warming from the steam rising from the solution in the flask. The flasks so prepared are then allowed to stand till perfectly cold before performing an experiment. When it was desired to perform an experiment the solution of the body intended as nucleus contained in the bulb tube is first crystallised by touching with a platinum wire, or preferably by introducing a crystal of the salt contained in the bulb. Crystallisation having thus taken place in the bulb, and the crystal added having become enclosed in the fresh deposit, the bulb is lowered into the liquid contained in the flask and allowed to remain there for some time to show that the disturbance produced by its introduction into the fluid does not excite crystallisation. The bulb is finally lightly broken under the fluid and the result observed. Bulbs also containing water, pieces of washed glass, &c., may be broken under the supersaturated solutions, to show that the disruption produced on breaking does not excite crystallisation.

A second method which may be employed for the introduction of nuclei is that introduced by Professor Liversidge.* This consists of a siphon tube, in which crystallisation is induced in the first limb and allowed gradually to pass over the bend and down to the point of the second limb.

The following table gives the results of many experiments on the action of isomorphous and also of dissimilar substances on supersaturated solutions of each other.

Isomorphous Sulphates on Magnesium Sulphate.

Substance in solution.	Substance added.	Result.	Remarks.
$MgSO_4.7H_2O$	$ZnSO_4.7H_2O$	Active	Crystallisation induced at once, the crystals forming long needles.
,,	$NiSO_4.7H_2O$	,,	
,,	$CoSO_4.7H_2O$	,,	Crystallisation induced after some time, the crystals forming attached to the nucleus.
,,	$FeSO_4.7H_2O$	,,	
,,	$MgSO_4.7H_2O$	,,	
,,	$NiSO_4.6H_2O$	,,	Crystallisation induced after some time, the crystals attached to the nucleus generally being truncated needles.
,,	$FeSO_4.xH_2O$	,,	
,,	$CoSO_4.xH_2O$	,,	

Dissimilar Bodies on Magnesium Sulphate.

$MgSO_4.7H_2O$	$MgK_2(SO_4)_2.6H_2O$	Inactive.	
,,	$Na_2SO_4.10H_2O$	,,	
,,	$Na_2S_2O_3.5H_2O$	,,	
,,	$NaCl$	,,	
,,	Glass	,,	

Isomorphous Salts on Sodium Sulphate.

$Na_2SO_4.10H_2O$	$Na_2SeO_4.10H_2O$	Active	Crystallisation immediate.
,,	$Na_2CrO_4.10H_2O$	,,	,, ,,

* Proc. Roy. Soc., xx. 497.

Dissimilar Bodies on Sodium Sulphate.

Substance in solution.	Substance added.	Result.	Remarks.
$Na_2SO_4.10H_2O$	$MgSO_4.7H_2O$	Inactive.	
,,	Na_2SeO_4	,,	
,,	$Na_2S_2O_3.5H_2O$	,,	
,,	$Na_2HPO_4.10H_2O$	,,	
,,	KCl	,,	
,,	NaCl	,,	
,,	$KClO_3$	,,	
,,	$NaI.4H_2O$	,,	
,,	Glass	,,	

Chrome Alum and Iron Alum on Common Alum.

$AlK(SO_4)_2.12H_2O$	$CrK(SO_4)_2.12H_2O$	Active	The chrome alum solution was prepared by saturating at 70°, and then allowing it to crystallise in the bulb tube.
,,	$FeK(SO_4)_2.12H_2O$	,,	

Bodies of the same form or belonging to the same system, but not similarly constituted on Alum.

$AlK(SO_4)_2.12H_2O$	NaCl (cubes)	Inactive.	
,,	FeS_2 (cubes)	,,	
,,	Fe_3O_4 (octahedra)	,,	

Hydro-disodic Phosphate and Hydro-disodic Arsenate.

$Na_2HPO_4.12H_2O$	$Na_2HAsO_4.12H_2O$	Active	Crystallisation immediate and very rapid.

When two salts which are not isomorphous are in a supersaturated solution together a separation of one or other of the salts may be effected within certain limits. Thus in a mixture of sodium sulphate and nickel sulphate, the former may be crystallised by touching with a crystal of the salt, and in allowing the mixture to remain at rest for a few minutes the liquor containing the nickel sulphate may be entirely poured off from the crystals of sodium sulphate.

A similar phenomenon may be seen by preparing in a long glass cylinder two supersaturated solutions one above the other, the lower one being sodium thiosulphate dissolved in its water of crystallisation, the upper one sodium acetate dissolved in a quarter its weight of water. When the whole has cooled down under a stopper of cotton wool a crystal of sodium thiosulphate may be introduced; this will pass through the acetate solution without solidifying it, but will cause the immediate solidification of the thiosulphate solution.

The following phenomena may be seen in experiments carried out on mixtures of dissimilar salts.

A. When the mixture consists of two salts which are not isomorphous.

(1) Sudden crystallisation may take place, gradually spreading through the solution on the addition of a nucleus, causing a deposition of the body belonging to the nucleus only.

(2) That when sudden crystallisation takes place, causing the deposition of both salts, there is a preponderance of the salt of the same nature as the nucleus.

(3) That the nucleus may remain growing slowly in the solution, becoming increased by a deposition of the salt of the same nature as the nucleus.

B. When the mixture consists of two isomorphous salts.

(1) Sudden crystallisation may occur, giving a deposition of both salts, apparently in the proportions in which they exist in solution.

(2) That when slow crystallisation takes place, the nucleus increases by a deposition of the least soluble salt, showing that in mixed supersaturated solutions a gradation of phenomena may be experienced, passing from those shown in the crystallisation of a true supersaturated solution to those shown in the crystallisation of an ordinary saturated solution.

Passing from the action of nuclei on supersaturated solutions of mixtures of dissimilar salts, it is interesting to examine the action of the different constituents on supersaturated solutions of double salts. The following table gives the results of certain experiments with such double salts:—

Substance in solution.	Nucleus added.	Result.
$HgCl_2(NH_4Cl)_2,3H_2O$	$HgCl_2$ (prismatic)	Active.
,,	$HgCl_2$ (deposited from hot solution).	Both active and inactive.
,,	NH_4Cl	Inactive.
$HgBr_2(NH_4Br)_2,3H_2O$	$HgBr_2$ (deposited in the cold).	Active.
,,	$HgBr_2$ (deposited from hot solution).	Both active and inactive.
,,	$(NH_4)Br$	Inactive.
$HgI_2(KI)_2$	HgI_2 (needle-shaped crystals).	Active.
,,	KI	Inactive.
$AlK(SO_4)_2,12H_2O$	$Al_23(SO_4),18H_2O$	,,
,,	K_2SO_4	,,
$NaNH_4HAsO_4.4H_2O$	$Na_2HAsO,12H_2O$	,,
,,	$(NH_4)_2HAsO_4,Aq_2$	,,

From these experiments it will be seen that in the case of the double salts formed by the halogen acids, certain of the component salts are capable of inducing crystallisation, but in the case of the salts formed from acids of higher basicity the component salts are incapable of causing that particular disruption of the solution.

Having now treated this question experimentally, it may be of advantage to examine it shortly from a theoretical point of view to see if any explanation may be offered at least with our present knowledge of these sudden changes from liquid to solid.

At the present time it would be rash to attempt a complete answer to the questions—

(a) What fully takes place when a salt dissolves?

(b) Why some salts always separate out when their hot solutions are cooled: and conversely why certain ones remain dissolved under the same conditions?

In order that a salt may dissolve in water we must suppose some attraction between the molecules of the salt undergoing solution, and the molecules of the water. With most salts, the power of the water to dissolve them is increased with rise of temperature, and this rise in temperature means increase in the active movement of the water and the salt molecules, and therefore greater facility for them to come near enough to one another for their mutual attractions to be exerted.

Then why do not all salts dissolve more in hot than in cold water? all do not do so, as you have seen in my diagram during lecture. This leads us to following up the completion of this attraction; namely, the combination of the water and the salt molecules which I have alluded to in my lecture under the term hydration.

We must suppose that some hydrates exist at a higher temperature than others, and this is borne out by experiment. In the case of salts whose hydrates exist only at the lower temperatures, the effect of raising the temperature would be merely to increase the vibratory movements and so shake asunder the water and salt molecules, the dehydrated salt naturally separating out, and therefore we could not expect more to go into solution by merely heating the liquid.

With regard to the second point. We must remember that there exists a strong attraction between the individual molecules which compose the salt. Taking then the case of a salt-like potassium chlorate, which in the solid state contains no water attached to it. We dissolve it in hot water, and on cooling, much of the salt separates out. We can suppose that the attraction of the water for the salt and the active movement produced by the rise in temperature overcome this attraction of salt molecule for salt molecule, but as the solution cools, this exercises its full force, and crystallisation ensues.

Now, taking the instances where the salt remains in solution even after cooling, but in much larger quantities than can be obtained by treating the solid salt with water at that temperature; this being what I have called "suspended crystallisation."

Let us consider the case of sodium sulphate as perhaps the most familiar instance. You have seen a large volume of that salt suddenly solidify on the introduction of the proper nucleus. Now why did not that salt behave like the potassium chlorate instead of remaining in solution in the liquid after cooling?

It has been suggested that this *super*-saturated solution is merely a saturated solution of the anhydrous salt. That may or may not be so. If it be the case, we can suppose that the molecules of the salt and water are prevented from arranging themselves in their normal order and proportions, and so there is a kind of strain throughout the liquid. This can only be overcome by something which will disturb the molecules sufficiently to bring about the necessary rearrangement.

Taking the instance of the suspended solidification of water cooled below its freezing point; we know that only the disruptive effect of shaking is required; but with sodium sulphate in water and many others, no amount of shaking, as we have seen, is sufficient. Some stronger force is required; this stronger force is found so far as we know at present only in the attraction for the salt of the body itself, or of some substance having the same crystalline form and a similar chemical composition, as has been already shown.

I say advisedly at present, because in my opinion it has not yet been conclusively proved, that no form of what we should call simply mechanical disturbance may not bring about sudden crystallisation in these so-called supersaturated solutions. Indeed, there is an interesting experiment which seems to foreshadow such a possibility. By dropping a single carefully washed crystal of alum into a supersaturated solution of that salt, we notice a very interesting phenomenon. The whole surface of the solution is covered with small crystals separated by definite and considerable intervals, and it appears as if the mere mechanical disturbance produced by the first crystal attracting to itself similar molecules, had caused the union of other molecules to form crystals in the remoter parts of the liquid. This is, I think, a very interesting case, which if studied with other similar instances may throw additional light on the causes of such crystallisation. If we suppose that the salt exists in solution as a hydrate, that is, in actual combination with water, we can imagine that each individual molecule of the hydrate attracts each other one, and is attracted by it equally; so if one molecule were to move towards another, it would be restrained by its neighbour, and that in its turn by those near it, and so a state of equilibrium would come about. However, whatever may be the condition of the salt in solution, the same cause, namely, the attraction of similar molecules, appears always to render the equilibrium unstable.

[J. M. T.]

Friday, February 18, 1887.

SIR FREDERICK ABEL, C.B. D.C.L. F.R.S. Manager and Vice-President, in the Chair.

WILLIAM CROOKES, Esq. F.R.S. V.P.C.S. *M.R.I.*

Genesis of the Elements.

IN the very words selected to denote the subject I have the honour of bringing before you, I have raised a question which may be regarded as heretical. At the time when our modern conception of chemistry first dawned upon the scientific mind, the average chemist as a matter of course accepted the elements as ultimate facts. He regarded his elements as absolutely simple, incapable of transmutation or decomposition, each a kind of barrier behind which we could not penetrate. If closely pressed he said that they were self-existent from all eternity, or that they had been individually created just as we now find them at the present day. Or he might argue that the origin of the elements did not in the least concern us, and was, indeed, a question lying outside the boundaries of science.

But in these our times of restless inquiry we cannot help asking what are these elements, whence do they come, what is their signification? We cannot but feel that unless some approach to an answer to these questions can be found, our chemistry, after all, is something profoundly unsatisfactory. These elements perplex us in our researches, baffle us in our speculations, and haunt us in our very dreams. They stretch like an unknown sea before us—mocking, mystifying, and murmuring strange revelations and possibilities.

If I venture to say that our commonly received elements are *not* simple and primordial, that they have *not* arisen by chance or have *not* been created in a desultory and mechanical manner but have been evolved from simpler matters—or perhaps indeed from one sole kind of matter—I do but give formal utterance to an idea which has been, so to speak, for some time "in the air" of science. Chemists, physicists, philosophers of the highest merit declare explicitly their belief that the seventy (or thereabouts) elements of our text-books are not the pillars of Hercules which we must never hope to pass.

Did time allow I might quote utterances of Dalton, of Professor Faraday, of Dr. Gladstone, of the late Sir Benjamin Brodie, of Professor Graham, of Dr. Mills, of Professor Stokes, of Mr. Norman Lockyer, all pointing in the same direction and all showing that in the course of their researches these servants of Science have been led to think that these same elements are not the final outcome—the be-all and the end-all of chemistry.

The law of Prout, and still more the better established and far-reaching periodic law of Newlands (since developed by Professors Mendeleeff, Meyer, and Carnelley), seem to presuppose the existence of a genetic relation among the elements.

Philosophers in the present as in the past,—men who certainly have not worked in the laboratory,—have reached the same view from another side. Thus Mr. Herbert Spencer records his conviction that "the chemical atoms are produced from the true or physical atoms by processes of evolution under conditions which chemistry has not yet been able to produce."

And the poet has forestalled the philosopher. Milton ('Paradise Lost,' Book V.) makes his Archangel Raphael say to Adam, instinct with the evolutionary idea, that the Almighty had created

> "one first matter all,
> Indued with various forms, various degrees
> Of substance."

If we can show how the so-called chemical elements might have been generated we shall be able to fill up a formidable gap in our knowledge of the universe. We have a preponderance of cumulative evidence to prove that both heavenly bodies and living organisms have been formed by evolution. We are seeking now to extend this law to the so-called elements, to the first principles of which stars and organisms alike consist.

If we survey the distribution of the chemical elements we find two very distinct cases. On the one hand we see bodies grouped in definite proportions with other bodies from which they differ exceedingly and to which they are held by affinity, more or less strong. To obtain either of two such bodies in a separate state, that affinity, as every student of chemistry knows, must be overcome. Instances of such association are too common and abundant to need mention. In such cases each of the bodies grouped together has fairly marked properties. One of them, moreover, for the most part has an atomic weight very different from that of the other.

In the second case we find bodies associated with other bodies more or less closely allied to themselves. They are not held together by any decided affinity; they are not combined in definite proportions, and their atomic weights are often almost identical. If we wish to obtain one or more of these bodies in a separate state, the difficulty encountered lies not in the strength of the affinities to be overcome but in the circumstance that whatever reagent we employ acts upon one of the substances in nearly the same manner as it does upon the other. Hence, to obtain one body of this kind entirely separate is an exceedingly tedious and difficult task. Nay, we are sometimes at a loss to decide whether we have before us a really simple body or a mixture of bodies whose properties are almost identical.

The most striking instance of such association is found in the metals of the so-called *rare earths*. These bodies form but a very trifling portion of the earth's crust. They are chiefly met grouped together in a few minerals, such as samarskite and gadolinite, which, so far, have been found in but few localities, and even in those are far from common. These earths form a group to themselves; chemically, they are so much alike that it taxes the utmost skill of the chemist to effect even a partial separation, and their history is so obscure that we do not yet know the number of them.

It will not be necessary here to explain in detail the process of chemical fractionation adopted for the separation of the rarer earths, since it could interest only the chemical specialist; moreover, it has been fully described in a paper I read before the British Association at Birmingham.

Stated in the briefest way the operation consists in fixing upon some chemical reaction in which there is the most likelihood of a difference in the behaviour of the elements under treatment, even though the difference be slight, and effecting such treatment incompletely, so that only a certain fraction of the total bases present is separated: the object being to get part of the material in an insoluble and the remainder in a soluble state.

Let us suppose that we have in solution two earths almost identical in their properties, but differing slightly, almost imperceptibly, in basicity. We add to the solution of the earths, which must be very dilute, weak ammonia to such an amount only that it precipitates one-half of the bases present. The dilution must be so great that a considerable time must elapse before the liquid shows a turbidity, and several hours will have to pass over before the action of the ammonia is complete. The liquid is then filtered, by which process we have the earths divided into two parts, no longer identical in their composition. We can easily see that there is now a slight difference in the basic value of the two portions of earths; the portion in solution being, though by a scarcely perceptible amount, more basic than that which the ammonia has precipitated. This minute difference is made to accumulate systematically until it becomes perceptible either by chemical or physical tests.

The accompanying diagram (Fig. 1), illustrates the scheme of fractionation. Starting from zero at the apex the precipitates all pass to the left and the filtrates to the right. Each circle represents a flask containing the solution under treatment, and the two arrows from each circle show the path pursued respectively by the precipitate and filtrate.

Such is the general outline of the process. But, as has been already intimated, the methods of separation suitable for different groups of earths vary. Where the constituents of yttrium and samarium are concerned, nothing seems available but straightforward fractionation continued month after month and year after year.

The further question whether an earth we have separated is really simple or is still a mixture has again to be decided by yet another process, to be effected only in a very high vacuum. To understand this process it is necessary to make an apparent total digression.

It seems, perhaps, strange to speak of exhausting the air in hollow bulbs and tubes until there is left in them only the one-millionth

Fig. 1.

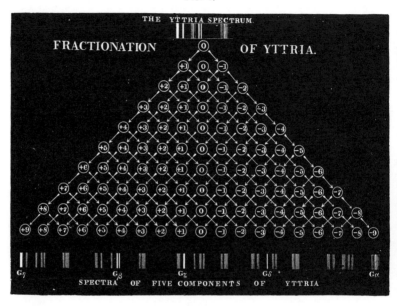

part of an atmosphere. It is only in modern times that atmospheric air has come to be regarded as matter. To this day a bottle or a jar is said to be "empty" if it contains no liquid or solid body, the air with which it is filled being completely ignored. According to the same common idea, how empty then must a vessel be when the air it contains is reduced to the one-millionth part of its original quantity! That something still remains is, however, proved by the fact that I have succeeded in reducing the pressure down to one fifty-millionth of an atmosphere. What this number represents will be better understood if I say that, given a barometric column one hundred miles in height, the remaining pressure would be equal only to about the tenth of an inch. Even this high degree of exhaustion by no means represents an absolute vacuum. I have in this glass tube perhaps the nearest approach to perfect emptiness yet artificially obtained. Its internal capacity is 5 cubic centimetres, and it is exhausted to the one fifty-millionth part of an atmosphere. It still contains 100,000000,000000 molecules. The internal space, therefore, is far, very far, from being absolutely void of matter.

The vacuum most suitable for experiments on these earths is one of about the millionth of an atmosphere. In a vacuum of this degree we find under the action of the induction-spark certain substances phosphorescing or behaving very differently from what they would if similarly treated at a lower vacuum or at the ordinary pressure of the atmosphere. When thus treated, the examination of the spectra of the phosphorescing earths furnishes what I have ventured to call the radiant matter test.

After a time, on examining the series of yttrium earths in the lowest line of flasks, their phosphorescent spectra are found to have become modified in the relative intensities of some of their lines. Ultimately different portions of the fractionated yttria give the five-spectra approximately shown at the foot of the diagram; whilst samaria also appears capable of being split up into two or perhaps three constituents.

These bodies, it must be clearly understood, are not impurities which may be removed, yttrium or samarium remaining in a pure state after their elimination. On the contrary, the molecule we formerly knew as yttrium has undergone a veritable splitting up into its constituents.

These constituents I have not as yet formally baptised. For more convenient reference and discussion I have provisionally ticketed them, as shown in the following Table.

TABLE I.

Position of Lines in the Spectrum.	Scale of Spectroscope.	Mean Wave-length of Line or Band.	$\frac{1}{\lambda^2}$	Provisional Name.	Probability.
Bright lines in—					
Violet	8·515	456	4809	$S\gamma$	New, or ytterbium.
Deep blue	8·931	482	4304	$G\alpha$	New.
Greenish blue (mean of a close pair)	9 650	545	3367	$G\beta$	New, or the $Z\beta$ of M. de Boisbaudran.
Green	9·812	564	3144	$G\gamma$	New.
Citron	9·890	574	3035	$G\delta$	New, or the $Z\alpha$ of M. de Boisbaudran.
Yellow	10·050	597	2806	$G\epsilon$	New.
Orange	10·129	609	2693	$S\delta$	New.
Red	10·185	619	2611	$G\zeta$	New.
Deep red	10·338	647	2389	$G\eta$	New.

I will rapidly sketch the most salient features of the rare earths when submitted to the radiant matter test. Some remain unaffected and thus are referred at once to a distinct group. Others have the curious property of preventing the induction-spark passing, and so simulating a non-conducting vacuum, when there is really plenty of residual gas present. The rare earth thoria possesses in the highest degree this obstructive property. Before me I have an exhausted tube having two sets of poles sealed in it, one set at each end. The size and distance apart of these poles are exactly the same in each case. At one end of the tube I have put some thorium sulphate, at the other end I have put yttrium sulphate. The exhaustion is now proceeding by aid of the Sprengel pump. I attach the wires of the induction coil to the poles at the thorium end, and, as you see, no current will pass; rather than pass through the tube, the spark prefers to strike across the spark-gauge in air—a striking distance of 37 millimetres,—showing an electromotive force of 34,040 volts. Now, without doing anything to affect the degree of exhaustion, I transfer the wires of the induction coil from the thorium to the yttrium end, and the spark passes at once. To balance the spark in air I must push the wires of the gauge together, till they are only 7 millimetres apart, equivalent to an electromotive force of 6440 volts : the fact of whether thoria or yttria is under the poles making a difference of 27,600 volts in the conductivity of the tube. The explanation of this eccentric action of thoria is not yet quite clear. From the great difference in the phosphorescence of the two earths, it is evident that the passage of electricity through these tubes is not so much dependent on the degree of exhaustion as upon the phosphorogenic property of the body opposite the poles.

Other earths become very phosphorescent, and their power of retaining residual phosphorescence differs greatly among themselves. This property we shall presently see is one of some importance. To examine this persistence of luminosity I have devised an instrument similar to Becquerel's phosphoroscope, but acting electrically instead of by means of direct light. It consists of an opaque disc, 30 inches in diameter, pierced with six openings near the edge. By means of a multiplying wheel and pulley the disc can be set in rapid rotation. At each revolution a stationary object behind one of the apertures is alternately exposed and hidden six times. A commutator forms part of the axis of the disc, and by connecting it with the wires from a battery, rotation of the disc produces alternate makes and breaks in the current. This primary current is then connected with the induction coil, from which the secondary current passes through the vacuum tube containing the earth under examination. When a phosphorescent body such as yttria is examined, if the wheel is turned slowly no light is seen when looked at from the front, as the current does not begin till the obscuration of the tube by an intercepting segment, and ends before the earth comes into view. When, however, the wheel is quickly turned, the residual phosphorescence lasts long enough to bridge over the brief interval between the cessation of the spark and the entry of the phosphorescent body into the field of view, and it is seen to glow with a faint light which becomes brighter as the speed of the wheel increases.

I will first put the phosphorescent earth glucina in the phosphoroscope. This phosphoresces of a bright blue colour, but the residual glow is so short that, with the highest speed of which the instrument is capable, you see no light whatever. In contrast I now put in a compound of the earth strontia. This also glows with a rich blue colour, showing in the spectroscope a continuous spectrum with a great concentration of light in the blue and violet. In the phosphoroscope the colour of the glow is bright green, showing in the spectroscope a continuous spectrum, with the red and blue ends cut off.

Alumina in the radiant matter tube glows with a rich crimson light. I will put some rubies—a crystalline form of alumina—in the phosphoroscope. Here the persistence of luminosity is so great that the red light is visible with the slowest speed, and with a high velocity the residual glow is nearly as strong as when the rubies are out of the instrument. Shakespeare, who is supposed to have mastered all knowledge, had he seen these rubies could hardly have described them more precisely than in the lines from 'Julius Cæsar':—

" . . . with unnumbered sparks
They are all fire, and every one doth shine."

Another distinctive phenomenon is that the earths of one group, yttrium and samarium, when submitted to the induction discharge *in vacuo*, yield discontinuous spectra.

These spectra are extremely complicated and change in their details in a puzzling manner. For many years I have persistently groped on in almost hopeless endeavour to get a clue to the meaning which I felt convinced was locked up in these systems of bands and lines. It was impossible to divest myself of the conviction that I was looking at a series of autograph inscriptions from the molecular world, evidently of intense interest, but written in a strange and baffling tongue. For a long time all attempts to decipher these mysterious signs were fruitless.

The meaning of the strongly-marked symbolic lines had first to be ascertained. After continued efforts I had to be content with roughly translating one group of coloured symbols as "yttrium" and another group as "samarium," disregarding the fainter lines, shadows, and wings frequently common to both. Constant practice in the decipherment has now given me fuller insight into what I may call the grammar of these hieroglyphic inscriptions. Every line and shadow of a line, each faint wing attached to a strong band, and every variation in intensity of the shadows and wings among themselves, has now a definite meaning which can be translated into the common symbolism of chemistry.

This leads us to what I may call the history of yttrium. Twelve months ago the name yttrium conveyed to all chemists a perfectly definite meaning. It was supposed to be an elementary or simple body, having a fixed atomic weight, $88 \cdot 9$, and its principal properties had been duly determined. Its phosphorescent spectrum gave a definite system of coloured bands, such as you see in the drawing before you (Fig. 2). Broadly speaking, there is a deep red band, a very luminous citron-coloured band, a pair of greenish-blue bands, and a blue band. These bands, it is true, varied slightly in relative intensities and in sharpness with almost every sample of yttria

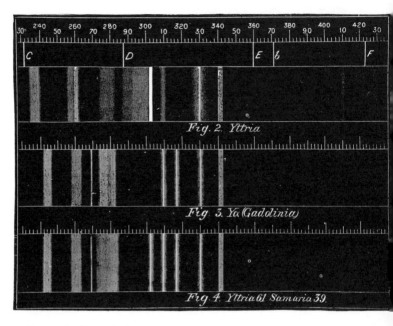

Fig. 2. Yttria

Fig. 3. Ya (Gadolinia)

Fig. 4. Yttria 61 Samaria 39.

I examined; yet the general character of the spectrum remained unchanged, and I habitually looked upon this spectrum as characteristic of yttrium; all the bands being visible when the earth was present in quantity, whilst only the strongest of all—the citron band—was visible when traces, such as millionths, were present. But that the whole system of bands spelled yttrium, and nothing but yttrium, I was firmly convinced.

The differences in the spectra of yttrium prepared from different sources are most distinctly seen on comparing the spectrum of yttrium from samarskite with that from gadolinite, hielmite, monazite, xenotime, fluocerite, euxenite, cerite, arrhenite, &c. Still, in spite of these slight differences, the several yttriums are practically all the same thing, and, as I have said, every living chemist a year ago would have regarded them as identical. But they have since yielded to persistent chemical fractionation, and I now call them old yttrium.

One property, above all others, is relied on by chemists as an indisputable proof of the identity of any particular chemical element.

When the vapour of an element is rendered incandescent by the electric spark, the characteristic system of lines in its spectrum is regarded as unalterable, and is looked upon as a certain proof that this special element is under examination. However much chemical or other tests fail to show the presence of a given element, the indications of its lines in the spectroscope are regarded as infallible. Spectrum analysis is the court of final appeal, whose decision no chemist has yet had the hardihood to dispute.

By way of illustration I will project on the screen the very characteristic system of lines given by yttrium when ignited by the electric spark,—a system, be it remembered, having no connection whatever with the peculiar phosphorescent spectrum yielded by yttrium. The coloured diagram gives as accurate a representation of the spark spectrum of yttrium as can be drawn by hand. Omitting minor lines you will notice two very strong groups of lines in the red and orange. These lines have been always regarded as the characteristic test for yttrium; the presence of these groups proves the presence of yttrium, and their absence proves its absence.

I now project the electric spark spectrum of $G\delta$ as pure as I have been able to prepare it. $G\delta$ is one of the bodies which by long and tedious fractionation I have separated from yttrium; it occurs at one extreme end of the fractioning, and differs not only from the parent yttrium in its phosphorescent spectrum, but by virtue of the process adopted for its isolation, it must likewise differ in chemical properties. But what tale does the spectrum tell? It tells us there is absolutely no difference between this spectrum and that given by old yttrium.

I now pass to the other end of the fractionation of yttrium, where a body, $G\eta$, concentrates giving a totally different phosphorescent spectrum to that given by $G\delta$. And it also differs chemically from old yttrium, and in a more marked manner from its brother, $G\delta$, at the other extremity of the fractionation. Look at its spark spectrum! It is perfectly identical both with old yttrium and with $G\delta$, and when I examine these three spectra in my laboratory with all the appliances for exact measurement, the whole system of lines is still identical.

What inference can be drawn from these results? Is discredit to be thrown on spectrum analysis? Is the superstructure which has been so laboriously raised upon its indications to fall to the ground? By no means. Spectrum analysis and its grand generalisations are on as firm a foundation as ever. I see two possible explanations of the facts I have brought before you. According to one hypothesis research has somewhat enlarged the field lying between the indications given by ordinary coarse chemistry and the searching scrutiny of the prism. Our notions of a chemical element have expanded. Hitherto the molecule has been regarded as an aggregate of two or more atoms, and no account has been taken of the architectural design on which these atoms have been joined. We may consider that the structure of a chemical element is more complicated than has hitherto been supposed. Between the molecules we are accustomed to deal with in chemical reactions and ultimate atoms as first created, come smaller molecules or aggregates of physical atoms; these sub-molecules differ one from the other, according to the position they occupied in the yttrium edifice.

Perhaps this hypothesis can be simplified if we imagine yttrium to be represented by a five-shilling piece. By chemical fractionation I have divided it into five separate shillings, and find that these shillings are not counterparts, but like the carbon atoms in the benzol ring, have the impress of their position, 1, 2, 3, 4, 5, stamped on them. These are the analogues of my $G\alpha$, $G\beta$, &c. If I now bring in a much more powerful and searching agent—if I throw my shillings into the melting-pot or dissolve them chemically—the mint stamp disappears and they all turn out to be silver. I submit my yttrium, or my $G\alpha$, $G\beta$, &c., to the intense heat of the electric spark, the little differences of molecular arrangement vanish, and the atoms of which the molecules of yttrium, $G\alpha$ and $G\beta$, are alike composed, reveal their presence in identical spectra.

An alternative theory commends itself to chemists, to the effect that the nine bodies shown in the above table (Table I.), are new chemical elements differing from yttrium and samarium in basic powers and several other chemical and physical properties, but not sufficiently to enable us to effect any but a slight separation. One of these bodies, $G\delta$, gives the phosphorescent citron line, and also the brilliant electric spectrum I have just exhibited. The other eight do not give electric spectra which can be recognised in the presence of a small quantity of $G\delta$, whilst the electric spectrum of $G\delta$ is so sensitive that it shines out in undiminished brilliancy even when the quantity present is extremely minute. In the process of fractionation, $G\alpha$, $G\beta$, $G\delta$, &c., are spread out and more or less separated from one another, yet the separation is imperfect at the best, and at any part there is enough $G\delta$ to reveal its presence by the sensitive electric spark test. The arguments in favour of each theory are strong and pretty evenly balanced. The compound molecule explanation is a good working hypothesis, which I think may account for the facts, while it does not postulate the rather heroic alternative of calling into existence eight or nine new elements to explain the phenomena. However, I submit it only as an hypothesis. If further research shows the new element theory is more reasonable, I shall be the first person to accept it.*

---

* Neither of these theories agrees with that of my distinguished friend M. Lecocq de Boisbaudran, who also has worked on these earths for some time. He considers that what I have called old yttrium is a true element, characterised by the spark spectrum already exhibited, but not giving a phosphorescent spectrum *in vacuo*. The bodies giving the phosphorescent spectra he considers to be impurities in yttrium. These he says are two in number, and he has provisionally named them $Z\alpha$ and $Z\beta$. By a method of his own, differing from mine, M. de Boisbaudran obtains fluorescent spectra of these bodies; but their

I now will introduce to you a substance which has been to me what the celebrated Rosetta stone was to the interpreters of Egyptian inscriptions. I received it from M. de Marignac, and it was nothing more than a small specimen of a new earth which he had obtained and had named provisionally $Y\alpha$.

In the radiant-matter tube this earth gives a bright spectrum as in the diagram before you (Fig. 3).

If we compare this spectrum with that ascribed to "old yttrium" (Fig. 2) we see that, omitting minor details, $Y\alpha$ is yttrium with the characteristic citron band left out and the green and orange bands of samarium added. Now look at the following diagram (Fig. 4), which represents the spectrum of a mixture of 61 parts of yttrium and 39 parts of samarium. It is almost to its minutest details identical with the spectrum of $Y\alpha$, but the citron band is as prominent as any other band. Hence $Y\alpha$ is shown to consist of samarium, with the greenish blue of yttrium and some of the other yttrium bands added to it. It proves, further, that the citron band which I had hitherto regarded as one of the essential bands of the yttrium spectrum can be entirely removed, whilst another characteristic yttrium group, the double green band, can remain with heightened brilliancy.

If now it were possible to remove the citron band-forming body from this mixture, I should leave $Y\alpha$ behind; I should, in fact, have recomposed $Y\alpha$ from its elements. I have no doubt whatever that this will ultimately be accomplished, but the preliminary work of fractionation is tedious to the last degree, and for its completion would occupy a space of time in comparison with which the life of man is all too brief.

Whilst I have not yet chemically removed the citron-forming constituent, I can physically suppress the citron band and show an artificial spectrum, imitating in the closest degree the natural spectrum of $Y\alpha$.

By means of the electrical phosphoroscope I am enabled to catch the spectrum of an earth immediately after it has suffered molecular bombardment in the vacuum. In this way I get the spectrum of the residual phosphorescence, and I have found that not all the constituents of these earths emit residual phosphorescence for the same duration of time.

When a little strontium is added to the yttrium-samarium mixture, the effect in the phosphoroscope is to suppress the residual phosphor-

---

fluorescent bands are extremely hazy and faint, rendering identification difficult. Some of them fall near lines in the spectra of my $G\beta$ and $G\delta$. At first sight it might appear that his and my spectra were due to the same bodies, but, according to M. de Boisbaudran, the chemical properties of the earths producing them are widely distinct. Those giving phosphorescent lines by my method occur at the yttrium extremity of the fractionation, where his fluorescent bands are scarcely shown at all; whilst his fluorescent phenomena are at their maximum quite at the terbium end of the fractionation, where no yttrium can be detected even by the direct spark, and where my phosphorescent lines are almost absent.

escence of $G\delta$—the citron band—and to enhance the phosphorescence of $G\beta$, the double green band, and the imitation of the $Y\alpha$ spectrum is complete.

I must here call attention to the experiments of Prof. A. E. Nordenskiöld, in the *Comptes Rendus* of the French Academy of Sciences for November 2nd, 1886. This eminent savant is working in the same direction as myself, with results which decidedly corroborate my experiments. He has taken the crude mixture of yttria, erbia, ytterbia, &c., just as it is precipitated from the minerals containing these rare earths. This mixture, for brevity's sake, he calls gadolinia, and he finds that this gadolinia, though palpably a compound body, has always a constant atomic weight, whatever the mineral from which it has been extracted. Or, to use Prof. Nordenskiöld's own words, "*Oxide of gadolinium, though it is not the oxide of a simple body, but a mixture of three isomorphous oxides* (even when it is derived from totally different minerals found in localities far apart from each other) *possesses a constant atomic weight.*" Therefore, as he significantly observes, " *We are in presence of a fact altogether new in chemistry.* For the first time we are confronted with the fact that three isomorphous substances, of a kind that chemists are still compelled to regard as elements, occur in nature not only always together, but in the same proportions. It seems that chemists here find themselves face to face with a problem analogous to that presented to astronomers in the origin of the minor planets."

These facts throw a new light upon certain important chemical questions. For the old yttrium passed muster as an element. It had a definite atomic weight, it entered into combination with other elements, and could be again separated from them as a whole. But now we find that excessive and systematic fractionation has acted the part of a chemical "sorting Demon," distributing the atoms of yttrium into groups, with certainly different phosphorescent spectra, and presumably different atomic weights, though, from the usual chemical point of view, all these groups behave alike. Here, then, is a so-called element whose spectrum does not emanate equally from all its atoms; but some atoms furnish some, other atoms others, of the lines and bands of the compound spectrum of the element. Hence the atoms of this element differ probably in weight, and certainly in the internal motions they undergo.

This is unlikely to be an isolated case. We may assume that the principle is of general application to all the elements. In some, possibly in all elements, the whole spectrum does not emanate from all their atoms, but different spectral rays may come from different atoms, and in the spectrum as we see it all these partial spectra are present together. This may be interpreted to mean that there are definite differences in the internal motions of the several groups of which the atoms of a chemical element consist. For example, we must now be prepared for some such events as that the seven series of bands in the absorption-spectrum of iodine may prove not all to emanate from every molecule, but that some of these molecules emit some of these series, others others, and in the jumble of all these molecules, to which is given the name "iodine vapour," the whole seven series are contributors.

Another important inference to be drawn from the facts is that yttrium atoms, though differing, do not differ continuously, but *per saltum*. We have evidence of this in the fact that the spectroscopic bands characteristic of each group are distinct from those of other groups, and do not pass gradually into them. We must accordingly expect, in the present state of science, that this is probably the case with the other elements. And the atoms of a chemical element being known to differ in one respect may differ in other respects, and presumably do somewhat differ in mass.

Returning, after this digression, to the idea of heavy and light atoms, we see how well this hypothesis accords with the new facts here brought to light. From every chemical point of view the stable molecular group, yttrium, behaves as an element. To split up yttrium requires not only enormous time and material, but the existence of a test by means of which the constituents of yttrium are capable of recognition. Had we tests as delicate for the constituent molecular groups of calcium, this element also might be resolved into simpler groupings. It is one thing, however, to find out means of separating bodies which we know to be distinct and to have colour or spectrum reactions to guide us at every step ; it is quite another thing to separate colourless bodies which are almost identical both in chemical reaction and atomic weight, especially if we have no suspicion that the body we examine is a mixture.

Again, it seems as if bodies we have been accustomed to regard as absolutely simple and elementary may be split up in different directions according to the means we bring to bear upon them. Until very lately our text-books made mention of an element under the name of didymium. With some trouble it had been separated from its accompanying bodies lanthanum and cerium. Its properties had been examined, and no one doubted its distinct and elementary character. It was viewed according to one of the common definitions of an element, as " a something to which we can add, but from which we can take nothing." When, behold ! Dr. Auer von Welsbach, examining this supposed simple body in a novel manner, succeeded in decomposing it into two simpler bodies, which he called neodymium and praseodymium ; and later researches, in which I have had a share, show that even neodymium and praseodymium are not the simplest bodies into which didymium can be dissected.

But it may be asked, What is the bearing of all this upon the great question of the genesis of the elements ? Have we chemists merely discovered some new "elements," or found out that a body hitherto held to be simple is in reality complex ? We have, I submit, done something decidedly different. If a metal which is found to have a fixed atomic weight is discovered to be a compound or a mixture, our best test for recognising an element, so-called, has melted away! Hitherto it has been considered that if the atomic weight of a metal, determined by different observers, setting out from different compounds, was always found to be constant (within, of course, the limits of experimental error), then such metal must rightly take rank among the simple or elementary bodies. We learn from Nordenskiöld's gadolinium that this is no longer the case. Again, we have here wheels within wheels. Gadolinium is not an element, but a compound, or rather, perhaps, a mixture of yttrium, erbium, and ytterbium. We have shown that yttrium is a complex of five or more new constituents. And who shall venture to gainsay that each of these constituents, if attacked in some different manner, and if the results were submitted to a test more delicate and searching than the radiant-matter test, might not be still further divisible? Where, then, is the actual ultimate element? As we advance it recedes like the tantalising mirage lakes and groves seen by the tired and thirsty traveller in the desert. Are we in our quest for truth to be thus deluded and baulked? The very idea of an element, as something absolutely primary and ultimate, seems to be growing less and less distinct.

But we have by no means done with the rare earths and their lessons. How is it that these bodies are found, as we actually find them, associated in certain rare minerals such as samarskite and gadolinite, but occurring only in a few localities? This fact is hard to account for on the ordinary theories of the origination of the elements.

I venture provisionally to conclude that our so-called elements or simple bodies are, in reality, compound molecules. To form a conception of their genesis I must beg you to carry your thoughts back to the time when the visible universe was "without form and void," and to watch the development of matter in the states known to us from an antecedent something. What existed anterior to our elements, before matter as we now have it, I propose to name *protyle*.*

---

* We require a word, analogous to protoplasm, to express the idea of the original primal matter existing before the evolution of the chemical elements. The word I have ventured to use for this purpose is compounded of πρό (*earlier than*) and ὕλη (*the stuff of which things are made*). The word is scarcely a new coinage, for in the 'Wisdom of Solomon' (xi., v. 17) we read:—" Thy almighty hand, that created the world—ἐξ ἀμόρφου ὕλης—out of formless stuff," the word here rendered "stuff" being in the original ὕλης, from which I have ventured to coin the word "protyle." Six hundred years ago Roger Bacon wrote in his *De Arte Chymiæ*, "The elements are made out of ὕλη, and every element is converted into the nature of another element." Professor Huxley reminds me that ὕλη, in the general sense of material substance, was first used by Aristotle, in whose works it is of very frequent occurrence. In fact the fundamental distinction in his Physical Philosophy is between ὕλη, or *matter*, and εἶδος, or *form*, which last pretty nearly answers to what we should call the sum total of the qualities, powers, and tendencies of a thing—or of forces as the cause of these. In the metaphysics and elsewhere Aristotle distinguishes (1) Πρώτη ὕλη, " Materia

But how can we suppose the protyle, or fire-mist, converted into the atomic condition? In amorphous matter we recognise a tendency to aggregation not to be identified with gravitation, since it is manifested among finely-divided matter, whether suspended in a medium of a specific gravity superior, equal, or inferior to its own. This agglutinative action is familiar to observers of natural phenomena. Clouds contracting to that appearance known as a mackerel sky; particles of carbon floating in the air, collecting, and ultimately falling as "blacks"; chemical precipitates, at first finely amorphous, but gradually becoming flocky, granular, and crystalline; vortex rings, suddenly quickening out of amorphous smoke;—all these, and many more, exemplify that universal formative principle in nature which I suggest first made itself manifest in the condensation of protyle into atomic matter.

A few weeks ago, in this theatre, Sir William Thomson asked you to travel back with him an imaginary excursion of about twenty million years. He pictured to you the moment immediately before the birth of our sun, when the Lucretian atoms rushed from all parts of space with velocities due to mutual gravitation, and, clashing together, formed in a few hours an incandescent fluid mass, the nucleus of a solar system with thirty million years of life in it. I will ask you to accompany me to a period even more remote,—to the very beginnings of time, before even the chemical atoms had consolidated from the original *protyle*. Let us imagine that at this primal stage all was in an ultra-gaseous state—a state differing from anything we can now conceive in the visible universe.

Now unless the expression "fire-mist" and the supposition that pristine matter was once in an intensely heated condition * are quite misleading and baseless, we have to deal with a process analogous to cooling. This operation, probably internal, reduces the temperature of the cosmic *protyle* to a point at which the first step in granulation takes place; matter as we know it comes into existence, and atoms are formed. As soon as an atom is formed out of *protyle* it is a store of energy, kinetic (from its internal motions), and potential (from its tendency to coalesce with other atoms by gravitation or chemically). To obtain this energy the neighbouring *protyle* must be laid under contribution, i. e. must be refrigerated by it, thereby accelerating the subsequent formation of other atoms. With the birth of gravitating

---

Prima," or matter undifferentiated into elements, without form, in fact, and consequently $αγνωστος$, unknown and unknowable, and (2) $ἐσχατη$ $ὕλη$. secondary or formed matter, such as earth, or metal, or water, or any other raw material with which we are familiar.

* I am constrained to use words expressive of high temperature; but I confess I am unable clearly to associate with *protyle* the idea of hot or cold. *Temperature, radiation*, and *free cooling* seem to require the periodic motions that take place in the chemical atoms; and the introduction of centres of periodic motion into protyle would involve its being so far changed into chemical atoms. Probably the first operation was more analogous to the formation of vortex rings than to a reduction of temperature.

matter, rushing suddenly together from every point of space, we thus get Sir. William Thomson's incandescent mass which is presently to cool down into a solar system. We cannot tell if electricity existed prior to the origin of the atomic condition of matter, but with the formation of atomic matter the other forms of energy which require matter in order to manifest themselves, begin to act, amongst others that form of energy which has for one of its factors that which we now speak of as atomic weight.

We have now to seek how protyle was converted not into one only kind of matter but into many. If we recognise that it contained within itself the potentiality of all atomic weights, how did these potentialities become actual? We may here call to mind the suggestion of Dr. E. J. Mills, that our elements are the result of successive polymerisations during the cooling process. We shall also derive much assistance from a method of illustrating the periodic law proposed by my friend Professor Emerson Reynolds, of the University of Dublin.

I must call your attention to a diagram (Fig. 5) in which I have slightly modified the original design of Professor Reynolds. I have represented the pendulum swing as gradually declining in amplitude according to a mathematical law. I have further interposed between cerium and lead another half-swing of the pendulum. This renders the oscillations more symmetrical and brings gold, mercury, thallium, lead, and bismuth to the side where they are fully in harmony with members of previous groups.

The chemical elements are arranged in order, according to their atomic weights, on the centre vertical line which is divided into equal parts.

Following the curve from hydrogen downwards, we see that the elements forming the eighth group of Mendeleeff's arrangement are situate near three of the ten nodal points. This eighth group is divided into the three triplets—iron, nickel, and cobalt; rhodium, ruthenium, and palladium; iridium, osmium, and platinum.

These bodies are interperiodic because their atomic weights exclude them from the small periods into which the other elements fall, and because their chemical relations with some members of the neighbouring groups show that they are probably interperiodic in the sense of being in transition stages.

Notice how accurately the series of like bodies fits into this scheme. Beginning at the top, run the eye down analogous positions in each oscillation, taking either the electro-positive or electro-negative swings. (See Table, p. 54.)

Notice, also, how orderly the metals discovered by spectrum analysis fit in their places—gallium, indium, and thallium; rubidium and cæsium.

The symmetry of nearly all this series proclaims at once that we are working in the right direction. Much also may be learned from the anomalies here visible. A few bodies, such as didymium, erbium,

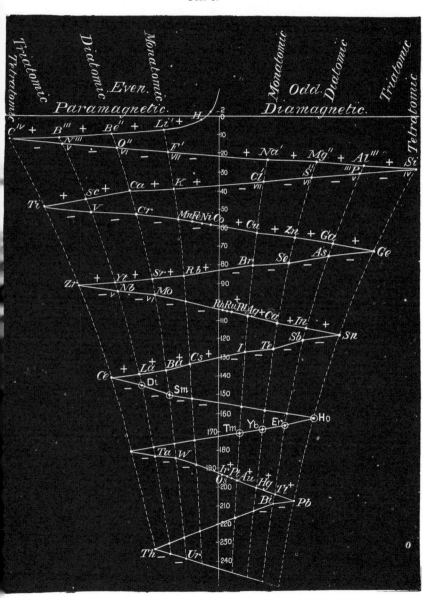

Fig. 5.

thulium, and ytterbium, are out of place, and require to have their atomic weights redetermined.

TABLE II.

VERTICAL SERIES OF THE ELEMENTS.

EVEN.							ODD.						
iv.	iii.	iii. v.	ii.	ii. vi.	i.	i. vii.	i. vii.	i.	ii. vi.	ii.	iii. v.	iii.	iv.
C	B	N	Be	O	Li	F	Cl	Na	S	Mg	P	Al	Si
Ti	Sc	V	Ca	Cr	K	Mn	Br	Cu	Se	Zn	As	Ga	Ge
Zr	Yt	Nb	Sr	Mo	Rb	—	I	Ag	Te	Cd	Sb	In	Sn
Ce	La	—	Ba	Sm	Cs	—	—	—	—	—	—	—	—
—	—	—	—	—	—	—	—	Au	—	Hg	Bi	Tl	Pb

The more I ponder over the arrangement of this zigzag curve, the more I become convinced that he who fully grasps its meaning holds the key to unlock some of the deepest mysteries of creation. As Mr. Browning puts it in his 'Parleyings'—"'Tis Man's to explore Up and down, inch by inch, with the taper his reason." Let us batter at the door of the Unknown and do our utmost to get a glimpse of some few of the secrets so darkly hidden.

Let us return in imagination to pre-geological ages, before the sun himself had been aggregated from the original protyle. We require two very reasonable postulates; let there be granted an antecedent form of energy having periodic cycles of ebb and swell, rest and activity. Let there also be granted an internal action akin to cooling operating slowly in the protyle. The first-born element would, in its simplicity, be most nearly allied to protyle. This is hydrogen, of all known bodies the simplest in structure and of the lowest atomic weight. For some time hydrogen would be the only existing form of matter (in our sense of the term). Between hydrogen and the next formed element there would be a gap in time, and in the interval the element standing next in order of simplicity would gradually be approaching its birth-point. In this interval we may suppose that the evolutionary process soon to determine the birth of a new element would fix likewise its atomic weight, its affinities, and its chemical position.

In this genesis of the elements the longer the time taken up in the cooling-down process, during which the hardening of protyle into atoms takes place, the more sharply defined would be the resulting elements; whilst the more rapid and the more irregular the cooling, the more closely the resulting bodies would fade into each other by almost imperceptible degrees. Thus we may conceive that the succession of events which gave rise to such groups as platinum, osmium, and iridium,—palladium, ruthenium, and rhodium,—iron, nickel, and cobalt,—might have produced only one element in each of these three groups if the process had been greatly prolonged. And conversely, had the rate of cooling been much more rapid, elements might have originated still more nearly identical than are nickel and cobalt. Thus may have arisen the closely allied elements of the cerium, yttrium, and similar groups. In fact, we may regard the collocation of the minerals of the class of samarskite and gadolinite as a kind of cosmical lumber-room where elements in a state of arrested development—the unconnected missing links of inorganic Darwinism—are gathered together.

Any well-defined element may be likened to a platform of stability, connected by ladders of unstable bodies. In the first coalescence of the primitive stuff there would be formed the smallest atoms; these would then unite, forming larger groups; the gaps between the several stages would gradually be bridged over and the stable element appropriate to that stage would absorb, so to speak, the unstable rungs of the ladder which led up to it. It may be questioned whether there is an absolute uniformity in the mass of every ultimate atom even of one and the same chemical element. Probably our atomic weights merely represent a mean value around which the actual atomic weights of the atoms vary within certain narrow limits. When, therefore, we say that, e.g. the atomic weight of calcium is 40, the actual fact may well be, that whilst the majority of the calcium atoms really have the atomic weight of 40, some are represented by $39 \cdot 9$ or $40 \cdot 1$, a smaller number by $39 \cdot 8$ or $40 \cdot 2$, and so on. The properties which we perceive in any element are thus the mean of a number of atoms differing among themselves very slightly, but still not identical.* Is this the true meaning of Newton's "old worn particles?"

That this speculation, hazardous as it may seem, is in some respects supported by the experimental results above described will, I think, be admitted. It seems to me that the hypothesis I have just suggested, if taken in conjunction with the diagram, Fig. 5, enables us to proceed a step or two further along the track of the evolution of the elements. We may trace in the undulating curve the action of two forms of energy, the one acting vertically and the other vibrating to and fro like a pendulum. Let the vertical line represent temperature gradually sinking through an unknown number of degrees from the dissociation-point of the first-formed element downwards to the dissociation-point of the last member of the scale.

But what form of energy is figured by the oscillating line? We see it swinging to and fro to points equidistant from a neutral centre. We see this divergence from neutrality confer atomicity of one, two, three, or four degrees as the distance from the centre increases to one, two, three, or four divisions. We see the approach to or the retrocession from this same neutral line deciding the electro-negative

---

* I venture to suggest that the heavier and lighter atoms formed from the protyle may have been partially sorted out by a process in nature somewhat analogous to the fractionation which has been already described. Such a sorting out would be effected chiefly whilst atomic matter was condensing from the primal state; but it may also have been carried on during geological ages in the wet way by successive solutions and re-precipitations of the various earths.

or electro-positive character of each element; those on the retreating half of the swing being positive, and those on the approaching half negative. In short, we are led to suspect that this oscillating power must be closely connected with the imponderable matter, essence, or source of energy we call electricity.

Let us now return to the period just preceding the birth of the first element. Before that time matter as it now is manifested did not exist. We can no more conceive of matter without energy than of energy without matter; indeed from one point of view the two are convertible terms. Let us assume that simultaneously with the creation of atoms all those attributes which enable us to discriminate one form of matter from another, start into being endowed with energy.

Our pendulum begins its swing from the electro-positive side; lithium, next to hydrogen in the simplicity of its atomic weight, is now formed, followed by glucinum, boron, and carbon. Each element, at the moment of birth, takes up definite quantities of electricity, and on these quantities its atomicity depends.* Thus are fixed the types of the monatomic, diatomic, triatomic, and tetratomic elements.

It has been pointed out by Dr. Carnelley that "those elements belonging to the even series of the periodic classification are always paramagnetic, whereas the elements belonging to the odd series are always diamagnetic." Now in our curve the even series to the left, so far as has been ascertained, are paramagnetic, whilst, with a few exceptions, all to the right are diamagnetic. The strongly magnetic group, iron, manganese, nickel, and cobalt, lie close together on the proper side. But the interperiodic groups, of which palladium and platinum are respectively examples, are supposed to be feebly magnetic. If this can be verified they form exceptions which have yet to be explained. Oxygen, which weight for weight is even more strongly magnetic than iron, lies near the beginning of the curve, whilst at the opposite end come the powerfully diamagnetic metals, bismuth and thallium.

We come now to the return or negative part of the swing; nitrogen appears and shows instructively how position governs the mean dominant atomicity. Nitrogen occupies a position immediately

---

* "Nature presents us with a single definite quantity of electricity. . . . For each chemical bond which is ruptured within an electrolyte a certain quantity of electricity traverses the electrolyte, which is the same in all cases."—G. JOHNSTONE STONEY, "On the Physical Units of Nature."—*British Association Meeting*, 1874, Section A. *Phil. Mag.*, May, 1881.

"The same definite quantity of either positive or negative electricity moves always with each univalent ion, or with every unit of affinity of a multivalent ion."—HELMHOLTZ, Faraday Lecture, 1881.

"Every monad atom has associated with it a certain definite quantity of electricity; every dyad has twice this quantity associated with it; every triad three times as much, and so on."—O. LODGE, "On Electrolysis," *British Association Report*, 1885.

below boron, a triatomic element, and, therefore, nitrogen is likewise triatomic. But nitrogen also follows upon carbon, a tetratomic body, and occupies the fifth position if we count from the place of origin. Now these seemingly opposing tendencies are beautifully harmonised by the endowment of nitrogen with a double atomicity, its atom being capable of acting either as a tri- or as a pentatomic element. With oxygen (di- and hexatomic) and fluorine (mon- and heptatomic) the same law holds good, and one half-oscillation of the pendulum is completed. Passing the neutral line again, we find successively formed the electro-positive bodies sodium (monatomic), magnesium (diatomic), aluminium (triatomic), and silicon (tetratomic).

Here we may notice a curious coincidence; at the beginning of this part of the curve stands carbon, the most ubiquitous element in the organic world. At the end, in opposition, stands silicon, the most commonly occurring element in the inorganic sphere. Further, as we move towards the median line, carbon is successively followed by nitrogen, oxygen, and fluorine, all entering into organic compounds and all gaseous in the free state. If we work back from silicon we find aluminium, magnesium, and sodium, all much less disposed to volatility, and all very prominent members of the mineral kingdom.

The first complete swing of the pendulum is accomplished by the birth of the three electro-negative elements, phosphorus, sulphur, and chlorine; all three, like the corresponding elements on the opposite homeward swing, having at least a double atomicity, depending on position.

Let us pause and examine the results. We have now formed the elements of water, of air, of ammonia, of carbonic acid, of plant and animal life; we have phosphorus for the brain, salt for the sea, clay and sand for the solid earth; two alkalies, an alkaline earth, an earth, along with their carbonates, borates, nitrates, fluorides, chlorides, sulphates, phosphates, and silicates, sufficient, it may be said, for animal and vegetable life, and for a world not so very different from that in which we live and move.

Again let us follow our pendulum. After the formation of chlorine this pendulum touches the neutral line, and is in the same position as in the beginning. Had everything remained as at first the next element to appear would again have been lithium, and the original cycle would have been eternally repeated, producing again and again the same fifteen elements. The conditions, however, are no longer the same: time has elapsed and the form of energy represented by the vertical line has declined; in other words, the temperature has sunk, and the first element to come into existence when the pendulum starts for its second oscillation is not lithium, but the metal next allied to it in the series, i.e. potassium, which may be regarded as the lineal descendant of lithium, with the same hereditary tendencies, but with less molecular mobility and a higher atomic weight.

Pass along the curve and in nearly every case the same law holds good. Thus the last element of the first complete vibration is chlorine. In the corresponding place in the second vibration we have not an exact repetition of chlorine but the very similar body bromine, and when the same position recurs for the third time we see iodine. I need not multiply examples. I may, however, point out that we have here a phenomenon which reminds us of alternating or cyclical generation in the organic world, or we may perhaps say of atavism, a recurrence to ancestral types, somewhat modified.

In this evolutionary scheme it cannot be expected that the potential elements should all be equal to each other. On the contrary, many degrees of stability will be represented, and if we look with a scrutinising eye we shall see our old friend the "missing link," coarse enough to be detected in the groups comprising such bodies as iron, nickel, and cobalt; palladium, ruthenium, and rhodium; iridium, osmium, and platinum : whilst in a more subtile form these missing links present themselves as representatives of the differences which I have suggested between the atoms of the same chemical element.

On the even or paramagnetic half of the swing the energy appears to have acted in a very irregular manner, whilst on the odd, or diamagnetic half, there is considerable regularity. Thus, between the extreme odd elements, silicon (28), germanium (73), tin (118), a missing element (163), and lead (208) there is a difference of exactly 45 units, rendering this half of the curve remarkably symmetrical. On the even side the differences are 36, 42, 51, 39 and 53 (assuming an atomic weight of 180 for a missing element between cerium and thorium). At first sight these differences appear to follow no law, but they gain interest when we see that the mean differences of these figures is $44 \cdot 2$—almost exactly the same as that on the odd side of the curve.

From this uniformity of difference—actual on the one side and average on the other—we may fairly infer that whilst on the odd side there has been little or no variation in the force symbolised by the vertical line, minor irregularities have been the rule on the even side. Or, in other words, the fall of temperature has been very uniform on the odd side—where, accordingly, we see that every original element represents a well marked group, sodium, magnesium, aluminium, silicon, phosphorus, and chlorine ; whilst on the even side the temperature has fallen with considerable fluctuations, thus preventing the formation here of any well-marked groups of elements, excepting those of which lithium and glucinum are the leading types.

Having thus detected irregularities in the fall of temperature in the protyle, we may next ask is there any fluctuation in the force represented by the pendulum-movement? This movement I have assumed to be connected with electrical energy. The earliest-formed elements are those in which chemical energy is at a maximum; as we descend the scale the affinities become feebler and the chemism grows more and more sluggish. In part this change may be due to the circumstance that the elements generated at a reduced temperature no longer possess great molecular mobility. But it is also extremely probable that the chemism-forming energy is itself dying out like the fires of the cosmic furnace. I have attempted to symbolise this gradual fading by a decrease in the amplitude of vibration.

The figures representing the scale of atomic weights may be supposed to represent, inversely, the scale of a gigantic pyrometer plunged into a cauldron where the elements of suns and worlds are undergoing formation. As the heat sinks, the elements generated increase in density and atomic weight. Below the formation-point of uranium the temperature will probably permit of the earlier-born elements forming combinations among themselves, and we shall witness, e. g. the birth of water, and the formation of those known compounds the dissociation of which is not beyond the powers of our terrestrial sources of heat.

Turning to the upper portion of the diagram we see that there is little room for elements of a lower atomic weight than hydrogen. But let us pass "through the looking-glass" and cross the zero line. What shall we find on the other side? Dr. Carnelley asks for an element of negative atomic weight; and here is ample room and verge enough for a shadow series of such unsubstantialities, leading, perhaps, to that "Unseen Universe" which two eminent physicists have discussed. Helmholtz says that electricity is probably as atomic as matter;* is electricity one of the negative elements? and the luminiferous ether another? Matter, as we now know it, does not here exist; and the forms of energy which are apparent in the motions of matter are as yet only latent possibilities.

A genesis of the elements such as is here sketched would not be confined to our little solar system, but would probably follow the same general sequence of events in every centre of energy now visible as a star.

It may be said that so far I have proved nothing. But I may submit that at least I have shown the improbability of the persistence of the ultimate character, and the eternal self-existence, the fortuitous origin, and the simultaneous creation of the elements. The analogy of these elements with the organic radicles, and still more with living organisms, constrains us to suspect that they are compound bodies, springing from a process of evolution. We have drawn corroborative evidence from the distribution and the association of the rare earths, evidence which seems to be converging to the

---

* "If we accept the hypothesis that the elementary substances are composed of atoms, we cannot avoid concluding that electricity also, positive as well as negative, is divided into definite elementary portions, which behave like atoms of electricity."—HELMHOLTZ, Faraday Lecture, 1881.

point of assuming a direct character. Led by the great law of continuity I have ventured to suggest a process by which our elements may have been originated. I dare not say *must* have been originated, for no one can be better aware than I am how much remains to be done before this great, this fundamental question can be finally solved. I earnestly hope that others will take up the task, and that chemistry, like biology, may find its Darwin.

If we consider the position we occupy with reference to the primary questions of chemistry, we might compare research to a game of chess. Man, the investigator, is playing, not with Satan for his soul, but with Nature for knowledge and power. Each element has its allotted moves on the great board of the universe; some of them dependent solely on themselves, and others on the interaction of the adjacent elements. Some of our elements may be compared to pawns, others to knights, bishops, or castles. The game is fearfully unequal. Our antagonist knows the power and the limitations of every piece, all the laws of the game, all possible moves, and is merciless in exacting penalty for errors. We experimentalists know nothing but what we have learned in countless losing games. But our knowledge is increasing. Nature no longer gives us fool's mate. The struggle becomes more obstinate, more exciting, we come upon new gambits, new combinations, and though still checkmated at the last, we take a few pawns, perhaps even a piece or two. Such partial successes were achieved when Lavoisier introduced the use of the balance and developed the theory of combustion; when Dalton put forward the atomic theory; when Davy decomposed the alkalies; when Wöhler effected the synthesis of urea; and when Faraday first liquefied a gas. On such and many similar occasions I can imagine our antagonist becoming thoughtful.

But suppose we one day win the game; that we find out what these obstinate elements really are, that we learn how they came into being, and wherefore their number, their properties, and their mutual relations are such as we find them? We shall then know, *à priori*, what we have now to find out by special experiment; we shall foresee the results of every conceivable reaction, and our theories will legitimate themselves by the power of prediction. To attain such knowledge seems to me the grand task of the chemistry of the coming age.

If you think I have given too free rein to the " scientific imagination " you will, I hope, forgive me as one who at least does not despair of the future of our Science.

[W. C.]

Friday, February 25, 1887.

WILLIAM HUGGINS, Esq. D.C.L. LL.D. F.R.S. Manager and Vice-President, in the Chair.

CAPTAIN W. DE W. ABNEY, R.E. F.R.S. *M.R.I.*

Sunlight Colours.

SUNLIGHT is so intimately woven up with our physical enjoyment of life that it is perhaps not the most uninteresting subject that can be chosen for what is—perhaps somewhat pedantically—termed a Friday evening " discourse." Now, no discourse ought to be possible without a text on which to hang one's words, and I think I found a suitable one when walking with an artist friend from South Kensington Museum the other day. The sun appeared like a red disk through one of those fogs which the east wind had brought, and I happened to point it out to him. He looked, and said, " Why is it that the sun appears so red?" Being near the railway station, whither he was bound, I had no time to enter into the subject, but said if he would come to the Royal Institution this evening I would endeavour to explain the matter. I am going to redeem that promise, and to devote at all events a portion of the time allotted to me in answering the question why the sun appears red in a fog. I must first of all appeal to what every one who frequents this theatre is so accustomed to, viz. the spectrum; I am going not to put it in the large and splendid stripe of the most gorgeous colours before you with which you are so well acquainted, but my spectrum will take a more modest form of purer colours some twelve inches in length.

I would ask you to notice which colour is most luminous. I think that no one will dispute that in the yellow we have the most intense luminosity, and that it fades gradually in the red on the one side and in the violet on the other. This then may be called a qualitative estimate of relative brightnesses; but I wish now to introduce to you what was novel last year, a quantitative method of measuring the brightness of any part.

Before doing this I must show you the diagram of the apparatus which I shall employ in some of my experiments.

R R are rays (Fig. 1) coming from the arc light, or, if we were using sunlight, from a heliostat, and a solar image is formed by a lens, L_1, on the slit s_1 of the collimator c. The parallel rays produced by the lens L_2 are partially refracted and partially reflected. The former pass through the prisms $P_1 P_2$, and are focused to form a spectrum by a lens, L_3, on D, a movable ground-glass screen. The rays are collected by a lens, L_4, tilted at an angle as shown, to form a white image of the near surface of the second prism on F.

Passing a card with a narrow slit S_2, cut in it in front of the spectrum, any colour which I may require can be isolated. The consequence is that, instead of the white patch upon the screen, I

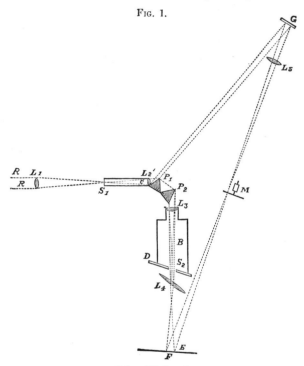

Fig. 1.

Colour Photometer.

have a coloured patch, the colour of which I can alter to any hue lying between the red and the violet. Thus, then, we are able to get a real patch of very appropriately homogeneous light to work with, and it is with these patches of colour that I shall have to deal. Is there any way of measuring the brightness of these patches? was a question asked by General Festing and myself. After trying various plans, we hit upon the method I shall now show you, and if any one works with it he must become fascinated with it on account of its almost childish simplicity—a simplicity, I may remark, which it took us some months to find out. Placing a rod before the screen, it casts a black shadow surrounded with a coloured background. Now I may cast another shadow from a candle or an incandescence lamp, and the two shadows are illuminated, one by the light of the coloured patch and the other by the light from an incandescence lamp which I am using to-night. [Shown.] Now one stripe is evidently too dark. By an arrangement which I have of altering the resistance interposed between the battery and the lamp, I can diminish or increase the light from the lamp, first making the shadow it illuminates too light and then too dark compared with the other shadow which is illuminated by the coloured light. Evidently there is some position in which the shadows are equally luminous. When that point is reached, I can read off the current which is passing through the lamp, and having previously standardised it for each increment of current, I know what amount of light is given out. This value of the incandescence lamp I can use as an ordinate to a curve, the scale number which marks the position of the colour in the spectrum being the abscissa. This can be done for each part of the spectrum, and so a complete curve can be constructed which we call the illumination curve of the spectrum of the light under consideration.

Now, when we are working in the laboratory with a steady light, we may be at ease with this method, but when we come to working with light such as the sun, in which there may be constant variation owing to passing, and maybe usually imperceptible, mist, we are met with a difficulty; and in order to avoid this, General Festing and myself substituted another method, which I will now show you. We made the comparison light part of the light we were measuring. Light which enters the collimating lens partly passes through the prisms and is partly reflected from the first surface of the prism; that we utilise, thus giving a second shadow. The reflected rays from P_1 fall on G, a silver-on-glass mirror. They are collected by L_5, and form a white image of the prism also at F. The method we can adopt of altering the intensity of the comparison light is by means of rotating sectors, which can be opened or closed at will, and the two shadows thus made equally luminous. [Shown.] But although this is an excellent plan for some purposes, we have found it better to adopt a different method. You will recollect that the brightest part of the spectrum is in the yellow, and that it falls off in brightness on each side, so, instead of opening and closing the sectors, they are set at fixed intervals, and the slit is moved in front of the spectrum, just making the shadow cast by the reflected beam too dark or too light, and oscillating between the two till equality is discovered. The scale number is then noted, and the curve constructed as before. It must be remembered that, on each side of the yellow, equality can be established.

This method of securing a comparison light is very much better for sun work than any other, as any variation in the light whose spectrum is to be measured affects the comparison light in the same degree. Thus, suppose I interpose an artificial cloud before the slit of the spectroscope, having adjusted the two shadows, it will be seen that the passage of steam in front of the slit does not alter the relative intensities; but this result must be received with caution. [The lecturer then proceeded to point out the contrast colours that the shadow of the rod illuminated by white light assumed.]

I must now make a digression. It must not be assumed that every one has the same sense of colour, otherwise there would be no colour-blindness. Part of the researches of General Festing and myself have been on the subject of colour-blindness, and these I must briefly refer to. We test all who come by making them match the luminosity of colours with white light, as I have now shown you; and as a colour-blind person has only two fundamental colour perceptions instead of three, his matching of luminosities is even more accurate than is that made by those whose eyes are normal or nearly normal. It is curious to note how many people are more or less deficient in colour-perception. Some have remarked that it is impossible that they were colour-blind, and would not believe it, and sometimes we have been staggered at first with the remarkable manner in which they recognised colour to which they ultimately proved deficient in perception. For instance, one gentleman when I asked him the name of a red colour patch, said it was sunset colour; he then named green and blue correctly, but when I reverted to the red patch he said green. On testing further he proved totally deficient in the colour-perception of red, and with a brilliant red patch he matched almost a black shadow. The diagram shows you the relative perceptions in the spectrum of this gentleman and myself. There are others who only see three-quarters, others half, and others a quarter the amount of red that we see, whilst some see none. Others see less green and others less violet, but I have met with no one that can see more than myself or General Festing, whose colour-perceptions are almost identical. Hence we have called our curve of illumination the "normal curve."

We have tested several eminent artists in this manner, and about one-half of the number have been proved to see only three-quarters of the amount of red which we see. It might be thought that this would vitiate their powers of matching colour, but it is not so. They paint what they see, and although they see less red in a subject, they see the same deficiency in their pigments; hence they are correct. If totally deficient, the case of course would be different.

Let us carry our experiments a step further, and see what effect what is known as a turbid medium has upon the illuminating value of different parts of the spectrum. I have here water which has been rendered turbid in a very simple manner. In it has been very cautiously dropped an alcoholic solution of mastic. Now mastic is practically insoluble in water, and directly the alcoholic solution comes in contact with the water it separates out in very fine particles, which, from their very fineness, remain suspended in the water. I propose now to make an experiment with this turbid water.

I place a glass cell containing water in front of the slit, and on the screen I throw a patch of blue light. I now change it for turbid water in a cell. This thickness much dims the blue; with a still greater thickness the blue has almost gone. If I measure the intensity of the light at each operation, I shall find that it diminishes according to a certain law, which is of the same nature as the law of absorption. For instance, if one inch diminishes the light one-half, the next will diminish it half of that again, the next half of that again, whilst the fourth inch will cause a final diminution of the total light of one-sixteenth. If the first inch allows only one-quarter of the light, the next will only allow one-sixteenth, and the fourth inch will only permit 1/256 part to pass. Let us, however, take a red patch of light and examine it in the same way. We shall find that, when the greater thickness of the turbid medium we used when examining the blue patch of light is placed in front of the slit, much more of this light is allowed to pass than of the blue. If we measure the light we shall find that the same law holds good as before, but that the proportion which passes is invariably greater with the red than the blue. The question then presents itself: Is there any connection between the amounts of the red and the blue which pass? Lord Rayleigh, some years ago, made a theoretical investigation of the subject; but, as far as I am aware no definite experimental proof of the truth of the theory was made till it was tested last year by General Festing and myself. His law was that for any ray, and through the same thickness, the light transmitted varied inversely as the fourth power of the wave-length. The wave-length 6000 lies in the red, and the wave-length 4000 in the violet. Now 6000 is to 4000 as 3 to 2, and the fourth powers of these wave-lengths are as 81 to 16, or as about 5 to 1. If, then, the four inches of our turbid medium allowed three-quarters of this particular red ray to be transmitted, they would only allow $(\frac{3}{4})^5$, or rather less than one-fourth, of the blue ray to pass. Now this law is not like the law of absorption for ordinary absorbing media, such as coloured glass for instance, because here we have an increased loss of light running from the red to the blue, and it matters not how the medium is made turbid, whether by varnish, suspended sulphur, or what not. It holds in every case, so long as the particles which make the medium turbid are small enough; and please to recollect that it matters not in the least whether the medium which is rendered turbid is solid, liquid, or air. Sulphur is yellow in mass, and mastic varnish is nearly white, whilst tobacco-smoke when condensed is black, and very minute particles of water are colourless: it matters not what the colour is, the loss of light is *always* the same. The result is simply due to the scattering of light by fine particles, such particles being small in dimensions compared with a wave of light. Now, in this trough is suspended 1/1000 of a cubic inch of mastic varnish, and the water in it measures about 100 cubic inches, or is 100,000 times more in bulk than the varnish. Under a microscope of ordinary power it is impossible to distinguish any particles of varnish: it looks like a homogeneous fluid, though we know that mastic will not dissolve in water. Now a wave-length in the red is about 1/40,000 of an inch, and a little calculation will show that these particles are well within the necessary limits. Prof. Tyndall has delighted audiences here with an exposition of the effect of the scattering of light by small particles in the formation of artificial skies, and it would be superfluous for me to enter more into that. Suffice it to say that when particles are small enough to form the artificial blue sky they are fully small enough to obey the above law, and that even larger particles will suffice. We may sum up by saying that very fine particles scatter more blue light than red light, and that consequently more red light than blue light passes through a turbid medium, and that the rays obey the law prescribed by theory. I will exemplify this once more by using the whole spectrum and placing this cell, which contains hyposulphite of soda in solution in water, in front of the slit. By dropping in hydrochloric acid, the sulphur separates out in minute particles; and you will see that, as the particles increase in number, the violet, blue, green, and yellow disappear one by one and only red is left, and finally the red disappears itself.

Now let me revert to the question why the sun is red at sunset. Those who are lovers of landscape will have often seen on some bright summer's day that the most beautiful effects are those in which the distance is almost of a match to the sky. Distant hills which when viewed close to are green or brown, when seen some five or ten miles away appear of a delicate and delicious, almost of a cobalt, blue colour. Now, what is the cause of this change in colour? It is simply that we have a sky formed between us and the distant ranges, the mere outline of which looms through it. The shadows are softened so as almost to leave no trace, and we have what artists call an atmospheric effect. If we go into another climate, such as Egypt or amongst the high Alps, we usually lose this effect. Distant mountains stand out crisp with black shadows, and the want of atmosphere is much felt. [Photographs showing these differences were shown.] Let us ask to what this is due. In such climates as England there is always a certain amount of moisture present in the atmosphere, and this moisture may be present as very minute particles of water—so minute indeed that they will not sink down in an atmosphere of normal density—or as vapour. When present as vapour the air is much more transparent, and it is a common expression to use, that when distant hills look "so close" rain may be expected shortly to follow, since the water is present in a state to precipitate in larger particles; but when present as small particles of water the hills look very distant, owing to what we may call the haze between us and them. In recent weeks every one has been able to see very multiplied effects of such haze. The ends of long streets, for instance, have been scarcely visible though the sun may have been shining, and at night the long vistas of gas lamps have shown light having an increasing redness as they became more distant. Every one admits the presence of mist on these occasions, and this mist must be merely a collection of intangible and very minute particles of suspended water. In a distant landscape we have simply the same or a smaller quantity of street-mists occupying, instead of perhaps 1000 yards, ten times that distance. Now I would ask, What effect would such a mist have upon the light of the sun which shone through it?

It is not in the bounds of present possibility to get outside our atmosphere and measure by the plan I have described to you the different illuminating values of the different rays, but this we can do:—First, we can measure these values at different altitudes of the sun, and this means measuring the effect on each ray after passing through different thicknesses of the atmosphere, either at different times of day, or at different times of the year, about the same hour. Second, by taking the instrument up to some such elevation as that to which Langley took his bolometer at Mount Whitney, and so to leave the densest part of the atmosphere below us. Now, I have adopted both these plans. For more than a year I have taken measurements of sunlight in my laboratory at South Kensington, and I have also taken the instrument up to 8000 feet high in the Alps, and made observations *there*, and with a result which is satisfactory in that both sets of observations show that the law which holds with artificially turbid media is under ordinary circumstances obeyed by sunlight in passing through our air: which is, you will remember, that more of the red is transmitted than of the violet, the amount of each depending on the wave-length. The luminosity of the spectrum observed

Fig. 2.

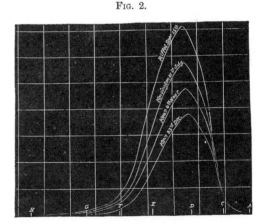

Relative Luminosities.

at the Riffel I have used as my standard luminosity, and compared all others with it. The result for four days you see in the diagram.

I have diagrammatically shown the amount of different colours which penetrated on the same days, taking the Riffel as ten. It will be seen that on December 23 we have really very little violet and less than half the green, although we have four-fifths of the red.

The next diagram before you shows the minimum loss of light which I have observed for different air thicknesses. On the top we

FIG. 3.

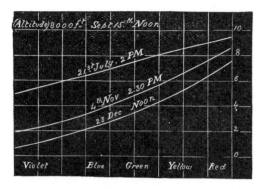

Proportions of Transmitted Colours.

have the calculated intensities of the different rays outside our atmosphere. Thus we have that through one atmosphere, and two, three, and four; and you will see what enormous absorption there is in the blue end at four atmospheres. The areas of these curves, which give the total luminosity of the light, are 761, 662, 577, 503, and 439; and if observed as astronomers observe the absorption of light, by means of stellar observations, they would have had the values, 761, 664, 578, 504, and 439—a very close approximation one to the other.

Next notice in the diagram that the top of the curve gradually inclines to go to the red end of the spectrum as you get the light transmitted through more and more air, and I should like to show you that this is the case in a laboratory experiment. Taking a slide with a wide and long slot in it, a portion is occupied by a right-angled prism, one of the angles of 45° being towards the centre of the slot. By sliding this prism in front of the spectrum I can deflect outwards any portion of the spectrum I like, and by a mirror can reflect it through a second lens, forming a patch of light on the screen overlapping the patch of light formed by the undeflected rays. If the two patches be exactly equal, white light is formed. Now, by placing a rod as before in front of the patch, I have two coloured stripes in a white field, and though the background remains of the same intensity of white, the intensities of the two stripes can be altered by moving the right-angled prism through the spectrum. The two stripes are now apparently equally luminous, and I see the point of equality is where the edge of the right-angled prism is in the green. Placing a narrow cell filled with our turbid medium in front of the slit, I find that the equality is disturbed, and I have to allow more of the yellow to come into the patch formed by the blue end of the spectrum, and consequently less of it in the red end. I again establish equality. Placing a thicker cell in front, equality is again disturbed, and I have to have less yellow still in the red half, and more in the blue half. I now remove the cell, and the inequality of luminosity is still more glaring. This shows, then, that the rays of maximum luminosity must travel towards the red as the thickness of the turbid medium is increased.

The observations at 8000 feet, here recorded, were taken on September 15 at noon, and of course in latitude 46° the sun could not be overhead, but had to traverse what would be almost exactly equivalent to the atmosphere at sea-level. It is much nearer the calculated intensity for no atmosphere intervening, than it is for one atmosphere. The explanation of this is easy. The air is denser at sea-level than at 8000 feet up, and the lower stratum is more likely to hold small water particles or dust in suspension than is the higher.

For, however small the particles may be, they will have a greater tendency to sink in a rare air than in a denser one, and less water vapour can be held per cubic foot. Looking, then, from my laboratory at South Kensington, we have to look through a proportionately larger quantity of suspended particles than we have at a high altitude when the air thicknesses are the same; and consequently the absorption is proportionately greater at sea-level than at 8000 feet high. This leads us to the fact that the real intensity of illumination of the different rays outside the atmosphere is greater than it is calculated from observations near sea-level. Prof. Langley, in this theatre, in a remarkable and interesting lecture, in which he described his journey up Mount Whitney to about 12,000 feet, told us that the sun was really blue outside our atmosphere, and at first blush the amount of extra blue which he deduced to be present in it would, he thought, make it so; but though he surmised the result from experiments made with rotating disks of coloured paper, he did not, I think, try the method of using pure colours, and consequently, I believe, slightly exaggerated the blueness which would result. I have taken Prof. Langley's calculations of the increase of intensity for the different rays, which I may say do not quite agree with mine, and I have prepared a mask which I can place in the spectrum giving the different proportions of each ray as calculated by him, and this when placed in front of the spectrum will show you that the real colour of sunlight outside the atmosphere, as calculated by Langley, can scarcely be called bluish. Alongside I place a patch of light which is very closely the colour of sunlight on a July day at noon in England. This comparison will enable you to gauge the blueness, and you will see that it is not very blue, and, in fact, not bluer perceptibly than that we have at the Riffel, the colour of the sunlight at which place I show in a similar way. I have also prepared some screens to show you the value of sunlight after passing through five and ten atmospheres. On an ordinary clear day you will see what a yellowness there is in the colour. It seems that after a certain amount of blue is present in white light the addition of more makes but little difference in the tint. But these last patches show that the light which passes through the atmosphere when it is feebly charged with particles does not induce the red of the sun as seen through a fog. It only requires more suspended particles in any thickness to induce it.

In observations made at the Riffel, and at 14,000 feet, I have found that it is possible to see far into the ultra-violet, and to distinguish and measure lines in the sun's spectrum which can ordinarily only be seen by the aid of a fluorescent eye-piece or by means of photography. Circumstantial evidence tends to show that the burning of the skin, which always takes place in these high altitudes in sunlight, is due to the great increase in the ultra-violet rays. It may be remarked that the same kind of burning is effected by the electric arc light, which is known to be very rich in these rays.

Again, to use a homely phrase, "You cannot eat your cake and have it." You cannot have a large quantity of blue rays present in your direct sunlight, and have a luminous blue sky. The latter must always be light scattered from the former. Now, in the high Alps you have, on a clear day, a deep blue-black sky, very different indeed from the blue sky of Italy or of England; and as it is the sky which is the chief agent in lighting up the shadows, not only in those regions do we have dark shadows on account of no intervening—what I will call—mist, but because the sky itself is so little luminous. In an artistic point of view this is important. The warmth of an English landscape in sunlight is due to the highest lights being yellowish, and to the shadows being bluish from the sky-light illuminating them. In the high Alps the high lights are colder, being bluer, and the shadows are dark, and chiefly illuminated by reflected direct sunlight. Those who have travelled abroad will know what the effect is. A painting in the Alps, at any high elevation, is rarely pleasing, although it may be true to Nature. It looks cold, and somewhat harsh and blue.

In London we are often favoured with easterly winds, and these, unpleasant in other ways, are also destructive of that portion of the sunlight which is the most chemically active on living organisms. The sunlight composition of a July day may, by the prevalence of an easterly wind, be reduced to that of a November day, as I have proved by actual measurement. In this case it is not the water particles which act as scatterers, but the carbon particles from the smoke.

Knowing, then, the cause of the change in the colour of sunlight, we can make an artificial sunset, in which we have an imitation light passing through increasing thicknesses of air largely charged with water particles. [The image of a circular diaphragm placed in front of the electric light was thrown on the screen in imitation of the sun, and a cell containing hyposulphite of soda placed in the beam. Hydrochloric acid was then added: as the fine particles of sulphur were formed, the disk of light assumed a yellow tint, and as the decomposition of the hyposulphite progressed, it assumed an orange and finally a deep red tint.] With this experiment I terminate my lecture, hoping that in some degree I have answered the question I propounded at the outset: why the sun is red when seen through a fog.

[W. DE W. A.]

Friday, March 25, 1887.

Sir Frederick Abel, C.B. D.C.L. F.R.S. Manager and Vice-President, in the Chair.

The Right Hon. Lord Rayleigh, M.A. D.C.L. LL.D. F.R.S. *M.R.I.*

On the Colours of Thin Plates.

The physical theory, as founded by Young and perfected by his successors, shows how to ascertain the composition of the light reflected from a plate of given material and thickness when the incident light is white; but it does not, and cannot tell us, except very roughly, what the *colour* of the light of such composition will be. For this purpose we must call to our aid the theory of compound colours, and such investigations as were made by Maxwell upon the chromatic relations of the spectrum colours themselves. Maxwell found that on Newton's chromatic diagram the curve representative of the spectrum takes approximately the simple form of two sides of a triangle, of which the angular points represent a definite red, a definite green, and a definite violet. The statement implies that yellow is a compound colour, a mixture of red and green.

In illustration of this fact, an experiment was shown in which a compound yellow was produced by absorbing agents. An infusion of litmus absorbs the yellow and orange rays; a thin layer of bichromate of potash removes the blue. Under the joint operation of these colouring matters the spectrum is reduced to its red and green elements, as may be proved by prismatic analysis; but, if the proportions are suitably chosen, the colour of the mixed light is yellow or orange. When the slit of the usual arrangement is replaced by a moderately large circular aperture, the prism throws upon the screen two circles of red and green light, which partially overlap. Where the lights are separated, the red and green appear; where they are combined, the resultant colour is yellow.

On the basis of Maxwell's data it is possible to calculate the colours of thin plates and to exhibit the results in the form of a curve upon Newton's diagram. The curve starts at a definite point, corresponding to an infinitely small thickness of the plate. This point is somewhat upon the blue side of white. As the thickness increases the curve passes very close to white, a little upon the green side. It then approaches the side of the triangle, indicating a full orange; and so on. In this way the colours of the various orders of Newton's scale are exhibited and explained. The principal discrepancy between the curve and the descriptions of previous observers relates to the precedence of the reds of the first and second orders. The latter has usually been considered to be the superior, while the diagram supports the claim of the former. The explanation is to be found in the inferior brightness (as distinguished from purity) of the red of the first order and its consequent greater liability to suffer by contamination with white light. Such white light, foreign to the true phenemenon, is always present when the thin plate is a plate of air enclosed between glass lenses. To make the comparison fairly, a soap film must be used, or recourse may be had to the almost identical series of colours presented by moderately thin plates of doubly refracting crystals when traversed by polarised light. Under these circumstances the red of the first order is seen to be equal or superior to that of the second order.

[RAYLEIGH.]

Friday, April 1, 1887.

Sir Frederick Abel, C.B. D.C.L. F.R.S. Manager and Vice-President, in the Chair.

Professor Dewar, M.A. F.R.S. *M.R.I.*

Light as an Analytic Agent.

Analytical research by means of the agency of light is usually carried out, by observing the effects produced on light of known quality by transmission through or reflection from matter; or on the other hand, by stimulating matter by the agency of heat or electricity to evolve light, and thereby noting its special qualities.

In previous lectures I have given an account of the results of a series of experiments, undertaken in concert with my colleague Prof. Liveing, with the object of elucidating certain obscure spectroscopic phenomena, and I propose this evening to extend the record of our work.

Not long since Berthelot published the results of some investigations of the rate of propagation of the explosion of mixtures of oxygen with hydrogen and other gases. He found that in a mixture of oxygen and hydrogen in the proportions in which they occur in water, the explosion progressed along a tube at the rate of 2841 metres per second; not far from the velocity of mean square for hydrogen particles, on the dynamic theory of gases, at a temperature of 2000°.

This is a velocity which, though very far short of the velocity of light, bears a ratio to it which cannot be called insensible. It is in fact about $\frac{1}{105000}$ part of it. Hence if the explosion were advancing towards the eye, the waves of light would proceed from a series of particles lit up in succession at this rate. This would be equivalent to a shortening of the wave-length of the light by about $\frac{1}{105000}$ part; and in the case of the yellow sodium lines would produce a shift of the lines towards the more refrangible side of the spectrum by a distance of about $\frac{1}{107}$ of the space between the two lines. It would require an instrument of very high dispersive power and sharply defined lines to make such a displacement appreciable. With lines of longer wave-length than the yellow sodium lines, the displacement would be proportionately greater. Further, if a receding explosion could be observed simultaneously with an advancing explosion, the relative shift of the line would be doubled, one image of the line observed being thrown as much towards the less-refrangible side of its proper position as the other was thrown towards the more-refrangible side. The two images of the red line of lithium would in this way be separated by a distance of about $\frac{1}{8}$ of a unit of Angström's scale; a quantity quite appreciable, though much less than the distance between the components of b_3, and about equal to the distance of the components of the less refrangible of the pair of lines E. We thought therefore that we might test theory by experiment.

A preliminary question had, however, to be answered. What lines could be seen in the flash of the exploding gases? We were pretty certain that the hydrogen lines could not be seen, but that probably we might get sufficient dust of sodium compounds floating in the gas to make the sodium lines visible. A preliminary observation was made on the flash of mixed hydrogen and oxygen in a Cavendish's eudiometer, which showed not only the yellow sodium lines, but the orange and green bands of lime and the indigo line of calcium all very brightly, as well as other lines not identified. The flash is very instantaneous, but nevertheless produces a strong impression on the eye; and by admitting the light of a flame into the spectroscope at the same time as that of the flash, the identity of the lines was established. That sodium should make itself seen was not surprising, but that the spectrum of lime should also be so bright had not been anticipated. At first we thought that some spray of the water over which the gases were confined must have found its way into the eudiometer; but subsequent observations led us rather to suppose that the lime was derived from the glass of the eudiometer. The lime-spectrum made its appearance when the eudiometer was quite clean and dry, and when the gases had been standing over water for a long time.

To obtain the high dispersion requisite, as already explained, we made use of one of Rowland's magnificent gratings, with a ruled surface of $3\frac{1}{8}$ by $2\frac{1}{8}$ inches, and the lines 14,438 to the inch. One telescope fitted with a collimating eye-piece served both as collimator and observing-telescope; and by this means we were able to use the spectra of the third and fourth orders with good effect.

Observations were made with this instrument on explosions in an iron tube shown in section in Fig. 1, half an inch in diameter, fitted at the end with a thick glass plate (a), held on by a screw-cap (c) and made tight with leaden washers. Small lateral tubes (d, d), at right angles to the main tube, were brazed into it near the two ends, for the purpose of connecting it with the air-pump, admitting the gases, and firing them. For this last purpose a platinum wire (b) fused into glass was cemented into the small tube, so that an electric spark could be passed from the wire to the side of the small tube when the gases were to be exploded.

To bring out the lithium lines, a small quantity of lithium carbonate in fine powder was blown into the tube before the cap with glass plate was screwed on. Powder was used because we supposed that it must be loose dust which would be lighted up by the explosion. The lithium lines came out bright enough, and it was unnecessary to put in any more lithium for any number of explosions. The tube was of course quite wet after the first explosion from the water formed, but the lithium lines were none the less strong. Indeed, after the tube had been very thoroughly washed out, the lithium lines continued to be visible at each explosion, though less brightly than at first. A good deal of continuous spectrum accompanies the flash which, from the overlapping of spectra of different orders, makes observations difficult, so a screen of red glass was used to cut this off when the lithium red line was under observation. In any case, however, close observation of the flashes is very trying, from the suddenness with which the illumination appears, and the briefness of its duration. At first we compared the lithium line given by the flash of the exploding gases with that produced by the flame of a small Bunsen burner in which a bead of fused lithium carbonate was held, both being in the field at once. While the flame-line was sharply defined, the flash-line had a different character, and was always diffuse at the edges; so that it was not possible in this way to substantiate the minute difference of wave-length indicated by theory, though the flash-line certainly seemed a little the more refrangible of the two.

We then tried taking the explosion in a tube bent round so as to be returned upon itself, the two parts of the tube being parallel to each other, and the glass ends side by side (Fig. 2). The axis of the

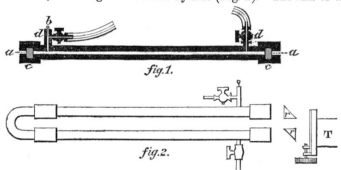

collimator (T) was made to coincide with that of one limb of the tube, so that the flash in that limb was seen directly; and by means of two reflecting prisms (r, r) the light from the other limb was thrown into the slit, and the two images were seen simultaneously one above the other. As the gas was ignited from one end of the tube, the flash was seen receding in one limb, approaching in the other, so that the displacement of the two lines would be doubled. Still we were unable to substantiate any relative displacement of the lines on account of their breadth and diffuse character. By washing out the tube the breadth of the lines was considerably reduced, but they remained diffuse at the edges, and baffled any observation sufficiently accurate to establish a displacement. Certainly there appeared to be a very slight displacement, but it was not so definite that one could be sure of it.

These observations, however, led us to some other interesting results. In the first place, one of the two images of the lithium line almost always was reversed—that is, showed a dark line down the middle. This was the line given by the flash approaching the slit. The receding flash in the other limb of the tube gave as broad a bright line, but it had no dark line in its middle. This observation was made a great many times; and the fact of the reversal established independently in the case of some other metallic lines by means of photographs. These reversals show that in the wave of explosion the temperature of the gases does not reach its maximum all at once; but the front of the wave is cooler than the part which follows and absorbs some of its radiation, while the rear of the wave does not produce the same effect. One would suppose that there must be cooler lithium-vapour in the rear of the wave as well as in its front; but it is possible that the absorption produced by it extends over the whole width of the line, and not only over a narrow strip in the middle. For we observed that when a little lithium carbonate was freshly put into the tube, the red line was so much expanded as to fill the whole field of view—that is to say, it was some ten or twelve times as wide as the distance between the two yellow lines of sodium; but by washing out the tube with water (that is, by reducing the quantity of lithium present in the tube), the line could be reduced in width until it was no wider than one-tenth of the distance between the two sodium lines. This seems to prove that the breadth of the line is directly dependent on the amount of lithium present.

M. Fievez has, in a recent publication ('Bulletins de l'Académie royale de Belgique'), concluded, from observations on sodium, that the widening of the lines is solely due to elevation of temperature. The flash of the exploding gases cannot be raised in temperature by the presence of a minute quantity more of a lithium compound; so that in our case the widening cannot be ascribed to anything but the increase in the quantity of lithium present, or to some consequence of that increase. It is not improbable that the amount of lithium vaporised in the front of the wave of explosion is less than in the following part, and hence the absorption-line is not so wide as the bright line behind it, while in the rear of the wave the absorption extends over the whole width of the bright band, and so is not so easily noticed. Only twice amongst many observations was any reversal of the lithium line seen in the receding wave of explosion.

On observing the flash with a spectroscope of small dispersion instead of that with the grating, the continuous spectrum was very bright, but the metallic lines stood out still brighter; not only the red line of lithium, but the orange, the green, and the blue lines were very bright, and continued so when the pressure of the gases before explosion was reduced from one atmosphere to one-third of an atmosphere. The violet line was not seen, but it may have been so much expanded as to be lost in the continuous spectrum; for it showed in a photograph afterwards taken. Other lines were, however, seen—the sodium yellow lines, the calcium indigo line, a group of other blue lines, and a group of green lines, amongst which one line was conspicuous, and this line, by comparison with the solar spectrum, was identified with E. We had not expected to see any lines of iron, as iron and its compounds give no lines in the flame of a Bunsen burner, and we supposed that it would only be volatilised at a much higher temperature. But the appearance of E suggested that other of the green and blue lines might be due to iron; so we proceeded to compare the positions of these lines with those of the electric spark between iron electrodes. For this purpose one of the spark-lines was first brought carefully on to the pointer, or cross wires, in the eye-piece of the observing telescope, and then, the passage of the spark being stopped, the flash of the exploding gases was observed. It was not difficult to see whether any line was on the pointer; and the observation could be repeated as many times as was desired without any shift of the apparatus. Nine of the most conspicuous green and yellowish-green lines in the flash were thus identified with lines of iron. For the blue and violet we adopted the photographic method as much less trying to the eyesight. Eight to twelve flashes were taken in succession without any shift of the apparatus, so as to accumulate their effects on the photographic plate. Eight flashes were found enough in general to produce a good impression, and more than twelve could not well be taken without turning out the water which accumulated in the tube, and cleaning the glass which closed its end. After the flashes had been taken, but without shifting the photographic plate, the slit of the spectroscope was partly covered, and the electric spark between iron points passed in front of the slit. We had thus on the plate the photograph of the flash as well as of the spark. Fourteen more lines in the indigo and violet were thus identified with iron lines; and on extending the photographs into the ultra-violet, and substituting quartz lenses and prisms for the glass ones hitherto used, a much larger number of lines were identified. There could be no doubt, then, that we had iron vapour in the flash. We supposed that it must be derived from dust of oxide shaken by the explosion off the sides of the tube, and we had the tube bored out clean and bright like a gun-barrel. This made no diminution in the brightness or number of the lines; and we came to the conclusion that the explosion detached particles of iron from the tube, and converted them into vapour. This was confirmed by finding that, however carefully the tube had been cleaned, the glass ends always became clouded with a rusty deposit after ten or twelve flashes. Altogether 68 lines of iron have been identified in the flash, of which about 40 lie in the ultra-violet between H and O. Only one iron line above O has been definitely identified, and that in only a few photographs. It is T.

As iron gave so many lines in the flash it was reasonable to suppose that more volatile metals would give their lines too. Linings of thin sheet copper, lead, cadmium, zinc, aluminium, and tin were successively put into the tube, and their effects on the flash observed. Copper gave one strong line in the green (wave-length $5104\cdot9$), but no other line in the visible part of the spectrum. In the ultra-violet two strong lines between Q and R came out in the photographs, frequently as reversed lines. Some of the photographs showed also something of the shaded bands in the blue which are ascribed to the oxide of copper. The green line of copper had been observed in the flash before the copper lining was put into the tube; and we concluded that the copper was derived from the brass with which the small lateral tubes were fastened into the large tube, or that the iron of the tube contained a little copper. When the leaden lining was used, only one visible line of lead was developed, and that was the strong violet line, but two ultra-violet lines between M and N were strongly depicted on the photographic plates. The violet line of lead had also been observed in many of the photographs taken before the leaden lining was introduced. This we ascribed to the leaden washers used to make the glass or quartz plates air-tight. The line was greatly increased in strength by the leaden lining. The zinc lining gave no visible line at all, notwithstanding the easy volatility of the metal; and in the ultra-violet it gave only a very doubtful impression of one of the lines near P. The cadmium, aluminium, and tin linings gave no lines at all. Zinc dust put into the tube gave no zinc lines, merely increased the continuous spectrum, and speedily rendered the quartz end opaque.

A clean wire of magnesium put into the tube gave the b group of lines, but no other line. No trace of the blue line, so conspicuous in the flame of burning magnesium, nor of the triplets near L and S, nor of the very strong line, the strongest of all in the arc, at wavelength 2852. b_1 and b_2 were well seen; but as b_4 is an iron line, as well as a magnesium line, and the iron line was visible in the flash before the magnesium wire was introduced, we cannot be sure whether the magnesium line, as well as the iron line, was present in the flash. Magnesia did not develop any line at all; merely augmented the continuous spectrum.

Compounds of sodium, such as the carbonate and chloride, introduced in powder gave the ultra-violet line between P and Q strongly, frequently reversed; but no other line except of course D. Potassium compounds developed, often reversed, the pair of violet lines, and also the ultra-violet pair near O, but no others.

A strip of silver developed two ultra-violet lines, one on either side of P; but we could not detect in the flash the well-known green lines of that metal. When powder of silver oxalate was introduced, the yellowish-green line (w.l. 5464) was seen at the first explosion but not afterwards. As silver oxalate is itself an explosive compound, decomposing with an evolution of heat, it is reasonable to ascribe the appearance of this line at the first explosion to the extra temperature so engendered.

Strips of copper, electroplated with nickel, brought out almost all the strong nickel lines in the ultra-violet between K and Q; 25 were photographed. When nickel oxalate was put in so as to give a powder of metallic nickel after the first explosion, the same lines were developed, and three additional lines in the ultra-violet. Only one line was seen in the visible part of the spectrum, and that was the yellowish-green line (w.l. 5476).

Copper wires electroplated with cobalt gave in the flash 22 lines in the violet and ultra-violet, between G and P; no lines beyond those limits. Cobalt oxalate gave no more.

No other metal gave anything like so many lines as iron, nickel, and cobalt; and it is remarkable that almost all the lines of these metals developed in the flash lie in the same region between G and P.

We expected that manganese would have given several lines in the flash; but it was not so. Neither metallic manganese, nor any of several compounds which we tried, gave us any lines of that metal except the violet triplet, and this was generally given by the iron tube alone, and was merely stronger for the manganese put in. The green channellings characteristic of manganese, and ascribed to the oxide, were, however, well seen when metallic manganese was used.

Chromium, introduced as bichromate of ammonia, which of course became chromium oxide at the first flash, gave three triplets in the green, the indigo, and the ultra-violet near N respectively, but no other lines.

Bismuth, antimony, and arsenic gave no lines, nor did mercury spread over a sheet of copper lining the tube. Several metals were tried as amalgams spread over such a piece of copper, but with no fresh results, except in the case of thallium, which gave the green line strongly, the strong line between L and M, and two lines between N and O.

On the whole it does not appear that the form in which the metal is introduced into the tube makes much difference. The merest traces of those which gave lines were sufficient. Generally when a metal had been put into the tube, its lines continued to show after the strip or lining had been removed. Thus, after the nickel strips had been taken out, and the tube cleaned out as completely as it could be mechanically, the nickel lines still came out in the flash, and the same was the case with other metals.

The strongest part of the water-spectrum, from s to near R, generally impressed itself more or less on the photographic plate; but, with the exception of T, which was only developed once or twice, no lines made their appearance in the region more refrangible than s.

Thus far the experiments had been made with the gases at the atmospheric pressure, or nearly so, before ignition. The proportions of hydrogen and oxygen were nearly two to one; but an excess of either gas to the extent of one-fifth did not sensibly affect the results.

Other explosive mixtures were tried. Carbonic oxide with

oxygen, and marsh-gas with oxygen, developed in general the same lines as the hydrogen mixture, but gave a much brighter continuous spectrum. Sulphuretted hydrogen, arseniuretted hydrogen, and antimoniuretted hydrogen, exploded with oxygen, also gave very bright continuous spectra, but no lines attributable to sulphur, arsenic, or antimony.

We have also tried explosions at higher pressures; mixtures of hydrogen, carbonic oxide, and marsh-gas respectively, with oxygen, were compressed into the tube by a condensing syringe until the pressure reached two and a half atmospheres, and in some cases three and a half atmospheres. The general effect of increasing the pressure was to strengthen very much the continuous spectrum, and also to intensify the bright lines, so that photographs could be taken with a smaller number of explosions. The lines previously observed to be reversed were more strongly reversed, but no new lines which we can attribute to the metals employed were noticed. No iron line more refrangible than T showed itself in the photographs. But a banded spectrum, of which traces had been noticed in the flash of the gases at lower pressure, came out decidedly. This spectrum occupies the region between P and R; it is not a regularly channelled spectrum, though probably under higher dispersion it would resolve itself into groups of lines like the water-spectrum. In fact it seems to us most probable that it is a development of the water spectrum, dependent on the pressure.

It seems very remarkable that metals so little volatile as iron, nickel, and cobalt should develop so many lines[*] in the flash, while more volatile metals show few or no lines. We do not know that any lines attributed to the metals, as distinct from their compounds, which have been observed in the gas-flame cannot be seen also in the flash of the exploding gases, unless they be the blue lines of zinc which Lecoq de Boisbaudran has seen faintly in the gas-flame when zinc chloride was introduced. These are, however, so faint in the flame, that they might easily escape notice in the much stronger continuous spectrum of the flash. But iron, nickel, and cobalt show no lines of those metals in a gas-flame. Mitscherlich ('Ann. de Phys. u. Chem.' Bd. 121, St. 3), by mixing vapour of ferric chloride with the hydrogen burnt in an oxyhydrogen-jet, obtained a number of the lines of iron. These form three groups—one below D, one near E, and one near G. The last two groups have a general correspondence with the lines developed in the explosions in the visible part of the spectrum; but exact identification is not possible with his figure. Of other metals he seems also to have found the same lines in the oxyhydrogen-jet which we have seen in the explosions, but with additional lines in several cases. Thus he found three zinc and as many cadmium lines, two of mercury, four of copper, and so on.

Gouy ('Comptes Rendus,' lxxxiv. 1877, p. 232) has observed in

---

[*] For detailed list of these lines see 'Proc. Roy. Soc.' vol. xxxvi. pp. 473-5.

the inner green cone of a modified Bunsen burner fed with gas mixed with spray of iron-salts, four green lines of iron which we did not find in the flash. He saw two of the blue lines, but not the other lines which we have noticed. In like manner with cobalt, he observed two feeble blue rays which we did not see in the explosions; also one zinc, one cadmium, and one silver line which we did not see; and he did not notice the green copper line which we always have seen in the explosions. In other cases he has noticed the same lines that we have noticed.

Comparing the spectrum of the explosions with that of iron wire burnt in a jet of coal-gas fed with oxygen, they may be called identical. We find in them generally the same lines and the same relative strengths of the lines. For instance, in the explosion-spectrum the strength of the groups of lines on either side of M and the line at wave-length $3859 \cdot 2$ is decidedly greater as compared with the other lines than it is in the arc-spectrum of iron. It is the same in that of iron burnt as above mentioned. T, however, comes out more strongly in the last-mentioned spectrum than in the explosions.

German-silver wire burnt in the coal-gas and oxygen jet gave the same nickel and copper lines as were developed in the explosions. Silver wire gave in the same jet the two silver lines near P, but no channelled spectrum. Spray of cobalt chloride gave also the same lines as in the explosions, with a few additional; while spray of manganese chloride gave the strong manganese triplet at wave-length about 2800, more refrangible than anything observed in the explosions, besides the usual violet triplet.

On the whole the spectra produced by the jet of coal-gas and oxygen are very similar to those of the explosions as far as the metallic lines go; they exhibit a few more lines, or it may be these are more easily observed.

Of the green and blue lines of iron seen by us in explosions nine are registered by Watts as occurring in the flame of a Bessemer converter; or at least the lines he gives are so near that we cannot doubt their identity.

When we come to make a comparison with the spectrum of the spark-discharge from a solution of ferric chloride, the differences become more marked. Not only are there many more lines in the spark-spectrum, but the relative intensities of those lines which are common to both spark and explosion are very different, and two of the iron lines seen in the explosions appear to be absent from the spark. The differences between the spectrum of the spark taken from a liquid electrode and that given by solid electrodes has usually been attributed to the lower temperature of the former; but the absence from the former spectrum of the line at wave-length 4132, and the feebleness of the line at wave-length 4143, both strong lines in the arc and in the explosions, as well as in the spark between solid electrodes, seem to indicate that the differences of spark-spectra are not simply due to differences of temperature. In fact we know so little about the mechanism, so to speak, of the changes of electric energy into heat, and of heat into radiation, that there is no good reason for assuming that the energy which takes the form of radiation in the electric discharge through a gas must first take the form of the motion of translation of the particles on which temperature depends. The gas may, for a short time, be intensely luminous at a very low temperature; and if the impulses which give rise to the vibratory movements of the particles be of different characters, the characters of the vibrations also may differ within certain limits.

Leaving, however, the realms of speculation, we may mention that we have before observed the spectrum of iron at a temperature intermediate between that of the oxyhydrogen-jet and that of the electric arc.

Some time since ('Proc. R. S.' xxxiv. p. 119, and 'Proc. Camb. Phil. Soc.' iv. p. 256) we described the spectrum proceeding from the interior of a carbon tube strongly heated by the electric arc playing on the outside. This spectrum approaches more nearly to that of the arc inasmuch as it shows all, or almost all, the iron lines given by the arc between F and O, and the aluminium pair between H and K; but it resembles the explosion-spectrum in the relative strength of some of the iron lines, and in the absence of almost all iron lines between O and T. The iron lines seen reversed against the hot walls of the carbon-tube correspond with the strongest of the explosion-lines; the strong lines near M and a little below L in the explosions being those most strongly reversed in the photographs of the carbon-tube. The greater completeness and extent of the iron spectrum, as well as the presence of the aluminium lines, which are entirely wanting in the explosion-spectrum, indicate that the temperature of the tube was higher than that of the explosion. That iron, nickel, and cobalt are volatile in some degree at the temperature of the explosion appears to be proved, and makes the appearance of iron lines at the very apices of solar prominences, as observed by Young, less astounding than it seemed to be at first sight. The ascending current of gas making the prominence may very well carry iron vapour with it; or we may not unreasonably suppose that there is meteoric dust containing iron everywhere in the outer atmosphere of the sun, which becomes volatilised, and emits the radiation observed, when it is heated up by the hot current of the prominence. What the temperature of such a current may be we cannot well gauge, but it is high enough to give the hydrogen-spectrum, of which no trace has been observed in the flash of the explosions or in the oxy-hydrogen-jet. The temperature of the explosions we know with tolerable accuracy, at least when the gases are at atmospheric pressure to begin with. Bunsen ('Phil. Mag.' 1867, p. 494) found the pressure of the explosion was for hydrogen and oxygen $9 \cdot 6$ atmospheres, and for carbonic oxide and oxygen $10 \cdot 3$ atmospheres, and he calculated the corresponding temperatures to be $2844°$ and $3033°$.

Recently published observations by Berthelot and Vieille ('Comptes Rendus,' xcviii. 1884, p. 548) put the pressure of explosion of oxygen and hydrogen at 9·8 atmospheres, and of carbonic oxide and oxygen at 10·1, and the corresponding temperatures 3240° and 3334°. The pressures determined by the two observers agree closely, and the calculated temperatures are not very discordant. On the whole, we cannot be wrong in assuming the temperature of the exploding gases to be about 3000°; and we see that at this degree such metals as iron, nickel, and cobalt are vaporous, and emit many characteristic rays, and that by far the greatest part of these rays lie between narrow limits of refrangibility, G and P. Even for other metals there is a predominance of rays in the same part of the spectrum. The lines of lead, potassium, and manganese, three out of four lines of thallium, and two-thirds of those of chromium, observed in the explosions, fall within the same region. It must not be inferred that these facts indicate the limit of the rate of oscillation which can be set up in consequence of an elevation of temperature to 3000°, because we know that the spectrum of the lime-light extends much further. But it might be possible to establish a sort of spectroscopic scale of temperatures if the lines which are successively developed as the temperature rises were carefully noted. Thus the appearance of the iron line T seems to synchronise with temperature of about 3000°. The lithium blue line is invisible in the flame of an ordinary Bunsen burner, but is just visible at the temperature of the inner green cone formed by reducing the proportion of gas to air in such a burner, while in the exploding gas the green line too is seen. It seems to need a temperature above 3000° to get the aluminium lines at H. Probably no line is ever abruptly brought out at a particular temperature—it will always be gradually developed as the temperature rises; yet the development may be rapid enough to give an indication which may be useful in default of means of more exact measurement. In former papers treating of spectroscopic problems ('Proc. Roy. Soc.' vol. xxxiv. p. 130, and xxix. p. 489) we have more than once adverted to the necessity of the study of the spectra both of flames and of the electric discharge under modified conditions of pressure. The projected experiments on the arc in lime-crucibles have not yet been carried out; but the present is a first instalment of a study of flame-spectra under such conditions.

[J. D.]

Friday, April 29, 1887.

HENRY POLLOCK, Esq. Treasurer and Vice-President,
in the Chair.

PROFESSOR H. S. HELE SHAW.

The Rolling Contact of Bodies.

WHEN two solid bodies roll upon each other, points in the surface of one successively come into contact with corresponding points in the surface of the other in a way which differs essentially from that which occurs in sliding contact, and it is the nature of this rolling-contact that the lecturer proposed to discuss in an experimental manner.

In the first place, it is well to understand clearly the nature of the relative motion of the two points which come into contact when the surfaces are such that no appreciable distortion of them takes place, and for this purpose one of the two bodies must be at rest. These may respectively be taken as the plane surface of the ground and a circular disk rolling upon it. An approximate representation of this motion is given by the end of the spokes of a wheel without its tyre. In this case it is seen that a point of the rolling body, when it is just coming into contact with the fixed surface, does so in a direction at right angles to the surface at rest, and also leaves it in the same direction. This action is very similar in kind to that which occurs with the continuous circle formed by the tyre. The path of a point in the rim can be drawn in a way visible to the audience by means of a piece of apparatus consisting of two circular glass plates held together by a hollow brass spindle in which slides an arm carrying a brush. The brush traces the well-known cycloid, of which the only portion now to be considered is that where it directly approaches the surface beneath. This part is perpendicular to that surface, and when epicycloids are drawn, by rolling the disk upon the arc of a circle, the same fact is brought out.

One body may, however, not merely roll upon another, and a normal pressure be exerted, but they may exert a tangential force upon each other. It is convenient to keep these two cases separate; examples of them being respectively the wheels of a railway carriage and those of the locomotive which draws it along. It is to be noted that the object in the former case is to permit one body to move relatively to another without permitting sliding contact of their surfaces, whilst, in the latter case, in addition to this, the object is to obtain such motion. There are, however, many cases in which it is merely the motion of a body about one point which is required, such as when motion is transmitted from the edge of one rotating disk to another, and then this distinction still more closely holds, as the normal pressure is only obtained so as to insure the necessary tangential resistance. Thus the objects of rolling motion may be classed as being—

(1) To *allow* the relative motion of one body to another with which it is in contact without permitting relative motion of that part of their surfaces in actual contact.

(2) To *obtain* the relative motion of such parts of the surfaces of bodies as are not in contact by means of statical contact of the parts which are.

The lecturer then proceeded to consider the practical proofs of the smallness of the resistance to rolling in cases where the distortion of the surfaces in contact is very small, as illustrated by the small tractive force required for heavy bodies properly mounted on wheels or on roller-bearings; mentioning the case of a 12-horse-power engine, the shaft of which continued to rotate for three-quarters of an hour after the motive power was withdrawn; and another case, of a turntable weighing 14 tons, which was kept in motion by a weight of $3\frac{1}{2}$ pounds acting upon it by means of a cord passing over a pulley. The small distortion of such surfaces when transmitting motion requiring expenditure of energy to maintain, was next made clear by giving certain facts as to the accuracy with which one surface was developed or measured out upon another. An account was given of experiments made with apparatus specially prepared by the lecturer to investigate this point. This apparatus consisted of two accurately turned brass disks properly mounted upon a frame, and the relative positions of these disks could be interchanged so that any minute differences in their peripheries could be detected. The experiments, which were very difficult to carry out accurately, showed that under the best circumstances, motion with an error of only 1 in 300,000 of the distance passed over could be obtained. This accurate measuring out of the surfaces one upon another was employed in various ways for purposes of measurement, and these, by means of models and diagrams, were briefly explained.

Although the foregoing facts prove that, under suitable conditions, distortion at the points of contact is very small, yet some resistance at these points *always* occurs, because no bodies are perfectly hard; and the nature of this distortion and consequent resistance was next discussed.

The explanation of the resistance opposed by a soft surface to a hard body rolling upon it, as first given by Prof. Osborne Reynolds, was applied by the lecturer to account for a very remarkable effect produced in the disk, globe, and cylinder integrator of Prof. James Thomson. This effect, which was the turning of the cylinder when the sphere was rolled along it in a horizontal direction, was reproduced by means of a large model. The action of a soft body rolling upon a hard surface was next considered, with the result of showing that the same reasoning would not account for the turning of the cylinder in the same direction as before with the above model, and the lecturer then proceeded, by means of diagrams, to offer an explanation of this and other phenomena. The various effects obtained with bodies of different relative degrees of hardness were discussed at length, but figures would be needed to make these points clear. Finally, an explanation was given of the cause of an error which always appeared in a certain important class of integrators caused by the slipping of the edge of a disk over a surface on which it rolled in circumstances under which it had apparently never been suspected that slipping did actually take place. This the lecturer had been enabled to discover and measure by means of a special piece of apparatus, a model of which was exhibited and the effects shown by its means.

The facts and reasoning, which were given in the lecture, all related to the rolling contact of bodies, and the lecturer ventured to think that, imperfect as the treatment of the subject had been, it was one of such importance, not merely from the point of view of the practical applications he had mentioned, but in its scientific aspect, dealing as it did from a novel point of view with the nature and properties of solid bodies, as to be worthy of being thus brought before the Royal Institution.

Friday, January 20, 1888.

WILLIAM HUGGINS, Esq. D.C.L. LL.D. F.R.S. Vice-President,
in the Chair.

The Right Hon. LORD RAYLEIGH, M.A. D.C.L. LL.D. F.R.S. *M.R.I.*
PROFESSOR OF NATURAL PHILOSOPHY, R.I.

Diffraction of Sound.

THE interest of the subject which I propose to bring before you this evening turns principally upon the connection or analogy between light and sound. It has been known for a very long time that sound is a vibration; and every one here knows that light is a vibration also. The last piece of knowledge, however, was not arrived at so easily as the first; and one of the difficulties which retarded the acceptance of the view that light is a vibration was that in some respects the analogy between light and sound seemed to be less perfect than it should be. At the present time many of the students at our schools and universities can tell glibly all about it; yet this difficulty is one not to be despised, for it exercised a determining influence over the great mind of Newton. Newton, it would seem, definitely rejected the wave theory of light on the ground that according to such a theory light would turn round the corners of obstacles, and so abolish shadows, in the way that sound is generally supposed to do. The fact that this difficulty seemed to Newton to be insuperable is, from the point of view of the advancement of science, very encouraging. The difficulty which stopped Newton two centuries ago is no difficulty now. It is well known that the question depends upon the relative wave-lengths in the two cases. Light-shadows are sharp under ordinary circumstances, because the wave-length of light is so small: sound-shadows are usually of a diffused character, because the wave-length of sound is so great. The gap between the two is enormous. I need hardly remind you that the wave-length of C in the middle of the musical scale is about 4 feet. The wave-length of the light with which we are usually concerned, the light towards the middle of the spectrum, is about the forty-thousandth of an inch. The result is that an obstacle which is immensely large for light may be very small for sound, and will therefore behave in a different manner.

That light-shadows are sharp is a familiar fact, but as I can prove it in a moment I will do so. We have here light from the electric arc thrown on the screen; and if I hold up my hand thus we have a sharp shadow at any moderate distance, which shadow can be made sharper still by diminishing the source of light. Sound-shadows, as I have said, are not often sharp; but I believe that they are sharper than is usually supposed, the reason being that when we pass into a sound-shadow—when, for example, we pass into the shade of a large obstacle, such as a building—it requires some little time to effect the transition, and the consequence is that we cannot make a very ready comparison between the intensity of the sound before we enter and its diminution afterwards. When the comparison is made under more favourable conditions, the result is often better than would have been expected. It is, of course, impossible to perform experiments with such obstacles before an audience, and the shadows which I propose to show you to-night are on a much smaller scale. I shall take advantage of the sensitiveness of a flame such as Professor Tyndall has often used here—a flame sensitive to the waves produced by notes so exceedingly high as to be inaudible to the human ear. In fact, all the sounds with which I shall deal to-night will be inaudible to the audience. I hope that no quibbler will object that they are therefore not sounds: they are in every respect analogous to the vibrations which produce the ordinary sensations of hearing.

I will now start the sensitive flame. We must adjust it to a reasonable degree of sensitiveness. I need scarcely explain the mechanism of these flames, which you know are fed from a special gasholder supplying gas at a high pressure. When the pressure is too high, the flame flares on its own account (as this one is doing now), independently of external sound. When the pressure is somewhat diminished, but not too much so—when the flame "stands on the brink of the precipice," were, I think, Tyndall's words—the sound pushes it over, and causes it to flare; whereas, in the absence of such sound, it would remain erect and unaffected. Now, I believe, the flame is flaring under the action of a very high note that I am producing here. That can be tested in a moment by stopping the sound, and seeing whether the flame recovers or not. It recovers now. What I want to show you, however, is that the sound shadows may be very sharp. I will put my hand between the flame and the source of sound, and you will see the difference. The flame is at present flaring; if I put my hand here, the flame recovers. When the adjustment is correct, my hand is a sufficient obstacle to throw a most conspicuous shadow. The flame is now in the shadow of my hand, and it recovers its steadiness: I move my hand up, the sound comes to the flame again, and it flares. When the conditions are at their best, a very small obstacle is sufficient to make the entire difference, and a sound shadow may be thrown across several feet from an obstacle as small as the hand. The reason of the divergence from ordinary experience here met with is, that while the hand is a fairly large obstacle in comparison with the wave-length of the sound I am here using, it would not be a sufficiently large obstacle in comparison with the wave-lengths with which we have to do in ordinary life and in music.

Everything then turns upon the question of the wave-length. The wave-length of the sound that I am using now is about half an inch. That is its complete length, and it corresponds to a note that would be very high indeed on the musical scale. The wave-length of middle C being four feet, the C one octave above that is two feet; two octaves above, one foot; three octaves above, six inches; four octaves, three inches; five octaves, one and a half inch; six octaves, three-quarters of an inch; between that and the next octave, that is to say, between six and seven octaves above middle C, is the pitch of the note that I was just now using. There is no difficulty in determining what the wave-length is. The method depends upon the properties of what are known as stationary sonorous waves as opposed to progressive waves. If a train of progressive waves are caused to impinge upon a reflecting wall, there will be sent back or reflected in the reverse direction a second set of waves, and the co-operation of these two sets of waves produces one set or system of stationary waves, the distinction being that, whereas in the one set the places of greatest condensation are continually changing and passing through every point, in the stationary waves there are definite points for the places of greatest condensation (nodes), and other distinct and definite (loops) for the places of greatest motion. The places of greatest variation of density are the places of no motion: the places of greatest motion are places of no variation of density. By the operation of a reflector, such as this board, we obtain a system of stationary waves, in which the nodes and loops occupy given positions relatively to the board.

You will observe that as I hold the board at different distances behind, the flame rises and falls—I can hardly hold it still enough. In one position the flame rises, further off it falls again; and as I move the board back the flame passes continually from the position of the node—the place of no motion—to the loop or place of greatest motion and no variation of pressure. As I move back the aspect of the flame changes; and all these changes are due to the reflection of the sound-waves by the reflector which I am holding. The flame alternately ducks and rises, its behaviour depending upon the different action of the nodes and loops. The nodes occur at distances from the reflecting wall, which are even multiples of the quarter of a wave-length; the loops are, on the other hand, at distances from the reflector which are odd multiples, bisecting therefore the positions between the loops. I will now show you that a very slight body is capable of acting as a reflector. This is a screen of tissue paper, and the effect will be apparent when it is held behind the flame and the distances are caused to vary. The flame goes up and down, showing that a considerable proportion of the sonorous intensity incident upon the paper screen is reflected back upon the flame; otherwise the exact position of the reflector would be of no moment. I have here, however, a different sort of reflector. This is a glass plate—I use glass so that those behind may see through it—and it will slide upon a stand here arranged for it. When put in this position the flame is very little affected; the place is what I call a node—a place where there is great pressure variation, but no vibratory velocity. If I move the glass back, the flame becomes vigorously excited; that position is a loop. Move it back still more and the flame becomes fairly quiet; but you see that as the plate travels gradually along, the flame goes through these evolutions as it occupies in succession the position of a node or the position of a loop. The interest of this experiment for our present purpose depends upon this —that the distances through which the glass plate, acting as a reflector, must be successively moved in order to pass the flame from a loop to the next loop, or from a node to the consecutive node, is in each case half the wave length; so that by measuring the space through which the plate is thus withdrawn one has at once a measurement of the wave length, and consequently of the pitch of the sound, though one cannot hear it.

The question of whether the flame is excited at the nodes or at the loops,—whether at the places where the pressure varies most or at those where there is no variation of pressure, but considerable motion of air—is one of considerable interest from the point of view of the theory of these flames. The experiment could be made well enough with such a source of sound as I am now using; but it is made rather better by using sounds of a lower pitch and therefore of greater wave-length, the discrimination being then more easy. Here is a table of the distances which the screen must be from the flame in order to give the maximum and the minimum effect, the minimum being practically nothing at all.

TABLE OF MAXIMA AND MINIMA.

Max.	Min.
1·1	
	3·0
4·5	
	5·9
7·5	
	8·9
10·3	
	11·7
13·0	
	14·7
15·9	

The distance between successive maxima or successive minima is very nearly 3 (centims), and this is accordingly half the length of the wave.

But there is a further question behind. Is it at the loops or is it at the nodes that the flame is most excited? The table shows what the answer must be, because the nodes occur at distances from the screen which are even multiples, and the loops at distances which are odd multiples; and the numbers in the table can be explained in only one way—that the flame is excited at the loops corresponding to the odd multiples, and remains quiescent at the nodes corresponding to the even multiples. This result is especially remarkable, because the ear, when substituted for the flame, behaves in the exactly opposite manner, being excited at the nodes and not at the loops. The experiment may be tried with the aid of a tube, one end of which is placed in the ear, while the other is held close to the burner. It is then found the ear is excited the most when the flame is excited least, and *vice versâ*. The result of the experiment shows, moreover, that the manner in which the flame is disintegrated under the action of sound is not, as might be expected, symmetrical in regard to the axis of the flame. If it were symmetrical, it would be most affected by the symmetrical cause, namely, the variation of pressure. The fact being that it is most excited at the loop, where there is the greatest vibratory velocity, shows that the method of disintegration is unsymmetrical, the velocity being a directed quantity. In that respect the theory of these flames is different from the theory of the water-jets investigated by Savart, which resolve themselves into detached drops under the influence of sonorous vibration. The analogy fails at this point, and it has been pressed too far by some experimenters on the subject. Another simple proof of the correctness of the result of our experiment is that it makes all the difference which way the burner is turned in respect of the direction in which the sound-waves are impinging upon it. If the phenomenon were symmetrical, it would make no difference if the flame were turned round upon its vertical axis. But we find that it does make a difference. This is the way in which I was using the flame, and you see that it is flaring strongly. If I now turn the burner round through a right angle, the flame stops flaring. I have done nothing more than turn the burner round and the flame with it, showing that the sound-waves may impinge in one direction with great effect, and in another direction with no effect. The sensitiveness occurs again when the burner is turned through another right angle; after three right angles there is another place of no effect; and after a complete revolution of the flame the original sensitiveness recurs. So that if the flame were stationary, and the sound-waves came, say, from the north or south, the phenomena would be exhibited; but if they came from the east or west, the flame would make no response.

This is of convenience in experimenting, because, by turning the burner round, I make the flame almost insensitive to a sound, and I am now free to show the effect of any sound that may be brought to it in the perpendicular direction. I am going to use a very small reflector—a small piece of looking-glass. Wood would do as well; but looking-glass facilitates the adjustment, because my assistant, by seeing the reflection, will be able to tell me when I am holding it in the best position. Now, the sound is being reflected from the bit of glass, and is causing the flame to flare, though the same sound, travelling a shorter distance and impinging in another direction, is incompetent to produce the result (Fig. 1).

I am now going to move the reflector to and fro along the line perpendicular to that joining the source and the burner, all the while maintaining the adjustment, so that from the position of the source of sound the image of the flame is seen in the centre of the mirror. Seen from the source, it is still as central as before, but it has lost its effect, and as I move it to and fro I produce cycles of effect and no

FIG. 1.

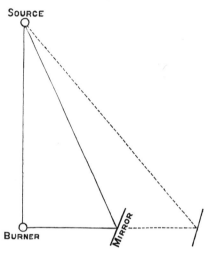

effect. What is the cause of this? The question depends upon something different from what I have been speaking of hitherto; and the explanation is, that we are here dealing with a diffraction phenomenon. The mirror is a small one, and the sound-waves which it reflects are not big enough to act in the normal manner. We are really dealing with the same sort of phenomena as arise in optics when we use small pin-holes for the entrance of our light. It is not very easy to make the experiment in the present form quite simple, because the mirror would have to be withdrawn, all the while maintaining a somewhat complicated adjustment. In order to raise the question of diffraction in its simplest shape, we must have a direct course for the sound between its origin and the place of observation, and interpose in the path a screen perforated with such holes as we desire to try.

The screen I propose to use is of glass. It is a practically perfect obstacle for such sounds as we are dealing with; but it is perforated here with a hole (20 cm. diameter), rendered more evident to those at a distance by means of a circle of paper pasted round it. The edge of the hole corresponds to the inner circumference of the paper. We shall thus be able to try the effect of different sized apertures, all the other circumstances remaining unchanged. The experiment is rather a difficult one before an audience, because everything turns on getting the exact adjustment of distances relatively to the wavelength. At present the sound is passing through this comparatively large hole in the glass screen, and is producing, as you see, scarcely any effect upon the flame situated opposite to its centre. But if (Fig. 2) I diminish the size of the hole by holding this circle of zinc (perforated with a hole 14 cm. in diameter) in front of it, it is seen that, although the hole is smaller, we get a far greater effect. That

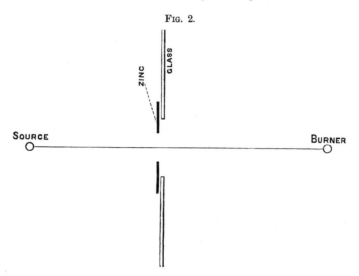

Fig. 2.

is a fundamental phenomenon in diffraction. Now I reopen the larger hole, and the flame becomes quiet. So that it is evident that in this case the sound produces a greater effect in passing through a small hole than in passing through a larger one. The experiment may be made in another way, by obstructing the central in place of the marginal part of the aperture in the glass. When I hold this unperforated disc of zinc (14 cm. in diameter) centrically in front, we get a greater effect than when the sound is allowed to pass through both parts of the aperture. The flame is now flaring vigorously under the action of the sonorous waves passing the marginal part of the aperture, whereas it will scarcely flare at all under the action of waves passing through both the marginal and the central hole.

This is a point which I should like to dwell upon a little, for it lies at the root of the whole matter. The principle upon which it depends is one that was first formulated by Huygens, one of the leading names in the development of the undulatory theory of light. In this diagram (Fig. 3) is represented in section the different parts of the obstacle. C represents the source of sound, B represents the flame, and A P Q is the screen. If we choose a point P on this screen, so that the whole distance from B to C, reckoned through P, viz. B P C, exceeds the shortest distance B A C by exactly half the wave-length of the sound, then the circular area, whose radius is A P, is the first zone. We take next another point, Q, so the whole distance B Q C exceeds the previous one by half a wave-length. Thus we get the

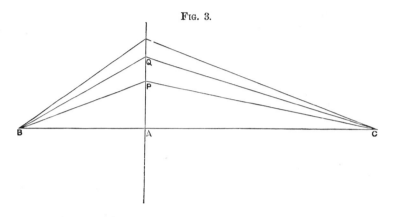

Fig. 3.

second zone represented by P Q. In like manner, by taking different points in succession such that the last distance taken exceeds the previous one every time by half a wave-length, we may map out the whole of the obstructing screen into a series of zones called Huygens' zones. I have here a material embodiment of that notion, in which the zones are actually cut out of a piece of zinc. It is easy to prove that the effects of the parts of the wave traversing the alternate zones are opposed, that whatever may be the effect of the first zone, A P, the exact opposite will be the effect of P Q, and so on. Thus, if A P and P Q are both allowed to operate, while all beyond Q is cut off, the waves will neutralise one another, and the effect will be immensely less than if A P or P Q operated alone. And that is what you saw just now. When I used the inner aperture only, a comparatively loud sound acted upon the flame. When I added to that inner aperture the additional aperture P Q, the sound disappeared, showing that the effect of the latter was equal and opposite to that of A P, and that the two neutralised each other.

[If $AC = a$, $AB = b$, $AP = x$, wave-length $= \lambda$, the value of x for the external radius of the nth zone is

$$x^2 = n \lambda \frac{a+b}{ab},$$

or, if $a = b$,

$$x^2 = \tfrac{1}{2} n \lambda a.$$

With the apertures used above, $x^2 = 49$ for $n = 1$; $x^2 = 100$, for $n = 2$; so that

$$\lambda a = 100,$$

the measurements being in centimetres. This gives the suitable distances, when λ is known. In the present case $\lambda = 1 \cdot 2$, $a = 83$.]

Closely connected with this there is another very interesting experiment, which can easily be tried, and which has also an important optical analogy. I mean the experiment of the shadow thrown by a circular disc. If a very small source of light be taken —such a source as would be produced by perforating a thin plate in the shutter of the window of a dark room with a pin and causing the rays of the sun to enter horizontally—and if we interpose in the path of the light a small circular obstacle and then observe the shadow thrown in the rear of that obstacle, a very remarkable peculiarity manifests itself. It is found that in the centre of the shadow of the obstacle, where the darkness might be expected to be greatest, there is, on the contrary, no darkness at all, but a bright spot, a spot as bright as if no obstacle intervened in the course of the light. The history of this subject is curious. The fact was first observed by Delisle in the early part of the eighteenth century, but the observation fell into oblivion. When Fresnel began his important investigations, his memoir on diffraction was communicated to the French Academy and was reported on by the great mathematician Poisson. Poisson was not favourably impressed by Fresnel's theoretical views. Like most mathematicians of the day, he did not take kindly to the wave theory; and in his report on Fresnel's memoir, he made the objection that if the method were applied, as Fresnel had not then done, to investigate what should happen in the shadow of a circular obstacle, it brought out this paradoxical result, that in the centre there would be a bright point. This was regarded as a *reductio ad absurdum* of the theory. All the time, as I have mentioned, the record of Delisle's observations was in existence. The remarks of Poisson were brought to the notice of Fresnel, the experiment was tried, and the bright point was rediscovered, to the gratification of Fresnel and the confirmation of his theoretical views. I don't propose to attempt the optical experiment now, but it can easily be tried in one's own laboratory. A long room or passage must be darkened: a fourpenny bit may be used as the obstacle, strung up by three hairs attached by sealing-wax. When the shadow of the obstacle is received on a piece of ground glass, and examined from behind with a magnifying lens, the bright spot will be seen without much difficulty. But what I propose to show you is the corresponding phenomenon in the case of sound. Fresnel's reasoning is applicable, word for word, to the phenomena we are considering just as much as to that which he, or rather Poisson, had in view. The disc (Fig. 4), which I shall hang up now between the source of sound and the flame, is of glass.

It is about 15 inches in diameter. I believe the flame is flaring now from being in the bright spot. If I make a small motion of the disc I shall move the bright spot and the effect will disappear. I am pushing the disc away now, and the flaring has stopped. The flame

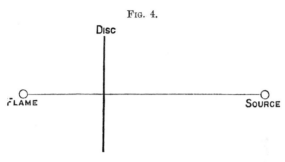

Fig. 4.

is still in the shadow of the disc, but not at the centre. I bring the disc back again, and when the flame comes into the centre it flares again vigorously. That is the phenomenon which was discovered by Delisle and confirmed by Arago and Fresnel, but mathematically it was suggested by Poisson.

Poisson's calculation related only to the very central point in the axis of the disc. More recently the theory of this experiment has been very thoroughly examined by a German mathematician, Lommel; and I have exhibited here one of the curves given by him embodying the results of his calculations on the subject (Fig. 5).

The abscissæ, measured horizontally, represent distances drawn outwards from the centre of the shadow O; the ordinates measure the intensity of the light at the various points. The maximum intensity O A is at the centre. A little way outwards at B the intensity falls almost, but not quite, to zero. At C there is a revival of intensity, indicating a bright ring; and further out there is a succession of subordinate fluctuations. The curve on the other side of O A would of course be similar. This curve corresponds to the distances and proportions indicated. a is the distance between the source of sound and the disc; b is the distance between the disc and the flame, the place where the intensity is observed. The numbers given are taken from the notes of an experiment which went well. If we can get our flame to the right point of sensitiveness we may succeed in bringing into view not only the central spot, but the revived sound which occurs after you have got away from the central point and have passed through the ring of silence. There is the loud central point. If I push the disc a little we enter the ring of silence B;* a little further, and the flame flares again, being now at C.

---

* With the data given above the diameter of the silent ring is two-thirds of an inch.

Although we have thus imitated the optical experiment, I must not leave you under the idea that we are working under the same conditions that prevail in optics. You see the diameter of my disc is 15 inches, and the length of my sound-wave is about half an inch.

Fig. 5.

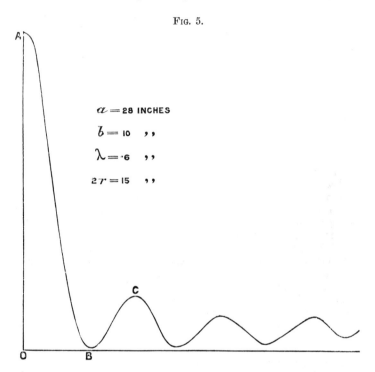

My disc is therefore about 30 wave-lengths in diameter, whereas the diameter of a disc representing 30 wave-lengths of light would be only about $\frac{1}{1000}$ inch. Still the conditions are sufficiently alike to get corresponding effects, and to obtain this bright point in the centre of the shadow conspicuously developed.

I will now make an experiment illustrating still further the principle of Huygens' zones, which I have already roughly sketched. I indicated that the effect of contiguous zones was equal and opposite, so that the effect of each of the odd zones is one thing, and of the even zones the opposite thing. If we can succeed in so preparing a screen as to fit the system of zones, allowing the one set to pass, and at the same time intercepting the other set, then we shall get a great effect at the central point, because we shall have removed those parts which, if they remained, would have neutralised the remaining parts.

Such a system has been cut out of zinc, and is now hanging before you. When the adjustments are correct there will be produced, under the action of that circular grating, an effect much greater than would result if the sound-waves were allowed to pass on without any obstruction. The only point difficult of explanation is as to what happens when the system of zones is complete, and extends to infinity, viz. when there is no obstruction at all. In that case it may be proved that the aggregate effect of all the zones is, in ordinary cases, half the effect that would be produced by any one zone alone, whereas if we succeed in stopping out a number of the alternate zones, we may expect a large multiple of the effect of one zone. The grating is now in the right position, and you see the flame flaring strongly, under the action of the sound-waves transmitted through these alternate zones, the action of the other zones being stopped by the interposition of the zinc. But the interest of the experiment is principally in this, that the flame is flaring *more* than it would do if the grating were removed altogether. There is now, without the grating, a very trivial flaring;[*] but when the grating is in position again—though a great part of the sound is thereby stopped out—the effect is far more powerful than when no obstruction intervened. The grating acts, in fact, the part of a lens. It concentrates the sound upon the flame, and so produces the intense magnification of effect which we have seen.

[The exterior radius of the nth zone being x, we have, from the formula given above:

$$\frac{1}{a} + \frac{1}{b} = \frac{n\lambda}{x^2};$$

so that if a and b be the distances of the source and image from the grating, the relation required to maintain the focus is as usual,

$$\frac{1}{a} + \frac{1}{b} = \frac{1}{f},$$

where f, the *focal length* is given by:

$$f = \frac{x^2}{n\lambda}.$$

In the actual grating, eight zones (the first, third, fifth, &c.) are occupied by metal. The radius of the first zone, or central circle, is 3 inches, so that $x^2/n = 9$. The focal length is necessarily a function of λ. In the present case $\lambda = \frac{1}{2}$ inch nearly, and therefore $f = 18$ inches. If a and b are the same, each must be made equal to 36 inches.]

[RAYLEIGH.]

---

[*] Under the best conditions the flame is absolutely unaffected.

Friday, April 27, 1888.

WILLIAM HUGGINS, Esq. D.C.L. LL.D. F.R.S. Vice-President,
in the Chair.

JAMES WIMSHURST, Esq. *M.R.I.*

Electrical Influence Machines.

I HAVE the honour this evening of addressing a few remarks to you upon the subject of influence machines, and the manner in which I propose to treat the subject is to state as shortly as possible, first, the historical portion, and afterwards to point out the prominent characteristics of the later and more commonly known machines. The diagrams upon the screen will assist the eye to the general form of the typical machines, but I fear that want of time will prevent me from explaining each of them.

In 1762 Wilcke described a simple apparatus which produced electrical charges by influence, or induction, and following this the great Italian scientist, Alexander Volta, in 1775 gave the electrophorus the form which it retains to the present day. This apparatus may be viewed as containing the germ of the principle of all influence machines yet constructed.

Another step in the development was the invention of the doubler by Bennet in 1786. He constructed metal plates which were thickly varnished, and were supported by insulating handles, and which were manipulated so as to increase a small initial charge. It may be better for me to here explain the process of building up an increased charge by electrical influence, for the same principle holds in all of the many forms of influence machines.

This Volta electrophorus, and these three blackboards, will serve for the purpose. I first excite the electrophorus in the usual manner, and you see that it then influences a charge in its top plate; the charge in the resinous compound is known as negative, while the charge induced in its top plate is known as positive. I now show you by this electroscope that these charges are unlike in character. Both charges are, however, small, and Bennet used the following system to increase them.

Let these three boards represent Bennet's three plates. To plate No. 1 he imparted a positive charge, and with it he induced a negative charge in plate No. 2. Then with plate No. 2 he induced a positive charge in plate No. 3. He then placed the plates Nos. 1 and 3 together, by which combination he had two positive charges within practically the same space, and with these two charges he induced a double charge in plate No. 2. This process was continued until the desired degree of increase was obtained. I will not go through the process of actually building up a charge by such means, for it would take more time than I can spare.

In 1787 Carvallo discovered the very important fact, that metal plates when insulated, always acquire slight charges of electricity; following up those two important discoveries of Bennet and Carvallo, Nicholson in 1788 constructed an apparatus, having two discs of metal insulated and fixed in the same plane. Then by means of a spindle and handle, a third disc, also insulated, was made to revolve near to the two fixed discs, metallic touches being fixed in suitable positions. With this apparatus he found that small residual charges might readily be increased. It is in this simple apparatus that we have the parent of influence machines (Diagram 1), and as it is now a hundred years since Nicholson described this machine in the Phil. Trans., I think it well worth showing a large sized Nicholson machine at work to-night.

In 1823 Ronalds described a machine in which the moving disc was attached to and worked by the pendulum of a clock. It was a modification of Nicholson's doubler, and he used it to supply electricity for telegraph working. For some years after these machines were invented no important advance appears to have been made, and I think this may be attributed to the great discoveries in galvanic electricity which were made about the commencement of this century by Galvani and Volta, followed in 1831 to 1857 by the magnificent discoveries of Faraday in electro-magnetism, electro-chemistry, and electro-optics, and no real improvement was made in influence machines till 1860, in which year Varley patented a form of machine shown in Diagram 2. It also was designed for telegraph working.

In 1865 the subject was taken up with vigour in Germany by Toepler, Holtz, and other eminent men. The most prominent of the machines made by them are figured in the Diagrams 3 to 6, but time will not admit of my giving an explanation of the many points of interest in them; it being my wish to show you at work such of the machines as I may be able, and to make some observations upon them.

In 1866 Bertsch invented a machine, but not of the multiplying type; and in 1867 Sir William Thomson invented the form of machine shown in Diagram 7, which, for the purpose of maintaining a constant potential in a Leyden jar, is exceedingly useful.

The Carré machine was invented in 1868, and in 1880 the Voss machine was introduced, since which time the latter has found a place in many laboratories. It closely resembles the Varley machine in appearance, and the Toepler machine in construction.

In condensing this part of my subject, I have had to omit many prominent names and much interesting subject matter, but I must state that in placing what I have before you, many of my scientific friends have been ready to help and to contribute, and, as an instance of this, I may mention that Professor Sylvanus P. Thompson at once placed all his literature and even his private notes of reference at my service.

I will now endeavour to point out the more prominent features of the influence machines which I have present, and, in doing so, I must ask a moment's leave from the subject of my lecture to show you a small machine made by that eminent worker, Faraday, which, apart from its value as his handiwork, so closely brings us face to face with the imperfect apparatus with which he and others of his day made their valuable researches.

The next machine which I take is a Holtz. It has one plate revolving, the second plate being fixed. The fixed plate, as you see, is so much cut away, that it is very liable to breakage. Paper inductors are fixed upon the back of it, while opposite the inductors, and in front of the revolving plate, are combs. To work the machine (1) a specially dry atmosphere is required ; (2) an initial charge is necessary ; (3) when at work the amount of electricity passing through the terminals is great ; (4) the direction of the current is apt to reverse ; (5) when the terminals are opened beyond the sparking distance the excitement rapidly dies away ; (6) it does not part with free electricity from either of the terminals singly.

It has no metal on the revolving plates, nor any metal contacts ; the electricity is collected by combs which take the place of brushes, and it is the break in the connection of this circuit which supplies a current for external use. On this point I cannot do better than quote an extract from page 339 of Sir William Thomson's Papers on Electrostatics and Magnetism, which runs : "Holtz's now celebrated electric machine, which is closely analogous in principle to Varley's of 1860, is, I believe, a descendant of Nicholson's. Its great power depends upon the abolition by Holtz of metallic carriers and metallic make-and-break contacts. It differs from Varley's and mine by leaving the inductors to themselves, and using the current in the connecting arc."

In respect to the second form of Holtz machine (Fig. 4) I have very little information, for since it was brought to my notice nearly six years ago I have not been able to find either one of the machines or any person who had seen one. As will be seen by the diagram it has two discs revolving in opposite directions, it has no metal sectors and no metal contacts. The "connecting arc circuit" is used for the terminal circuit. Altogether I can very well understand and fully appreciate the statement made by Professor Holtz in '*Uppenborn's Journal*' of May 1881, wherein he writes "that for the purpose of demonstration I would rather be without such machines."

The first type of Holtz machine has now in many instances been made up in multiple form, within suitably constructed glass cases, but when so made up great difficulty has been found in keeping each of the many plates to a like excitement. When differently excited the one set of plates furnished positive electricity to the comb, while the next set of plates gave negative electricity—as a consequence no electricity passed the terminals.

To overcome this objection, to dispense with the dangerously cut plates, and also to better neutralise the revolving plate, throughout its whole diameter, I made a large machine having twelve discs 2 feet 7 inches in diameter, and in it I inserted plain rectangular slips of glass between the discs, which might readily be removed; these slips carried the paper inductors. To keep all the paper inductors on one side of the machine to a like excitement, I connected them together by a metal wire. The machine so made worked splendidly, and your late secretary, Mr. Spottiswoode, sent on two occasions to take note of my successful modifications. The machine is now ten years old, but still works splendidly. I will show you a smaller sized one at work.

The next machine on which I make observations, is the Carré. It consists essentially of a disc of glass which is free to revolve without touch or friction. At one end of a diameter it moves near to the excited plate of a frictional machine, while at the opposite end of the diameter is a strip of insulating material, opposite which, and also opposite the excited amalgam plate, are combs for conducting the induced charges, and to which the terminals are metallically connected; the machine works well in ordinary atmosphere, and certainly is in many ways to be preferred to the simple frictional machine. In my experiments with it I found that the quantity of electricity might be more than doubled by adding a segment of glass between the amalgam cushions and the revolving plate. The current in this type of machine is constant.

The Voss machine has one fixed plate and one revolving plate. Upon the fixed plate are two inductors, while on the revolving plate are six circular carriers. Two brushes receive the first portions of the induced charges from the carriers, which portions are conveyed to the inductors. The combs collect the remaining portion of the induced charge for use as an outer circuit, while the metal rod with its two brushes neutralises the plate surface in a line of its diagonal diameter. When at work it supplies a considerable amount of electricity. It is self-exciting in ordinary dry atmosphere. It freely parts with its electricity from either terminal, but when so used the current frequently changes its direction, hence there is no certainty that a full charge has been obtained, nor whether the charge is of positive or negative electricity.

I next come to the type of machine with which I am more closely associated, and I may preface my remarks by adding that the invention sprang solely from my experience gained by constantly using and experimenting with the many electrical machines which I possessed. It was from these I formed a working hypothesis which led me to make the small machine now before you. The machine is unaltered. It excited itself when new with the first revolution. It so fully satisfied me with its performance that I had four others made, the first of which I presented to this Institution. Its construction is of the simplest character. The two discs of glass revolve near to each other, and in opposite directions. Each disc carries metallic sectors; each disc has its two brushes supported by metal rods, the rods to the two plates forming an angle of 90° with each other. The external circuit is independent of the brushes, and is formed by the combs and terminals.

The machine is self-exciting under all conditions of atmosphere, owing probably to each plate being influenced by, and influencing in turn its neighbour, hence there is the minimum surface for leakage. When excited the direction of the current never changes; this circumstance is due probably to the circuit of the metallic sectors and the make-and-break contacts always being closed, while the combs and the external circuit are supplemental, and for external use only. The quantity of electricity is very large and the potential high. When suitably arranged the length of spark produced is equal to nearly the radius of the disc. I have made them from 2 inches to 7 feet in diameter, with equally satisfactory results. The Diagram No. 9 shows the distribution of the electricity upon the plate surfaces, when the machine is fully excited. The inner circle of signs corresponds with the electricity upon the front surface of the disc. The two circles of signs between the two black rings refer to the electricity between the discs, while the outer circle of signs corresponds with the electricity upon the outer surface of the back disc. The diagram is the result of experiments which I cannot very well repeat here this evening, but in support of the distribution shown on the diagram I will show you two discs at work made of a flexible material, which when driven in one direction, close together at the top and bottom, while in the horizontal diameter they are repelled. When driven in the reverse direction the opposite action takes place.

I have also experimented with the cylindrical form of the machine; the first of these I made in 1882, and it is before you. The cylinder gives inferior results to the simple discs, and is more complicated to adjust. You notice I neither use nor recommend vulcanite, and it is perhaps well to caution my hearers against the use of that material for the purpose, for it warps with age, and when left in the daylight it changes and becomes useless.

I have now only to speak of these larger machines. They are in all respects made up with the same plates, sectors, and brushes as were used by me in the first experimental machines, but for convenience sake they are fitted in numbers within a glass case.

This machine has eight plates of 2 feet 4 inches diameter; it has been in the possession of the Institution for about three years.

This large machine, which has been made for this lecture, has twelve discs, each 2 feet 6 inches in diameter. The length of spark from it is $13\frac{5}{8}$ inches.

During the construction of the machine every care was taken to avoid electrical excitement in any of its parts, and after its completion several friends were present to witness the fitting of the brushes and the first start. When all was ready the terminals were connected to an electroscope, and the handle was moved so slowly that it occupied thirty seconds in moving one-half revolution, and at that point violent excitement appeared.

The machine has now been standing with its handle secured for about eight hours; no excitement is apparent, but still it may not be absolutely inert; of this each one present must judge, but I will connect it with this electroscope, and then move the handle slowly, so that you may see when the excitement commences and judge of its absolutely reliable behaviour as an instrument for public demonstration. I may say that I have never under any condition found this type of machine to fail in its performance.

I now propose to show you the beautiful appearances of the discharge, then the length of sparks, which appear to be almost continuous, and then in order that you may judge of the relative capabilities of each of these three machines, we will work them all at the same time.

The large frictional machine which is in use for this comparison belongs to this Institution. It was made for Napoleon in 1822, and its great power is so well known to you that a better standard could not be desired.

These five Leyden jars are of equal size; I will connect one of them only to the large frictional machine, while I connect two jars to each of the two large machines of the influence type. The difference in power of the machines is then seen to be very marked. The exhibition may be considered as a miniature thunderstorm with almost no intermission between the lightning flashes.

In conclusion I may be permitted to say that it is fortunate I had not read the opinions of Sir William Thomson and Professor Holtz, as quoted in the earlier part of my lecture, previous to my own practical experiments. For had I read such opinions from such authorities I should probably have accepted them without putting them to practical test. As the matter stands I have done those things which they said I ought not to have done, and I have left undone those things which they said I ought to have done, and by so doing I think you must freely admit, that I have produced an electric generating machine of great power, and have placed in the hands of the physicist, for the purposes of public demonstration, or original research, an instrument more reliable than anything hitherto produced.

[J. W.]

Friday, June 8, 1888.

Sir Frederick Bramwell, D.C.L. F.R.S. Honorary Secretary and Vice-President, in the Chair.

Professor Dewar, M.A. F.R.S. *M.R.I.*

Phosphorescence and Ozone.

In spectroscopic observations the experimenter is often much puzzled by the phenomena presented in high vacua, and the perplexity is largely due to the fact that we are unacquainted with the chemical changes which take place under such conditions. Special apparatus has to be devised for the purpose of attempting to solve some of these questions. Friction, heat, light, and electricity, will stimulate certain bodies, and cause them to become phosphorescent, and cooling the body may prevent the continuance of the luminosity. Again, by cooling the centre of a plate which has been coated with sulphide of calcium, light will make it phosphorescent everywhere but in the place it has been cooled. Heat increases the luminosity at first, but it afterwards dies out more quickly than where the plate has not been heated.

Geissler was the first to discover that phosphorescence is sometimes set up in residual gases in vacuum tubes. This was illustrated by sending a discharge through a series of vacuum bulbs, in which the traces of gas remained luminous for about five seconds after the discharge had ceased; when one of the bulbs was heated, on passing the discharge once more, that bulb alone remained dark. Becquerel and others investigated these phenomena; some of the inquirers came to the conclusion that they were produced only by oxygen compounds; others thought them to be due to some drying agent used in the construction of the bulbs.

Ozone is a very unstable body, which cannot be kept unless produced at a low temperature; its boiling-point is about $-100°$ C., and at this temperature it is a blue liquid which exhibits high absorbent powers in the luminous part of the spectrum. At low temperatures substances may be dissolved in it, with which it explodes at high temperatures; bisulphide of carbon is one of these substances. On a former occasion I have shown that at $-150°$ C. phosphorus does not combine with liquid oxygen, neither does sodium nor potassium, so that the absence of chemical combination between ozone and oxidisable substances is another proof of the negation of chemical combination at low temperatures. Smell is one of the most delicate tests of the presence of ozone, but inapplicable in the instance of the contents of a vacuum tube; the investigator has then to resort to chemical means and the study of the absorption spectra. In making ozone from oxygen, low pressures and the presence of moisture favour the action, and that such conditions should favour chemical changes is contrary to what might have been expected.

In order to carry out the following experiments a good and powerful air-pump is required, and the Institution is fortunate in possessing a specially constructed instrument generously presented by the inventor, William Anderson, Esq. M.I.C.E. Director-General of Ordnance Factories.

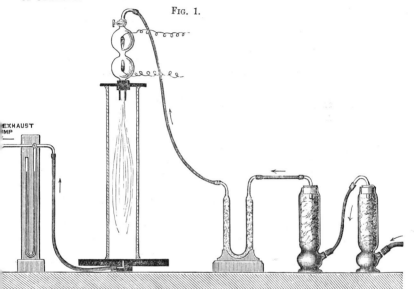

Phosphorescent Gases Apparatus.

The more essential part of the apparatus is represented in Fig. 1. Common air is first dried and purified by passing through one vessel containing calcium chloride, and another containing caustic potash; the latter absorbs the carbonic acid. The air is next filtered by passage through a U-tube filled with cotton wool, after which it enters through a carefully adjusted small tap, the two-bulbed vacuum tube represented in the cut. The narrow channel between the bulbs is necessary; the glow is concentrated thereby, and this seems to have something to do with the effects obtained. It makes no difference whether platinum, charcoal, or aluminium poles be used inside the vacuum tube. The lower part of the tube opens into a tall glass vessel, connected below with the exhaust pump and a mercurial pressure gauge.

When the current of highly attenuated air blows downwards through the vacuum tube (which is surrounded by a box to prevent any light being seen from the electrical discharge) a luminous glow, about two feet in length, resembling to the eye the tail of a comet, appears in the large glass vessel below. This phosphorescent glow is connected in some way with ozone, as it occurs only with oxygen compounds; impure air is fatal to success in the experiments, the glow being very sensitive to traces of organic matter, especially to the vapour of essential oils and substances which have a smell. Hydrogen extinguishes the light; pure oxygen increases it, and makes the glow shorter, with a tendency to break up at the lower end; carbonic acid gives a glow not so bright as air, and ozone is produced from its decomposition. Nitrous oxide gives a very bright brush. The phosphorescence disappears at once when a pocket-handkerchief containing any odoriferous matter is brought near the air inlet, and afterwards much time is lost in getting the apparatus to work as before. It is no easy matter to get the brushes back again; this was first found out accidentally—for days and weeks they had been puzzled in the laboratory to understand why one tube worked better than another. Bisulphide of carbon is an organic body, and is the only one, so far discovered, which allows the glow to be obtained in its presence. That these downward luminous brushes contain ozone is proved by means of the iodide of potassium starch test (and others), which darken in the brushes, but are not acted upon when placed outside them near the inner sides of the lower glass vessel. By suddenly altering the rate of oxygen supply passing through the vacuum tube most beautiful effects of apparent explosions of phosphorescent glows can be produced. It is remarkable that the rate of oxygen or air supply can be so regulated that the luminous emission seems to come from a steady current of gas passing down the middle of the tube, of almost uniform diameter, and blending very little with the surrounding space.

It is usually supposed that ozone is destroyed by heat, and can only be produced at a low temperature, yet in these vacuum tubes it is produced at a white heat. The piece of apparatus represented in Fig. 2 enables the chemist to demonstrate that ozone can be continuously produced by heating pure oxygen to a temperature of about $1600°$ C. The apparatus consists of a glass tube bent at its lower end, and passing up the centre of a tube of platinum; a little hole in the latter is placed just above the top orifice of the glass tube. The upper part of the apparatus is covered with an outer tube of platinum, which at the top very nearly touches the inner one. In action a regulated current of water flows up the inner platinum tube, then passes down the central glass tube, which is made longer than the platinum tube; consequently it sucks in and carries down air, which it draws through the little hole in the top of the inner platinum tube. The top of the outer tube is then raised to near the melting-point of platinum, by means of an oxyhydrogen flame, and the oxygen beneath raised to this temperature is suddenly withdrawn and cooled by the water current, and carried down to a collecting vessel for examination. When tested, it is found to contain ozone; hence ozone has two centres of stability, and one of them is at about the melting-point of platinum. In this experiment ozone is formed by the action of a high temperature, owing to the dissociation of the oxygen molecules and their partial recombination into the more complex

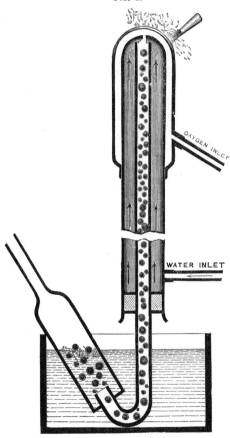

Fig. 2.

Ozone Apparatus.

molecules of ozone. We may conceive it not improbable that some of the elementary bodies might be formed somewhat like the ozone in the whole experiment, but at very high temperatures, by the collocation of certain dissociated constituents and with the simultaneous absorption of heat.

[J. D.]

Friday, February 8, 1889.

Sir Frederick Bramwell, Bart. D.C.L. F.R.S. Honorary Secretary
and Vice-President, in the Chair.

Sir William Thomson, D.C.L. LL.D. F.R.S. *M.R.I.*

Electrostatic Measurement.

A FUNDAMENTAL requisite of a measuring instrument is that its application to make a measurement shall not alter the magnitude of the thing measured. When this condition is not fulfilled (as is essentially the case with an electric measuring instrument not kept permanently in or on the electric circuit or system to which it is applied), it is the magnitude as influenced or modified by the measuring instrument which is actually measured, and the measurement is to be interpreted on this understanding, whatever may be the circumstances.

The nearest approach in electric measuring instruments to the fulfilment of this condition, of not altering the magnitude of the thing measured, is attained by the electrometer when applied to measure differences of potential between different points of a wire, or metallic mass of any shape, in which electricity is kept flowing by a battery or dynamo or other electromotive apparatus. The insulation of any practical electrometer is so nearly perfect that the conduction of electricity through the instrument does not sensibly diminish the difference of potentials of the points touched by the electrodes, and the consumption of energy is therefore practically *nil*. In this respect, therefore, the quadrant electrometer would be ideally perfect: but it is only available for potentials of a few volts, and in its most sensitive adjustment indicates about $\frac{1}{100}$ of a volt. The lecturer has therefore designed for ordinary use in connection with electric lighting and the other practical applications of electric energy, a series of instruments which will measure by electrostatic force potentials of from 40 volts to 50,000 volts. The construction of the various types of this series was fully explained.

The standardisation of these instruments up to 200 or 300 volts is made exceedingly easy, by aid of his centiampère balance and continuous rheostat, with a voltaic battery of any kind, primary or secondary, capable of giving a fairly steady current of $\frac{1}{10}$ of an ampère through it and the platinoid resistance in series with it. The accuracy of the electro-magnetic standardisation, within the range of the direct application of this method, is quite within $\frac{1}{20}$ per cent. A method of multiplication by aid of condensers, which was explained, gives an accuracy quite within $\frac{1}{5}$ per cent. for the measurement in volts up to

2000 or 3000 volts; and with not much less accuracy, by aid of an intermediate electrometer, up to 50,000 volts.

He also explained, and illustrated by a drawing, an absolute electrometer which he had constructed for the purpose of measuring "v," the number of electrostatic units of potential or electromotive force in the electro-magnetic unit of potential. This number "v" is essentially a velocity, and experiments have proved it to be so nearly equal to the velocity of light that from all the direct observations hitherto made we cannot tell whether it is a little greater than, or a little less than, or absolutely equal to, the velocity of light.

The determination was made by comparing the electro-magnetic with the electrostatic value, in C. G. S. units, as given by the balance, for a potential of 10,000 volts: but hitherto he has not been able to make sure of the absolute accuracy of the electrostatic balance to closer than $\frac{1}{2}$ per cent.

The results of a great number of measurements which had been made in the Physical Laboratory of the University of Glasgow during the previous two months gave the required number, "v," within $\frac{1}{2}$ per cent. of 300,000 kilometres per second; the velocity of light is known to be within $\frac{1}{4}$ per cent. of 300,000 kilometres per second. Results of previous observers for determining "v" had almost absolutely proved it at least as close an agreement with the 300,000,000 metres. He expressed his obligations to his assistants and students in the Physical Laboratory of Glasgow University, Messrs. Meikle, Shields, Sutherland, and Carver, who worked with the greatest perseverance and accuracy, in the laborious and often irksome observations by which he had attempted to determine "v" by the direct electrometer method, as exactly as, or more exactly than, it has been determined by other observers and other methods.

Note added March 14th, 1889.

The measurement of "v" by Sir William Thomson and Profs. Ayrton and Perry, communicated to the British Association at Bath, was too small (292) on account of the accidental omission of a correction regarding the effective area of the attracted disk in the absolute electrometer. When this correction is applied their result is brought up to 298, which exactly agrees with Profs. Ayrton and Perry's previous determination by another method, in Japan. Prof. J. J. Thomson's result is 296·3. It is understood that Rowland has found 299. The result of Sir William Thomson's latest observations, founded wholly on the comparison of electrometric and electro-magnetic determinations of potential in absolute measure, is 30·1 legal ohms, or 30·04 Rayleigh ohms. Assuming, as is now highly probable, that the Rayleigh ohm is considerably nearer than the legal ohm to the true ohm, the result for "v" is 300,400,000 metres per second. Sir William does not consider that this result can be trusted as demonstrating the truth within $\frac{1}{3}$ per cent.

Friday, February 15, 1889.

WILLIAM CROOKES, Esq. F.R.S. Vice-President, in the Chair.

PROFESSOR A. W. RÜCKER, M.A. F.R.S. *M.R.I.*

Electrical Stress.

THE subject of the discourse was brought before the members of the Royal Institution some years ago by Mr. Gordon. In the interval a considerable amount of work has been done upon it, both in England and Germany, and many experiments have been devised to illustrate it. Some of the more striking of these, though of great interest to the student, are rarely or never shown in courses of experimental lectures. The lecturer and Mr. C. V. Boys, F.R.S., last year devised a set of apparatus which has made the optical demonstration of electrical stress comparatively easy, and most of the results obtained by Kerr and Quincke can now be demonstrated to audiences of a considerable size. Before discussing this portion of his subject the lecturer introduced it by an explanation of principles on which the experiments are founded.

Magnetic lines of force can easily be mapped out by iron filings, but the exhibition of electrical lines of force in a liquid is a more complex matter. In the first place, if two oppositely electrified bodies are introduced into a liquid which is a fairly good non-conductor, convective conduction is set up. Streams of electrified liquid pass from the one to the other. The highly refracting liquid phenyl thiocarbamide appears to be specially suitable for experiments on this subject. If an electrified point is brought over the surface a dimple is formed which becomes deeper as the point approaches it. At the instant at which the needle touches the liquid the dimple disappears, but a bubble of air from the lower end frequently remains imprisoned in the vortex caused by the downward rush of the electrified liquid from the point. It oscillates a short distance below the point, and indicates clearly the rapid motions which are produced in the fluid in its neighbourhood. When the needle is withdrawn a small column of liquid adheres to it. This effect is, however, seen to greater advantage if a sphere about 5 mm. in diameter is used instead of the needle-point. When this is withdrawn a column of liquid about 5 mm. high and 2 mm. in diameter is formed between the sphere and the surface. A similar experiment was made by Faraday on a much larger scale with oil of turpentine, and he detected the existence of currents which are in accord with the view that the unelectrified liquid flows up the exterior of the cylinder, becomes electrified by contact, and is repelled down its axis. In view of this explanation, and the movements assumed can be clearly seen in the phenyl thiocarbamide, the performance of the experiment on a small scale is not without interest. The possibility of the formation of such violent up-and-down currents in so small a space must depend upon a very nice adjustment between the properties of the liquid and the forces in play. It is also obvious that such movements of the liquid must be a disturbing element in any attempt to make the lines of electric force visible.

Again, if a solid powder be suspended in a liquid into which electrified solids are introduced, it tends to accumulate round one of the poles. This subject has been investigated by W. Holtz. Sometimes the powder appears to move in a direction opposed to that in which the liquid is streaming. Sometimes two powders will travel towards different poles.

If powdered antimony sulphide be placed in ether, it settles at the bottom of the liquid, and if either two wires insulated with glass up to their points, or two vertical plates be used as electrodes, and slightly electrified, the solid particles arrange themselves along the lines of force. If the electrification be increased, they cluster round the positive pole. On suddenly reversing the electrification by means of a commutator, they stream along lines of force to the pole from which they were previously repelled. Other methods of obtaining the lines of force have been devised. They can, for instance, be shown by crystals of sulphate of quinine immersed in turpentine.

The tendency of the lines of force to separate one from the other was illustrated by Quincke's experiment. A bubble of air is formed in bisulphide of carbon between two horizontal plates. It is in connection with a small manometer, and when the plates are oppositely excited, the electrical pressure acting at right angles to the lines of force, being greater in the liquid than in air, compels the bubble to contract and depresses the manometer.

Kerr's experiments depend upon the fact that, since the electrical stress is a tension along the lines of force, and a pressure at right angles to them, a substance in which such a stress is produced assumes a semicrystalline condition in the sense that its properties along, and perpendicular to, the lines of force are different. Light is therefore transmitted with different velocities according as the direction of vibration coincides with, or is perpendicular to, these lines; and the familiar phenomena of the passage of polarised light through crystals may be imitated by an electrically stressed liquid.

The bisulphide of carbon used must be dry, and, to make the phenomena clearly visible, it is necessary that the light should travel through a considerable thickness. Thus, to represent the stress between two spheres, elongated parallel cylinders should be used, the axes of which are parallel to the course of the rays of light. These appear on the screen as two dark circles. Between crossed Nicols, the planes of polarisation of which are inclined at 45° to the horizontal, the field is dark until the cylinders are electrified, when light is restored in the space between them.

If parallel plates with carefully rounded edges, and about 2 millimetres apart, are used, the colours of Newton's rings appear in turn, the red of the third order being sometimes reached. If one plate is convex towards the other, the colours of the higher orders appear in the middle, and travel outwards as the stress is increased. The experiments may be varied by using two concentric cylinders, or two sheets of metal bent twice at right angles to represent a section through a Leyden jar. In the first case a black cross is formed; and in the second, black brushes unite the lower angles of the images of the edges of the plates. By the interposition of a piece of selenite, which shows the blue of the second order, two of the quadrants contained between the arms of the cross become green, and the others red. In like manner the horizontal and vertical spaces between the inner and outer coatings of the "jar" become differently coloured.

There are several phenomena connected with the stress in insulators which present considerable difficulties. Thus it is found impossible to restore the light between crossed Nicols by subjecting a solid placed between them to electrical stress in a uniform field. That the non-uniformity of the field has nothing to do with the phenomenon in liquids, though at first disputed, is now generally admitted. It may be readily proved by means of a Franklin's pane, of which half is pierced with windows. The glow is much weakened by thus replacing a uniform by a non-uniform field.

Again, though most dielectrics when placed in an electric field expand, the fatty oils contract. Prof. J. J. Thomson has recently pointed out that this indicates that another set of strains are superposed upon those assumed in the ordinary explanations of these phenomena, and by which they may be neutralised or overcome.

In experiments with carbon bisulphide it is necessary to take every precaution against fire. For this purpose the cell which contains the liquid should be immersed in a larger cell, so that if—as sometimes happens—the passage of a spark cracks the glass the liquid may flow into a confined space. This should stand in a tray with turned-up edges, and an extinguisher of tin plate should be at hand to place over the whole apparatus. No Leyden jars should be included in the electrical circuit. The difficulties which formerly arose in the exhibition of experiments in statical electricity owing to the presence of moisture in the air of a lecture-room are now immensely reduced by the Wimshurst machine, which works with unfailing certainty under adverse conditions. A new and very beautiful machine was kindly lent by Mr. Wimshurst for the purposes of the lecture.

[A. W. R.]

Friday, March 8, 1889,

Sir James Crichton Browne, M.D. LL.D. F.R.S.
Vice-President, in the Chair.

Professor Oliver Lodge, LL.D. F.R.S.

The Discharge of a Leyden Jar.

It is one of the great generalisations established by Faraday, that all electrical charge and discharge is essentially the charge and discharge of a Leyden jar. It is impossible to charge one body alone. Whenever a body is charged positively, some other body is *ipso facto* charged negatively, and the two equal opposite charges are connected by lines of induction. The charges are, in fact, simply the ends of these lines, and it is as impossible to have one charge without its correlative as it is to have one end of a piece of string without there being somewhere, hidden it may be, split up into strands it may be, but somewhere existent, the other end of that string.

This I suppose familiar fact that all charge is virtually that of a Leyden jar being premised, our subject for this evening is at once seen to be a very wide one, ranging in fact over the whole domain of electricity. For the charge of a Leyden jar includes virtually the domain of electrostatics; while the discharge of a jar, since it constitutes a current, covers the ground of current electricity all except that portion which deals with phenomena peculiar to steady currents. And since a current of electricity necessarily magnetises the space around it, whether it flow in a straight or in a curved path, whether it flow through wire or burst through air, the territory of magnetism is likewise invaded; and inasmuch as a Leyden jar discharge is oscillatory, and we now know the vibratory motion called light to be really an oscillating electric current, the domain of optics is seriously encroached upon.

But though the subject I have chosen would permit this wide range, and though it is highly desirable to keep before our minds the wide-reaching import of the most simple-seeming fact in connection with such a subject, yet to-night I do not intend to avail myself of any such latitude, but to keep as closely and distinctly as possible to the Leyden jar in its homely and well-known form, as constructed out of a glass bottle, two sheets of tinfoil, and some stickphast.

The act of charging such a jar I have permitted myself now for some time to illustrate by the mechanical analogy of an inextensible endless cord able to circulate over pulleys, and threading in some portion of its length a row of tightly-gripping beads which are connected to fixed beams by elastic threads.

The cord is to represent electricity; the beads represent successive strata in the thickness of the glass of the jar, or, if you like, atoms of dielectric or insulating matter. Extra tension in the cord represents negative potential, while a less tension (the nearest analogue to pressure adapted to the circumstances) represents positive

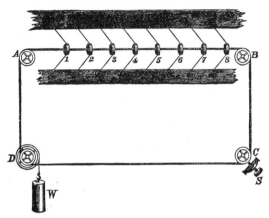

Mechanical analogy of a circuit partly *dielectric*; for instance, of a charged condenser. A is its positive coat, B its negative.

potential. Forces applied to move the cord, such as winches or weights, are electromotive forces; a clamp or fixed obstruction represents a rheostat or contact-breaker; and an excess or defect of cord between two strata of matter represents a positive or a negative charge.

The act of charging a jar is now quite easily depicted as shown in the diagram.

To discharge the jar one must remove the charging E.M.F. and unclamp the screw, i.e. close the circuit. The stress in the elastic threads will then rapidly drive the cord back, the inertia of the beads will cause it to overshoot the mark, and for an instant the jar will possess an inverse charge. Back again the cord swings, however, and a charge of same sign as at first, but of rather less magnitude, would be found in the jar, if the operation were now suspended. If it be allowed to go on, the oscillations gradually subside, and in a short time everything is quiescent, and the jar is completely discharged.

All this occurs in the Leyden jar, and the whole series of oscillations, accompanied by periodic reversal and re-reversal of the charges of the jar, is all accomplished in the incredibly short space of time occupied by a spark.

Consider now what the rate of oscillation depends on. Manifestly on the elasticity of the threads and on the inertia of the matter which is moved. Take the simplest mechanical analogy, that of the vibration of a loaded spring, like the reeds in a musical box. The stiffer the spring, and the less the load, the faster it vibrates. Give a mathematician these data, and he will calculate for you the time the spring takes to execute one complete vibration, the "period" of its swing. [Loaded lath in vice.]

The electrical problem and the electrical solution are precisely the same. That which corresponds to the flexibility of the spring, is in electrical language called static capacity, or, by Mr. Heaviside, permittance. That which corresponds to the inertia of ordinary matter is called electro-magnetic inertia, or self-induction, or, by Mr. Heaviside, inductance.

Increase either of these, and the rate of oscillation is diminished. Increasing the static capacity corresponds to lengthening the spring; increasing the self-induction corresponds to loading it.

Now the static capacity is increased simply by using a larger jar, or by combining a number of jars into a battery in the very old established way. Increase in the self-induction is attained by giving the discharge more space to magnetise, or by making it magnetise a given space more strongly. For electro-magnetic inertia is wholly due to the magnetisation of the space surrounding a current, and this space may be increased or its magnetisation intensified as much as we please.

To increase the space we have only to make the discharge take a long circuit instead of a short one. Thus we may send it by a wire all round the room, or by a telegraph wire all round a town, and all the space inside it and some of that outside will be more or less magnetised. More or less, I say, as it is a case of less rather than more. Practically very little effect is felt except close to the conductor, and accordingly the self-induction increases very nearly proportionally to the length of the wire, and not in proportion to the area inclosed: provided also the going and return wires are kept a reasonable distance apart, so as not to encroach upon each other's appreciably magnetised regions.

But it is just as effective, and more compact, to intensify the magnetisation of a given space by sending the current hundreds of times round it instead of only once; and this is done by inserting a coil of wire into the discharge circuit.

Yet a third way there is of increasing the magnetisation of a given space, and that is to fill it with some very magnetisable substance such as iron. This, indeed, is a most powerful method under many circumstances, it being possible to increase the magnetisation and therefore the self-induction or inertia of the current some 5000 times by the use of iron.

But in the case of the discharge of a Leyden jar, iron is of no advantage. The current oscillates so quickly that any iron introduced into its circuit, however subdivided into thin wires it may be, is protected from magnetism by inverse currents induced in its outer skin, as your Professor of Natural Philosophy* has shown, and accordingly it does not get magnetised; and so far from increasing the inductance of the discharge circuit it positively diminishes it by the reaction effect of these induced currents: it acts, in fact, much as a mass of copper might be expected to do.

The conditions determining rate of oscillation being understood we have next to consider what regulates the damping out of the vibrations, i. e. the total duration of the discharge.

Resistance is one thing. To check the oscillations of a vibrating spring you apply to it friction, or make it move in a viscous medium, and its vibrations are speedily damped out. The friction may be made so great that oscillations are entirely prevented, the motion being a mere dead-beat return to the position of equilibrium; or, again, it may be greater still, and the motion may correspond to a mere leak or slow sliding back, taking hours or days for its accomplishment. With very large condensers, such as are used in telegraphy, this kind of discharge is frequent, but in the case of a Leyden jar discharge it is entirely exceptional. It can be caused by including in the circuit a wet string, or a capillary tube full of distilled water, or a slab of wood, or other atrociously bad conductor of that sort; but the conditions ordinarily associated with the discharge of a Leyden jar, whether it discharge through a long or a short wire, or simply through its tongs, or whether it overflow its edge or puncture its glass, are such as correspond to oscillations, and not to leak. [Discharge jar first through wire and next through wood.]

When the jar is made to leak through wood or water the discharge is found to be still not steady: it is not oscillatory indeed, but it is intermittent. It occurs in a series of little jerks, as when a thing is made to slide over a resined surface. The reason of this is that the terminals discharge faster than the circuit can supply the electricity, and so the flow is continually stopped and begun again.

Such a discharge as this, consisting really of a succession of small sparks, may readily appeal to the eye as a single flash, but it lacks the noise and violence of the ordinary discharge; and any kind of moving mirror will easily analyse it into its constituents and show it to be intermittent. [Shake a mirror, or waggle head, or opera-glass.]

It is pretty safe to say, then, that whenever a jar discharge is not oscillatory it is intermittent, and when not intermittent is oscillatory. There is an intermediate case when it is really deadbeat, but it could only be hit upon with special care, while its occurrence by accident must be rare.

So far I have only mentioned resistance or friction as the cause of the dying out of the vibrations; but there is another cause, and that a most exciting one.

The vibrations of a reed are damped partly indeed by friction and

---

* Lord Rayleigh.

imperfect elasticity, but partly also by the energy transferred to the surrounding medium and consumed in the production of sound. It is the formation and propagation of sound waves which largely damp out the vibrations of any musical instrument. So it is also in electricity. The oscillatory discharge of a Leyden jar disturbs the medium surrounding it, carves it into waves which travel away from it into space: travel with a velocity of 185,000 miles a second: travel precisely with the velocity of light. [Tuning-fork.]

The second cause, then, which damps out the oscillations in a discharge circuit is *radiation: electrical* radiation if you like so to distinguish it, but it differs in no respect from ordinary radiation (or "radiant heat" as it has so often been called in this place); it differs in no respect from Light except in the physiological fact that the retinal mechanism, whatever it may be, responds only to waves of a particular, and that a very small, size, while radiation in general may have waves which range from 10,000 miles to a millionth of an inch in length.

The seeds of this great discovery of the nature of light were sown in this place: it is all the outcome of Faraday's magneto-electric and electrostatic induction: the development of them into a rich and full-blown theory was the greatest part of the life-work of Clerk-Maxwell: the harvest of experimental verification is now being reaped by a German. But by no ordinary German. Dr. Hertz, now Professor in the Polytechnicum of Karlsruhe, is a young investigator of the highest type. Trained in the school of Helmholtz, and endowed with both mathematical knowledge and great experimental skill, he has immortalised himself by a brilliant series of investigations which have cut right into the ripe corn of scientific opinion in these islands, and by the same strokes as have harvested the grain have opened up wide and many branching avenues to other investigators.

At one time I had thought of addressing you this evening on the subject of these researches of Hertz, but the experiments are not yet reproducible on a scale suited to a large audience, and I have been so closely occupied with some not wholly dissimilar, but independently conducted, researches of my own—researches led up to through the unlikely avenue of lightning-conductors—that I have had as yet no time to do more than verify some of them for my own edification.

In this work of repetition and verification Prof. Fitzgerald has, as related in a recent number of NATURE (February 21, p. 391), probably gone further; and if I may venture a suggestion to your Honorary Secretary, I feel sure that a discourse on Hertz's researches from Prof. Fitzgerald next year would be not only acceptable to you, but would be highly conducive to the progress of science.

I have wandered a little from my Leyden jar, and I must return to it and its oscillations. Let me very briefly run over the history of our knowledge of the oscillatory character of a Leyden jar discharge. It was first clearly realised and distinctly stated by that excellent experimentalist, Joseph Henry, of Washington, a man not wholly unlike Faraday in his mode of work, though doubtless possessing to a less degree that astonishing insight into intricate and obscure phenomena; wanting also in Faraday's circumstantial advantages.

This great man arrived at a conviction that the Leyden jar discharge was oscillatory by studying the singular phenomena attending the magnetisation of steel needles by a Leyden jar discharge, first observed in 1824 by Savary. Fine needles, when taken out of the magnetising helices, were found to be not always magnetised in the right direction, and the subject is referred to in German books as anomalous magnetisation. It is not the magnetisation which is anomalous, but the currents which have no simple direction; and we find in a memoir published by Henry in 1842, the following words :—

"This anomaly, which has remained so long unexplained, and which, at first sight, appears at variance with all our theoretical ideas of the connection of electricity and magnetism, was, after considerable study, satisfactorily referred by the author to an action of the discharge of the Leyden jar, which had never before been recognised. The discharge, whatever may be its nature, is not correctly represented (employing for simplicity the theory of Franklin) by the single transfer of an imponderable fluid from one side of the jar to the other; the phenomenon requires us to admit *the existence of a principal discharge in one direction and then several reflex actions backward and forward each more feeble than the preceding, until the equilibrium is obtained.* All the facts are shown to be in accordance with this hypothesis, and a ready explanation is afforded by it of a number of phenomena, which are to be found in the older works on electricity, but which have until this time remained unexplained."*

The italics are Henry's. Now if this were an isolated passage it might be nothing more than a lucky guess. But it is not. The conclusion is one at which he arrives after a laborious repetition and serious study of the facts, and he keeps the idea constantly before him when once grasped, and uses it in all the rest of his researches on the subject. The facts studied by Henry do in my opinion support his conclusion, and if I am right in this it follows that he is the original discoverer of the oscillatory character of a spark, although he does not attempt to state his theory. That was first done, and completely done, in 1853, by Sir William Thomson; and the progress of experiment by Feddersen, Helmholtz, Schiller, and others has done nothing but substantiate it.

The writings of Henry have been only quite recently collected and published by the Smithsonian Institution of Washington in accessible form, and accordingly they have been far too much ignored.

---

* 'Scientific Writings of Joseph Henry,' vol. i. p. 201. Published by the Smithsonian Institution, Washington, 1886.

The two volumes contain a wealth of beautiful experiments clearly recorded, and well repay perusal.

The discovery of the oscillatory character of a Leyden jar discharge may seem a small matter but it is not. One has only to recall the fact that the oscillators of Hertz are essentially Leyden jars—one has only to use the phrase " electro-magnetic theory of light"—to have some of the momentous issues of this discovery flash before one.

One more extract I must make from that same memoir by Henry,* and it is a most interesting one; it shows how near he was, or might have been, to obtaining some of the results of Hertz; though if he had obtained them, neither he nor any other experimentalist could possibly have divined their real significance.

It is, after all, the genius of Maxwell and of a few other great theoretical physicists whose names are on everyone's lips† which endows the simple induction experiments of Hertz and others with such stupendous importance.

Here is the quotation :—

" In extending the researches relative to this part of the investigations, a remarkable result was obtained in regard to the distance at which induction effects are produced by a very small quantity of electricity; a single spark from the prime conductor of a machine, of about an inch long, thrown on to the end of a circuit of wire in an upper room, produced an induction sufficiently powerful to magnetise needles in a parallel circuit of iron placed in the cellar beneath, at a perpendicular distance of 30 feet, with two floors and ceilings, each 14 inches thick, intervening. The author is disposed to adopt the hypothesis of an electrical *plenum* [in other words, of an ether], and from the foregoing experiment it would appear that a single spark is sufficient to disturb perceptibly the electricity of space throughout at least a cube of 400,000 feet of capacity; and when it is considered that the magnetism of the needle is the result of the difference of two actions, it may be further inferred that the diffusion of motion in this case is almost comparable with that of a spark from a flint and steel in the case of light."

Comparable it is, indeed, for we now know it to be the self-same process.

One immediate consequence and easy proof of the oscillatory character of a Leyden jar discharge is the occurrence of phenomena of sympathetic resonance.

Everyone knows that one tuning-fork can excite another at a

---

* Loc. cit., p. 204.

† And of one whose name is not yet on everybody's lips, but whose profound researches into electro-magnetic waves have penetrated further than anybody yet understands into the depths of the subject, and whose papers have very likely contributed largely to the theoretical inspiration of Hertz—I mean that powerful mathematical physicist, Mr. Oliver Heaviside.

reasonable distance if both are tuned to the same note. Everyone knows, also, that a fork can throw a stretched string attached to it into sympathetic vibration if the two are tuned to unison or to some simple harmonic. Both these facts have their electrical analogue. I have not time to go fully into the matter to-night, but I may just mention the two cases which I have myself specially noticed.

A Leyden jar discharge can so excite a similarly-timed neighbouring Leyden jar circuit as to cause the latter to burst its dielectric if thin and weak enough. The well-timed impulses accumulate in the neighbouring circuit till they break through a quite perceptible thickness of air.

Put the circuits out of unison by varying the capacity or by including a longer wire in one of them; then, although the added wire be a coil of several turns, well adapted to assist mutual induction as ordinarily understood, the effect will no longer occur, until the capacity is suitably diminished and the synchronism thus restored.

That is one case, and it is the electrical analogue of one tuning-fork exciting another. It is too small at present to show here satisfactorily, for I only recently observed it, but it is exhibited in the library at the back.

The other case, analogous to the excitation of a stretched string of proper length by a tuning-fork, I published last year under the name of the experiment of the recoil kick, where a Leyden jar circuit sends waves along a wire connected by one end with it, which waves splash off at the far end with an electric brush or long spark.

I will show merely one phase of it to-night, and that is the reaction of the impulse accumulated in the wire upon the jar itself, causing it to either overflow or burst. (Sparks of gallon or pint jar made to overflow by wire round room.*)

The early observations by Franklin on the bursting of Leyden jars, and the extraordinary complexity or multiplicity of the fracture that often results, are most interesting. His electric experiments as

---

* During the course of this experiment, the gilt paper on the wall was observed by the audience to be sparkling, every gilt patch over a certain area discharging into the next, after the manner of a spangled jar. It was probably due to some kind of sympathetic resonance. Electricity splashes about in conductors in a surprising way everywhere in the neighbourhood of a discharge. For instance, a telescope in the hand of one of the audience was reported afterwards to be giving off little sparks at every discharge of the jar. Everything which happens to have a period of electric oscillation corresponding to some harmonic of the main oscillation of a discharge is liable to behave in this way. When light falls on an opaque surface it turns into some other form of energy. What the audience saw was probably the result of waves of electrical radiation being quenched or reflected by the walls of the room, and generating electric currents in the act. It is these electric surgings which render such severe caution necessary in the erection of lightning-conductors.

This explanation is merely tentative. I have had no time to investigate the matter locally.

well as Henry's well repay perusal, though, of course, they belong to the infancy of the subject.

He notes the striking fact that the bursting of a jar is an extra occurrence, it does not replace the ordinary discharge in the proper place, it accompanies it; and we now know that it is precipitated by it, that the spark occurring properly between the knobs sets up such violent surgings that the jar is far more violently strained than by the static charge or mere difference of potentials between its coatings; and if the surgings are at all even roughly properly timed, the jar is bound to either overflow or burst.

Hence a jar should always be made without a lid, and with a lip protruding a carefully considered distance above its coatings: not so far as to fail to act as a safety valve, but far enough to prevent overflow under ordinary and easy circumstances.

And now we come to what is after all the main subject of my discourse this evening, viz. the optical and audible demonstration of the oscillations occurring in the Leyden jar spark. Such a demonstration has, so far as I know, never before been attempted, but if nothing goes wrong we shall easily accomplish it.

And first I will do it audibly. To this end the oscillations must be brought down from their extraordinary frequency of a million or hundred thousand a second to a rate within the limits of human audition. One does it exactly as in the case of the spring—one first increases the flexibility and then one loads it. [Spark from battery of jars and varying sound of same.]

Using the largest battery of jars at our disposal, I take the spark between these two knobs—not a long spark, $\frac{1}{4}$ inch will be quite sufficient. Notwithstanding the great capacity, the rate of vibration is still far above the limit of audibility, and nothing but the customary crack is heard. I next add inertia to the circuit by including a great coil of wire, and at once the spark changes character, becoming very shrill but an unmistakable whistle, of a quality approximating to the cry of a bat. Add another coil, and down comes the pace once more, to something like 5000 per second, or about the highest note of a piano. Again and again I load the circuit with magnetisability, and at last the spark has only 500 vibrations a second, giving the octave, or perhaps the double octave, above the middle C.

One sees clearly why one gets a musical note: the noise of the spark is due to a sudden heating of the air; now if the heat is oscillatory, the sound will be oscillatory too, but both will be an octave above the electric oscillation, if I may so express it, because two heat-pulses will accompany every complete electric vibration, the heat production being independent of direction of current.

Having thus got the frequency of oscillation down to so manageable a value, the optical analysis of it presents no difficulty: a simple looking-glass waggled in the hand will suffice to spread out the spark into a serrated band, just as can be done with a singing or a sensitive flame, a band too of very much the same appearance.

Using an ordinary four-square rotating mirror driven electromagnetically at the rate of some two or three revolutions per second, the band is at the lowest pitch seen to be quite coarsely serrated; and fine serrations can be seen with four revolutions per second in even the shrill whistling sparks.

The only difficulty in seeing these effects is to catch them at the right moment. They are only visible for a minute fraction of a revolution, though the band may appear drawn out to some length. The further away the spark is from the mirror, the more drawn out it is, but also the less chance there is of catching its image.

With a single observer it is easy to arrange a contact maker on the axle of the mirror which shall bring on the discharge at the right place in the revolution, and the observer may then conveniently watch for the image in a telescope or opera-glass, though at the lower pitches nothing of the kind is necessary.

But to show it to a large audience various plans can be adopted. One is to arrange for several sparks instead of one; another is to multiply images of a single spark by suitably adjusted reflectors, which if they are concave will give magnified images; another is to use several rotating mirrors; and indeed I do use two, one adjusted so as to suit the spectators in the gallery.

But the best plan that has struck me is to combine an intermittent and an oscillatory discharge. Have the circuit in two branches, one of high resistance so as to give intermittences, the other of ordinary resistance so as to be oscillatory, and let the mirror analyse every constituent of the intermittent discharge into a serrated band. There will thus be not one spark, but several successive sparks, close enough together to sound almost like one, separate enough in the rotating mirror to be visible on all sides at once, and each one analysed into its component alternations.

But to achieve this one must have great exciting power. In spite of the power of this magnificent Wimshurst machine, it takes some time to charge up our great Leyden battery, and it is tedious waiting for each spark. A Wimshurst does admirably for a single observer, but for a multitude one wants an instrument which shall charge the battery not once only but many times over, with overflows between, and all in the twinkling of an eye.

To get this I must abandon my friend Mr. Wimshurst, and return to Michael Faraday. In front of the table is a great induction coil; its secondary has the resistance needed to give an intermittent discharge. The quantity it supplies at a single spark will fill our jars to overflowing several times over. The discharge circuit and all its circumstances shall remain unchanged. [Excite jars by coil.]

Running over the gamut with this coil now used as our exciter instead of the Wimshurst machine—everything else remaining exactly as it was—you hear the sparks give the same notes as before, but with a slight rattle in addition, indicating intermittence as well as alternation. Rotate the mirror, and everyone should see one or other of the serrated bands of light at nearly every break of the primary current of the coil. [Rotating mirror to analyse sparks.]

The musical sparks which I have now shown you were obtained by me during a special digression * which I made while examining the effect of discharging a Leyden jar round heavy glass or bisulphide of carbon. The rotation of the plane of polarisation of light by a steady current, or by a magnetic field of any kind properly disposed with respect to the rays of light, is a very familiar one in this place. Perhaps it is known also that it can be done by a Leyden jar current. But I do not think it is; and the fact seems to me very interesting. It is not exactly new—in fact, as things go now it may be almost called old, for it was investigated six or seven years ago by two most highly skilled French experimenters, Messrs. Bichat and Blondlot.

But it is exceedingly interesting as showing how short a time, how absolutely no time, is needed by heavy glass to throw itself into the suitable rotatory condition. Some observers have thought they had proved that heavy glass requires time to develop the effect, by spinning it between the poles of a magnet and seeing the effect decrease; but their conclusions cannot be right, for the polarised light follows every oscillation in a discharge, the plane of polarisation being waved to and fro as often as 70,000 times a second in my own observation.

Very few persons in the world have seen the effect. In fact, I doubt if anyone had seen it a month ago except Messrs. Bichat and Blondlot. But I hope to make it visible to most persons here, though I hardly hope to make it visible to all.

Returning to the Wimshurst machine as exciter, I pass a discharge round the spiral of wire inclosing this long tube of CS_2, and the analysing Nicol being turned to darkness, there may be seen a faint—by those close to not so faint, but a very momentary—restoration of light on the screen at every spark. (CS_2 tube experiment on screen.)

Now I say that this light restoration is also oscillatory. One way of proving this fact is to insert a biquartz between the Nicols. With a steady current it constitutes a sensitive detector of rotation, its sensitive tint turning green on one side and red on the other. But with this oscillatory current a biquartz does absolutely nothing. (Biquartz.)

That is one proof. Another is that rotating the analyser either way weakens the extra brightening of the field, and weakens it equally either way.

But the most convincing proof is to reflect the light coming

---

* Most likely it was a conversation which I had with Sir Wm. Thomson, at Christmas, which caused me to see the interest of getting slow oscillations. **My attention has mainly been directed to getting them quick.**

through the tube upon our rotating mirror, and to look now, not at the spark, or not only at the spark, but at the faint band into which the last residue of light coming through polariser and tube and analyser is drawn out. (Analyse the light in rotating mirror.)

At every discharge this faint streak brightens in places into a beaded band: these are the oscillations of the polarised light: and when examined side by side they are as absolutely synchronous with the oscillations of the spark itself as can be perceived.

Rotating the analysing Nicol a little, one sees every alternate bead grow fainter, while the other alternate ones brighten; thus directly establishing the fact of alternations, as distinct from intermittences. A certain definite rotation will obliterate one set altogether, and make the beading appear twice as coarse, as if it belonged to the octave below. [For further details see 'Philosophical Magazine' for April, 1889.]

Out of a multitude of phenomena connected with the Leyden jar discharge I have selected a few only to present to you here this evening. Many more might have been shown, and great numbers more are not at present adapted for presentation to an audience, being only visible with difficulty and close to.

An old and trite subject is seen to have in the light of theory an unexpected charm and brilliancy. So it is with a great number of other old familiar facts at the present time.

The present is an epoch of astounding activity in physical science. Progress is a thing of months and weeks, almost of days. The long line of isolated ripples of past discovery seem blending into a mighty wave, on the crest of which one begins to discern some oncoming magnificent generalisation. The suspense is becoming feverish, at times almost painful. One feels like a boy who has been long strumming on the silent key-board of a deserted organ, into the chest of which an unseen power begins to blow a vivifying breath. Astonished, he now finds that the touch of a finger elicits a responsive note, and he hesitates, half delighted, half affrighted, least he be deafened by the chords which it would seem he can now summon forth almost at will.

[O. L.]

Friday, April 12, 1889.

SIR FREDERICK BRAMWELL, Bart. D.C.L. F.R.S. Honorary Secretary and Vice-President, in the Chair.

The Right Hon. LORD RAYLEIGH, M.A. D.C.L. LL.D. F.R.S. *M.R.I.*
PROFESSOR OF NATURAL PHILOSOPHY, R.I.

Iridescent Crystals.

THE principal subject of the lecture is the peculiar coloured reflection observed in certain specimens of chlorate of potash. Reflection implies a high degree of discontinuity. In some cases, as in decomposed glass, and probably in opals, the discontinuity is due to the interposition of layers of air; but, as was proved by Stokes, in the case of chlorate crystals the discontinuity is that known as twinning. The seat of the colour is a very thin layer in the interior of the crystal and parallel to its faces.

The following laws were discovered by Stokes:—

(1) If one of the crystalline plates be turned round in its own plane, without alteration of the angle of incidence, the peculiar reflection vanishes twice in a revolution, viz. when the plane of incidence coincides with the plane of symmetry of the crystal. [Shown.]

(2) As the angle of incidence is increased the reflected light becomes brighter and rises in refrangibility. [Shown.]

(3) The colours are not due to absorption, the transmitted light being strictly complementary to the reflected.

(4) The coloured light is not polarised. It is produced indifferently, whether the incident light be common light or light polarised in any plane, and is seen whether the reflected light be viewed directly or through a Nicol's prism turned in any way. [Shown.]

(5) The spectrum of the reflected light is frequently found to consist almost entirely of a comparatively narrow band. When the angle of incidence is increased, the band moves in the direction of increasing refrangibility, and at the same time increases rapidly in width. In many cases the reflection appears to be almost total.

In order to project these phenomena a crystal is prepared by cementing a smooth face to a strip of glass, whose sides are not quite parallel. The white reflection from the anterior face of the glass can then be separated from the real subject of the experiment.

A very remarkable feature in the reflected light remains to be noticed. If the angle of incidence be small, and if the incident light be polarised in or perpendicularly to the plane of incidence, the reflected light is polarised in the *opposite* manner. [Shown.]

Similar phenomena, except that the reflection is white, are exhibited by crystals prepared in a manner described by Madan. If the crystal be treated beyond a certain point the peculiar reflection disappears, but returns upon cooling. [Shown.]

In all these cases there can be little doubt that the reflection takes place at twin surfaces, the theory of such reflection* reproducing with remarkable exactness most of the features above described. In order to explain the vigour and purity of the colour reflected in certain crystals, it is necessary to suppose that there are a considerable number of twin surfaces disposed at approximate equal intervals. At each angle of incidence there would be a particular wave length for which the phases of the several reflections are in agreement. The selection of light of a particular wave length would thus take place upon the same principle as in diffraction spectra, and might reach a high degree of perfection.

In illustration of this explanation an acoustical analogue is exhibited. The successive twin planes are imitated by parallel and equidistant discs of muslin (Figs. 1 and 2) stretched upon brass rings

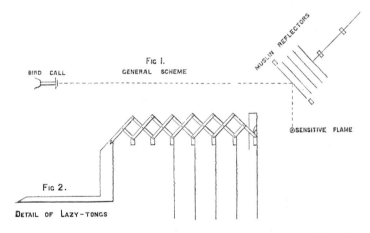

and mounted (with the aid of three lazy-tongs arrangements), so that there is but one degree of freedom to move, and that of such a character as to vary the interval between the discs without disturbing their equidistance and parallelism.

The source of sound is a bird-call, giving a pure tone of high pitch (inaudible), and the percipient is a high pressure flame issuing from a burner so oriented that the direct waves are without influence

---

* 'Phil. Mag.' Sept. 1888.

upon the flame.* But the waves reflected from the muslin arrive in the effective direction, and if of sufficient intensity induce flaring. The experiment consists in showing that the action depends upon the distance between the discs. If the distance be such that the waves reflected from the several discs co-operate,† the flame flares, but for intermediate adjustments recovers its equilibrium. For full success it is necessary that the reflective power of a single disc be neither too great nor too small. A somewhat open fabric appears suitable.

It was shown by Brewster that certain natural specimens of Iceland spar are traversed by thin twin strata. A convergent beam, reflected at a nearly grazing incidence from the twin planes, depicts upon the screen an arc of light, which is interrupted by a dark spot corresponding to the plane of symmetry. [Shown.] A similar experiment may be made with small rhombs in which twin layers have been developed by mechanical force after the manner of Reusch.

The light reflected from fiery opals has been shown by Crookes to possess in many cases a high degree of purity, rivalling in this respect the reflection from chlorate of potash. The explanation is to be sought in a periodic stratified structure. But the other features differ widely in the two cases. There is here no semicircular evanescence, as the specimen is rotated in azimuth. On the contrary, the coloured light transmitted perpendicularly through a thin plate of opal undergoes no change when the gem is turned round in its own plane. This appears to prove that the alternate states are not related to one another as twin crystals. More probably the alternate strata are of air, as in decomposed glass. The brilliancy of opals is said to be readily affected by atmospheric conditions.

---

* See 'Proc. Roy. Inst.' Jan. 1888.
† If the reflection were perpendicular, the interval between successive discs would be equal to the half wave-length, or to some multiple of this.

Friday, May 3, 1889.

SIR FREDERICK BRAMWELL, Bart. D.C.L. F.R.S. Honorary Secretary and Vice-President, in the Chair.

SIR HENRY ROSCOE, M.P. D.C.L. LL.D. V.P.R.S.

Aluminium.

CHEMISTS of many lands have contributed to our knowledge of the metal aluminium. Davy, in 1807, tried in vain to reduce alumina by means of the electric current. Oerstedt, the Dane, in 1824, pointed out that the metal could be obtained by treating the chloride with an alkali metal; this was accomplished in Germany by Wöhler in 1827, and more completely in 1845, whilst in 1854, Bunsen showed how the metal can be obtained by electrolysis. But it is to France, by the hands of Henri St. Claire Deville, in the same year, that the honour belongs of having first prepared aluminium in a state of purity, and of obtaining it on a scale which enabled its valuable properties to be recognised and made available, and the bar of "silver-white metal from clay," was one of the chemical wonders in the first Paris Exhibition of 1855. Now England and America step in, and I have this evening to relate the important changes which further investigation has effected in the metallurgy of aluminium. The process suggested by Oerstedt, carried out by Wöhler, and modified by Deville, remains in principle unchanged. The metal is prepared, as before, by a reduction of the double chloride of aluminium and sodium, by means of metallic sodium in presence of cryolite; and it is therefore not so much a description of a new reaction as of improvements of old ones of which I have to speak.

I may perhaps be allowed to remind my hearers that more than 33 years ago, Mr. Barlow, then secretary to the Institution, delivered a discourse, in the presence of M. Deville, on the properties and mode of preparation of aluminium, then a novelty. He stated that the metal was then sold at the rate of 3*l.* per ounce, and the exhibition of a small ingot, cast in the laboratory by M. Deville, was considered remarkable. As indicating the progress since made, I may remark that the metal is now sold at 20*s.* per lb., and manufactured by the ton, by the Aluminium Company, at their works at Oldbury, near Birmingham. The improvements which have been made in this manufacture by the zeal and energy of Mr. Castner, an American metallurgist, are of so important a character, that the process may properly be termed the Deville-Castner process.

The production of aluminium previous to 1887, probably did not exceed 10,000 lbs. per annum, whilst the price at that time was very high. To attain even this production required that at least 100,000 lbs. of double chloride, and 40,000 lbs. of sodium should be manufactured annually. From these figures an idea of the magnitude of the undertaking assumed by the Aluminium Company may be estimated, when we learn that they erected works having an annual

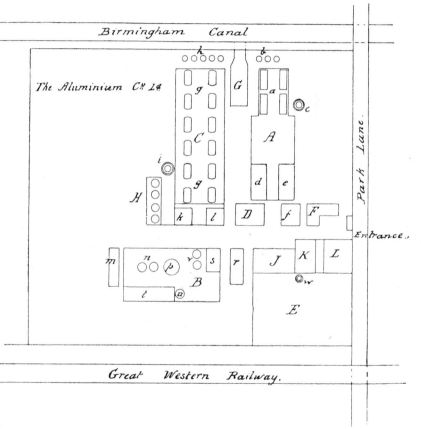

producing capacity of 100,000 lbs. of aluminium. To accomplish this, required not only that at least 400,000 lbs. of sodium, 800,000 lbs. of chlorine, and 1,000,000 lbs. of double chloride, should be annually manufactured, but in addition that each of these materials should be produced at a very low cost, in order to enable the metal to be sold at 20*s*. per lb.

Annexed is a sketch plan of the works, which now cover a space of nearly five acres. They are divided into five separate departments, viz., 1st, sodium, A ; 2nd, chlorine, B ; 3rd, chloride, C ; 4th, aluminium, D ; and 5th, foundry, rolling, wire mills, &c., E.

In each department an accurate account is kept of the production each day, the amount of material used, the different furnaces and apparatus in operation, &c. In this manner it has been found possible to ascertain each day exactly how the different processes are progressing, and what effect any modification has, either on cost, quantity, or quality of product. By this means a complicated chemical process is reduced to a series of very simple operations, so that whilst the processes are apparently complicated and difficult to carry out successfully, this is not the case now that the details connected with the manufacture have been perfected, and each operation carried on quite independently until the final materials are brought together for the production of the aluminium.

Manufacture of Sodium.

The first improvement occurs in the manufacture of sodium by what is known as the " Castner Process." The successful working of this process marks an era in the production of sodium, as it not only has greatly cheapened the metal, but has enabled the manufacture to be carried out upon a very large scale with little or no danger. Practically, the process consists in heating fused caustic soda in contact with carbon whilst the former substance is in a perfectly liquid condition. By the process in vogue before the introduction of this method, it was always deemed necessary that special means should be taken to guard against actual fusion of the mixed charges, which, if it were to take place, would to a large extent allow the alkali and reducing material to separate. Thus having an infusible charge to heat, requiring the employment of a very high temperature for its decomposition, the iron vessels must be of small circumference to allow the penetration of the heat to the centre of the charge without actually melting the vessel in which the materials are heated. By the new process, owing to the alkali being in a fused or perfectly liquid condition in contact directly with carbon, the necessity of this is avoided, and consequently, the reduction can be carried on in large vessels at a comparatively low temperature. The reaction taking place may be expressed as follows:—

$$3NaHO + C = Na_2CO_3 + 3H + Na.$$

The vessels in which the charges of alkali and reducing material are heated are of egg-shaped pattern, about 18 inches in width at their widest part and about 3 feet high, and are made in two portions, the lower one being actually in the form of a crucible, while the upper one is provided with an upright stem and a protruding hollow arm. This part of the apparatus is known as the cover. In commencing the operation, these covers are raised in the heated furnace through apertures provided in the floor of the heated chamber, and are then fastened in their place by an attachment adjusted to the stem; the hollow arm extends outside the furnace. Directly below each aperture in the bottom of the furnace are situated the hydraulic lifts; attached to the top of which are the platforms upon which are placed the crucibles to be raised into the furnace. Attached to the hydraulic lifts are the usual reversing valves for lowering or raising, and the platform is of such a size as, when raised, completely to fill the bottom aperture of the furnace. The charged crucible, being placed upon the platform, is raised into its position, the edges meeting those of the cover, forming an air-tight joint which prevents the escape of gas and vapour from the vessel during reduction, except by the hollow arm provided for this purpose. The natural expansion of the iron vessels is accommodated by the water-pressure in the hydraulic lifts, so that the joint of the cover and crucible are not disturbed until it is intended to lower the lift for the purpose of removing the crucible.

The length of time required for the first operation of reduction and distillation is about two hours. At the end of this time the crucibles are lowered, taken from the platforms by a large pair of tongs on wheels, carried to a dumping pit, and thrown on their side. The residue is cleaned out, and the hot pot, being again gripped by the tongs, is taken back to the furnace. On its way, the charge of alkali and reducing material is thrown in. It is again placed on the lift and raised in position against the edges of the cover. The time consumed in making the change is $1\frac{1}{2}$ minute, and it only requires about seven minutes to draw, empty, recharge, and replace the five crucibles in each furnace. In this manner the crucibles retain the greater amount of their heat, so that the operation of reduction and distillation now only requires one hour and ten minutes. Each of the four furnaces, of five crucibles each, when in operation, are drawn alternately, so that the process is carried on night and day.

Attached to the protruding hollow arm from the cover are the condensers, which are of a peculiar pattern specially adapted to this process, being quite different from those formerly used. They are about 5 inches in diameter, and nearly 3 feet long, and have a small opening in the bottom about 20 inches from the nozzle. The bottom of these condensers is so inclined that the metal condensed from the vapour issuing from the crucible during reduction, flows down and out into a small pot placed directly below this opening. The uncondensed gases escape from the condenser at the further end, and burn with the characteristic sodium flame. The condensers are also provided with a small hinged door at the further end, by means of which the workmen from time to time may look in to observe how the distillation is progressing. Previous to drawing the crucibles from the furnace for the purpose of emptying and recharging, the small pots each containing the distilled metal are removed, and empty ones substituted. Those removed each contain on an average about 6 lbs. of metal, and are taken directly to the sodium casting shop, where it is melted and cast, either into large bars ready to be used for making aluminium, or in smaller sticks to be sold.

Special care is taken to keep the temperature of the furnaces at about 1000° C., and the gas and air valves are carefully regulated, so as to maintain as even a temperature as possible. The covers remain in the furnace from Sunday night to Saturday afternoon, and the crucibles are kept in use until they are worn out, when new ones are substituted without interrupting the general running of the furnace. A furnace in operation requires 250 lbs. of caustic soda every one hour and ten minutes, and yields in the same time 30 lbs. of sodium, and about 240 lbs. of crude carbonate of soda. With the four furnaces at work 120 lbs. of sodium can be made every 70 minutes, or over a ton in the 24 hours. The residual carbonate, on treatment with lime in the usual manner, yields two-thirds of the original amount of caustic operated upon. The sodium, after being cast, is saturated with kerosene oil, and stored in large tanks holding several tons, placed in rooms specially designed both for security against either fire or water.

Chlorine Manufacture.

This part of the works is connected with the adjacent works of Messrs. Chance Bros. by a large gutta-percha pipe, by means of which from time to time hydrochloric acid is supplied direct into the large storage cisterns, from which it is used as desired for making the chlorine. For the preparation of the chlorine gas needed in making the chloride, the usual method is employed; that is, hydrochloric acid and manganese dioxide are heated together, when chlorine gas is evolved with effervescence, and is led away by earthenware and lead pipes to large lead-lined gasometers, where it is stored.

The materials for the generation of the chlorine are brought together in large tanks, or stills, built up out of great sandstone slabs, having rubber joints, and the heating is effected by the injection of steam. The evolution of gas, at first rapid, becomes gradually slower, and at last stops; the hydrochloric acid and manganese dioxide being converted into chlorine and manganous chloride. This last compound remains dissolved in the "spent still liquor" and is reconverted into manganese dioxide, to be used over again, by Weldon's Manganese Recovery Process. Owing to the difficulty of keeping up a regular supply of chlorine under a constant pressure directly from the stills, in order that the quantity passed into the sixty different retorts in which the double-chloride is made can be regulated and fed as desired, four large gasometers were erected. Each of these is capable of holding 1,000 cubic feet of gas, and is completely lined with lead, as are all the connecting mains, &c., this being the only available metal which withstands the corrosive action of chlorine. The gasometers are filled in turn from the stills, the chlorine consumed being taken direct from a gasometer under a regular pressure until it is exhausted; the valves being changed, the supply is taken from another holder, the emptied one being refilled from the still.

Manufacture of the Double Chloride.

Twelve large regenerative gas furnaces are used for heating, and in each of these are fixed five horizontal fire-clay retorts about 10 ft. in length, into which the mixture for making the double chloride is placed. These furnaces have been built in two rows, six on a side, the clear passage-way down the centre of the building, which is about 250 ft. long, being 50 ft. in width. Above this central passage is the staging carrying the large lead-mains for the supply of the chlorine coming from the gasometers. Opposite each retort, and attached to the main, are situated the regulating valves, connected with lead and earthenware pipes, for the regulation and passage of the chlorine to each retort. The valves are of peculiar design, and have been so constructed that the chlorine is made to pass through a certain depth of liquid, which not only by opposing a certain pressure allows a known quantity of gas to pass in a given time, but also prevents any return from the retort into the main, should an increase of pressure be suddenly developed in the retorts.

The mixture with which the retorts are charged is made by grinding together hydrate of alumina, salt, and charcoal. This mixture is then moistened with water, which partially dissolves the salt, and thrown into a pug mill of the usual type for making drain pipes, excepting that the mass is forced out into solid cylindrical lengths upon a platform alongside of which a workman is stationed with a large knife, by means of which the material is cut into lengths of about 3 inches each. These are then piled on top of the large furnaces to dry. In a few hours they have sufficiently hardened to allow of their being handled. They are then transferred to large wagons, and are ready to be used in charging the retorts.

The success of this process is in a great measure dependent—1st, on the proportionate mixture of materials; 2nd, on the temperature of the furnace; 3rd, on the quantity of chlorine introduced in a given time; and 4th, on the actual construction of the retorts. I am, however, not at liberty to discuss the details of this part of the process, which have only a commercial interest. In carrying on the operation, the furnaces or retorts, when at the proper temperature, are charged by throwing in the balls until they are quite full, the fronts are then sealed up, and the charge allowed to remain undisturbed for about four hours, during which time the water of the alumina hydrate is completely expelled. At the end of this time the valves on the chlorine main are opened, and the gas is allowed to pass into the charged retorts. In the rear of each retort, and connected therewith by means of an earthenware pipe, are the condenser boxes, which are built in brick. These boxes are provided with openings or doors, and also with earthenware pipes connected with a small flue for carrying off the uncondensed vapours to the large chimney. At first the chlorine passed into each retort is all absorbed by the charge, and only carbonic oxide escapes into the open boxes, where it burns. After a certain time, however, dense fumes are evolved, and the boxes are then closed, while the connecting pipe between the box and the small flue serves to carry off the uncondensed vapours to the chimney.

The reaction which takes place is as follows:—

$$Al_2O_3 + 2NaCl + 3C + 6Cl = 2AlCl_3NaCl + 3CO.$$

The chlorine is passed in for about 72 hours in varying quantity, the boxes at the back being opened from time to time by the workmen to ascertain the progress of the distillation. At the end of the time mentioned the chlorine valves are closed and the boxes at the back of the furnace are all thrown open. The crude double chloride, as distilled from the retorts, condenses in the connecting pipe and trickles down into the boxes, where it solidifies in large irregular masses. The yield from a bench of five retorts will average from 1,600 to 1,800 lbs., which is not far from the theoretical quantity. After the removal of the crude chloride from the condenser boxes, the retorts are opened at their charging end, and the residue, which consists of a small quantity of alumina, charcoal, and salt, is raked out and remixed in certain proportions with fresh material, to be used over again. The furnace is immediately re-charged and the same operations repeated, so that from each furnace upwards of 3,500 lbs. of chloride are obtained weekly. With ten of the twelve furnaces always at work the plant is easily capable of producing 30,000 lbs. of chloride per week, or 1,500,000 lbs. per annum.

Owing to the presence of iron, both in the materials used (viz., charcoal, alumina, &c.) and in the fireclay composing the retorts, the distilled chloride always contains a varying proportion of this metal in the form of ferrous and ferric chlorides. When it is remembered that it requires 10 lbs. of this chloride to produce 1 lb. of aluminium by reduction, it will be quite apparent how materially a very small percentage of iron in the chloride will influence the quality of the resulting metal. I may say that, exercising the utmost care as to the purity of the alumina and the charcoal used, and after having the retorts made of special fireclay containing only a very small percentage of iron, it was found almost impossible to produce upon a large scale a chloride containing less than 0·3 per cent. of iron.

This crude double chloride, as it is now called at the works, is highly deliquescent, and varies in colour from a light yellow to a dark red. The variation in colour is not so much due to the varying percentage of iron contained as to the relative proportion of ferric or ferrous chlorides present, and although a sample may be either very dark or quite light, it may still contain only a small percentage of iron if it be present as ferric salt, or a very large percentage if it is in the ferrous condition. Even when exercising all possible precautions, the average analysis of the crude double chloride shows about 0·4 per cent. of iron. The metal subsequently made from this chloride therefore never contained much less than about 5 per cent. of iron, and, as this quantity greatly injures the capacity of aluminium for drawing into wire, rolling, &c., the metal thus obtained required to be refined. This was successfully accomplished by Mr. Castner and his able assistant Mr. Cullen, and for some time all the metal made was refined, the iron being lowered to about 2 per cent.

The process, however, was difficult to carry out, and required careful manipulation, but as it then seemed the only remedy for effectively removing the iron, it was adopted and carried on for some time quite successfully, until another invention of Mr. Castner rendered it totally unnecessary. This consisted in purifying the double chloride before reduction. I cannot now explain this process, but I am able to show some of the product. This purified chloride, or pure double chloride, is, as you see, quite white, and is far less deliquescent than the crude, so that it is quite reasonable to infer that this most undesirable property is greatly due to the former presence of iron chlorides. I have seen large quantities containing upwards of $1\frac{1}{2}$ per cent. of iron, or 150 lbs. to 10,000 of the chloride, completely purified from iron in a few minutes, so that, whilst the substance before treatment was wholly unfit for the preparation of aluminium, owing to the presence of iron, the result was, like the sample exhibited, a mass containing only 1 lb. of iron in 10,000, or 0·01 per cent. The process is extremely simple, and adds little or no appreciable cost to the final product. After treatment, this pure chloride is melted in large iron pots and run into drums similar to those used for storing caustic soda. As far as I am aware, it was generally believed to be an impossibility to remove the iron from anhydrous double chloride of aluminium and sodium, and few if any chemists have ever seen a pure white double chloride.

Aluminium Manufacture.

I now come to the final stage of the process, viz., the reduction of the pure double chloride by sodium. This is effected, not in a tube of Bohemian glass, as shown in Mr. Barlow's lecture in 1856, but in a large reverberatory furnace, having an inclined hearth about 6 feet square, the inclination being towards the front of the furnace, through which are several openings at different heights. The pure chloride is ground together with cryolite in about the proportions of two to one, and is then carried to a 'staging erected above the reducing furnace. The sodium, in large slabs or blocks, is run through a machine similar to an ordinary tobacco-cutting machine, where it is cut into small thin slices; it is then also transferred to the staging above the reducing furnace.

Both materials are now thrown into a large revolving drum, when they become thoroughly mixed. The drum being opened and partially turned, the contents drop out into a car on a tramway directly below. The furnace having been raised to the desired temperature, the dampers of the furnace are all closed to prevent the access of air, the heating gas also being shut off. The car is then moved out on the roof of the furnace until it stands directly over the centre of the hearth. The furnace roof is provided with large hoppers, and through these openings the charge is introduced as quickly as possible. The reaction takes place almost immediately, and the whole charge quickly liquefies. At the end of a certain time the heating gas is again introduced and the charge kept at a moderate temperature for about two hours. At the end of this period the furnace is tapped by driving a bar through the lower opening, which has previously been stopped with a fire-clay plug, and the liquid metal run out in a silver stream into moulds placed below the opening. When the metal has all been drawn off, the slag is allowed to run out into small iron wagons and removed. The openings being again plugged up, the furnace is ready for another charge. From each charge, composed of about 1200 lbs. of pure chloride, 600 lbs. of cryolite, and 350 lbs. of sodium, about 115 to 120 lbs. of aluminium is obtained.

The purity of the metal entirely depends upon the purity of the chloride used, and without exercising more than ordinary care the metal tests usually indicate a purity of metal above 99 per cent. On the table is the metal run from a single charge, its weight is 116 lbs., and its composition, as shown by analysis, is $99 \cdot 2$ aluminium, $0 \cdot 3$ silicon, and $0 \cdot 5$ iron. This I believe to be the largest and the purest mass of metal ever made in one operation.

The result of eight or nine charges are laid on one side, and then melted down in the furnace to make a uniform quality, the liquid metal, after a good stirring, being drawn off into moulds. These large ingots, weighing about 60 lb. each, are sent to the casting shop, there to be melted and cast into the ordinary pigs, or other shapes, as may be required for the making of tubes, sheets, or wire, or else used directly for making alloys of either copper or iron.

The following table shows approximately the quantity of each material used in the production of one ton of aluminium:—

Metallic sodium	6,300 lbs.
Double chloride	22,400 „
Cryolite	8,000 „
Coal	8 tons.

To produce 6,300 lbs. of sodium is required:—

Caustic soda	44,000 lbs.
Carbide made from pitch, 12,000 lbs., and iron turnings, 1,000 lbs.	7,000 „
Crucible castings	2½ tons.
Coal	75 „

For the production of 22,400 lbs. double chloride is required:—

Common salt	8,000 lbs.
Alumina hydrate	11,000 „
Chlorine gas	15,000 „
Coal	180 tons.

For the production of 15,000 lbs. of chlorine gas is required:—

Hydrochloric acid	180,000 lbs.
Limestone dust	45,000 „
Lime	30,000 „
Loss of manganese	1,000 „

(These figures were rendered more evident by the aid of small blocks, each cut a given size so as to represent the relative weights of the different materials used to produce one unit of aluminium.)

It might seem, on looking over the above numbers, as if an extraordinary amount of waste occurred, and as if the production is far below that which ought to be obtained, but a study of the figures will show that this is not the case. I would wish to call attention to one item in particular, viz. fuel, it having been remarked that the consumption of coal must prevent cheap production. I think when it is remembered that coal, such as used at the works, cost only 4s. per ton, while the product is worth 2240l. per ton, the cost of coal is not an item of consequence in the cost of production. The total cost of the coal to produce one ton of metal being 50l.; the actual cost for fuel is less than sixpence for every pound of aluminium produced. The ratio of cost of fuel to value of product is indeed less than is the case in making either iron or steel. In concluding my remarks as to the method of manufacture and the process in general, I may add that I do not think it is too much to expect, in view of the rapid strides already made, that in the future, further improvements and modifications will enable aluminium to be produced and sold even at a lower price than appears at present possible.

Properties of Aluminium.

In its physical properties aluminium widely differs from all the other metals. Its colour is a beautiful white, with a slight blue tint. The intensity of this colour becomes more apparent when the metal has been worked, or when it contains silicon or iron. The surface may be made to take a very high polish, when the blue tint of the metal become manifest, or it may be treated with caustic soda and then nitric acid, which will leave the metal quite white. The extensibility or malleability of aluminium is very high, ranking with gold and silver if the metal be of good quality. It may be beaten out into thin leaf quite as easily as either gold or silver, although it requires more careful annealing.

It is extremely ductile and may be easily drawn, especial care only being required in the annealing.

The excessive sonorousness of aluminium is best shown by example (large suspended bar being struck). Faraday has remarked, after experiments conducted in his laboratory, that the sound produced by an ingot of aluminium is not simple, and one may distinguish the two sounds by turning the vibrating ingot.

After being cast it has about the hardness of pure silver, but may be sensibly hardened by hammering.

Its tensile strength varies between 12 and 14 tons to the inch (test sample which was shown having been broken at 13 tons or 27,000 lbs.), ordinary cast iron being about 8 tons. Comparing the strength of aluminium in relation to its weight, it is equal to steel of 38 tons tensile strength. The specific gravity of cast aluminium is $2 \cdot 58$, but after rolling or hammering this figure is increased to about $2 \cdot 68$.

The specific gravity of aluminium being 1, copper is $3 \cdot 6$, nickel $3 \cdot 5$, silver 4, lead $4 \cdot 8$, gold $7 \cdot 7$.

The fusibility of aluminium has been variously stated as being between that of zinc and silver, or between 600 and 1000° C.

As no reliable information has ever been made public on this subject, my friend, Professor Carnelley, undertook to determine it. I was aware, from information gained at the works at Oldbury, that a small increase in the percentage of contained iron materially raised its point of fusion, and it has been undoubtedly due to this cause that such wide limits are given for the melting point. Under these circumstances two samples were forwarded for testing, of which No. 1, containing $\frac{1}{2}$ per cent. of iron, had a melting point of 700° C.; whereas No. 2, containing 5 per cent. of iron, does not melt at 700° and only softens somewhat above that temperature but undergoes incipient fusion at 730°.

According to Faraday, aluminium ranks very high among metallic conductors of heat and electricity, and he found that it conducted heat better than either silver or copper. The specific heat is also very high, which accounts for length of time required for an ingot of the metal to either melt or get cold after being cast.

Chemically, its properties are well worthy of study.

Air, either wet or dry, has absolutely no effect on aluminium at the ordinary temperature, but this property is only possessed by a very pure quality of metal, and the pure metal in mass undergoes only slight oxidation even at the melting point of platinum.

Thin leaf, however, when heated in a current of oxygen, burns with a brilliant, bluish white light. (Experiment shown). If the metal be pure, water has no effect on it whatever, even at a red-heat. Sulphur and its compounds also are without action on it, while, under the same circumstances, nearly all metals would be discoloured with great rapidity. (Experiment shown using silver and aluminium under the same conditions.)

Dilute sulphuric acid and nitric acid, both diluted and concentrated, have no effect on it, although it may be dissolved in either hydrochloric acid or caustic alkali. Heated in an atmosphere of chlorine it burns with a vivid light, producing aluminium chloride. (Experiment shown). In connection with the subject it may be of interest to state the true melting point of the double chloride of aluminium and sodium, which has always been given at 170° to 180° C., but which Mr. Baker, the chemist to the works, finds lies between 125° and 130° C.

Uses of Aluminium.

Its uses, unalloyed, have heretofore been greatly restricted. This is, I believe, alone owing to its former high price, for no metal possessing the properties of aluminium could help coming into larger use if its cost were moderate. Much has been said as to the impossibility of soldering it being against its popular use, but I believe that this difficulty will now soon be overcome. The following are a few of the purposes to which it is at present put: telescope tubes, marine glasses, eye glasses and sextants, especially on account of its lightness. Fine wire for the making of lace, embroidery, &c. Leaf in the place of silver leaf, sabre sheaths, sword handles, &c., statuettes and works of art, jewellery and delicate physical apparatus, culinary utensils, harness fittings, metallic parts of solders' uniforms, dental purposes, surgical instruments, reflectors (it not being tarnished by the products of combustion), photographic apparatus, aeronautical and engineering purposes, and especially for the making of alloys.

Alloys of Aluminium.

The most important alloys of aluminium are those made with copper. These alloys were first prepared by Dr. Percy, in England, and now give promise of being largely used. The alloy produced by the addition of 10 per cent. of aluminium to copper, the maximum amount that can be used to produce a satisfactory alloy, is known as aluminium bronze. Bronzes, however, are made which contain smaller amounts of aluminium, possessing in a degree the valuable properties of the 10 per cent. bronze. According to the percentage of aluminium up to 10 per cent., the colour varies from red gold to pale yellow. The 10 per cent. alloy takes a fine polish, and has the colour of jewellers' gold. The 5 per cent. alloy is not quite so hard, the colour being very similar to that of pure gold. I am indebted to Prof. Roberts Austen for a splendid specimen of crystallised gold, as also for a mould in which the gold at the mint is usually cast, and in this I have had prepared ingots of the 10 and 5 per cent. alloy, so that a comparison may be made of the colour of these with a gold ingot cast in the same mould, for the loan of which I have to thank Messrs. Johnson, Matthey, & Co., all of which are before you.

I have also ingots of the same size, of pure aluminium, from which an idea of the relative weights of gold and aluminium may be obtained.

To arrive at perfection in the making of these alloys, not only is it required that the aluminium used should be of good quality, but also that the copper must be of the very best obtainable. For this purpose only the best brands of Lake Superior copper should be used. Inferior brands of copper or any impurities in the alloy give poor results. The alloys all possess a good colour, polish well, keep their colour far better than all other copper alloys, are extremely malleable and ductile, can be worked either hot or cold, easily engraved, the higher grades have an elasticity exceeding steel, are easily cast into complicated objects, do not lose in remelting, and are possessed of great strength, dependent, of course, on the purity and percentage of contained aluminium. The 10 per cent. alloy, when cast, has a tensile strength of between 70,000 and 80,000 lbs. per square inch, but when hammered or worked, the test exceeds 100,000 lbs. (A sample shown broke at 105,000 lbs.).

An attempt to enumerate either the present uses or the possible future commercial value of these alloys is beyond my present purpose. I may, however, remark that they are not only adapted to take the place of bronze, brass, and steel, but they so far surpass all of those metals, both physically and chemically, as to make their extended use assured. (Sheets, rods, tubes, wire, and ingots shown.)

But even a more important use of aluminium seems to be its employment in the iron industry, of which it promises shortly to become a valuable factor, owing to certain effects which it produces when present, even in the most minute proportions. Experiments are now being carried on at numerous iron and steel works, in England, on the Continent, and in America. The results so far attained are greatly at variance, for whilst in the majority of cases the improvements made have encouraged the continuance of the trials, in others the result has not been satisfactory. On this point I would wish to say to those who may contemplate making use of aluminium in this direction, that it would be advisable before trying their experiments to ascertain whether the aluminium alloy they may purchase actually contains any aluminium at all, for some of the so-called aluminium alloys contain little or no aluminium, and this may doubtless account for the negative results obtained. Again, others contain such varying proportions of carbon, silicon, and other impurities, as to render their use highly objectionable.

It seems to be a prevailing idea with some people, that because aluminium is so light compared with iron, that they cannot be directly alloyed, and furthermore, that for the same reason, alloys made by the direct melting together of the two metals would not be equal to an alloy where both metals are reduced together. Now, of course, this is not the case, and the statement has been put forward by those who were only able to make the alloys in one way.

Aluminium added to molten iron and steel lowers their melting points, consequently increases the fluidity of the metal, and causes it to run easily into moulds and set there, without entrapping air and other gases, which serve to form blow-holes and similar imperfections. It is already used by a large number of steel founders, and seems to render the production of sound steel castings more certain and easy than is otherwise possible.

One of the most remarkable applications of this property which aluminium possesses of lowering the melting-point of iron has been made use of by Mr. Nordenfelt in the production of castings of wrought iron.

Aluminium forms alloys with most other metals, and although each possess peculiar properties which in the future may be utilised, at present they are but little used.

In conclusion, I beg to call your attention to the wood models on the table, one being representative of aluminium, the other aluminium bronze. The originals of these models are now in the Paris Exhibition, each weighing 1000 lbs. With regard to aluminium bronze, I cannot speak positively, but the block of pure aluminium is undoubtedly the largest casting ever made in this most wonderful metal. I have to thank the Directors of the Aluminium Company, and especially Mr. Castner, for furnishing me with the interesting series of specimens of raw and manufactured metal for illustrating my discourse.

[H. E. R.]

Friday, May 10, 1889.

WILLIAM CROOKES, Esq. F.R.S. Vice-President, in the Chair.

PROFESSOR DEWAR, M A. F.R.S. *M.R.I.*

Optical Properties of Oxygen and Ozone.

IN the course of experiments on the spectra of gases at high pressures, Professor Liveing and I have made observations on the absorption-spectrum of oxygen which confirm and extend the observations of Egoroff and Jansen. The interest of this spectrum is so great, on account of the important part which oxygen plays in our world, and its free condition in our atmosphere, that it deserves a separate notice.

In order to include the ultra-violet rays in our observations we have had to contrive windows of quartz to the apparatus containing the gases. A strong steel tube, 165 centimetres long and 5 centimetres wide, was fitted with gun-metal ends, bearing by curved

FIG. 1.

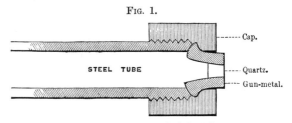

Section through one end of the tube.

surfaces upon the conical openings of the tube, and forced home by powerful screw-caps. Each gun-metal end was pierced centrally by a conical opening fitted with a quartz stopper, 2·1 centimetres thick, and of the same diameter, with plane polished ends. A small amount of wax was interposed between the stopper and the gun-metal for the purpose of ensuring a uniform bearing for the quartz, which is very brittle. Trial proved that the tube thus fitted would sustain, without leakage, a pressure of upwards of 260 atmospheres. The tube had, besides, near each end, a screw-plug valve for admitting the gases. About the centre of the tube was placed a quartz lens, rather less in diameter than the tube, held in place by three springs which pressed against the walls of the tube. This lens had a focal length of about 46 centimetres; so that when a source of light was placed about 10 centimetres from one end of the tube, an image of it was formed on the slit of the spectroscope at about the same distance from the other end of the tube, and thereby loss of light, so far as it was due to the distance of the source, was reduced to a minimum.

Ordinary oxygen was let into the tube from an iron bottle until the pressure reached 85 atmospheres, and on viewing an arc light through the tube the following absorptions were visible:—

(1) A very dark band sharply defined on its more refrangible side, gradually fading out on its less refrangible side, and divided into two parts by a streak of light, occupying the position of A of the solar spectrum.

(2) A much weaker, but precisely similar band in the position of B of the solar spectrum.

(3) A dark band very diffuse on both edges, extending from about λ 6360 to λ 6225, with a maximum intensity at about λ 6305.

(4) A still darker band a little above D, beginning with a diffuse edge at about λ 5810, rapidly coming to a maximum intensity at about λ 5785, and then gradually fading on the more refrangible side, and disappearing at about λ 5675.

(5) A faint narrow band in the green at about λ 5350.

(6) A strong band in the blue, diffuse on both sides, extending from about λ 4795 to λ 4750.

When photographs were taken of the ultra-violet part of the spectrum of the arc and of the iron spark, the gas appeared to be quite transparent for violet and ultra-violet rays up to about λ 2745.

From that point the light gradually diminished, and beyond λ 2664 appeared to be wholly absorbed.

The pressure of the oxygen in the tube was then increased to 140 atmospheres. This had the effect of increasing sensibly the darkness of all the bands above described; but brought out no new bands, except a faint band in the indigo at about λ 4470. In the ultra-violet the absorption appeared to be complete for all rays beyond about λ 2704.

The foregoing observations were made with a spectroscope of small dispersion. We next brought to bear on the spectrum a large instrument with one of Rowland's gratings. Even with the high dispersion of this instrument, the bands at A could not be resolved into lines; they remained two diffuse bands; though the red potassium-lines, which were produced by sprinkling the electrode of the arc with a potassium-salt, were sharply defined and widely separated. None of the other bands were resolvable into lines. This we attribute to the density of the gas, by which the lines are expanded so as to obliterate the interspaces; and this supposition is confirmed by the observation of Ångström, that the band in the solar spectrum which appears to be identical with that observed by us a little above D, was resolved into fine lines when the sun was high, but appeared as a continuous band when the sun was near the horizon.

On letting down the pressure the bands were all weakened; A, though weaker, became more sharply defined at the more refrangible edge. The faint band in the indigo λ 4470 remained just visible until the pressure fell below 110 atmospheres. At 90 atmospheres A and B were still well seen and sharp, but all the other bands weaker. B remained visible until the pressure fell to 40 atmospheres. A was then still well seen, the band just above D very faint, and the others almost gone. At 30 atmospheres A was still easily seen, and there was a trace of the band above D. At 25 atmospheres this band had gone, but A remained visible until the pressure fell to less than 20 atmospheres. Hence an amount of oxygen not greater than that contained in a column of air 150 metres long at ordinary pressure, is sufficient to produce a visible absorption at A. The quantity of oxygen in the tube at the highest pressure we used falls, however, far short of the quantity traversed by the solar rays in passing through the atmosphere when the sun is vertical.

It will be noted that the bands, if we except the faint two in the green and indigo respectively, appear to be identical with those terrestrial bands in the solar spectrum which Ångström found to be as strong when the air was dried by intense frost as at other times. At least the positions of the maxima agree closely, and that near D shows the same peculiarity in having its maximum near the less refrangible end. We did not, however, observe α, which would be fainter than B, and if, like A and B, unresolvable, would be lost in the diffuse band which covers that region. The bands above numbered 3, 4, 5, 6, agree also with those observed by Olszewski,[*] to be produced by a layer of liquid oxygen 12 millimetres thick. The point also at which the absorption of the ultra-violet rays begins, agrees with that at which the absorption by ozone begins, as observed by Hartley[†]; but the oxygen, as we used it, did not appear to transmit the more refrangible rays beyond 2320, which seem to pass through ozone. Egoroff[‡] found that A remained visible when he looked through 80 metres of atmosphere, but 3 kilogrammes of atmosphere failed to produce α.

When the pressure in our tube was reduced, a cloud was always formed which rendered the contents of the tube nearly opaque; the faint light which was then transmitted had always a green tinge.

It is remarkable that the compounds of oxygen do not show any similar absorptions. Ångström thought it improbable that oxygen should have a spectrum of such a character, since he failed to obtain an emission spectrum resembling it; and suggested that the absorptions might be due to carbonic acid gas or to ozone, or possibly to oxygen in the state in which it becomes fluorescent.[§] Neither carbonic acid gas nor nitrous oxide, at a pressure of 50 atmospheres in our tube, show any sensible absorption in the visible spectrum;

---

[*] Wied. 'Ann.,' xxxiii. p. 570.
[†] 'Journ. Chem. Soc.,' xxxix. p. 57.
[‡] 'Comptes Rendus,' ci. p. 1144.
[§] 'Spect. Norm.,' p. 41.

and the absorption of the ultra-violet rays by the latter gas begins at a higher point, namely about λ 2450, than that of uncombined oxygen. In fact, we see the anomalies of the selective absorption by compounds as compared with that of their elements when we take the case of water, which has a remarkable transparency for those ultra-violet rays for which oxygen is opaque.

These observations show that all stellar spectra observed in our atmosphere, irrespective of the specific ultra-violet radiation of each star, must be limited to wave-lengths not less than λ 2700, unless we can devise means to eliminate the atmospheric absorption by observations at exceedingly high altitudes.

We have extended our observations to much longer columns of oxygen. A steel tube 18 metres long (see Fig. 2) was fitted with the same quartz ends as had been used with the shorter tube, and with two quartz lenses symmetrically placed inside the tube, one near

FIG. 2.

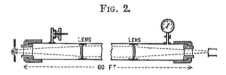

Section of Steel Tube.

each end, so that when an arc lamp was placed about 14 centimetres from one end of the tube, the image of it was formed on the slit of the spectroscope at the same distance from the other end.

When the tube was filled with air only at ordinary pressure, no absorptions could be detected, but when the air was replaced by oxygen at the pressure of the atmosphere the absorption of A was just visible, though neither B nor any other absorption-band could be traced. As the pressure of the oxygen was increased, A became much darker and more distinct, and B came out sharply defined. The absorption-band about λ 5785 was next seen, and the dark bands about λ 6300 and λ 4770, were just visible when the pressure reached 20 atmospheres.

At a pressure of 30 atmospheres A was very black, B also strong and sharply defined, and the forementioned bands were all quite strong and had the same general characters as when seen through the shorter tube; the band about λ 5350 also could be seen, but there was only a bare trace of that in the indigo about λ 4470. At 60 atmospheres these last two absorptions could be well seen, all the other bands were very strong, B still quite sharp, but A somewhat obscured by a general absorption at the red end. At 90 atmospheres this general absorption at the red end seemed to extend to about one-third of the distance between A and B; but A could still be seen, when the slit was wide, as a still darker band on a dark red background; B was still sharp, and the other absorptions all strengthened

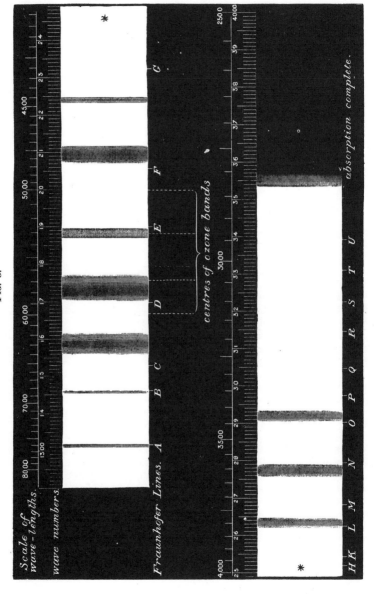

Fig. 3.

and somewhat expanded. The diffuse edges of several bands now extended from about —(1) λ 6410 to 6190, (2) λ 5865 to 5635, (3) λ 5350 to 5280, (4) λ 4820 to 4710, λ 4480 to 4455.

Photographs taken when the pressure of the oxygen was 90 atmospheres show a faint absorption-band about L of the solar spectrum, a stronger band extending from about λ 3600 to 3640, a broad diffuse band about the place of the solar line O, and complete absorption above P. The accompanying diagram, Fig. 3, represents the absorption of 18 metres of ordinary oxygen at a pressure of about 97 atmospheres.

The absorbent column in the tube at the highest pressure used contained a mass of oxygen about equal to that in a vertical column of the earth's atmosphere of the same section as the tube; but the intensity of the bands produced by the compressed gas was far greater than that of the corresponding bands in the solar spectrum with a low sun. When the arc light was replaced by a piece of white paper reflecting light from the sky through the tube, it appeared to the naked eye to have a faint blue tint, similar to that of liquid oxygen, which, comparing our observation with Olszewski's, seems to have the same absorptive powers as the dense gas, if we except A. This exception is probably only apparent, and due to the difficulty of observing A under the circumstances of Olszewski's experiment.

The greatly increased intensity of the absorption-bands at high pressures bears out Jansen's observation, that in this group the absorption is proportional to the product of the thickness of the absorbent stratum into the square of its density, while the absorptions to which A and B belong vary directly as the density.

The appearance, on looking through the tube when gas at high pressure is streaming into it, is very much like that of a black and a colourless liquid, which do not mix, being stirred together, and the tube soon ceases to transmit any light. Transparency returns as the density becomes uniform. Currents produced by heating the tube at one or two points produce a similar effect, and show that such currents in the atmosphere of a star may stop all rays coming from its interior.

We hope before long to get the tube fitted with rock-salt ends and lenses, and to determine the total absorption of radiation by similar masses of oxygen, nitrogen, and hydrogen.

[J. D.]

Friday, May 17, 1889.

JOHN RAE, M.D. LL.D. F.R.S. Vice-President, in the Chair.

PROFESSOR SILVANUS P. THOMPSON, B.A. D.Sc. *M.R.I.*

*Optical Torque.**

SEVENTY-EIGHT years have elapsed since the first discovery, by Arago, of the remarkable chromatic effects produced by slices of quartz crystals upon light, previously polarised, which was caused to traverse them. These effects were shown, one year later, by Biot, to be caused by a peculiar action of the quartz in rotating the plane of polarisation; the amount of the rotation being different for lights of different colours. Ever since then, the rotation of the plane of polarisation of light has been a topic familiar to physicists. It has stimulated the devotee of research to an endless variety of experiments and suggestive speculations: it has lured on the mathematician to problems which tax his utmost skill: it has afforded to the lecturer an array of beautiful and striking illustrations. Here, in this place, made classical by the researches and expositions of Thomas Young, of Michael Faraday, and of William Spottiswoode, and last, but not least, by the labours of those eminent men whom we rejoice still to number amongst the living—here, I say, on this classic ground, the rotation of the plane of polarisation of light is almost a household word, and its phenomena are amongst the most familiar. We know now that not only certain actual crystals, such as quartz, bromate of soda, and cinnabar, rotate the plane of polarisation, but that many non-crystalline bodies—liquids, such as turpentine, oil of lemons, solutions of sugar and of various alkaloids, and even certain vapours, such as that of camphor—possess the same property.

In 1845, at the very culminating point of his unique career of research, Faraday opened a new field of inquiry, linking together for the first time the science of optics with that of magnetism, by his discovery that the rotation of the plane of polarisation of light could be effected by the application of magnetic forces. This effect he observed first in his peculiar "heavy-glass," when it lay in a powerful magnetic field. Subsequently he found other bodies to possess similar properties: some of these being magnetic liquids, such as solutions of iron, others being diamagnetic. Time will only permit me in passing to refer to the researches of Verdet, and those of Lord Rayleigh and of Mr. Gordon upon the numerical values of the

---

* The blocks of the woodcuts illustrating this discourse have been kindly lent by the publishers of *Nature*.

magneto-optic rotation in these substances. H. Becquerel has extended them to gases, and has shown how the magnetism of the earth rotates the plane of polarisation of the light which, previously polarised by reflection from the aërial particles which give the sky its blue tint, passes earthward through the oxygen of the air.

Other experimenters have dealt with the rotatory effects (whether crystalline, molecular, or magnetic) in relation to lights of different colours, and have studied the dispersion which arises from the greater actual angle of optical torsion which is produced upon waves of short wave-length (violet and blue) than that which is produced under the influence of equal rotatory forces upon the waves of longer wave-length (red and orange). It has also been demonstrated that the plane of polarisation of waves of invisible light, whether those of the infra-red, or those of the ultra-violet species, if they have been previously polarised, can be rotated just as can that of waves of visible light.

In 1877, Dr. Kerr, of Glasgow, discovered a point which Faraday had sought for, but fruitlessly—namely, that in the act of reflection at the pole or surface of a magnet, there is a rotation of the plane of polarisation of light. This discovery was completed in 1884 by Kundt, of Strasburg, by the further demonstration, also dimly foreseen by Faraday, that a magneto-optic rotation of the plane of polarisation is caused by the passage of previously polarised light through a normally magnetised film of iron so thin as to be transparent.

Lastly, in this brief enumeration, we were shown a month ago, by Oliver Lodge, how the magnetic impulses generated by the rapid oscillatory discharges of the Leyden jar can produce corresponding rapid oscillatory rotation in the plane of polarisation of the waves of previously polarised light.

You will not have failed to notice the cumbrous phrase which, whether in speaking of the purely optical effects (of quartz, or sugar, or turpentine), or in speaking of the magneto-optic effects of more recent discovery, I have employed to connote a very simple fact. You may have wondered that any lover of simple English speech should indulge in such sesquipedalian words.

Of course, at this period of the nineteenth century it is no longer open to debate that light consists of waves. The plane of polarisation of the waves of light is the plane of polarisation of the light itself. The rotation of the plane of polarisation is the rotation of the polarised waves, and therefore of the polarised light itself. Yet I must draw attention to the fact that in all the array of discoveries which I have enumerated, that which had been observed was the rotation— whether by crystalline, molecular, or magnetic means—not of natural light, but of light which had by some means been previously polarised. It was not known to Arago or to Biot, to Fresnel, to Faraday, nor even to Spottiswoode or to Maxwell, that natural unpolarised light could be rotated. They may have inferred so, but it was not in their time even demonstrable that a beam of circularly-polarised light could be rotated upon itself in the same sense as that in which a beam of plane-polarised light can be rotated.

That light of any and every kind, however completely polarised or devoid of that which is called polarisation, can be, and in fact is, rotated when it passes across a slice of quartz or along a magnetic field, is a wider generalisation of more recent date; but one of the reality of which I hope to convince you before the warning finger of the clock puts a period to my discourse.

In order the better to enable this audience to comprehend the ultimate significance of this discovery, I must claim the indulgence of those amongst them who are already familiar with the subject of the polarisation of light, whilst I go back to the most simple elementary matters. Having illustrated the fundamental facts about the plane of polarisation of light and its twisting, I shall then go on to methods of precisely measuring the amount of optical torsion produced by the various substances under various conditions. And after dealing with the magnetic as well as the crystalline and molecular methods of producing optical torsion in the case of light that has been previously polarised into a given plane, I shall be in a position to speak of the nature of the torque,* or twisting force, which in the several cases produces the torsion; and shall finally endeavour to indicate the scope of the researches by which it is now definitely ascertained that the very same optical forces which are capable of impressing a rotation upon light which has been artificially polarised into a definite plane, are also capable of impressing a rotation upon natural non-polarised light.

At the outset, to elucidate to any who may not comprehend the meaning of the term polarisation as applied to wave-motion, I will show a simple apparatus, constructed from my designs by Mr. Groves. In this there are two sets of movable beads, fixed upon stems which pass into a box containing a piece of mechanism actuated by means of a handle. These beads, when I turn the handle, oscillate to and fro in definite directions, and, by their successive motions, give rise to progressive waves. One set of beads, tinted red, executes movements in a plane inclined 45° to the right, another set, silvered, simultaneously executes movements at 45° to the left. There are therefore here two waves, the planes of polarisation of their movements being at right angles to one another. Their velocity of march is equal; but in this model, as a matter of fact, their phases differ by one-quarter—that is to say, each successive wave of the one set is always a quarter of a wave-length behind the corresponding wave of the other set. [Model exhibited.]

---

* The convenient term *Torque* was first proposed by Prof. James Thomson, of Glasgow, for the older and more cumbrous phrase "moment of couple," or "angular force." Its general acceptance by engineers justifies the extension of the term to optics. As a mechanical torque is that which produces or tends to produce mechanical torsion, so optical torque may be defined as that which produces or tends to produce optical torsion.

Now, in the case of waves of natural light from all ordinary sources—sun, stars, candles, gas-flames, or electric lights—the waves emitted are not found to be polarised. That is to say, their motions are not executed in any particular plane, nor even in any particular path of any kind; they appear to be absolutely heterogeneous at least so far as this, that no vibration of the millions of millions emitted in a second of time is followed by more (on the average) than about 50,000 vibrations of a similar sort, executed along a similar path—the plane of the polarisation, if any, changing after the lapse of such an incredibly short time that for most purposes the vibrations in different directions are as inextricably mixed as if they had all been simultaneously jumbled up. Since, then, natural light is non-polarised or miscellaneous, the production of polarised light must be brought about by the employment of polarising apparatus or agents which will so operate on or affect the mixed waves as to bring their vibrations into one direction—or, what amounts to the same thing, transmit the light whilst destroying or absorbing those parts of the vibrations which are executed *across* the desired line of vibration. So we have *polarisers* consisting of tourmaline slices; oblique bundles of thin glass plates; black-glass reflectors; and Nicol prisms cut from calc-spar. About the two latter I may be permitted a passing word presently. These objects polarise, i. e. turn into one plane, the vibrations of light falling upon them. A rough mechanical illustration may here be permitted me. A long indiarubber cord is passed through the open ends of a box provided with vertical partitions. Fig. 1 shows the arrangement. These partitions confine the motion of the cord, and effectually polarise the vibrations which I now impart to the cord by shaking the end of it to and fro. If the partitions are vertical, the box polarises, into vertical vibrations only, the miscellaneous vibrations which are sent to it. If rotated until its partitions are horizontal, it polarises the vibrations into a horizontal position.

Let us now turn to the optical analogue of this experiment. The large Nicol prism which I introduce into the field of the electric-light lantern, polarises the light, so that the vibrations are executed simply in an up-and-down direction. Your eye will not detect this, the motion being millions of times too rapid. To detect the direction an analyser is necessary. For this purpose a second apparatus of the same sort is used, for then, by crossing the positions of the two, the whole of the light is cut off; the second Nicol prism, if set so as to transmit only horizontal vibrations, cutting off the vertical vibrations that are sent through the first prism. So, while the first prism serves as a polariser, the second serves as an analyser to detect by cutting them off when turned to the proper position, the direction of the polarisation which had been previously impressed by the first prism.

Here I may illustrate the action of the analyser for determining the plane of polarisation of the vibrations by the extinction which it produces when turned to the crossed position. For this purpose I

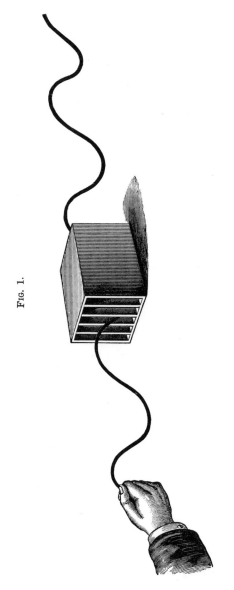

Fig. 1.

Box with partitions to illustrate polarisation of vibrations.

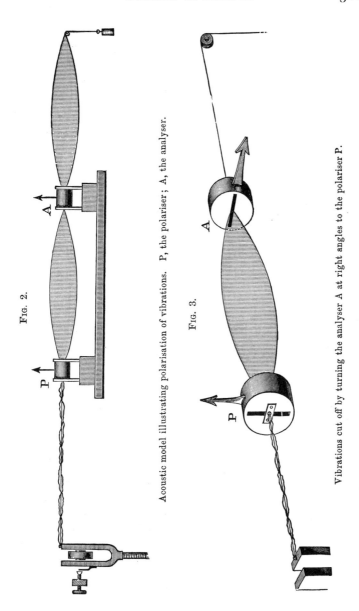

Fig. 2.

Acoustic model illustrating polarisation of vibrations. P, the polariser; A, the analyser.

Fig. 3.

Vibrations cut off by turning the analyser A at right angles to the polariser P.

have refined upon the box with partitions, using instead parallel plates of glass mounted in wooden cylinders, whilst for the cord swung by hand I am using Professor Schwedoff's device, and am producing the vibrations in this silken cord by means of an electrically-driven tuning-fork (Fig. 2). At the first nodal point of the stretched cord a pair of parallel glass plates act as a polariser, the cord beyond that point vibrating in the plane thus imposed upon it. I can alter this plane at will by rotating the polariser. This polariser, P, consisting of a pair of glass plates, is mounted in a cylindrical mount, and is provided with an arrow to indicate their direction. If now at any subsequent node I introduce a second such device, it will act as an analyser, A. This excellent suggestion is due to M. Macé de Lepinay. In Fig. 2 the polariser and analyser are parallel. You see (Fig. 3) how the vibration is extinguished when the positions of analyser and polariser are crossed. Half a degree of error in the position of the analyser produces something less than perfect extinction of the vibrations. Hence it is possible, by this analyser, to determine the plane of the vibrations to the accuracy of half a degree. I should say that the whole of this model has been constructed by my assistant Mr. Eustace Thomas.

Now let me show you the optical effect which corresponds to this. Placing a second Nicol prism as analyser in the path of the polarised waves, I turn it to the position where it cuts off the polarised light. The "*dark field*" so produced by the crossed Nicol prisms corresponds to the motionless cord beyond the crossed analyser of the acoustic apparatus.

Returning for a moment to two well-known forms of polarising apparatus, viz. the black glass reflector and the Nicol prism, I may be permitted to refer to some recent attempts to improve upon these devices.

The Nicol prism, as is well known, consists of a rhomb of Iceland spar cut into two pieces, which are reunited by a film of Canada balsam. As originally devised, it had oblique end faces (Fig. 4) and a comparatively narrow angle (19°) of aperture. These may be noticed in the small example which I here exhibit, which is an original constructed by William Nicol himself. It also has the disadvantage of giving a field in which the directions of the planes of polarisation are not strictly parallel to one another throughout its whole extent. Consequently there is never complete extinction of light all over the field at one time. Hartnack and others have attempted to remedy this by giving the prism a different form and using other materials than Canada balsam. I have from time to time made many attempts to improve upon the original construction.

Fig. 4.

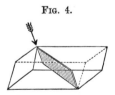

Nicol prism: original form: arrow shows crystallographic axis.

First, I have made the end faces principal planes of section (Fig. 5); secondly I have made the axis of vision cross the crystallographic axis at right angles, so getting a flatter field, a shorter length, a wider angle, and less loss of light by reflection. Mr. Ahrens, the prism-cutter, on whose able assistance I have relied during the last six or seven years in cutting these prisms, has aided me with his ingenuity in devising a method of cutting up the spar so as to give these advantages with a minimum waste of material. He has further devised a method of putting a polarising prism together in three

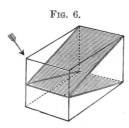

Fig. 5.   Fig. 6.

S. P. Thompson's modification of the Nicol prism.   Ahrens's triple prism.

instead of two pieces—illustrated in the diagram (Fig. 6)—which gives a still wider angle. The prism which I shall use as analyser in the next experiments is one of these forms.

Unfortunately at present there is a spar-famine, pieces of Iceland spar of a size and purity suitable for the making of large polarisers such as that I employ being not now procurable at any price. To avoid the excessive cost of large Nicols I have lately got Mr. Ahrens to construct for me a large reflection-polariser, on the plan of Delezenne, but modified by Mr. Ahrens in detail. In this prism the light is first turned to the proper polarising angle by a large total-reflection prism of glass, and then reflected back, parallel to its original path, by impinging upon a mirror of black glass covered by a single sheet of the thinnest patent plate glass to increase the intensity of the light. This form of polariser, depicted in Fig. 7, is quite equal for projection purposes to a Nicol prism of equal aperture, and is much less costly. This one has $2\frac{7}{8}$ inches clear aperture.

Having so far reviewed the apparatus for polarising and analysing I will return to the apparatus set with its prisms crossed, so that the analyser completely extinguishes the polarised light emitted from the polariser.

If in the space between polariser and analyser anything be introduced which can either resolve obliquely the polarised vibrations or twist them bodily round, then there will not be complete extinction; the amount of light passing the analyser depending in the one case on the obliquity of the resolution, in the other upon the degree to which the vibrations are twisted or rotated upon themselves.

The effect of oblique resolution I may illustrate by introducing a slice of tourmaline between the crossed Nicols, and rotating it till it stands at 45°; or, in the acoustic model, by introducing an oblique pair of guide pins.

The other case—namely, that of producing a bodily twist of the vibrations, rotating the plane of polarisation around the path of the wave—is not so easily illustrated by the model. But it is optically perfectly simple: all that is requisite is to introduce between the

Fig. 7.

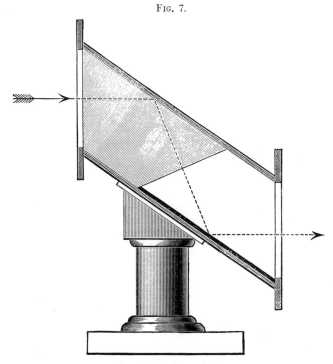

Ahrens's reflecting polariser.

crossed Nicols a thin slice of that crystal—namely, quartz—in which this effect of rotating the plane of polarisation was first observed.

I take a clear plate of quartz, just 1 millimetre in thickness, and interpose it between the crossed Nicol prisms. You will note how the introduction of this plate of quartz brings *some* light into view.

Suppose we now turn the analyser to try and obtain extinction: we get tinting. If we put in a coloured glass so as to work with one kind of light only, we shall get extinction at a particular angle. The table of data to which I invite your attention states this amount for the different colours.

OPTICAL TORSION PRODUCED BY PLATE OF QUARTZ.

	1 millimetre.	3·75 millimetres.
Red	19°	71°·2
Orange	21·5	80·6
Yellow	24	90
Green	29	108·7
Peacock	31	116·2
Blue	35·5	133·1
Violet	42·8	161

If we use a piece of quartz so thick that it rotates any particular tint just 90°, that tint will be cut off by the crossed analyser, and all others will—in greater or less proportion—be transmitted, so that the resulting tint will be complementary to that cut off. For example, a slice so thick as to twist yellow waves round 90° must be just 3·75 millimetres thick. (I may remark, for the benefit of those who think it easier to express this exact thickness in fractions of a British inch, that the quartz which rotates yellow light 90° must have a thickness equal to one-eighth, plus three-sixteenths of an eighth, plus one sixty-fourth of an eighth of an inch.) When such a quartz is placed between the crossed Nicols the light shown is yellow; but if placed between parallel Nicols (i. e. in the bright field) it shows a rich purplish-violet colour, the complementary of the yellow. This particular tint Biot found to be excessively sensitive, the smallest inaccuracy in adjustment between the prisms at once producing a change, the colour appearing too red or too blue, according to the directions in which the analyser has been turned out of exact adjustment. This tint is accordingly known as the "transition tint" or "sensitive tint," its accurate definition being due to the fact that the human eye is more sensitive to the presence or absence of the complementary yellow than to any other tint in the whole spectrum. If we take, however, a quartz plate twice as thick as this—namely, $7\frac{1}{2}$ millimetres thick—it will give the yellow light a torsion of 180°. Hence this thickness gives the purple transition tint in the dark field, and yellow in the bright field. A quartz plate $11\frac{1}{4}$ millimetres thick gives again a transition tint in the bright field. I shall recur presently to the question of the transition tints of the several orders.

One of the familiar facts in this subject is that there are two kinds of quartz crystals, optically alike in every other respect, differing only in this, that one kind produces a right-handed twist, the other kind a left-handed twist. All the pieces of quartz I have so far employed are right-handed specimens. I now introduce two small slices of crystal, each 3·75 millimetres thick, giving the yellow tint when the Nicols are exactly crossed, but you will notice that when we are using the right-handed crystal, the tint grows reddish as the analyser is turned towards the left, and greenish when the analyser is turned towards the right; whereas, when I substitute the left-handed slice, the tint grows greenish as the analyser is turned towards the left, and reddish when it is turned towards the right. If the analyser is turned through an exact right-angle, we get an extinction of the yellow light, the remaining blue and red rays combining to give us the purple transition tint.

You will have noticed that the way in which we have (approximately) measured the angle of rotation has been first to set the analyser to extinction, then to introduce the substance which has the property of rotating the beam, then to turn the analyser again to extinction, and read off its angle. For, of course, the angle through which the analyser is turned measures the angle through which the plane of polarisation has been turned.

It is possible, however, to show in the lantern something like a more obvious rotation of the light by introducing between the Nicols a crystal star, built up of radial pieces of mica, twenty-four in number (Fig. 8). You see in the bright field a white cross with black sectors at 45°. Or, in the dark field we have a black cross with vertical and horizontal arms, the sectors next to those that are black seeming dusky. If now I put in a quartz plate between the star and the analyser, you see the cross shift round, and it shows colours, because the blue rays are twisted round more than the green, the green than the yellow, the yellow than the red. Repeating the experiment with the 3·75 millimetre quartz which turns yellow waves round just 90°, we get this gorgeous radiation of colours, and our black cross is turned into a yellow one. With the 7·5 millimetre quartz, the black cross is replaced by one of "transition" tint.

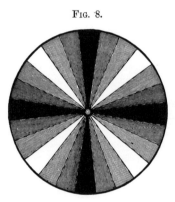

FIG. 8.

Mica disk of twenty-four rays, showing black cross in the dark field.

The black crosses seen in certain sections of natural crystals, sphæroliths, sections of stalactites, crystallisations of salicine and of Epsom salts, may also be used instead of the 24-rayed star of mica. But best of all I find to be the beautiful black cross which is seen by polarised light in the prepared crystalline lens taken from the eye of a fish. You notice how, when the fish-lens is projected and the quartz introduced, the cross turns round.

This is, however, a rough-and-ready way of displaying the rotation, and it is of vast practical importance that precise methods of measuring the angle of rotation should be available—of vast importance, because in several large industries this optical process is applied as a method of rapid analysis. I have named a solution of sugar as being an " active " substance. In the industry of sugar-refining, as in that of brewing, the strength of sugar in the liquids is directly measured by measuring its optical effect. Consequently there has been developed a special instrument, the *polarimeter*, for this express purpose.

I have here examples of several practical forms of polarimeters; there are diagrams of several more upon the walls.

The problem of finding the best polarimeter naturally leads to the inquiry what special means there are for making the observation of the angle more precise than by merely observing the extinction of the light, its restoration when the active substance is interposed, and the subsequent renewal of extinction when the analysing prism is turned.

Biot considered that much greater accuracy could be attained by watching for the restoration of the sensitive tint than by watching for the mere restoration of extinction of the light. Accordingly we will use the plate of quartz 7·5 millimetres thick, giving the purple tint, to enable us to measure the rotation produced by the tube of sugar solution which is now inserted in the beam of polarised light. You notice how the tint has changed. But I have only to turn the analyser to an amount equal to that to which the light has been twisted by the sugar, and again I obtain the sensitive transition tint.

The eye is not always, however, alive to minute changes of colour in a single coloured patch; it much more readily distinguishes a minute difference between two tints when both are present at once. Hence Soleil devised the well-known biquartz arrangement, consisting of two pieces of crystal, equal in thickness, but possessing opposite rotations. You will notice how the slightest inaccuracy in placing the analyser causes the two halves of the field to differ in tint. This is especially marked when the tint chosen is the transition purple.

It will be convenient here for me to refer to some researches, not yet published, which I have made, as to the various orders of transition tints, with the view of ascertaining which of them is the most sensitive —which of them, in fact, shows the greatest change of tint for the smallest amount of rotation. Reference to the diagram on the wall displaying Newton's tints will make clear what I mean by the transition tints of the several orders. The tints obtained from quartzes of varying thicknesses may be considered as approximately identical with the tints of Newton's rings, provided we remember that the air-film which gives any particular tint in Newton's rings is about 1/300,000 part as thick as the quartz which yields the corresponding tint in the polariscope. Better far than any painted diagram, because richer and purer, are the tints now thrown upon the screen by introducing into the field a thin wedge of selenite, displaying the whole of the colours of the first three orders of Newton's scale. You will notice the successive recurrence of purple tints, both in the colours seen in the bright field, and in those seen in the dark field.

First I will show you the transition tints of the first and second orders in the bright field. That of the second order is much less intense than that of the first; and yet it is very sensitive, turning to a green tint whilst the first order purple has only turned to a blue. On the other hand, with reversed rotation of the analyser it turns to red less rapidly than does the tint of the first order.

Next I take the transition tints of orders I., II., and III. in the dark field. These, though arranged, by means of superposed half-disks of "quarter-wave" plates, to be optically equivalent to biquartzes of two rotations, are really built up of selenite and mica. You will notice how the tint of order I. surpasses in sensitiveness both the others. I cannot here show you on the screen the means by which I have compared the tint of order I. in the dark field with that of order I. in the other set. Suffice it to say that I find the tint of order I. in the dark field—corresponding to 7·5 millimetres thickness—more sensitive than that of order I. in the bright field, which corresponds to 3·75 millimetres thickness.

A method which was at one time supposed to be more precise, was that of placing a spectroscope (or its prism) in front of the analyser, and watching the motion along the spectrum of the interference bands which are then seen. My three pieces of crystal remain. I introduce a slit in front of them, also a single film of quarter-wave mica, and then a prism to give the spectrum. This prism (Fig. 9), by the way, is a new sort of direct-vision prism, having a single very wide-angled prism of Jena glass inclosed in a cell with parallel ends containing cinnamic ether (first recommended by Wernicke), a liquid which has the same mean refractive power but widely different dispersion. It is preferable to bisulphide of carbon in several respects; its odour is a delicate reminiscence of cinnamon; it is barely volatile; and it is whiter than bisulphide. This prism, which is shown also in plan in Fig. 10, was constructed for me by Messrs. R. and J. Beck. It will be seen that the dark bands in the spectrum are nebulous and ill-defined. It was proposed to secure accuracy by turning the analyser until they shift along to a definite point. But their want of definition prevents precision. There is no advantage in using the higher orders of tints which give more bands; for, though the bands are certainly better defined, their progression across the spectrum for a given amount of rotation is proportionally smaller.

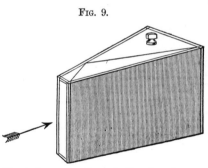

Fig. 9.

Direct-vision prism for projection of spectrum.

Another suggestion, due to Sénarmont, is to use two sets of superposed wedges of right- and left-handed quartz. Such you now see before you. Instead of starting with extinction you start with coinci-

Fig. 10.

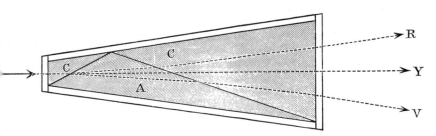

Direct-vision prism. A, wide-angled prism of Jena glass ; C, cinnamic ether.

dence between the upper and lower set of bands. Any rotation of the light shifts the bands, one set moving to left, the other to right. By turning the analyser through an equal angle coincidence is again obtained.

Another method, used by Wild in his polaristrobometer, is to produce the phenomenon known as Savart's bands (due to the introduction of two crossed slices of quartz cut at a particular angle). The bands disappear when the analyser is set in a particular direction. Anything that twists the plane of polarisation causes them to reappear; but they again fade out when the analyser is turned through an equal angle.

There is another method in exact polarimetry, due to Soleil, in which the optical torsion due to the sugar is counterbalanced or compensated by introducing a pair of sliding wedges of quartz of the opposite rotation. This device is known as a "compensator." By sliding the quartzes over one another a greater or less thickness of quartz is introduced at will. But I must not stop to illustrate this elegant device.

Yet one other method must be mentioned, and this is certainly the most preferable. It consists in aiding the eye to recognise with precision a particular degree of extinction, by the device, first suggested in 1856 by Pohl, of covering a portion of the visible field with something which slightly alters the initial plane of polarisation, so that complete blackness is not obtained at once over both parts of the field. A common device is to cover half the field with a slice of some thin crystal—mica or quartz—so that only one half can be perfectly black at any instant. As an example, here is the field covered half over with a plate of mica of the thickness known as half-wave. The result is that when one half of the field is black the other is light. Adjust the analyser now to equality. Now introduce something that rotates the light—say a tube with sugar solution in it. At once the balance is upset, and I must, in order to get equality of illumination, turn my analyser through an angle equal to that of the optical torsion.

Of the same class are the polarimeters with special prisms made in two parts slightly inclined to one another. The earliest of these was devised by the late Professor Jellett, of Dublin, and has been followed by imitations of the same plan by Cornu, by Lippich, and by Schmidt and Haensch. The beautiful "shadow polarimeter," by the latter firm, which I here exhibit, has the divided prism, and a quartz compensator.

I have suggested two simpler methods of accomplishing the same end. In the first place, I have proposed to use *twin-prisms*. These are made on a plan suggested to me by finding that Mr. Ahrens's method of cutting calc-spar for prisms was admirably adapted for making such prisms, either with wide or narrow angles between the respective planes of polarisation in the two parts of the visible field. Two such twin-prisms, one with $90°$, the other with $2\frac{1}{2}°$, between the prisms, are here on the table. In the second place, I have essayed a polarimeter, an example of which is before you, in which an arrangement of twin-mirrors (each set at the polarising angle, but slightly inclined to one another) is made to yield a half-shadow effect.

Before I leave the subject of quartz I must refer to the famous mathematical theory of Fresnel, who endeavoured to explain its action

FIG. 11.

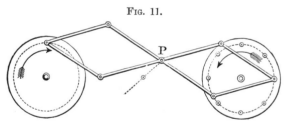

Model illustrating recomposition of rectilinear motion from two opposite circular motions.

upon light by supposing that the plane-polarised wave on entering it is split into two waves, consisting of oppositely circularly-polarised light, which traverse the crystal with different speeds. On emerging they recombine to form plane-polarised light, the plane of which, however, depends on the retardation of phase between the two components. I here introduce a mechanical model to illustrate one of the points in this theory—namely, the recombination of two circular motions to form a straight-line motion. These two disks (Fig. 11), which turn in opposite senses, but at equal rates, represent two circularly-polarised beams of light. The linkages, which connect two pins on these disks, compound their motions at the central point, P, which executes, as you see, a straight line. But now, suppose one of these circular motions to be retarded behind the other, an effect which I can imitate by shifting one of the pins to another position on the disk. Still the resultant motion is a straight line, but it is now executed in a direction oblique to the former. In other words, its plane has been rotated. Of course this model must not be taken as establishing the truth of Fresnel's ingenious theory; it is at best a rough kinematical representation of it.

We have, however, the puzzling fact still to account for that there should be two kinds of quartz crystals, right- and left-handed. Sir John Herschel first showed that natural crystals of quartz themselves often indicated their optical nature, by the presence of certain little secondary faces or facets which lay obliquely across the corners of the primary faces. These are indicated in the diagrams (Figs. 12 and 13), and may be seen in two of the specimens of quartz crystals

FIG. 12.   FIG. 13.

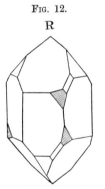
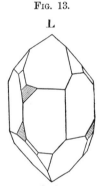

Quartz crystal, showing characteristic facets: right-handed.   Quartz crystal, showing characteristic facets: left-handed.

which lie upon the table. The largest of these is right-handed. The wider generalisations of Pasteur, respecting the crystalline form of optically active substances, show that those substances which exercise an optical torque, whether as crystals or in solution, belong to the class of forms which the crystallographer distinguishes as possessing non-superposable hemihedry. In other words, they all show *skew symmetry*, as if in the growth of them they had been built up in some screw-fashion around an axis, and must therefore be either right-handed or left-handed screws. By piling up a number of wooden slabs in skew-symmetric fashion, I am able roughly to illustrate (Figs. 14 and 15) the difference between the right-handed and the left-handed structure. It is a curious fact, if I am rightly informed, that down to the present date the only substances possessing this skew symmetry are natural substances; that those which the chemist can produce by artificial synthesis are all optically inactive. It is perhaps equally significant that as yet no inorganic substances have been found which will in the liquid state rotate the light. This appears to be a property possessed solely by certain compounds of carbon.

FIG. 14.   FIG. 15.

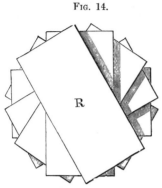 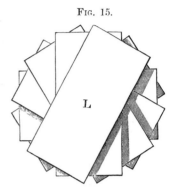

Skew-symmetrical arrangement: right-handed.   Skew-symmetrical arrangement: left-handed.

Quartz fused in the blowpipe or dissolved in potash shows no trace of rotatory power.

Yet we can have little doubt that this property is bound up in the yet unravelled facts of atomic and molecular structure. In the case of the liquids, such as turpentine, and sugar solution, there must be some skew symmetry in the grouping of atoms in the molecule to produce the result. In the case of quartz, there must be a skew in the building of the molecules; there must—to borrow a phrase from the architect—be an oblique *bonding* of the minute bricks of which its transparent mass is builded. Though we cannot even rebuild it from its solution, we know this must be so, for we can reproduce all the optical phenomena which it exhibits by an actual skew building of thin slices of another non-rotatory crystal. Here is an artificial object (I built it myself) constructed on Reusch's plan, from sixteen thin slips of mica built up in staircase fashion—right-handedly—one above the other, and set symmetrically at equal angles of 45° to one another, the whole set making a cork-screw of two complete turns. In the lantern it behaves just as a quartz of about 9 millimetres thickness would do. It even gives tolerably perfect rings, as quartz does, when viewed by convergent light.

I must now pass hastily onwards to the great discovery of Faraday. Here (Fig 16) is a magnetising coil of wire M, having about 8300 turns, and enclosed in an iron jacket. When it is traversed by a powerful electric current from the dynamo machine, it produces an intense magnetic field along its axis. In this axial position lies a bar of heavy glass, not quite so dense as that which Faraday himself used, but nearly so. The bar lies along the line of light from our lantern, but the polariser P (the Ahrens reflector, Fig. 7), and analyser A (the Ahrens triple spar prism, Fig. 6), are crossed, so that here is the dark field. On turning on the current, light is at once restored, being twisted to the right when the current circulates right-handedly. To measure the rotation, I must turn the analyser; and now I find that, owing to the greater rotation of blue waves than of red, complete extinction does not occur. Introducing a half-shadow plate, and using coloured glasses, it is very easy to verify the greater amount of rotation for blue light, and to show that reversing the current reverses the rotation. You will perhaps better understand it if I use (as in Fig. 16) the 24-ray star S, which I have previously employed. It is now obvious to you that there is a large rotation—over 50° in fact—which is reversed when I reverse the magnetising current. We have here the fundamental experiment of magneto-optics. But now we meet with another consideration. Reflect that the circulation of current, if it be taken as right-handed when regarded from one end of the coil, will be left-handed when regarded from the other end of the coil.

Fig. 16.

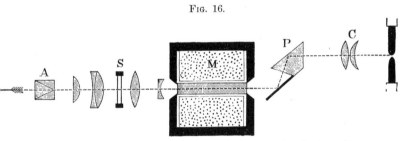

Projection of magnetic rotation of light. C, condensing lenses; P, reflecting polariser; magnetising coil surrounding bar of heavy glass; S, mica star of twenty-four rays; A, analyser (Ahrens's triple prism).

This is, therefore, no case of skew symmetry; it clearly indicates that something is going on in the glass which tends to twist the light quite irrespective of which way the light enters.

The next magneto-optic phenomenon is that discovered by Dr. Kerr of the rotation of the plane of polarisation by reflection at the surface of a magnet. To observe this at all requires good apparatus and a keen eye. So far as I am aware, it has never been projected on the screen. If I can succeed in doing so, it will only be because I have special means of the most favourable character for so doing. We withdraw the bar of heavy glass from the coil, and replace it (Fig. 17) by an iron core polished at its coned end. This will be intensely magnetised when the current is turned on.

Now we must throw the beam of light obliquely down the hollow of the coil, polarising it by one of my improved Nicol prisms P, as it goes down. After reflection it is focussed by a lens which sends it through the analysing prism A. You see the dim spot of reflected

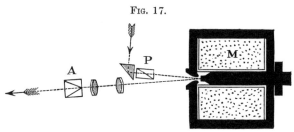

Fig. 17.

Apparatus for projecting rotation of plane of polarisation by reflection at pole of magnet. P, polariser; M, magnetising coil with coned iron core; A, analyser.

light upon the screen. Now for the current: "on," "off," "on," "off." Reversing its direction ought to double the amount of torsion.

Whilst Mr. Thomas is making the needful arrangements for the next experiment, I may mention that it was found by Kerr that the effect was approximately proportional to the magnetic induction through the iron. I have myself tried some further experiments: for example, using a bar of lodestone instead of an iron core. The light reflected from lodestone is also twisted. I should expect the ferro-aluminium alloy which Sir H. Roscoe showed us a fortnight ago to do the same thing, because that alloy is, as I have found, susceptible of magnetisation. But I should not expect manganese steel to rotate the light, because of its singularly non-magnetisable nature.

The experiment of Kundt, transmitting polarised light through a thin transparent film of iron, magnetised normally whilst the light is passing through it, is another difficult of repetition before an audience. The small disks here are covered with films of iron, kindly prepared for me by Mr. Crookes, by squirting them electrically in a high vacuum. But the thin ones barely transmit enough light to make the observation of the effect possible, even to the solitary observer. I have observed the effect projected on the screen, using this very coil and these transparent mirrors. It requires, however, an absolutely dark room, and is at best so faint that it would be hopeless to attempt to show it to a large audience. Professor Kundt has not only observed similar rotations in other magnetic films of nickel and cobalt but has even shown that the degree of rotation of the light is proportional not to the magnetising force, but to the resulting magnetic induction. This is a result of utmost importance in considering the theory of the phenomenon. He has further shown that whereas the magnetic rotations in elementary bodies, whether magnetic or diamagnetic, are in the same sense as that in which the current circulates, the magnetic rotations in compound magnetic bodies, such as a solution of sulphate of iron in water, are in the opposite sense.

These experiments with transparent mirrors of iron raise interesting speculations as to the probable nature of a transparent magnet, if such there could be. It is one of the cardinal points of Maxwell's celebrated electro-magnetic theory of light, that the better a body conducts electric currents, the greater is its tendency to absorb light and become opaque. Now, suppose it were possible to obtain a substance such as to possess greater electric conductivity in one direction than in another, such a substance ought to absorb those vibrations of light which are executed in the direction of the greater electric conductivity more than those in the direction at right angles. In other words, such a substance ought, like the tourmaline, to polarise light by absorption. Now, since the researches of Sir W. Thomson in 1856, we have known that the electric conductivity of iron is altered in the direction of the magnetic lines of force, when it is powerfully magnetised. More recently it has been discovered—I myself observed it in tinfoil, and announced the discovery to the Physical Society a few days before the announcement of the same fact by Righi—that non-magnetic metals alter their resistance in the magnetic field. Notably so do bismuth and tellurium. I had therefore conceived it possible that a film of iron or possibly of tellurium, if strongly magnetised in its own plane, might exhibit polar absorption and act like a tourmaline. Unfortunately, if the effect exists it is so faint as to be yet undiscovered, though I have made many efforts to find such. I have further tried to obtain a similar result by making a transparent magnet out of a film of magnetic oxide of iron, precipitated chemically. In this too I have not succeeded. I have tried to precipitate a transparent film of magnetic oxide in the midst of a transparent jelly. And I have mixed particles of precipitated oxide with melted gelatine so as to get a film. In this way I hoped to get, by placing the preparation in a strong magnetic field, a sort of magnetic structure which would operate upon waves of light. That such a task was not hopeless was shown by two facts: first, that many mere vegetable and animal structures can act as polarisers; and second, that a mere film of paint, such as indigo, can, if a proper mechanical drag is given to it so as to produce structure, also act as a polariser.

The film of indigo-carmine which I have here, acts nearly as strongly, though not quite as evenly, as a tourmaline slice, and costs but a fraction of a penny.

Well, my films of jelly enclosing particles of magnetic oxide of iron do faintly act on polarised light; but their action is not as marked as that of films of jelly enclosing actual small scraps of iron. This film, when placed across the poles of this electromagnet, between two crossed Nicols at 45°, shows an action when the magnet is turned on, as you see by the way in which it flashes into light in the dark field. When the jelly is fresh, and of the proper consistency, the action is very strong, but with the rather dry sample before you I fear we can only call the effect a *succès d'estime*.

Incidentally, in the course of these experiments on magnetic films,

I came across a new magnetic body unknown hitherto, I believe, to the chemist, namely, a magnetic double oxide of cobalt and iron—a ferroso-cobaltic oxide, I think—a black powder, a sample of which I have here.

It also occurred to me, as a matter of speculation, that if I could strongly magnetise a crystal of ferrous sulphate or nickelous sulphate, whilst viewing it by convergent polarised light, I might find some interesting phenomena, which should, if they existed, show some sort of a relation between the direction of the optic axis and that of the lines of the magnetic field. I thought that a longitudinal magnetisation might possibly set up a rotatory phenomenon like that in quartz in so far as to disturb the central field between the arms of the black cross; however, not by the most powerful magnetising could I discover any such effect. Again, I thought that by magnetising transversely to the optic axis I might possibly succeed in turning the uniaxial crystal into a biaxial, or producing by magnetism an effect resembling the action of heat on crystals of selenite. Owing probably to the small depth of any crystals that can be obtained, I have failed so far to obtain any such effect, though I am convinced that it must exist.

An effect precisely analogous to the magnetic effect which I vainly sought has, however, been lately discovered by Prof. Röntgen. I sought a distortion of the optic axis by transversely magnetising, and I sought it in crystals of sulphate of nickel; he has found a distortion of the optic axis by transversely electrifying, and he has found it in crystals of quartz.

Suppose a piece of quartz crystal is cut as a square prism, its long faces being principal planes of section respectively parallel to and at right angles to two of the natural faces of the hexagonal prism. Fig. 18 shows the form of the portion cut. The + and − signs in this

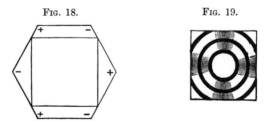

Fig. 18.   Fig. 19.

figure refer to the pyro-electric poles of the crystal. Such a piece viewed by convergent light shows the usual rings and black cross with a coloured centre (Fig. 19). If now two opposite faces be covered with tinfoil, and the crystal be electrified transversely, the rings are distorted into lemniscates, the direction of the distortion changing with the sign of the electrification. It is necessary to use a red glass, or still better sodium light, to observe the changes in form on reversing the sign of the charges. Figs. 20 and 21, 22 and 23, show the changes of form, but these sketches grossly exaggerate the effects. As you see upon the screen, when the charges imparted by this fine Wimshurst machine are rapidly reversed, there is a decided distortion of the rings, but it is small in amount.

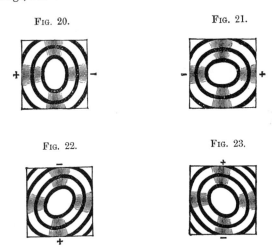

Fig. 20.   Fig. 21.

Fig. 22.   Fig. 23.

Returning to the phenomena of the rotation impressed by magnetism on polarised light, I may point out that the torque which a magnetic field exerts on the light waves appears to be really an action upon the matter through which the light-waves are passing. It is as though the magnetic field were really a portion of space rotating rapidly on itself, or perhaps as though the ether were there rotating, and that this rotation in some way dragged the particles of matter along with it. It has long been supposed necessary, in order to account for the refractive and dispersive properties of transparent bodies, to consider that their particles are in some way concerned in and partake of the vibrations going on in the ether within them or between their molecules. It is impossible to explain the phenomena of magneto-optic rotation by the supposition that any skew structure is imparted to the medium; for these phenomena unlike those of quartz, do not exhibit skew symmetry. There seems to be no other way of explaining the magneto-optic torsion of light than by supposing that the molecules of matter in the magnetic field are actually subjected to rotatory actions; as indeed was suggested long ago by Sir William Thomson.

However, there is room here not only for speculation but for experiment. Some day, when facts enough have been collected, we shall be ready to build thereon the wider generalisation which at present seems to escape us.

So far we have been applying an optical torque to previously polarised light, and producing a torsion of it. It remains for me yet to describe the means by which, in the hands of Professor Abbe and Professor Sohncke, it has been demonstrated that natural, non-polarised light is actually rotated when subjected to an optical torque.

The way of doing this is to make use of the principle of interference. Here is the slit from which a narrow beam of light-waves issues. At a point a little distance away is a Fresnel's biprism which splits up the light (without polarising it) into two beams, just as if we had two slits or sources of light. These two beams pass along, and meet upon this distant screen, and give us—what? A set of interference fringes having a bright line down the middle, because this part of the screen is exactly equidistant from the two sources of light.

But these dark interference fringes that lie right and left can only exist because, in the first place the vibrations have travelled unequal paths differing by an odd number of half wave-lengths; and secondly, because (owing to the method adopted of using two images of one slit) the phases of the emitted waves from the two sources are identical.

This being so, let us now introduce across the two interfering beams of light a special biquartz, made of right-and-left handed quartz of only 1·88 mm. thick. This will rotate—*if it rotates natural light at all*—the yellow light in one beam 45° to the right and that of the other beam 45° to the left. The angles will be a little more for green and blue, a little less for red and orange. Consequently we shall not get quite a perfect result for all kinds of colours. But for the main body of the light the result is this: that because the two beams have had their respective vibrations turned so that, whatever their primitive positions, they are now at right angles to one another, they cannot interfere. In other words, if it be true that the quartz rotates natural light, the interference bands will die out. [Experiment shown.]

Here I have the light passing through the biprism only, and giving us this narrow series of interference bands. You must notice carefully—with opera-glasses if you have them—the narrow bright and dark stripes. Now I shift this little diaphragm so that the light passes through the biquartz as well. Instead of sharp interference bands we have merely a dull line of nebulous light. The disappearance of the fringes proves that quartz does twist the non-previously polarised waves of light.

That the magnetic field can also exert a magnetic torque on non-polarised light is readily proved, at least when one already has the biquartz. Two strips of heavy glass of exactly equal length and similar quality, such as those I hold in my hand, must be introduced in the respective paths of the two beams: and one at least of them must be surrounded by a magnetising coil. The biquartz has wiped out the interference fringes; but on magnetising one of the two pieces of heavy-glass, or on magnetising the two in opposite senses, the interference bands can be made to reappear. It is in this way that Professor Sohncke's experiment—hardly suitable for a lecture theatre—was performed. It is in this way that we establish upon an experimental basis the fact that light itself, and not merely the plane of its polarisation, experiences an optical torsion when subjected to those forces which, whether crystalline, molecular, or magnetic, exert upon it an optical torque.

[S. P. T.]

Friday, May 31, 1889.

Sir Frederick Abel, C.B. D.C.L. F.R.S. Vice-President,
in the Chair.

D. Mendeléeff, Esq. LL.D.

professor of chemistry in the university of st. petersburg.

An Attempt to apply to Chemistry one of the Principles of Newton's Natural Philosophy.

Nature, inert to the eyes of the ancients, has been revealed to us as full of life and activity. The conviction that motion pervaded all things, which was first realised with respect to the stellar universe, has now extended to the unseen world of atoms. No sooner had the human understanding denied to the earth a fixed position and launched it along its path in space, than it was sought to fix immovably the sun and the stars. But astronomy has demonstrated that the sun moves with unswerving regularity through the star-set universe at the rate of about 50 kilometres per second. Among the so-called fixed stars are now discerned manifold changes and various orders of movement. Light, heat, electricity—like sound—have been proved to be modes of motion; to the realisation of this fact modern science is indebted for powers which have been used with such brilliant success, and which have been expounded so clearly at this lecture table by Faraday and by his successors. As in the imagination of Dante, the invisible air became peopled with spiritual beings, so before the eyes of earnest investigators, and especially before those of Clerk Maxwell, the invisible mass of gases became peopled with particles: their rapid movements, their collisions, and impacts became so manifest that it seemed almost possible to count the impacts and determine many of the peculiarities or laws of their collisions. The fact of the existence of these invisible motions may at once be made apparent by demonstrating the difference in the rate of diffusion through porous bodies of the light and rapidly moving atoms of hydrogen and the heavier and more sluggish particles of air. Within the masses of liquid and of solid bodies we have been forced to acknowledge the existence of persistent though limited motion of their ultimate particles, for otherwise it would be impossible to explain, for example, the celebrated experiments of Graham on diffusion through liquid and colloidal substances. If there were, in our times, no belief in the molecular motion in solid bodies, could the famous Spring have hoped to attain any result by mixing carefully dried powders of potash, saltpetre, and acetate of soda, in order to produce, by pressure, a chemical reaction between these substances through the interchange of their metals, and have derived, for the conviction of the incredulous, a mixture of two hygroscopic though solid salts—nitrate of soda and acetate of potash?

In these invisible and apparently chaotic movements, reaching from the stars to the minutest atoms, there reigns, however, a harmonious order which is commonly mistaken for complete rest, but which is really a consequence of the conservation of that dynamic equilibrium which was first discerned by the genius of Newton, and which has been traced by his successors in the detailed analysis of the particular consequences of the great generalisation, namely, relative immovability in the midst of universal and active movement.

But the unseen world of chemical changes is closely analogous to the visible world of the heavenly bodies, since our atoms form distinct portions of an invisible world, as planets, satellites, and comets form distinct portions of the astronomer's universe; our atoms may therefore be compared to the solar systems, or to the systems of double or of single stars, for example, ammonia (NH^3) may be represented in the simplest manner by supposing the sun nitrogen surrounded by its planets of hydrogen; and common salt ($NaCl$) may be looked upon as a double star formed of nitrogen and chlorine. Besides, now that the indestructibility of the elements has been acknowledged, chemical changes cannot otherwise be explained than as changes of motion, and the production by chemical reactions of galvanic currents, of light, of heat, of pressure, or of steam power, demonstrate visibly that the processes of chemical reaction are inevitably connected with enormous though unseen displacements, originating in the movements of atoms in molecules. Astronomers and natural philosophers, in studying the visible motions of the heavenly bodies and of matter on the earth, have understood and have estimated the value of this store of energy. But the chemist has had to pursue a contrary course. Observing in the physical and mechanical phenomena which accompany chemical reactions the quantity of energy manifested by the atoms and molecules, he is constrained to acknowledge that within the molecules there exist atoms in motion, endowed with an energy which, like matter itself, is neither being created nor is capable of being destroyed. Therefore, in chemistry, we must seek dynamic equilibrium not only between the molecules but also in their midst among their component atoms. Many conditions of such equilibrium have been determined, but much remains to be done, and it is not uncommon, even in these days, to find that some chemists forget that there is the possibility of motion in the interior of molecules, and therefore represent them as being in a condition of death-like inactivity.

Chemical combinations take place with so much ease and rapidity; possess so many special characteristics, and are so numerous, that their simplicity and order was for a long time hid from investigators. Sympathy, relationship, all the caprices or all the fancifulness of human intercourse, seemed to have found complete analogies, in chemical combinations, but with this difference, that the characteristics of the material substances—such as silver, for example, or of any other body—remain unchanged in every subdivision from the largest masses to the smallest particles, and consequently their characteristics must be a property of its particles. But the world of heavenly luminaries appeared equally fanciful at man's first acquaintance with it, so much so, that the astrologers imagined a connection between the individualities of men and the conjunctions of planets. Thanks to the genius of Lavoisier and of Dalton, man has been able, in the unseen world of chemical combinations, to recognise laws of the same simple order as those which Copernicus and Kepler proved to exist in the planetary universe. Man discovered, and continues every hour to discover *what* remains unchanged in chemical evolution, and *how* changes take place in combinations of the unchangeable. He has learned to predict, not only what possible combinations may take place, but also the very existence of atoms of unknown elementary bodies, and has besides succeeded in making innumerable practical applications of his knowledge to the great advantage of his race, and has accomplished this notwithstanding that notions of sympathy and affinity still preserve a strong vitality in science. At present we cannot apply Newton's principles to chemistry, because the soil is only being now prepared. The invisible world of chemical atoms is still waiting for the creator of chemical mechanics. For him our age is collecting a mass of materials, the inductions of well-digested facts, and many-sided inferences similar to those which existed for Astronomy and Mechanics in the days of Newton. It is well also to remember that Newton devoted much time to chemical experiments, and while considering questions of celestial mechanics, persistently kept in view the mutual action of those infinitely small worlds which are concerned in chemical evolutions. For this reason, and also to maintain the unity of laws, it seems to me that we must, in the first instance, seek to harmonise the various phases of contemporary chemical theories with the immortal principles of the Newtonian natural philosophy, and so hasten the advent of true chemical mechanics. Let the above considerations serve as my justification for the attempt which I propose to make to act as a champion of the universality of the Newtonian principles, which I believe are competent to embrace every phenomenon in the universe, from the rotation of the fixed stars, to the interchanges of chemical atoms.

In the first place I consider it indispensable to bear in mind that, up to quite recent times, only a one-sided affinity has been recognised in chemical reactions. Thus, for example, from the circumstance that red-hot iron decomposes water with the evolution of hydrogen, it was concluded that oxygen had a greater affinity for iron than for hydrogen. But hydrogen, in presence of red-hot iron scale, appropriates its oxygen and forms water, whence an exactly opposite conclusion may be formed.

During the last ten years a gradual, scarcely perceptible, but most important change has taken place in the views, and consequently in the researches of chemists. They have sought everywhere, and have always found systems of conservation or dynamic equilibrium substantially similar to those which natural philosophers have long since discovered in the visible world, and in virtue of which the position of the heavenly bodies in the universe is determined. There where one-sided affinities only were at first detected, not only secondary or lateral ones have been found, but even those which are diametrically opposite, yet among these, dynamical equilibrium establishes itself not by excluding one or other of the forces, but regulating them all. So the chemist finds in the flame of the blast furnace, in the formation of every salt, and, with especial clearness, in double salts, and in the crystallisation of solutions, not a fight ending in the victory of one side, as used to be supposed, but the conjunction of forces; the peace of dynamic equilibrium resulting from the action of many forces and affinities. Carbonaceous matters, for example, burn at the expense of the oxygen of the air, yielding a quantity of heat and forming products of combustion, in which it was thought that the affinities of the oxygen with the combustible elements were satisfied. But it appeared that the heat of combustion was competent to decompose these products, to dissociate the oxygen from the combustible elements, and therefore, to explain combustion fully, it is necessary to take into account the equilibrium between opposite reactions, between those which evolve, and those which absorb heat.

In the same way, in the case of the solution of common salt in water, it is necessary to take into account, on the one hand, the formation of compound particles generated by the combination of salt with water, and on the other the disintegration or scattering of the new particles formed, as well as of those originally contained. At present we find two currents of thought, apparently antagonistic to each other, dominating the study of solutions: according to the one, solution seems a mere act of building up or association; according to the other, it is only dissociation or disintegration. The truth lies, evidently, between these views; it lies, as I have endeavoured to prove by my investigations into aqueous solutions, in the dynamic equilibrium of particles tending to combine and also to fall asunder. The large majority of chemical reactions which appeared to act victoriously along one line have been proved capable of acting as victoriously even along an exactly opposite line. Elements which utterly decline to combine directly may often be formed into comparatively stable compounds by indirect means, as for example in the case of chlorine and carbon; and consequently the sympathies and antipathies, which it was thought to transfer from human relations to those of atoms, should be laid aside until the mechanism of chemical relations is explained. Let us remember, however, that chlorine, which does not form with carbon the chloride of carbon, is strongly absorbed, or, as it were, dissolved by carbon, which leads us to suspect incipient chemical action even in an external and purely surface contact, and involuntarily gives rise to conceptions of that unity of the forces of nature which has been so energetically insisted on by Sir William Grove and formulated in his famous paradox. Grove noticed that platinum, when fused in the oxyhydrogen flame, during which operation water is formed, when allowed to drop into water decomposes the latter and produces the explosive oxyhydrogen mixture. The explanation of this paradox, as of many others which arose during the period of chemical renaissance has led, in our time, to the promulgation by Henri St. Claire Deville of the conception of dissociation and of equilibrium, and has recalled the teaching of Berthollet, which, notwithstanding its brilliant confirmation by Heinrich Rose and Dr. Gladstone, had not, up to that period, been included in received chemical views.

Chemical equilibrium in general, and dissociation in particular, are now being so fully worked out in detail, and applied in such various ways, that I do not allude to them to develop, but only use them as examples by which to indicate the correctness of a tendency to regard chemical combinations from points of view differing from those expressed by the term hitherto appropriated to define chemical forces, namely, "affinity." Chemical equilibria dissociation, the speed of chemical reactions, thermo-chemistry, spectroscopy, and, more than all, the determination of the influence of masses and the search for a connection between the properties and weights of atoms and molecules; in one word, the vast mass of the most important chemical researches of the present day, clearly indicates the near approach of the time when chemical doctrines will submit fully and completely to the doctrine which was first announced in the Principia of Newton.

In order that the application of these principles may bear fruit it is evidently insufficient to assume that statical equilibrium reigns alone in chemical systems or chemical molecules: it is necessary to grasp the conditions of possible states of dynamical equilibria, and to apply to them kinetic principles. Numerous considerations compel us to renounce the idea of statical equilibrium in molecules, and the recent yet strongly supported appeals to dynamic principles constitute, in my opinion, the foundation of the modern teaching relating to atomicity, or the valency of the elements, which usually forms the basis of investigations into organic or carbon compounds.

This teaching has led to brilliant explanations of very many chemical relations and to cases of isomerism, or the difference in the properties of substances having the same composition. It has been so fruitful in its many applications and in the foreshadowing of remote consequences, especially respecting carbon compounds, that it is impossible to deny its claims to be ranked as a great achievement of chemical science. Its practical application to the synthesis of many substances of the most complicated composition entering into the structure of organised bodies, and to the creation of an unlimited number of carbon compounds, among which the colours derived from coal tar stand prominently forward, surpass the synthetical powers of Nature itself. Yet this teaching, as applied to the structure of carbon compounds, is not on the face of it directly applicable to the investigation of other elements, because in examining the first it is possible to assume that the atoms of carbon have always a definite and equal number of affinities, while in the combinations of other elements this is evidently inadmissible. Thus, for example, an atom of carbon yields only one compound with four atoms of hydrogen and one with four atoms of chlorine in the molecule, while the atoms of chlorine and hydrogen unite only in the proportions of one to one. Simplicity is here evident and forms a point of departure from which it is easy to move forward with firm and secure tread. Other elements are of a different nature. Phosphorus unites with three and with five atoms of chlorine, and consequently the simplicity and sharpness of the application of structural conceptions are lost. Sulphur unites only with two atoms of hydrogen, but with oxygen it enters into higher orders of combination. The periodic relationship which exists among all the properties of the elements, such, for example, as their ability to enter into various combinations, and their atomic weights, indicate that this variation in atomicity is subject to one perfectly exact and general law, and it is only carbon and its near analogues which constitute cases of permanently preserved atomicity. It is impossible to recognise as constant and fundamental properties of atoms, powers which, in substance, have proved to be variable. But by abandoning the idea of permanence, and of the constant saturation of affinities—that is to say, by acknowledging the possibility of free affinities—many retain a comprehension of the atomicity of the elements "under given conditions"; and on this frail foundation they build up structures composed of chemical molecules, evidently only because the conception of manifold affinities gives, at once, a simple statical method of estimating the composition of the most complicated molecules.

I shall enter neither into details, nor into the various consequences following from these views, nor into the disputes which have sprung up respecting them (and relating especially to the number of isomers possible on the assumption of free affinities), because the foundation or origin of theories of this nature suffers from the radical defect of being in opposition to dynamics. The molecule, as even Laurent expressed himself, is represented as an architectural structure, the style of which is determined by the fundamental arrangement of a few atoms, while the decorative details, which are capable of being varied by the same forces, are formed by the elements entering into the combination. It is on this account that the term "structural" is so appropriate to the contemporary views of the above order, and that the "constructors" seek to justify the tetrahedric, plane, or prismatic disposition of the atoms of carbon in benzole. It is evident that the consideration relates to the statical position of atoms and molecules and not to their kinetic relations. The atoms of the structural type are like the lifeless pieces on a chess board: they are endowed but with the voices of living beings, and are not those living beings themselves; acting, indeed, according to laws, yet each possessed of a store of energy, which, in the present state of our knowledge, must be taken into account.

In the days of Haüy, crystals were considered in the same statical and structural light, but modern crystallographers, having become more thoroughly acquainted with their physical properties and their actual formation, have abandoned the earlier views and have made their doctrines dependent on dynamics.

The immediate object of this lecture is to show that, starting with Newton's third law of motion, it is possible to preserve to chemistry all the advantages arising from structural teaching, without being obliged to build up molecules in solid and motionless figures, or to ascribe to atoms definite limited valencies, directions of cohesion, or affinities. The wide extent of the subject obliges me to treat only a small portion of it, namely of *substitutions*, without specially considering combinations and decompositions, and, even then, limiting myself to the simplest examples, which, however, will throw open prospects embracing all the natural complexity of chemical relations. For this reason, if it should prove possible to form groups similar, for example, to H^4 or CH^6 as the remnants of molecules CH^4 or C^2H^6 we shall not pause to consider them, because, as far as we know, they fall asunder into two parts, $H^2 + H^2$ or $CH^4 + H^2$, as soon as they are even temporarily formed, and are capable of separate existence, and therefore can take no part in the elementary act of substitution. With respect to the simplest molecules which we shall select—that is to say, those of which the parts have no separate existence, and therefore cannot appear in substitutions— we shall consider them according to the periodic law, arranging them in direct dependence on the atomic weight of the elements.

Thus, for example, the molecules of the simplest hydrogen compounds—

HF	H^2O	H^3N	H^4C
hydrofluoric acid	water	ammonia	methane

correspond to elements the atomic weights of which decrease consecutively,

$$F = 19, O = 16, N = 14, C = 12.$$

Neither the arithmetical order (1, 2, 3, 4 atoms of hydrogen) nor the total information we possess respecting the elements will permit us to interpolate into this typical series one more additional element; and therefore we have here, for hydrogen compounds, a natural base upon which are built up those simple chemical combinations which we take as typical. But even they are competent to unite with each other, as we see, for instance, in the property which hydrofluoric acid has of forming a hydrate, that is of combining with water; and the similar attribute of ammonia, resulting in the formation of a caustic alkali, NH^3H^2O, or NH^4OH.

Having made these indispensable preliminary observations, I may now attack the problem itself and attempt to explain the so-called structure, or rather construction of molecules, that is to say, their constitution and transformations without having recourse to the teaching of "structionists," but on Newton's dynamical principles.

Of Newton's three laws of motion, only the third can be applied directly to chemical molecules when regarded as systems of atoms among which it must be supposed that there exist common influences or forces, and resulting compounded relative motions. Chemical reactions of every kind are undoubtedly accomplished by changes in these internal movements, respecting the nature of which nothing is known at present, but the existence of which, the mass of evidence collected in modern times, forces us to acknowledge as forming part of the common motion of the universe, and as a fact further established by the circumstance that chemical reactions are always characterised by changes of volume or the relations between the atoms or the molecules. Newton's third law, which is applicable to every system, declares that, " action is always associated with reaction, and is equal to it." The brevity and conciseness of this axiom was, however, qualified by Newton in a more expanded statement, " the action of bodies one upon another are always equal, and in opposite directions." This simple fact constitutes the point of departure for explaining dynamic equilibrium, that is to say, systems of conservancy. It is capable of satisfying even the dualists, and of explaining, without additional assumptions, the preservation of those chemical types which Dumas, Laurent, and Gerhardt created unit types, and those views of atomic combinations which the structionists express by atomicity or the valency of the elements, and, in connection with them, the various numbers of affinities. In reality if a system of atoms or a molecule be given, then in it, according to the third law of Newton, each portion of atoms acts on the remaining portion in the same manner, and with the same force as the second set of atoms acts on the first. We infer directly from this consideration that both sets of atoms, forming a molecule, are not only equivalent with regard to themselves, as they must be according to Dalton's law, but also that they may, if united, replace each other. Let there be a molecule containing atoms A B C, it is clear that, according to Newton's law, the action of A on B C must be equal to the action of B C on A, and if the first action is directed on B C, then the second must be directed on A, and consequently then, where A can exist in dynamic equilibrium, B C may take its place and act in a like manner. In the same way the action of C is equal to the action of A B. In one word every two sets of atoms forming a molecule are equivalent to each other, and may take each other's place in other molecules, or, having the power of balancing each other, the atoms or their complements are endowed with the power of replacing each other. Let us call this consequence of an evident axiom "the principle of substitution," and let us apply it to those typical forms of hydrogen compounds, which we have already discussed, and which, on account of their simplicity and regularity, have served as starting points of chemical argument long before the appearance of the doctrine of structure.

In the type of hydrofluoric acid, HF, or in systems of double stars, are included a multitude of the simplest molecules. It will be sufficient for our purpose to recall a few: for example the molecules of chlorine, Cl^2, and of hydrogen, H^2, and hydrochloric acid, HCL, which is familiar to all in aqueous solution as spirit of salt, and which has many points of resemblance with HF, HB_2, HI. In these cases division into two parts can only be made in one way, and therefore the principle of substitution renders it probable that exchanges between the chlorine and the hydrogen can take place, if they are competent to unite with each other. There was a time when no chemist would even admit the idea of any such action; it was then thought that the power of combination indicated a polar difference of the molecules in combination, and this thought set aside all idea of the substitution of one component element by another.

Thanks to the observations and experiments of Dumas and Laurent fifty years ago, such fallacies were dispelled, and in this manner, this same principle of substitution was exhibited. Chlorine and bromine acting on many hydrogen compounds, occupy immediately the place of their hydrogen, and the displaced hydrogen, with another atom of chlorine or bromine, forms hydrochloric acid or bromide of hydrogen. This takes place in all typical hydrogen compounds. Thus chlorine acts on this principle on gaseous hydrogen—reaction, under the influence of light, resulting in the formation of hydrochloric acid. Chlorine acting on the alkalis, constituted similarly to water, and even on water itself—only, however, under the influence of light and only partially because of the unstability of HClO—forms by this principle bleaching salts, which are the same as the alkalis, but with their hydrogen replaced by chlorine. In ammonia and in methane, chlorine can also replace the hydrogen. From ammonia is formed in this manner the so-called chloride of nitrogen, NCl^3, which decomposes very readily with violent explosion on account of the evolved gases, and falls asunder as chlorine and nitrogen. Out of marsh gas, or methane, CH^4, may be obtained consecutively, by this method, every possible substitution, of which chloroform, $CHCl^3$, is the best known, and chloro-carbonic acid, CCl^4 the most instructive. But by virtue of the fact that chlorine and bromine act, in the manner shown, on the simplest typical hydrogen compounds, their action on the more complicated ones may be assumed to be the same. This can be easily demonstrated. The hydrogen of benzole, C^6H^6, reacts feebly under the influence of light on liquid bromine, but Gustavson has shown that the addition of the smallest quantity of metallic aluminium causes energetic action, and the evolution of large volumes of bromide of hydrogen,

If we pass on to the second typical hydrogen compound, that is to say water, its molecule, HOH, may be split up in two ways: either into an atom of hydrogen and a molecule of oxide of hydrogen, HO, or into oxygen, O, and two atoms of hydrogen, H; and therefore, according to the principle of substitution, it is evident that one atom of hydrogen can exchange with oxide of hydrogen, HO, and two atoms of hydrogen, H, with one atom of oxygen, O.

Both these forms of substitution will constitute methods of oxidation, that is to say, of the entrance of oxygen into the compound—a reaction which is so common in nature as well as in the arts, taking place at the expense of the oxygen of the air or by the aid of various oxidising substances or bodies which part easily with their oxygen. There is no occasion to reckon up the unlimited number of cases of such oxidising reactions. It is sufficient to state that in the first of these oxygen is directly transferred, and the position, the chemical function, which hydrogen originally occupied is, after the substitution, occupied by the hydroxyl. Thus ammonia, NH^3, yields hydroxylamine, $NH^2(OH)$, a substance which retains many of the properties of ammonia.

Methane and a number of other hydrocarbons yield, by substitution of the hydrogen by its oxide, methylic, $CH^3(OH)$, and other alcohols. The substitution of one atom of oxygen for two atoms of hydrogen is equally common with hydrogen compounds. By this means alcoholic liquids containing ethyl alcohol, or spirits of wine, $C^2H^5(OH)$, are oxidised till they become vinegar or acetic acid, $C^2H^3O(OH)$. In the same way caustic ammonia, or the combination of ammonia with water, NH^3H^2O, or $NH^4(OH)$, which contains a great deal of hydrogen, by oxidation exchanges four atoms of hydrogen for two atoms of oxygen, and become converted into nitric acid $NO^2(OH)$. This process of conversion of ammonia salts into saltpetre goes on in the fields every summer, and with especial rapidity in tropical countries. The method by which this is accomplished, though complex, though involving the agency of all-permeating microorganisms, is, in substance, the same as that by which alcohol is converted into acetic acid, or glycol, $C^2H^4(OH)^2$, into oxalic acid, if we view the process of oxidation in the light of the Newtonian principles.

But while speaking of the application of the principle of substitution to water, we need not multiply instances, but must turn our attention to two special circumstances which are closely connected with the very mechanism of substitutions.

In the first place, the replacement of two atoms of hydrogen by one atom of oxygen may take place in two ways, because the hydrogen molecule is composed of two atoms, and therefore, under the influence of oxygen, the molecule forming water may separate before the oxygen has time to take its place. It is for this reason that we find, during the conversion of alcohol into acetic acid, that there is an interval during which is formed aldehyde, C^2H^4O, which, as its very name implies, is "alcohol dehydrogenatum," or alcohol deprived of hydrogen. Hence aldehyde combined with hydrogen yields alcohol; and united to oxygen, acetic acid.

For the same reason there should be, and there actually are, intermediate products between ammonia and nitric acid, $NO^2(HO)$, containing either less hydrogen than ammonia, less oxygen than nitric acid, or less water than caustic ammonia. Accordingly we find, among the products of the de-oxidisation of nitric acid and the oxidisation of ammonia, not only hydroxylamine, but also nitrous oxide, nitrous and nitric anhydrides. Thus, the production of nitrous acid results from the removal of two atoms of hydrogen from caustic ammonia and the substitution of the oxygen for the hydrogen, $NO(OH)$; or by the substitution, in ammonia, of three atoms of hydrogen by hydroxyl, $N(OH)^3$, and by the removal of water; $N(OH)^3 - H^2O = NO(OH)$. The peculiarities and properties of nitrous acid, as, for instance, its action on ammonia and its conversion, by oxidation, into nitric acid, are thus clearly revealed.

On the other hand, in speaking of the principle of substitution as applied to water, it is necessary to observe that hydrogen and hydroxyl, H and OH, are not only competent to unite, but also to form combinations with themselves, and thus become H^2 and H^2O^2; and such are hydrogen and the peroxide thereof. In general, if a molecule AB exists, then molecules AA and BB can exist also. A direct reaction of this kind does not, however, take place in water, therefore undoubtedly, at the moment of formation hydrogen reacts on the peroxide of hydrogen, as we can show, at once, by experiment; and further because the peroxide of hydrogen, H^2O^2, exhibits a structure containing a molecule of hydrogen, H^2, and one of oxygen, O^2, either of which is capable of separate existence. The fact, however, may now be taken as thoroughly established, that, at the moment of combustion of hydrogen or of the hydrogen compounds, peroxide of hydrogen is always formed, and not only so, but in all probability its formation invariably precedes the formation of water. This was to be expected as a consequence of the law of Avogadro and Gerhardt, which leads us to expect this sequence in the case of equal interactions of volumes of vapours and gases; and in the peroxide of hydrogen we actually have such equal volumes of the elementary gases.

The instability of peroxide of hydrogen—that is to say, the ease with which it decomposes into water and oxygen, even at the mere contact of porous bodies—accounts for the circumstance that it does not form a permanent product of combustion, and is not produced during the decomposition of water. I may mention this additional consideration that, with respect to the peroxide of hydrogen, we may look for its effecting still further substitutions of hydrogen by means of which we may expect to obtain still more highly oxidised water-compounds, such as H^2O^3 and H^2O^4. These Schönbein and

Bunsen have long been seeking, and Berthelot is investigating them at this moment. It is probable, however, that the reaction will stop at the last compound, because we find that, in a number of cases, the addition of 4 atoms of oxygen seems to form a limit. Thus, OsO^4, $KClO^4$, $KMnO^4$, K^2SO^4, Na^3PO^4, and such like, represent the highest grades of oxidation.*

As for the last 40 years, from the times of Berzelius, Dumas, Liebig, Gerhardt, Williamson, Frankland, Kolbe, Kekulé, and Butlerow, most theoretical generalisations have centred round organic or carbon compounds, so we will, for the sake of brevity, leave out the discussion of ammonia derivatives, notwithstanding their simplicity in respect to the doctrine of substitutions; we will dwell more especially on its application to carbon compounds, starting from methane, CH^4, as the simplest of the hydrocarbons, containing in its molecule one atom of carbon. According to the principles enumerated we may derive from CH^4 every combination of the form CH^3X, CH^2X^2, CHX^3, and CX^4, in which X is an element, or radical, equivalent to hydrogen, that is say, competent to take its place or to combine with it. Such are the chlorine substitutes mentioned already, such is wood-spirit, $CH^3(OH)$, in which X is represented by the residue of water, and such are numerous other carbon derivatives. If we continue, with the aid of hydroxyl, further substitutions of the hydrogen of methane we shall obtain successively $CH^2(OH)^2$, $CH(OH)^3$, and $C(OH)^4$. But if, in proceeding thus, we bear in mind that $CH^2(OH)^2$ contains two hydroxyls in the same form as peroxide of hydrogen, H^2O^2 or $(OH)^2$, contains them—and moreover not only in one molecule, but together, attached to one and the same atom of carbon—so here we must look for the same decomposition as that which we find in peroxide of hydrogen, and accompanied also by the formation of water as an independently existing molecule;

---

* Because more than four atoms of hydrogen never unite with one atom of the elements, and because the hydrogen compounds (e. g. HCl, H^2S, H^3P, H^4Si) always form their highest oxides with four atoms of oxygen, and as the highest forms of oxides (OSO^4 RO^4) also contain four of oxygen, and eight groups of the periodic system, corresponding to the highest basic oxides R^2O, RO, R^2O^3, RO^2, R^2O^5, RO^3, R^2O^7, and RO^4, imply the above relationship, and because of the nearest analogues among the elements—such as Mg, Zn, Cd, and Hg; or Cr, Mo, W and U; or Si, Ge, Sn, and Pt; or F, Cl, Br, and J and so forth—not more than four are known, it seems to me that in these relationships there lies a deep interest and meaning with regard to chemical mechanics. But because, to my imagination, the idea of unity of design in Nature, either acting in complex celestial systems or among chemical molecules, is very attractive, especially because the atomic teaching at once acquires its true meaning, I will recall the following facts relating to the solar system. There are eight major planets, of which the four inner ones are not only separated from the four outer by asteroids, but differ from them in many respects, as for example in the smallness of their diameters and their greater density. Saturn with his ring has eight satellites, Jupiter and Uranus have each four. It is evident that in the solar systems also we meet with these higher numbers four and eight which appear in the combination of chemical molecules.

therefore $CH^2(OH)^2$ should yield, as it actually does, immediately water and the oxide of methylene, CH^2O, which is methane with oxygen substituted for two atoms of hydrogen. Exactly in the same manner out of $CH(OH)^3$ are formed water and formic acid, $CHO(OH)$, and out of $C(OH)^4$ is produced water and carbonic acid, or directly carbonic anhydride, CO^2, which will therefore be nothing else than methane with the double replacement of pairs of hydrogen by oxygen. As nothing leads to the supposition that the four atoms of hydrogen in methane differ one from the other, so it does not matter by what means we obtain any one of the combinations indicated—they will be identical; that is to say, there will be no case of actual isomerism, although there may easily be such cases of isomerism as have been distinguished by the term metamerism.

Formic acid, for example, has two atoms of hydrogen, one attached to the carbon left from the methane, and the other attached to the oxygen which has entered in the form of hydroxyl, and if one of them be replaced by some substance X it is evident that we shall obtain bodies of the same composition, but of different construction, or of different orders of movement among the molecules, and therefore endowed with other properties and reactions. If X be methyl, CH^3, that is to say, a group capable of replacing hydrogen because it is actually contained with hydrogen in methane itself, then by substituting this group for the original hydrogen, we obtain acetic acid, $CCH^3O(OH)$, out of formic, and by substitution of the hydrogen in its oxide or hydroxyl we obtain methyl formiate, $CHO(OCH^3)$. These bodies differ so much from each other physically and chemically that, at first sight, it is hardly possible to admit that they contain the same atoms in identically the same proportions. Acetic acid, for example, boils at a higher temperature than water, and has a higher specific gravity than it, while its metamer, formo-methylic ether, is lighter than water, and boils at 30°, that is to say, it evaporates very easily.

Let us now turn to carbon compounds containing two atoms of carbon to the molecule, as in acetic acid, and proceed to evolve them from methane by the principle of substitution. This principle declares at once that methane can only be split up in the four following ways:—

1. Into a group CH^3 equivalent with H. Let us call changes of this nature methylation.

2. Into a group CH^2 and H^2. We will call this order of substitutions methylenation.

3. Into CH and H^3, which commutations we will call acetylenation.

4. Into C and H^4, which may be called carbonisation.

It is evident that hydrocarbon compounds containing two atoms of carbon, can only proceed from methane, CH^4, which contains four atoms of hydrogen by the first three methods of substitution; carbonising would yield free carbon if it could take place directly, and if the molecule of free carbon—which is in reality very complex, that is to say strongly polyatomic, as I have long since been proving by various means—could contain only C^2 like the molecules O^2, H^2, N^2, and so on.

By methylation we should evidently obtain from marsh gas, ethane, $CH^3CH^3 = C^2H^6$.

By methylenation, that is by substituting group CH^2 for H^2, methane forms ethylene, $CH^2CH^2 = C^2H^4$.

By acetylenation, that is by substituting three atoms of hydrogen, H^3, in methane, by the remnant CH, we get acetylene $CHCH = C^2H^2$.

If we have applied the principles of Newton correctly, there should not be any other hydrocarbons containing two atoms of carbon in the molecule. All these combinations have long been known, and in each of them we can not only produce those substitutions of which an example has been given in the case of methane, but also all the phases of other substitutions, as we shall find from a few more instances, by the aid of which I trust that I shall be able to show the great complexity of those derivatives which, on the principle of substitution, can be obtained from each hydrocarbon. Let us content ourselves with the case of ethane, CH^3CH^3, and the substitution of the hydrogen by hydroxyl. The following are the possible changes.

1. $CH^3CH^2(OH)$: this is nothing more than spirit of wine, or ethyl alcohol, $C^2H^5(OH)$ or C^2H^6O.

2. $CH^2(OH)CH^2(OH)$: this is the glycol of Wurtz, which has shed so much light on the history of alcohol. Its isomer may be $CH^3CH(OH)^2$, but as we have seen in the case of $CH(OH)^2$, it decomposes giving off water, and forming aldehyde, CH^3CHO, a body capable of yielding alcohol by uniting with hydrogen and of yielding acetic acid by uniting with oxygen.

If glycol $CH^2(OH)CH^2(OH)$ loses its water, it may be seen at once that it will not now yield aldehyde, CH^3CHO, but its isomer, $\begin{matrix}CH^2CH^2\\O\end{matrix}$, the oxide of ethylene. I have here indicated in a special manner the oxygen which has taken the place of two atoms of the hydrogen of ethane taken from different atoms of the carbon.

3. $CH^3C(OH)^3$ decomposed as $CH(OH)^3$, forming water and acetic acid $OH^3CO(OH)$. It is evident that this acid is nothing else than formic acid, $CHO(OH)$, with its hydrogen replaced by methyl. Without examining further the vast number of possible derivatives, I will direct your attention to the circumstance that in dissolving acetic acid in water we obtain the maximum contraction and the greatest viscosity when to the molecule $CH^3CO(OH)$ is added a molecule of water, which is the proportion which would form the hydrate $CH^3C(OH)^3$. It is probable that the doubling of the molecule of acetic acid at temperatures approaching its boiling point has some connection with this power of uniting with one molecule of water.

4. $CH^2(OH)C(OH)^3$ is evidently alcoholic acid, and indeed this compound, after losing water, answers to glycolic acid, $CH^2(OH)CO$

(OH). Without investigating all the possible isomers, we will note only that the hydrate $CH(OH)^2CH(OH)^2$ has the same composition as $CH^2(OH)C(OH)^3$, and although corresponding to glycol, and being a symmetrical substance, it becomes on parting with its water aldehyde of oxalic acid, or the glyoxal of Debus, $CHOCHO$.

5. $CH(OH)^2C(OH^3)$, from the tendency of all the preceding, corresponds to glyoxylic acid, aldehyde acid, $CHOCO(OH)$, because the group $CO(OH)$, or carboxyl, enters into the compositions of organic acids, and the group CHO defines the aldehyde function.

6. $C(OH)^3C(OH)^3$ through the loss of $2H^2O$ yields the bibasic oxalic acid $CO(OH)CO(OH)$, which generally crystallises with $2H^2O$, following thus the normal type of hydration characteristic of ethane.*

Thus, by applying the principle of substitution, we can, in the simplest manner, derive not only every kind of hydrocarbon compound, such as the alcohols, the aldehyde alcohols, aldehydes, alcohol acids, and the acids, but also combinations analogous to hydrated crystals which usually are disregarded.

But even those unsaturated substances, of which ethylene, CH^2CH^2, and acetylene, $CHCH$, are types, may be evolved with equal simplicity. With respect to the phenomena of isomerism, there are many possibilities among the hydrocarbon compounds containing two atoms of carbon, and without going into details it will be sufficient to indicate that the following formulæ, though not identical, will be isomeric substantially among themselves:—CH^3CHX^2 and CH^2XCH^2X, although both contain $C^2H^4X^2$, or CH^2CX^2 and $CHXCHX$, although both contain $C^2H^2X^2$, if by X we indicate chlorine or generally an element capable of replacing one atom of hydrogen, or capable of uniting with it. To isomerism of this kind belongs the case of aldehyde and the oxide of ethylene, to which we have already referred, because both have the composition C^2H^4O.

---

* One more isomer, $CH^2CH(OH)$, is possible, that is secondary vinyl alcohol, which is related to ethylene, CH^2CH^2, but derived by the principle of substitution from CH^4. Other isomers of the composition C^2H^4O, such, for example, as $CHCH^3(OH)$, are impossible, because it would correspond to the hydrocarbon $CHCH^3 = C^2H^4$, which is isomeric with ethylene, and it cannot be derived from methane. If such an isomer existed, it would be derived from CH^2, but such products are up to the present unknown. In such cases the insufficiency of the points of departure of the statical structural teaching is shown. It first admits constant atomicity and then rejects it, the facts serving to establish either one or the other view; and therefore, it seems to me, that we must come to the conclusion that the structural method of reasoning, having done a service to science, has outlived the age, and must be regenerated as, in their time, was the teaching of the electro-chemists, the radicalists, and the adherents of the doctrine of types. As we cannot now lean on the views above stated, it is time to abandon the structural theory. They will all be united in chemical mechanics, and the principle of substitution must be looked upon only as a preparation for the coming epoch in chemistry, where such cases as the isomerism of fumaric and maleic acids, when explained dynamically, as proposed by Le Bel and Van't Hoff, may yield points of departure.

What I have said appears to me sufficient to show that the principle of substitution adequately explains the composition, the isomerism and all the diversity of combination of the hydrocarbons, and I shall limit the further development of these views to preparing a complete list of every possible hydrocarbon compound containing three atoms of carbon in the molecule. There are eight in all, of which only five are known at present.*

Among those possible for C^3H^6 there should be two isomers, propylene and trimethylene, and they are both already known. For C^3H^4 there should be three isomers: allylene and allene are known, but the third has not yet been discovered; and for C^3H^2 there should be two isomers, though neither of them are known as yet. Their composition and structure is easily deduced from ethane, ethylene, and acetylene, by methylation, methylenation, by acetylenation and by carbonisation.

1. $C^3H^8 = CH^3CH^2CH^3$ out of CH^3CH^3 by methylation. This hydrocarbon is named propane.

2. $C^3H^6 = CH^3CHCH^2$ out of CH^3CH^3 by methylenation. This substance is propylene.

3. $C^3H^6 = CH^2CH^2CH^2$ out of CH^3CH^3 by methylenation. This substance is trimethylene.

4. $C^3H^4 = CH^3CCH$ out of CH^3CH^3 by acetylenation or from CHCH by methylation. This hydrocarbon is named allylene.

5. $C^3H^4 = CHCH$ out of CH^3CH^3 by acetylenation, or from $\begin{array}{c}CH^2\\CH^2CH^2\end{array}$ by methylenation, because $\begin{array}{cc}CH^2CH = CHCH\\CH & CH^2\end{array}$. This body is as yet unknown.

6. $C^3H^4 = CH^2CCH^2$ out of CH^2CH^2 by methylenation. This hydrocarbon is named allene, or iso-allylene.

7. $C^3H^2 = CHCH$ out of CH^3H^3 by symmetrical carbonisation,
$\phantom{C^3H^2 = CHCH \text{ out of}} C$
or out of CH^2CH^2 by acetylenation. This compound is unknown.

8. $C^3H^2 = CC$ out of CH^3CH^3 by carbonisation, or out of $\begin{array}{c}CHCH\\CH^2\end{array}$ by methylenation. This compound is unknown.

If we bear in mind that for each hydrocarbon serving as a type in the above tables there are a number of corresponding derivatives, and that every compound obtained may, by further methylation, methylenation, acetylenation, and carbonisation, produce new hydrocarbons, and these may be followed by a numerous suite of derivatives and an immense number of isomeric bodies, it is possible to understand the limitless number of carbon compounds, although they all

---

* Conceding variable atomicity, the structurists must expect an incomparably larger number of isomers, and they cannot now decline to acknowledge the change of atomicity, were it only for the examples HgCl and $HgCl^2$, CO and CO^2, PCl^3 and PCl^5.

have the one substance, methane, for their origin. The number of substances is so enormous that it is no longer a question of enlarging the possibilities of discovery, but rather of finding some means of testing them, analogous to the well-known two which for a long time have served as gauges for all carbon compounds.

I refer to the law of even numbers and to that of limits, the first enunciated by Gerhardt forty years ago, with respect to hydrocarbons, namely, that their molecules always contain an even number of atoms of hydrogen. But by the method which I have used of deriving all the hydrocarbons from methane, CH^4, this law may be deduced as a direct consequence of the principle of substitutions. Accordingly, in methylation, CH^3 takes the place of H, and therefore CH^2 is added. In methylenation the number of atoms of hydrogen remains unchanged, and at each acetylenation it is reduced by two, and in carbonisation by four atoms, that is to say, an even number of atoms of hydrogen is always added or removed. And because the fundamental hydrocarbon, methane, CH^4, contains an even number of atoms of hydrogen, therefore all its derivative hydrocarbons will also contain even numbers of hydrogen, and this constitutes the law of even numbered parts.

The principle of substitutions explains with equal simplicity the conception of limiting compositions of hydrocarbons, C^nH^{2n+2}, which I derived, in 1861,* in an empirical manner from accumulated materials available at that time, and on the basis of the limits to combinations worked out by Dr. Frankland for other elements.

Of all the various substitutions the highest proportion of hydrogen is yielded by methylation, because in that operation alone does the quantity of hydrogen increase; therefore, taking methane as a point of departure, if we imagine methylation effected $(n-1)$ times we obtain hydrocarbon compounds containing the highest quantities of hydrogen. It is evident that they will contain

$$CH^4 + (n-1)\ CH^2,\ \text{or}\ C^nH^{2n+2},$$

because methylation leads to the addition of CH^2 to the compound.

It will thus be seen that by the principle of substitution—that is to say, by the third law of Newton—we are able to deduce, in the simplest manner, not only the individual composition, the isomerism, and relations of substances, but also the general laws which govern their most complex combinations, without having recourse either to statical constructions, to the definition of atomicities, to the exclusion of free affinities, or to the recognition of those single, double, or treble ties which are so indispensable to structurists in the explanation of the composition and construction of hydrocarbon compounds. And yet, by the application of the dynamic principles of Newton, we can

---

* 'Essai d'une théorie sur les limites des combinaisons organiques,' par D. Mendeléeff, 2/11 août 1861, 'Bulletin de l'Académie i. d. Sc. de St. Pétersbourg,' t. v.

attain to that chief and fundamental object—the comprehension of isomerism in hydrocarbon compounds, and the forecasting of the existence of combinations as yet unknown, by which the edifice raised by structural teaching is strengthened and supported. Besides, and I count this for a circumstance of special importance, the process which I advocate will make no difference in those special cases which have been already so well worked out, such as, for example, the isomerism of the hydrocarbons and alcohols, even to the extent of not interfering with the nomenclature which has been adopted, and the structural system will retain all the glory of having worked up, in a thoroughly scientific manner, the store of information which Gerhardt had accumulated about the middle of the fifties, and the still higher glory of establishing the rational synthesis of organic substances. Nothing will be lost to the structural doctrine, except its statical origin; and as soon as it will embrace the dynamic principles of Newton, and suffer itself to be guided by them, I believe that we shall attain, for chemistry, that unity of principle which is now wanting. Many an adept will be attracted to that brilliant and fascinating enterprise, the penetration into the unseen world of the kinetic relations of atoms, to the study of which the last twenty-five years have contributed so much labour and such high inventive faculties.

D'Alembert found in mechanics, that if inertia be taken to represent force, dynamic equations may be applied to statical questions which are thereby rendered more simple and more easily understood.

The structural doctrine in chemistry has unconsciously followed the same course, and therefore its terms are easily adopted; they may retain their present forms provided that a truly dynamical, that is to say, Newtonian meaning be ascribed to them.

Before finishing my task and demonstrating the possibility of adapting structural doctrines to the dynamics of Newton, I consider it indispensable to touch on one question which naturally arises, and which I have heard discussed more than once. If bromine, the atom of which is eighty times heavier than that of hydrogen, takes the place of hydrogen, it would seem that the whole system of dynamic equilibrium must be destroyed.

Without entering into the minute analysis of this question, I think it will be sufficient to examine it by the light of two well-known phenomena, one of which will be found in the department of chemistry, and the other in that of celestial mechanics, and both will serve to demonstrate the existence of that unity in the plan of creation, which is a consequence of the Newtonian doctrines. Experiments demonstrate that when a heavy element is substituted for a light one, in a chemical compound—an atom of magnesium in the oxide of that metal, for example, for mercury, the atom of which is $8\frac{1}{3}$ times heavier—the chief chemical characteristics or properties are generally though not always preserved.

The substitution of silver for hydrogen, than which it is 108 times heavier, does not affect all the properties of the substance, though it does some. Therefore chemical substitutions of this kind, the substitution of light for heavy atoms, need not necessarily entail changes in the original equilibrium; and this point is still further elucidated by the consideration that the periodic law indicates the degree of influence of an increment of weight in the atom as affecting the possible equilibria, and also what degree of increase in the weight of the atoms reproduces some, though not all, the properties of the substance.

This tendency to repetition, these periods, may be likened to those annual or diurnal periods with which we are so familiar on the earth. Days and years follow each other: but, as they do so, many things change; and in like manner chemical evolutions, changes in the masses of the elements, permit of much remaining undisturbed, though many properties undergo alteration. The system is maintained according to the laws of conservation in nature, but the motions are altered in consequence of the change of parts.

Next, let us take an astronomical case, such for example as the earth and the moon, and let us imagine that the mass of the latter is constantly increasing. The question is, what will then occur? The path of the moon in space is a wave-line similar to that which geometricians have named epicycloidal, or the locus of a point in a circle rolling round another circle. But in consequence of the influence of the moon, it is evident that the path of the earth itself cannot be a geometric ellipse, even supposing the sun to be immovably fixed; it must be an epicycloidal curve, though not very far removed from the true ellipse, that is to say, it will be impressed with but faint undulations. It is only the common centre of gravity of the earth and the moon which describes a true ellipse round the sun. If the moon were to increase, the relative undulations of the earth's path would increase in amplitude, those of the moon would also change, and when the mass of the moon had increased to an equality with that of the earth, the path would consist of epicycloidal curves crossing each other, and having opposite phases. But a similar relation exists between the sun and the earth because the former is also moving in space. We may apply these views to the world of atoms, and suppose that, in their movements, when heavy ones take the place of those that are lighter, similar changes take place provided that the system or the molecule is preserved throughout the change.

It seems probable that in the heavenly systems, during incalculable astronomical periods changes have taken place and are still going on similar to those which pass rapidly before our eyes during the chemical reaction of molecules and the progress of molecular mechanics, may—we hope will—in course of time, permit us to explain those changes in the stellar world which have more than once been noticed by astronomers, and which are now so carefully studied. A coming Newton will discover the laws of these changes. Those laws, when applied to chemistry, may exhibit peculiarities, but these will certainly be mere variations on the grand harmonious theme which reigns in nature. The discovery of the laws which produce this harmony in chemical evolutions will only be possible, it seems to me, under the banner of Newtonian dynamics which have so long waved over the domains of mechanics, astronomy, and physics. In calling chemists to take their stand under its peaceful and catholic shadow I imagine that I am aiding in establishing that scientific union which the managers of the Royal Institution wish to effect, who have shown their desire to do so by the flattering invitation which has given me—a Russian—the opportunity of laying before the countrymen of Newton an attempt to apply to chemistry one of his immortal principles.

[D. M.]

Friday, June 14, 1889.

WILLIAM HUGGINS, Esq. D.C.L. LL.D. F.R.S. Vice-President,
in the Chair.

C. V. BOYS, Esq. F.R.S.

Quartz Fibres.

IN almost all investigations which the physicist carries out in the laboratory, he has to deal with, and to measure with accuracy, those subtle, and, to our senses, inappreciable forces to which the so-called laws of nature give rise. Whether he is observing by an electrometer the behaviour of electricity at rest, or by a galvanometer the action of electricity in motion; whether in the tube of Crookes he is investigating the power of radiant matter, or by the famous experiment of Cavendish he is finding the mass of the earth—in these and in a host of other cases he is bound to measure, with certainty and accuracy, forces so small that in no ordinary way could their existence be detected; while disturbing causes, which might seem to be of no particular consequence, must be eliminated if his experiments are to have any value. It is not too much to say that the very existence of the physicist depends upon the power which he possesses of producing at will, and by artificial means, forces against which he balances those that he wishes to measure.

I had better, perhaps, at once indicate in a general way the magnitude of the forces with which we have to deal.

The weight of a single grain is not to our senses appreciable, while the weight of a ton is sufficient to crush the life out of any one in a moment. A ton is about 15,000,000 grains. It is quite possible to measure, with unfailing accuracy, forces which bear the same relation to the weight of a grain that a grain bears to a ton.

To show how the torsion of wires or threads is made use of in measuring forces, I have arranged what I can hardly dignify by the name of an experiment. It is simply a straw hung horizontally by a piece of wire. Resting on the straw is a fragment of sheet-iron weighing ten grains. A magnet, so weak that it cannot lift the iron, yet is able to pull the straw round through an angle so great that the existence of the feeble attraction is evident to every one in the room.

Now it is clear that if, instead of a straw moving over the table simply, we had here an arm in a glass case and a mirror to read the motion of the arm, it would be easy to observe a movement a hundred or a thousand times less than that just produced, and, therefore, to measure a force a hundred or a thousand times less than that exerted by this feeble magnet.

Again, if instead of wire as thick as an ordinary pin, I had used the finest wire that can be obtained, it would have opposed the movement of the straw with a far less force. It is possible to obtain wire ten times finer that this stubborn material, but wire ten times finer is much more than ten times more easily twisted. It is ten thousand times more easily twisted. This is because the torsion varies as the fourth power of the diameter, so we say $10 \times 10 = 100$; $100 \times 100 = 10,000$. Therefore, with the finest wire, forces 10,000 times feebler still could be observed.

It is, therefore, evident how great is the advantage of reducing the size of a torsion wire. Even if it is only halved, the torsion is

Scale of 1000ths of an inch for Figs. 1 to 7. The scale of Figs. 8 and 9 is much finer.

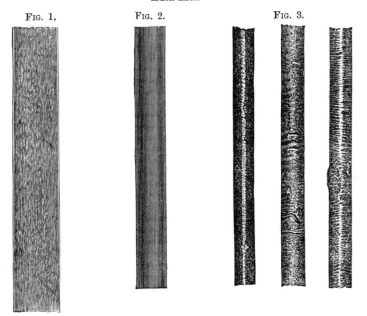

Fig. 1. Fig. 2. Fig. 3.

reduced sixteen-fold. To give a better idea of the actual sizes of such wires and fibres as are in use, I shall show upon the screen a series of photographs taken by Mr. Chapman, on each of which a scale of thousandths of an inch has been printed.

The first photograph (Fig. 1) is an ordinary hair—a sufficiently familiar object, and one that is generally spoken of as if it were rather fine. Much finer than this is the specimen of copper wire now on the screen (Fig. 2), which I recently obtained from Messrs. Nalder Brothers. It is only a little over one-thousandth of an inch in diameter. Ordinary spun glass, a most beautiful material, is about one-thousandth of an inch in diameter, and this would appear to be an ideal torsion thread (Fig. 3). Owing to its fineness, its torsion would be extremely small, and the more so because glass is more easily deformed than metals. Owing to its very great strength, it can carry heavier loads than would be expected of it. I imagine many physicists must have turned to this material in their endeavour to find a really delicate torsion thread. I have so turned, only to be disappointed. It has every good quality but one, and that is its imperfect elasticity. For instance, a mirror hung by a piece of spun glass is casting an image of a spot of light on the scale. If I turn the mirror, by means of a fork, twice to the right, and then turn it back again, the light does not come back to its old point of rest, but oscillates about a point on one side, which, however, is slowly changing, so that it is impossible to say what the point of rest really is. Further, if the glass is twisted one way first, and then the other way, the point of rest moves in a manner which shows that it is not influenced by the last deflection alone: the glass remembers what was done to it previously. For this reason spun glass is quite unsuitable as a torsion thread; it is impossible to say what the twist is at any time, and, therefore, what is the force developed.

FIG. 4.

So great has the difficulty been in finding a fine torsion thread, that the attempt has been given up, and in all the most exact instruments silk has been used. The natural cocoon fibres, as shown on the screen (Fig. 4), consist of two irregular lines gummed together, each about one two-thousandth of an inch in diameter. These fibres must be separated from one another and washed. Then each component will, according to the experiments of Gray, carry nearly 60 grains before breaking, and can be safely loaded with 15 grains. Silk is, therefore, very strong, carrying at the rate of from 10 to 20 tons to the square inch. It is further valuable in that its torsion is far less than that of a fibre of the same size of metal or even of glass, if such could be produced. The torsion of silk, though exceedingly small, is quite sufficient to upset the working of any delicate instrument, because it is never constant. At one time the fibre twists one way, and at another time another, and the evil effect can only be mitigated by using large apparatus in which strong forces are developed. Any attempt that may be made to increase the delicacy of apparatus by reducing their dimensions is at once prevented by the relatively great importance of the vagaries of the silk suspension.

The result, then, is this. The smallness, the length of period, and therefore delicacy, of the instruments at the physicist's disposal have until lately been simply limited by the behaviour of silk. A more perfect suspension means still more perfect instruments, and therefore advance in knowledge.

It was in this way that some improvements that I was making in an instrument for measuring radiant heat came to a deadlock about two years ago. I would not use silk, and I could not find anything else that would do. Spun glass, even, was far too coarse for my purpose; it was a thousand times too stiff.

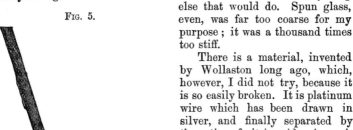

FIG. 5.

There is a material, invented by Wollaston long ago, which, however, I did not try, because it is so easily broken. It is platinum wire which has been drawn in silver, and finally separated by the action of nitric acid. A specimen about the size of a single line of silk is now on the screen, showing the silver coating at one end (Fig. 5).

As nothing that I knew of could be obtained that would be of use to me, I was driven to the necessity of trying by experiment to find some new material. The result of these experiments was the development of a process of almost ridiculous simplicity, which it may be of interest for me to show.

The apparatus consists of a small cross-bow, and an arrow made of straw with a needle point. To the tail of the arrow is attached a fine rod of quartz, which has been melted and drawn out in the oxyhydrogen jet. I have a piece of the same material in my hand, and now, after melting their ends and joining them together, an operation which produces a beautiful and dazzling light, all I have to do is to liberate the string of the bow by pulling the trigger with one foot, and then, if all is well, a fibre will have been drawn by the arrow, the existence of which can be made evident by fastening to it a piece of stamp-paper.

In this way threads can be produced of great length, of almost any degree of fineness, of extraordinary uniformity, and of enormous strength. I do not believe, if any experimentalist had been promised by a good fairy that he might have anything he desired, that he would have ventured to ask for any one thing with so many valuable properties as these fibres possess. I hope, in the course of this evening, to show that I am not exaggerating their merits.

In the first place, let me say something about the degree of fineness to which they can be drawn. There is now projected upon the screen a quartz fibre one five-thousandth of an inch in diameter (Fig. 6). This is one which I had in constant use in an instrument and carrying about 30 grains. It has a section only one-sixth of that of a single line of silk, and it is just as strong. Not being organic, it is in no way affected by changes of moisture and temperature, and so it is free from the vagaries of silk which give so much trouble. The piece used in the instrument was about 16 inches long. Had it been necessary to employ spun glass, which hitherto was the finest torsion material, then, instead of 16 inches, I should have required a piece 1000 feet long, and an instrument as high as the Eiffel tower to put it in.

There is no difficulty in obtaining pieces as fine as this yards long if required, or in spinning it very much finer. There is upon the screen a single line made by the small garden spider, and the size of this is perfectly evident (Fig. 7). You now see a quartz fibre far finer than this, or, rather, you see a diffraction phenomenon, for no true image is formed at all; but even this is a conspicuous object in comparison with the tapering ends, which it is absolutely impossible to trace in a microscope. The next two photographs, taken by Mr. Nelson, whose skill and resources are so famous, represent the extreme end of a tail of quartz, and though the scale is a great deal larger than that used in the other photographs, the end will be visible only to a few. Mr. Nelson has photographed here what it is absolutely impossible to see. What the size of these ends may be, I have no means of telling. Dr. Royston Piggott has estimated some of them at less than one-millionth of an inch, but whatever they are, they supply for the first time objects of extreme smallness, the form of which is certainly known, and therefore I cannot help looking upon them as more satisfactory tests for the microscope than diatoms and other things of the real shape of which we know nothing whatever.

Since figures as large as a million cannot be realised properly, it may be worth while to give an illustration of what is meant by a fibre one-millionth of an inch in diameter.

A piece of quartz an inch long and an inch in diameter would, if drawn out to this degree of fineness, be sufficient to go all the way round the world 658 times; or a grain of sand, just visible—that is, one-hundredth of an inch long and one-hundredth of an inch in diameter—would make 1000 miles of such thread. Further, the pressure inside such a thread, due to a surface tension equal to that of water, would be 60 atmospheres.

Going back to such threads as can be used in instruments, I have made use of fibres one ten-thousandth of an inch in diameter, and with these the torsion is 10,000 times less than that of spun glass.

As these fibres are made finer, their strength increases in proportion to their size, and surpasses that of ordinary bar steel, reaching, to use the language of engineers, as high a figure as 80 tons to the inch. Fibres of ordinary size have a strength of 50 tons to the inch.

While it is evident that these fibres give us the means of producing an exceedingly small torsion, and one that is not affected by weather, it is not yet evident that they may not show the same fatigue that makes spun glass useless. I have therefore a duplicate apparatus with a quartz fibre, and you will see that the spot of light comes back to its true place on the screen after the mirror has been twisted round twice.

I shall now for a moment draw your attention to that peculiar property of melted quartz that makes threads such as I have been describing a possibility. A liquid cylinder, as Plateau has so beautifully shown, is an unstable form. It can no more exist than can a pencil stand on its point. It immediately breaks up into a series of spheres. This is well illustrated in that very ancient experiment of shooting threads of resin electrically. When the resin is hot, the liquid cylinders which are projected in all directions break up into spheres, as you see now upon the screen. As the resin cools, they begin to develop tails; and when it is cool enough, i. e. sufficiently viscous, the tails thicken, and the beads become less, and at last uniform threads are the result. The series of photographs show this well.

There is a far more perfect illustration, which we have only to go into the garden to find. There we may see in abundance what is now upon the screen—the webs of those beautiful geometrical spiders. The radial threads are smooth, like the one you saw a few minutes ago, but the threads that go round and round, are beaded. The spider draws these webs slowly, and at the same time pours upon them a liquid, and still further to obtain the effect of launching a liquid cylinder in space, he, or rather she, pulls it out like the string of a bow, and lets it go with a jerk. The liquid cylinder cannot exist, and the result is what you now see upon the screen (Fig. 8). A more perfect illustration of the regular breaking up of a liquid cylinder, it would be impossible to find. The beads are, as Plateau showed they ought to be, alternately large and small, and their regularity is marvellous. Sometimes two still smaller beads are developed, as may be seen in the second photograph, thus completely agreeing with the results of Plateau's investigations.

I have heard it maintained that the spider goes round her web and places these beads there afterwards. But since a web with about 360,000 beads is completed in an hour—that is, at the rate of about 100 a second—this does not seem likely. That what I have said is true, is made more probable by the photograph of a beaded web that I have made myself by simply stroking a quartz fibre with a straw wetted with castor-oil (Fig. 9). It is rather larger than a spider line; but I have made beaded threads, using a fine fibre, quite indistinguishable from a real spider web, and they have the further similarity that they are just as good for catching flies.

FIG. 8. FIG. 9.

Now, going back to the melted quartz, it is evident that if it ever became perfectly liquid, it could not exist as a fibre for an instant. It is the extreme viscosity of quartz, at the heat even of an electric arc, that makes these fibres possible. The only difference between quartz in the oxyhydrogen jet, and quartz in the arc, is that in the first you make threads, and in the second are blown bubbles. I have in my hand some microscopic bubbles of quartz, showing all the perfection of form and colour that we are familiar with in the soap bubble.

An invaluable property of quartz is its power of insulating perfectly, even in an atmosphere saturated with water. The gold leaves now diverging, were charged some time before the lecture, and hardly show any change, yet the insulator is a rod of quartz only three-quarters of an inch long, and the air is kept moist by a dish of water. The quartz may even be dipped in the water, and replaced with the water upon it, without any difference in the insulation being observed.

Not only can fibres be made of extreme fineness, but they are wonderfully uniform in diameter. So uniform are they, that they perfectly stand an optical test so severe that irregularities invisible in any microscope would immediately be made apparent. Every one must have noticed, when the sun is shining upon a border of flowers and shrubs, how the lines which the spiders use as railways to travel upon from place to place glisten with brilliant colours. These colours are only produced when the fibres are sufficiently fine. If you take one of these webs and examine it in the sunlight, you will find that the colours are variegated, and the effect consequently is one of great beauty.

The quartz fibre of about the same size shows colours in the same way, but the tint is perfectly uniform on the fibre. If the colour of the fibre is examined with a prism, the spectrum is found to consist of alternate bright and dark bands. Upon the screen are photographs taken by Mr. Briscoe, a student in the laboratory at South Kensington, of the spectra of some of these fibres at different angles of incidence. It will be seen that coarse fibres have more bands than fine, and that the number increases with the angle of incidence of the light. There are peculiarities in the march of the bands as the angle increases which I cannot describe now. I may only say that they appear to move not uniformly but in waves, presenting very much the appearance of the legs of a caterpillar walking.

So uniform are the quartz fibres, that the spectrum from end to end consists of parallel bands. Occasionally a fibre is found which presents a slight irregularity here and there. A spider line is so irregular that these bands are hardly observable; but, as the photograph on the screen shows, it is possible to trace them running up and down the spectrum when you know what to look for.

To show that these longitudinal bands are due to the irregularities, I have drawn a taper piece of quartz by hand, in which the two edges make with one another an almost imperceptible angle, and the spectrum of this shows the gradual change of diameter by the very steep angle at which the bands run up the spectrum.

Into the theory of the development of these bands I am unable to enter; that is a subject on which your Professor of Natural Philosophy is best able to speak. Perhaps I may venture to express the hope, as the experimental investigation of this subject is now rendered possible, that he may be induced to carry out a research for which he is so eminently fitted.

Though this is a subject which is altogether beyond me, I have been able to use the results in a practical way. When it is required to place into an instrument a fibre of any particular size, all that has to be done is to hold the frame of fibres towards a bright and distant light, and look at them through a low-angled prism. The banded spectra are then visible, and it is the work of a moment to pick out one with the number of bands that has been found to be given by a fibre of the desired size. A coarse fibre may have a dozen or more, while such fibres as I find most useful have only two dark bands. Much finer ones exist, showing the colours of the first order with one dark band; and fibres so fine as to correspond to the white, or even the gray, of Newton's scale, are easily produced.

Passing now from the most scientific test of the uniformity of these fibres, I shall next refer to one more homely. It is simply this; the common garden spider, except when very young, cannot climb up one of the same size as the web on which she displays such activity. She is perfectly helpless, and slips down like a bead upon a wire. After vainly trying to make any headway, she finally puts her hands (or feet) into her mouth, and then tries again, with no better success. I may mention that a male of the same species is able to run up one of these with the greatest ease, a feat which may perhaps save the lives of a few of these unprotected creatures when quartz fibres are more common.

It is possible to make any quantity of very fine quartz fibre without a bow and arrow at all, by simply drawing out a rod of quartz over and over again in a strong oxyhydrogen jet. Then, if a stand of any sort has been placed a few feet in front of the jet, it will be found covered with a maze of thread, of which the photograph on the screen represents a sample. This is hardly distinguishable from the web spun by this magnificent spider in corners of greenhouses and such places. By regulating the jet and the manipulation, anything from one of these stranded cables to a single ultro-microscope line may be developed.

And now that I have explained that these fibres have such valuable properties, it will no doubt be expected that I should perform some feat with their aid which, up to the present time, has been considered impossible, and this I intend to do.

Of all experiments, the one which has most excited my admiration, is the famous experiment of Cavendish, of which I have a full-size model before you. The object of this experiment is to weigh the earth by comparing directly the force with which it attracts things with that due to large masses of lead. As is shown by the model, any attraction which these large balls exert on the small ones will tend to deflect this six-foot beam in one direction, and then if the balls are reversed in position, the deflection will be in the other direction. Now, when it is considered how enormously greater the earth is than these balls, it will be evident that the attraction due to them must in comparison be excessively small. To make this evident, the enormous apparatus you see had to be constructed, and then, using a fine torsion wire, a perfectly certain but small effect was produced. The experiment, however, could only be successfully carried out in cellars or specially protected places, because changes of temperature produced effects greater than those due to gravity.*

Now I have, in a hole in the wall, an instrument no bigger than a galvanometer, of which a model is on the table. The balls of the Cavendish apparatus, weighing several hundredweight each, are replaced by balls weighing $1\frac{3}{4}$ lb. only. The smaller balls of $1\frac{3}{4}$ lb. are replaced by little weights of 15 grains each. The 6-foot beam is replaced by one that will swing round freely in a tube three-quarters of an inch in diameter. The beam is, of course, suspended by a quartz fibre. With this microscopic apparatus, not only is the very feeble attraction observable, but I can actually obtain an effect eighteen times as great as that given by the apparatus of Cavendish, and, what is more important, the accuracy of observation is enormously increased.

The light from a lamp passes through a telescope lens, and falls on the mirror of the instrument. It is reflected back to the table, and thence by a fixed mirror to the scale on the wall, where it comes to a focus. If the mirror on the table were plane, the whole movement of the light would be only about eight inches, but the mirror is

---

* Dr. Lodge has been able, by an elaborate arrangement of screens, to make this attraction just evident to an audience.—C. V. B.

convex, and this magnifies the motion nearly eight times. At the present moment the attracting weights are in one extreme position, and the line of light is quiet. I will now move them to the other position, and you will see the result—the light slowly begins to move, and slowly increases in movement. In forty seconds it will have acquired its highest velocity, and in forty more, it will have stopped at 5 feet 8½ inches from the starting point, after which it will slowly move back again, oscillating about its new position of rest. It has moved up to and stopped exactly at the division indicated.

It is not possible at this hour to enter into any calculations; I will only say that the motion you have seen is the effect of a force of less than one ten millionth of the weight of a grain, and that with this apparatus I can detect a force two thousand times smaller still. There would be no difficulty even in showing the attraction between two No. 5 shot.

And now, in conclusion, I would only say that if there is anything that is good in the experiments to which I have this evening directed your attention, experiments conducted largely with sticks, and string and straw and sealing-wax, I may perhaps be pardoned if I express my conviction that in these days we are too apt to depart from the simple ways of our fathers, and instead of following them, to fall down and worship the brazen image which the instrument-maker hath set up.

[C. V. B.]

Note.—I have since learnt that in 1841 M. Gaudin melted quartz and drew it out by hand into threads. I have given an abstract of his experiments in the 'Electrical Review' of July 19th.